Inductively Coupled Plasma Emission Spectroscopy

CHEMICAL ANALYSIS

A SERIES OF MONOGRAPHS ON
ANALYTICAL CHEMISTRY AND ITS APPLICATIONS

Editors
P. J. ELVING, J. D. WINEFORDNER
Editor Emeritus: **I. M. KOLTHOFF**

Advisory Board

Fred W. Billmeyer, Jr.   Victor G. Mossotti
Eli Grushka   A. Lee Smith
Barry L. Karger   Bernard Tremillon
Viliam Krivan   T. S. West

VOLUME 90

A WILEY-INTERSCIENCE PUBLICATION

JOHN WILEY & SONS

New York / Chichester / Brisbane / Toronto / Singapore

Inductively Coupled Plasma Emission Spectroscopy

Part 1

METHODOLOGY, INSTRUMENTATION, AND PERFORMANCE

Edited by

P. W. J. M. Boumans

Philips Research Laboratories
Eindhoven, The Netherlands

A WILEY-INTERSCIENCE PUBLICATION

JOHN WILEY & SONS

New York / Chichester / Brisbane / Toronto / Singapore

Copyright © 1987 by John Wiley & Sons, Inc.

All rights reserved. Published simultaneously in Canada.

Reproduction or translation of any part of this work
beyond that permitted by Section 107 or 108 of the
1976 United States Copyright Act without the permission
of the copyright owner is unlawful. Requests for
permission or further information should be addressed to
the Permissions Department, John Wiley & Sons, Inc.

Library of Congress Cataloging in Publication Data:
Inductively coupled plasma emission spectroscopy.

  (Chemical analysis, ISSN 0069-2883; v. 90)
  "A Wiley-Interscience Publication."
  Includes index.
  Contents: pt. 1. Methodology, instrumentation, and
performance.
  1. Plasma spectroscopy.   I. Boumans, P. W. J. M.
(Paul Willy Joseph Maria)   II. Series.
QD96.P62I4   1987   543'.0873   86-18984
ISBN 0-471-09686-5 (v. 1)

Printed in the United States of America

10 9 8 7 6 5 4 3 2 1

CONTRIBUTORS

P. W. J. M. Boumans, Philips Research Laboratories, Eindhoven, The Netherlands

J. A. C. Broekaert, Institut für Spektrochemie und angewandte Spektroskopie, Dortmund, Federal Republic of Germany

H. Bubert, Institut für Spektrochemie und angewandte Spektroskopie, Dortmund, Federal Republic of Germany

W.-D. Hagenah, Institut für Spektrochemie und angewandte Spektroskopie, Dortmund, Federal Republic of Germany

G. M. Hieftje, Department of Chemistry, Indiana University, Bloomington, Indiana

J. W. Olesik, Department of Chemistry, University of North Carolina, Chapel Hill, North Carolina

PREFACE

This book is the first part of a two-volume treatise, *Inductively Coupled Plasma Emission Spectroscopy*, and addresses methodology, instrumentation, and performance. Part 2 deals with applications and fundamentals. The original intention was to cover the complete subject in a single volume. However, the explosive growth of inductively coupled plasma atomic emission spectroscopy (ICP–AES) in recent years and the scarcity of adequate and modern textbooks on emission spectroscopy indicated the need for a more comprehensive treatment. Therefore, this work was set up in its present form, with a fourfold aim:

1. To fill an essential gap in the AES literature
2. To provide a critical and tutorial survey of more than 20 years of research, development, and application in the field of ICPs and related plasma sources
3. To act as a handbook and textbook for the novice and the expert
4. To serve as an aide-memoire and major source of reference for broad groups of analytical spectroscopists, analytical chemists, physical chemists, and physicists including researchers, technicians, and applied analysts.

Part 1 is comprised of nine chapters. Five of them were written by me: Introduction to AES (Chapter 1), Plasma Sources other than ICPs (Chapter 2), ICPs (Chapter 3), Basic Concepts and Characteristics of ICP–AES (Chapter 4), and Line Selection and Spectral Interferences (Chapter 7). I shared the authorship with G. M. Hieftje in a chapter on torches (Chapter 5) and with J. A. C. Broekaert in a chapter on sample introduction (Chapter 6). The topic of spectrometers was entrusted to J. W. Olesik (Chapter 8), and detection and measurement to H. Bubert and W.-D. Hagenah (Chapter 9).

The heavy seal which my authorship set on Part 1 emphasizes my concern as editor to produce a book that treats the basic principles of emission spectroscopy in a tutorial, systematic, and consistent way. I have been conscious, however, that a consistent treatment of a subject by a single author entails the hazards of bias and shortsightedness. Therefore, I employed the scholarship, experience, and specific expertise of a group of authors to cover those topics that can be easily split off the main line without risking loss of consistency and coherence. These authors have been allotted only a relatively modest place in Part 1, but had the opportunity to set their stamp fully in Part 2, where my task was limited to that of a rather critical editor and text revisor.

The difference in expertise and scientific background among the authors has resulted in a balanced distribution of diverse, though consistent, views and opinions throughout the work as a whole. The participation of authors from various countries, mainly located in Western Europe and North America, has further contributed to an equilibrated treatment of the various topics. I have made efforts to formulate their assignments strictly and to inspect their texts carefully in order to prevent undue overlap. Only in those cases where the expression of different, but complementary visions appeared beneficial have I overlooked some overlap.

The thematic arrangement of the topics has resulted in a compilation of chapters that are more or less self-existent and may be consulted independently of each other. In the treatment of the topics, either a tutorial approach or a review character predominates, depending on the subject matter. In all instances, however, the literature has been comprehensively covered.

For practical reasons, the complete material was split into two parts, published as separate volumes, each being self-consistent within certain limits. Part 1 covers the basis of ICP-AES as an analytical method and deals with (1) the fundamental analytical concepts: detection limits, precision, accuracy, dynamic range, multielement capability, spectral interferences, and line selection; (2) the principles of the instrumentation: sample introduction devices, torches, spectrometers, and devices for detection and measurement. For obvious reasons a treatment of the principles of generators and a detailed survey of commercial instruments have been omitted.

Part 2 is comprised of (1) six chapters on applications, subdivided according to field: metals and industrial materials, geological, organics, agriculture and food, biological, and environmental; (2) four chapters dealing with fundamentals: aerosol generation, plasma modelling, plasma diagnostics, excitation mechanisms, and discharge characteristics; (3) a chapter on direct analysis of solids; (4) a chapter in which the capabilities of a variety of atomic spectroscopic methods for elemental trace analysis are assessed. Further details will be given in the Preface to Part 2.

The work is intended primarily for spectroscopists and analytical chemists in the field of ICP-AES. However, the way in which the subjects are addressed makes the information also of major interest to workers in related fields: AES using plasma sources other than ICPs, ICP atomic fluorescence spectrometry (AFS), and ICP mass spectrometry (MS). More generally, anyone wishing to become acquainted with the essentials of emission spectroscopy either will find much of the desired information directly in this work or will be guided by it to more specialized literature.

It is hoped that this treatise will be a useful source of reference to many people working in the fields of AES and ICP-AES concerned with research, development, or applications in universities, industrial organizations, or government institutes. Let the insights gained by consulting this work stimulate the further progress of emission spectroscopy.

I want to thank the authors for their contributions, their willingness to accept my suggestions, and their perseverance to carry this project appropriately to its end. I also want to thank the series editor, Professor J. D. Winefordner, for his patience and encouragement, and for his understanding of my position as editor and contributor.

P. W. J. M. BOUMANS

Eindhoven, The Netherlands
January 1987

CONTENTS

CHAPTER 1. INTRODUCTION TO ATOMIC EMISSION SPECTROMETRY 1
P. W. J. M. Boumans

CHAPTER 2. PLASMA SOURCES OTHER THAN INDUCTIVELY COUPLED PLASMAS 45
P. W. J. M. Boumans

CHAPTER 3. INDUCTIVELY COUPLED PLASMAS 69
P. W. J. M. Boumans

CHAPTER 4. BASIC CONCEPTS AND CHARACTERISTICS OF ICP-AES 100
P. W. J. M. Boumans

CHAPTER 5. TORCHES FOR INDUCTIVELY COUPLED PLASMAS 258
P. W. J. M. Boumans and G. M. Hieftje

CHAPTER 6. SAMPLE INTRODUCTION TECHNIQUES IN ICP-AES 296
J. A. C. Broekaert and P. W. J. M. Boumans

CHAPTER 7. LINE SELECTION AND SPECTRAL INTERFERENCES 358
P. W. J. M. Boumans

CHAPTER 8. SPECTROMETERS 466
J. W. Olesik

CHAPTER 9. DETECTION AND MEASUREMENT 536
H. Bubert and W.-D. Hagenah

INDEX 567

Inductively Coupled Plasma Emission Spectroscopy

CHAPTER 1

INTRODUCTION TO ATOMIC EMISSION SPECTROMETRY

P. W. J. M. BOUMANS

Philips Research Laboratories
Eindhoven, The Netherlands

1.1. Optical Atomic Spectrometry and Atomic Emission Spectrometry
1.2. Atomic Emission Spectra and Excitation Sources
1.3. Multielement Capability of AES
1.4. Spectroscopic Instrumentation for AES
    1.4.1. Introduction
    1.4.2. Polychromators
    1.4.3. Monochromators
    1.4.4. Spectrographs
    1.4.5. Progress Induced by Advent of ICP
1.5. Qualitative and Semiquantitative Emission Spectroscopic Analysis
1.6. Quantitative Analysis: Calibration and Interferences
    1.6.1. Basic Concepts
    1.6.2. Calibration Functions
    1.6.3. Interferences and Approaches to Circumvent Them
        1.6.3.1. Addition Technique
        1.6.3.2. Buffer Addition Technique
        1.6.3.3. Isoformation Technique
    1.6.4. Classification of Interferences
    1.6.5. Multicomponent System with Multiple Interferences
1.7. Sample Introduction and Excitation Source
    1.7.1. General
    1.7.2. Arc
    1.7.3. Spark
    1.7.4. Glow Discharge
    1.7.5. Hollow Cathode Discharge
    1.7.6. Laser
    1.7.7. Furnace
    1.7.8. Exploding Conductors
    1.7.9. Flames
    1.7.10. Plasma Sources
    1.7.11. Combined Sources

1.8. **Guide to the Literature on Emission Spectroscopic Analysis**
    1.8.1. Introduction
    1.8.2. Encyclopedical Information
    1.8.3. Classical Treatises on the Emission Spectroscopic Methodology
    1.8.4. Recent Works on Emission Spectroscopic Methodology
    1.8.5. Works Dealing with Fundamentals of Excitation Sources
    1.8.6. Books on Flame Spectrometry
    1.8.7. Bibliography of Sections 1.8.2 to 1.8.6.
    1.8.8. Abstracts, Compilations, and Regular Reviews
    1.8.9. Journals
References

1.1. OPTICAL ATOMIC SPECTROMETRY AND ATOMIC EMISSION SPECTROMETRY

Atomic emission spectrometry (AES), atomic absorption spectrometry (AAS), and atomic fluorescence spectrometry (AFS) are three branches of that part of analytical spectrometry which derives analytical information from atomic spectra in the optical region of the electromagnetic spectrum. This optical region is the ultraviolet (UV), the visible, and the near infrared. The atomic spectra in this region originate from energy transitions in the outer electronic shells of free atoms or ions.

Atomic emission spectrometry is often denoted "*optical* emission spectrometry" (OES), in particular because the acronym "AES" also refers to Auger electron spectroscopy. In this work we will use the terms "atomic emission spectrometry," or briefly "emission spectrometry," and the acronym "AES."

The three classes of radiation processes that form the basis of AES, AAS, and AFS are schematically depicted in Fig. 1.1 and elucidated in the figure caption.

In all three methods the sample must be atomized, that is, dissociated into free atoms and/or ions. In AES, this is accomplished in an excitation source. The latter, then, not only furnishes the energy for the atomization of the sample but also for the excitation of the free atoms and ions of the elements to be determined (analytes).

Inductively coupled plasmas (ICP) are chiefly used as excitation sources for AES. Although initial work on their use as atomizers for AFS showed some promise [1], it was not until recently that their application to AFS analysis was more fully exploited (Section 3.4). The use of ICPs as atomizers for AAS does not entail advantages over that of the common atomizers: flames and furnaces [2–9]. However, although Magyar and Aeschbach [10] found the sensitivity of AAS measurements along the diameter of an ICP to be ten times lower than in flame AAS, they considered the approach useful for its high selectivity and precision in the determination of metals in complex compounds, which are dif-

1.2. ATOMIC EMISSION SPECTRA AND EXCITATION SOURCES

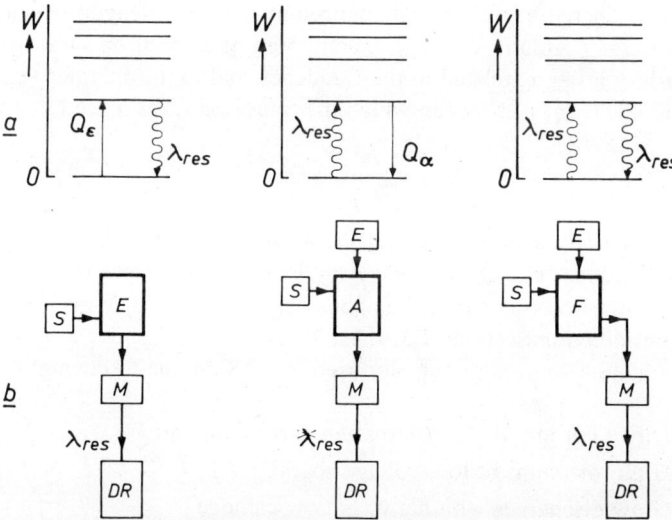

Figure 1.1. Illustration of the principles of AES, AAS, and AFS. Diagrams (*a*) represent energy levels of free atoms; block diagrams (*b*) show the experimental setup. W = energy, Q = energy transferred on collision, λ = wavelength, S = sample introduction, E = excitation source or external radiation source, A = absorption cell, F = fluorescence cell, M = radiation selector (spectrometer), DR = detector and readout unit. For simplicity diagrams (*a*) show only transitions between the atomic ground level and the first excited level (resonance transition). Left: in AES an amount of energy, Q_e, is transferred to the atom on collision with another particle resulting in excitation. Subsequent de-excitation by spontaneous emission leads to the emission of radiation of wavelength λ_{res}. Center: in AAS the atom is excited by absorption of radiation from an external source. The atom loses the excitation energy, Q_α, in collision with other particles. As a result, the intensity of the incident beam is attenuated. Right: in AFS the atom is excited by absorption of radiation from an external source in the same way as in AAS. However, de-excitation takes place by emission of radiation. This so-called fluorescence radiation is observed in a direction perpendicular to the incident beam.

ficult to dissociate in flames. High sensitivity was said to be generally not important in this type of application.

ICPs are also used as ion sources in mass spectrometry. This relatively new, but most promising analysis method will be only briefly discussed in this book (see Section 3.5).

1.2. ATOMIC EMISSION SPECTRA AND EXCITATION SOURCES

Optical emission spectra are observed by dispersing the radiation according to wavelength so that photons ($h\nu$) of different frequency (ν) appear in the focal plane of a spectroscopic apparatus as an array of monochromatic images of the entrance slit. These images are characterized by their wavelengths (λ) and called "spectral lines." Atomic spectra are composed of *discrete* spectral lines. The wavelength of a line is related to the frequency and to the energies (E_q, E_p) of the atomic levels (q, p) between which the transition takes place by

$$h\nu_{qp} = \frac{hc}{\lambda_{qp}} = E_q - E_p \qquad (1.1)$$

where c is the speed of light and h the Planck constant. Classical tables of atomic emission spectra list the wavelengths of spectral lines in Ångstrøm (1 Å = 10^{-10} m); in recent works the nanometer (nm) is used. Wavelength tables will be considered in Sections 7.3.4 and 7.4.4.

The excitation sources principally used in AES are the following:

1. Direct current (dc) or alternating current (ac) arcs
2. High, medium, or low voltage sparks
3. Glow discharges with flat or hollow cathode
4. Lasers
5. Flames
6. Inductively coupled plasmas (ICP)
7. Direct current plasmas (DCP)
8. Capacitively coupled microwave plasmas (CMP)
9. Microwave induced plasmas (MIP)
10. Furnaces
11. Exploding wires or foils

Sources (1) to (3) and (6) to (9) are electrical gas discharges, at atmospheric or reduced pressure, in which a gas is energized by an electric current. The gases of a flame are heated by the chemical energy released in the combustion process. Lasers primarily effect vaporization of a (solid) sample as a result of absorption of powerful laser radiation. For AES analysis one uses either the spectrum emitted by the hot vapor plume formed upon the laser impact, or the spectrum emitted after additional excitation of the plume by the direct action of a spark or transfer of the vapor to an ICP. In furnaces, atomization and excitation are achieved either both electrothermally or in two steps: atomization by resistive heating and excitation by an electric discharge. The exploding-wire technique uses the rapid Joule heating and explosive vaporization of a wire or foil that may occur when a high-current source, such as a capacitor charged to several thousand volts, is switched across a thin metallic wire or foil (cf. Section 1.7).

When an excitation source is used for spectrochemical analysis, it no longer stands on its own: essential is the introduction of samples. To that end, solid

1.2. ATOMIC EMISSION SPECTRA AND EXCITATION SOURCES

and liquid samples must be brought into a form so that they can be readily evaporated and atomized by interacting with the hot gases in the source. Proper control of the step "sample introduction" is vital so that the composition of the gas in the source will unambiguously reflect the sample composition. Only then will the spectrum provide a picture from which quantitative analytical information can be derived.

Sample introduction is the most troublesome part of an emission spectroscopic analysis. This difficulty is demonstrated not only by the wide diversity of sample introduction techniques, but also by the large variety of excitation sources. The reasons for this are the diversity of sample types and the fact that the sample introduction technique must be compatible with the properties of the source. Mutual consistency of source characteristics and sample introduction technique, in turn, is indispensable because sample volatilization and atomization cannot be easily separated from the subsequent steps: excitation and emission. This will become evident in the brief discussion of sample introduction techniques and excitation sources in Section 1.7 and in the comprehensive treatment of sample introduction in ICP-AES in Chapter 6 and Part 2, Chapters 7 and 8. Sample preparation is a separate topic to be dealt with in the chapters on applications (Part 2, Chapters 1-6).

A common feature of all excitation sources is the presence of a flowing or stationary gas: the discharge or support gas, the flames gases, the surrounding atmosphere or an artificial, controlled atmosphere. Once a sample has entered the excitation source, the sample constituents are dispersed in the gas as free atoms, ions, molecules, and possibly larger aggregates. This composite gas is denoted "plasma" if generated by an electric discharge, laser impact, or wire or foil explosion. Physically a gas is called "plasma" if the following conditions are fulfilled: (1) the dimensions must be much larger than the Debye length (λ_D); (2) the particle density should be so large that there are many particles present within a sphere of radius λ_D; (3) (quasi-)electroneutrality should prevail; and (4) interactions should take place mainly between charged particles mutually, or between charged and neutral particles.

The radiation emitted from an excitation source encompasses the contributions from all components: the gaseous atmosphere, the free atoms, ions and molecules of the sample, and any species formed between the constituents of the gaseous atmosphere mutually or with the sample constituents. The emission spectrum of the source then consists of line and band spectra superimposed on continua that result from (1) various types of interactions (including recombination) between free electrons and ions, (2) recombinative interactions between atoms, and (3) thermal emission from incandescent solid particles.

The information about the sample composition must be extracted from the composite radiation of the source, and is derived from atomic or ionic lines, or exceptionally from molecular bands, as is the case with fluorine, which can only be indirectly determined using bands of InF, MgF, or CaF because the resonance lines of F are located in the far vacuum ultraviolet (VUV) region (see, however, the text that follows).

A chemical element (M) of a sample will be present in an excitation source in the form of various species, for example, diatomic oxide molecules (MO), neutral atoms (M I), singly charged ions (M II), doubly charged ions (M III), and so on. Under the conditions prevailing in spectrochemical excitation sources one or two types of species (e.g., MO and M I, M I and M II, or M II and M III) will predominate, the most common being the presence of neutral atoms and singly charged ions, with a bias toward the former, for instance, in a dc arc, or the latter, for instance, in a spark at atmospheric pressure or an ICP.

Each of the species in the form of which a chemical element is present emits a characteristic spectrum from which one or more lines may be chosen to identify the chemical element (qualitative analysis) and to determine its concentration (quantitative analysis).

For the identification of an analyte, one establishes the presence of one or several "most sensitive" lines of the element, while ascertaining at the same time that the lines found at the characteristic wavelengths do not originate from other chemical elements that emit lines at wavelengths similar to or in the close vicinity of the most sensitive lines of the analyte ("line coincidence").

Quantitative analysis uses an empirically established relationship between the intensity of a spectral line of the analyte and its concentration in the sample: a calibration curve. This approach is necessary because the relationship between intensity and concentration cannot be calculated with sufficient accuracy. As for most physical methods of analysis, AES is not, therefore, an absolute method. Also, intensity is not measured absolutely but in relative units, for example, volts, amperes, or counts. Intensity of a spectral line is often referred to as "(analysis) signal" or "line signal."

The absolute intensity of a spectral line I_{qp} (W m^{-3} sr^{-1}) corresponding to a transition between levels q and p is given by

$$I_{qp} = \frac{1}{4\pi} A_{qp} n_{xq} h \nu_{qp} \qquad (1.2)$$

where A_{qp} (s^{-1}) is the transition probability and n_{xq} (m^{-3}) is the population of the upper level of the transition expressed as number density (often also referred to as "concentration in the source").

For a source in local thermal equilibrium (LTE) concentration n_{xq} is related to the concentration of species x by the Boltzmann equation:

$$n_{xq} = n_x \frac{g_q \exp(-E_q/kT)}{Z_x(T)} \qquad (1.3)$$

where E_q is the excitation energy (eV) or potential (V) of level q (Fig. 1.2), g_q the statistical weight of level q, T (K) the absolute temperature, Z_x the partition function of species x, and k the Boltzmann constant [11, 12].

1.2. ATOMIC EMISSION SPECTRA AND EXCITATION SOURCES

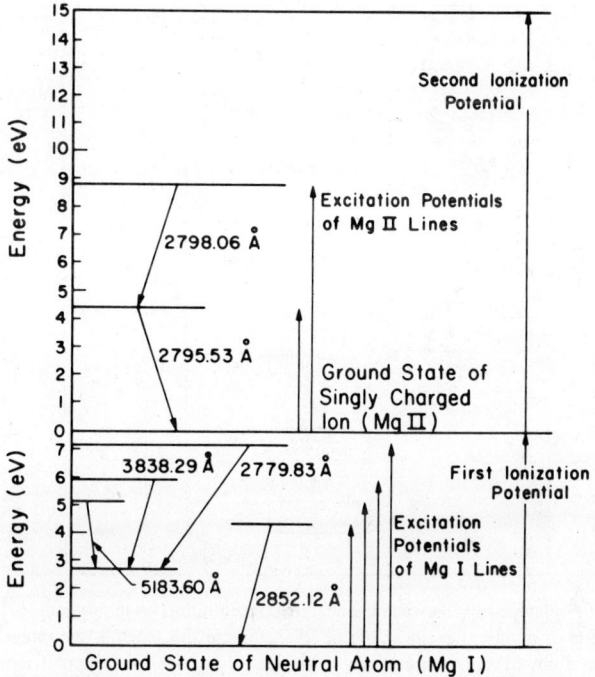

Figure 1.2. Simplified energy level diagram for Mg I and Mg II. Numbers near the arrows indicate the wavelengths of the corresponding spectral lines. Note that the excitation potentials are always taken with respect to the ground level of the corresponding species, thus for the atomic lines with respect to the ground level of the atom and for the ionic lines with respect to the ground level of the ion. (cf. Boumans [11].)

The concentrations of the various species n_x, n_y, n_z, . . . of a chemical element make up the total concentration (n) of the element:

$$n = n_x + n_y + n_z + \ldots \tag{1.4}$$

which is related to the total pressure (P) by

$$n = P/kT \tag{1.5}$$

The relative values of n_x, n_y, n_z, . . . depend on the dissociation energies of possible molecular species (in particular, monoxides) and on the ionization energies of the successive ionization stages of the element. The total concentration n (m^{-3}) is related to the concentration c (e.g., μg/g or μg/mL) of the element in the sample via complex relationships governed by evaporation, dissociation, and transport processes.

The preceding picture is schematically visualized in Fig. 1.3 for three LTE plasmas (1, 2, 3) characterized by temperatures T_1, T_2, and T_3 respectively. In

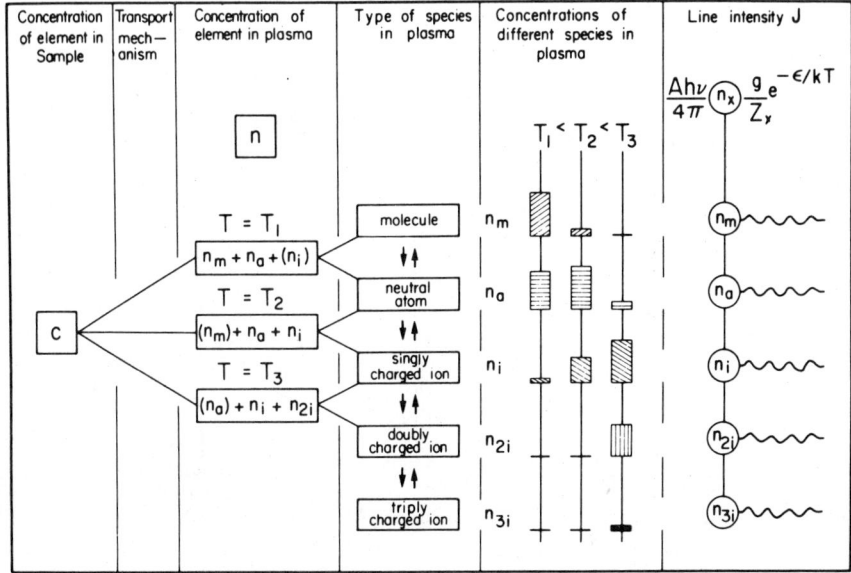

Figure 1.3. Interrelationships between the analyte concentration (c) in the sample, the corresponding concentration or number density (n) in the plasma, and the spectral-line intensities emitted at different temperatures by the various analyte species: molecules (n_m), neutral atoms (n_a), singly charged ions (n_i), doubly charged ions (n_{2i}), and triply charged ions (n_{3i}). (cf. Boumans [11].)

plasma 1 diatomic molecules and neutral atoms predominate and, therefore, the spectra of the neutral atoms will constitute the basic source of analytical inference; in plasma 2 the spectra of both neutral atoms and singly charged ions will form the source of information, and in plasma 3 this will be the spectra of singly and doubly charged ions.

Which set of spectral lines of a chemical element will be the most sensitive ones under particular conditions depends on

> Properties of the source, that is, in the case of LTE plasmas, the temperature and the gaseous atmosphere, and in the case of non-LTE plasmas, the types of processes that govern dissociation, ionization, recombination, and excitation, and the values of the parameters involved.
> Dissociation energy of the monoxide of the element.
> Ionization energies of successive ionization stages of the element (Table 1.1).
> Excitation energies (Fig. 1.2), the transition probabilities; and wavelengths of the lines.[1]

---

[1] An elementary treatment of the various relationships can be found in Section 6.3 of Boumans' chapter, "Excitation of Spectra" [11].

Table 1.1. Ionization Potentials (in V) of the Neutral Atoms (IP–I) and Singly Charged (IP–II) Ions, and the Atomic Masses (M) and Atomic Numbers (A) of the Elements [13, 14]

	IP–I	IP–II	M	A		IP–I	IP–II	M	A
Ac	6.9	12.1	227	89	Na	5.14	47.29	22.99	11
Ag	7.58	21.49	107.9	47	Nb	6.88	14.32	92.91	41
Al	5.99	18.83	26.98	13	Nd	5.49	10.72	144.2	60
Am	6.0	—	243	95	Ne	21.56	40.96	20.18	10
Ar	15.76	27.63	39.95	18	Ni	7.63	18.17	58.71	28
As	9.81	18.63	74.92	33	O	13.62	35.12	16.00	8
Au	9.22	20.5	197.0	79	Os	8.7	17	190.2	76
B	8.30	25.15	10.81	5	P	10.49	19.72	30.97	15
Ba	5.21	10.00	137.3	56	Pb	7.42	15.03	207.2	82
Be	9.32	18.21	9.01	4	Pd	8.34	19.43	106.4	46
Bi	7.29	16.69	209.0	83	Pm	5.55	10.90	145	61
Br	11.84	21.8	79.90	35	Po	8.42	—	210	84
C	11.26	24.38	12.01	6	Pr	5.42	10.55	140.9	59
Ca	6.11	11.87	40.08	20	Pt	9.0	18.56	195.1	78
Cd	8.99	16.91	112.4	48	Pu	5.8	—	244	94
Ce	5.47	10.85	140.1	58	Ra	5.28	10.15	226.0	88
Cl	12.97	23.81	35.45	17	Rb	4.18	27.28	85.47	37
Co	7.86	17.06	58.93	27	Re	7.88	16.6	186.2	75
Cr	6.77	16.50	52.00	24	Rh	7.46	18.08	102.9	45
Cs	3.89	25.1	132.9	55	Rn	10.75	—	222	86
Cu	7.73	20.29	63.55	29	Ru	7.37	16.76	101.1	44
Dy	5.93	11.67	162.5	66	S	10.36	23.33	32.06	16
Er	6.10	11.93	167.3	68	Sb	8.64	16.53	121.7	51
Eu	5.67	11.25	152.0	63	Sc	6.54	12.80	44.96	21
F	17.42	34.97	19.00	9	Se	9.75	21.19	78.96	34
Fe	7.87	16.18	55.85	26	Si	8.15	16.34	28.09	14
Ga	6.00	20.51	69.72	31	Sm	5.63	11.07	150.4	62
Gd	6.14	12.1	157.2	64	Sn	7.34	14.63	118.7	50
Ge	7.90	15.93	72.59	32	Sr	5.69	11.03	87.62	38
H	13.60	—	1.008	1	Ta	7.89	16.2	180.9	73
He	24.59	54.42	4.003	2	Tb	5.85	11.52	158.9	65
Hf	7.0	14.9	178.5	72	Tc	7.28	15.26	98.9	43
Hg	10.44	18.76	200.6	80	Te	9.01	18.6	127.6	52
Ho	6.02	11.80	164.9	67	Th	6.2	11.5	232.0	90
I	10.45	19.31	126.9	53	Ti	6.82	13.58	47.90	22
In	5.79	18.87	114.8	49	Tl	6.11	20.43	204.4	81
Ir	9.1	—	192.2	77	Tm	6.18	12.05	168.9	69
K	4.34	31.62	39.10	19	U	6.2	—	238.0	92
Kr	14.00	24.36	83.80	36	V	6.74	14.65	50.94	23
La	5.58	11.06	138.9	57	W	7.98	17.7	183.8	74
Li	5.39	75.64	6.941	3	Xe	12.13	21.21	131.3	54
Lu	5.43	13.9	175.0	71	Y	6.38	12.24	88.90	39
Mg	7.65	15.04	24.30	12	Yb	6.25	12.17	173.0	70
Mn	7.43	15.64	54.94	25	Zn	9.39	17.96	65.38	30
Mo	7.10	16.15	95.94	42	Zr	6.84	13.13	91.22	40
N	14.53	29.60	14.01	7					

Frequently a particular set of spectral lines (usually the resonance lines) of a species are the most sensitive lines of the *species* under all conditions, that is, irrespective of the source. This is so because excitation energy, transition probability, and frequency are invariable line characteristics, while the Boltzmann equation is often applicable, though with certain restrictions as to the meaning of the temperature occurring in it.

However, since the relative concentrations of the species originating from a chemical element may vary drastically from one type of source to another, the sets of most sensitive lines will vary accordingly. Thus, in sources where neutral atoms prevail (e.g., arcs) atomic lines are generally the most sensitive, whereas in sources with predominantly ionic species (e.g., sparks, ICPs) the ionic lines of many elements are the most sensitive.

Generally speaking, the most sensitive lines of the neutral atoms and singly charged ions of the majority of the chemical elements have wavelengths in the region between 190 and 600 nm [12], the exceptions being

$\lambda < 120$ nm: Ar, F, H, He, Ne

120 nm $< \lambda <$ 190 nm: Br, C, Cl, I, N, O, P, S, Xe

600 nm $< \lambda$: Cs, K, Li, Rb, Rn

The wavelength range accessible with a particular spectrometer depends on the type and blaze of the grating, the geometry of the instrument, the type(s) of detector (usually one or more photomultipliers (PMT), and the gaseous atmosphere in the light path (cf. Section 1.4 and Chapters 8 and 9). The range between about 160 and 900 nm can be covered in general without great difficulty, albeit that below 190 nm an instrument flushed with argon or nitrogen or vacuum apparatus is indispensable. The lower limit of 160 nm is dictated by the transmittance of the silica windows of the bialkali or solar blind (Cs–Te) PMTs commonly used in the UV region. PMTs with MgF_2 windows are also available and allow measurements down to 115 nm. Access to lower wavelengths is possible with windowless PMTs. Work in that far VUV region has been performed by a group of French spectroscopists who used specifically designed spectrometers and excitation sources ("gliding spark"). A recently described spectrometer [15] covers a wavelength region between 14 and 320 nm and enables the use of spectral lines of multiply ionized atoms of gases (O IV, N V), rare gases, halogens, metalloids (B III, C IV, Si IV, P IV, S VI), as well as metals [15, 16].

Work with particular emphasis on the other end of the wavelength range of atomic spectroscopy has been performed by Fry and his group [17]. This work concerns ICP excited near infrared atomic spectra of O, N, F, Cl, Br, S, and C, the selective gas chromatographic ICP determination of compounds containing oxygen and nitrogen atoms, and the use of a photodiode array spectrometer for near infrared multielement atomic spectroscopy.

1.3. MULTIELEMENT CAPABILITY OF AES

Multielement capability of an analysis method denotes the potential of the simultaneous determination of a large variety of chemical elements, in principle, any group of elements at random selected from the Periodic Table, in one and the same sample. A generalized treatment of the multielement capability of atomic methods including detection systems has been given by Winefordner et al. [18].

Multielement capability is inherent in AES techniques using an excitation source of sufficiently high temperature, that is ≳ 5000 K, so that atomization is (virtually) complete, and atomic and ionic species dominate. Suitable analysis lines can then be found for elements having widely differing ionization potentials and being present at widely differing concentrations [19]. For alkali metals, however, a temperature of 5000 K is too high: they are ionized to a large degree so that atomic line emission is weak, whereas ionic line emission is absent or negligible because alkali ions behave spectroscopically as noble gases. Although alkalis can be determined by many excitation sources, the best source for them is a flame, for example, a propane–air (2200 K) or acetylene–air (2500 K) flame.

For other elements, flames are predominantly used only as atomizers in AAS. To achieve a high degree of atomization for elements forming stable oxides, high-temperature flames such as the nitrous oxide–acetylene flame (3200 K) [20, 21] and the acetylene–oxygen flame [22] have been developed. These flames offered good perspectives as excitation sources for many elements [22–26]. The detection limits of some 40 elements are reasonable, whereas those for the remaining elements are poor. Therefore, the multielement capability of flames is essentially limited. The poor values are found for elements with resonance lines below about 350 nm (excitation energy ≳ 3.5 eV) and for elements that form stable compounds (oxides) and, thus, cannot be sufficiently atomized thermally at 3200 K. In addition, chemical interferences may be troublesome for these elements. For these various reasons hot flames have not gained popularity as excitation sources for AES and were actually ousted by "plasma sources," that is, electrical gas discharges, usually in an inert gaseous atmosphere, having the appearance of flames (DCP, ICP, CMP) but with a substantially higher temperature (≳ 5000 K) in the zone useful for analysis. These plasma sources combine the spatial and temporal stability of flames with the excitation characteristics of arcs and sparks, and permit direct analysis of liquids. In a similar way as in flame spectrometry (see Section 1.8) the liquid samples are nebulized and introduced as aerosols into the plasma. Owing to their high temperature, plasma sources have excellent multielement capabilities, but also many other desirable analytical features.

A penalty of the multielement capability of AES is the problem of line coincidences, which may cause severe troubles in trace analysis of samples whose major constituents have line-rich spectra (e.g., Fe, Ni, Co, Cr, Mn, Nb, Ta, Mo, V, W, Sc, Y, Zr, La, rare earths, platinum metals, Th, and U). Spectroscopic equipment providing high resolution (bandwidth $\approx$ 0.01 nm) is then often required in particular for rare earths and associated elements, and Th, and U. Separation of the traces from the matrix may be also necessary to achieve the desired detection limits.

The exploitation of the fundamentally existing multielement capability of AES requires the excitation source to be combined with such spectroscopic apparatus that intensities of spectral lines can be measured at several wavelengths using either a simultaneous or a rapid sequential mode of measurement.

1.4. SPECTROSCOPIC INSTRUMENTATION FOR AES

1.4.1. Introduction

In spite of many attempts to exploit solid-state devices, such as photodiode (PD) arrays and vidicon tubes (see Chapter 9), these efforts have not yet been successful enough to result in generally accepted instruments for AES. The prime limitation of the solid-state devices is the essentially poorer signal-to-noise ratio (SNR) in the UV, as compared to PMTs. Their overall dimensions and spatial resolution may impose further limitations.

A type of PMT known as "image dissector tube" showed much promise for rapid sequential detection when combined with an echelle spectrometer [27, 28]. This combination temporarily reached the stage of a commercial emission spectrometer ("IDES"), which, however, did not find widespread application.

The AES spectrometers commercially available at present all use photomultipliers as detectors. These instruments are either *polychromators* equipped with an array of exit slits and PMTs for simultaneous multielement analysis, or *monochromators* for sequential multielement analysis. In addition, spectrographs using photographic detection remain in use, in particular, for general survey analysis with dc arc, ICP, or DCP.

The multielement capability of the three classes of instruments degrades in the order spectrograph > polychromator > monochromator, but other criteria will often predominate if an instrument has to be chosen for a particular analytical task. Details on spectroscopic instrumentation will be treated in Chapters 8 and 9. A brief characterization with special reference to ICP–AES is as follows.

1.4.2. Polychromators

- Polychromators are basically designed for routine multielement analysis involving many elements per sample, but with little diversity among the samples.
- They are inflexible as to line choice; a large diversity of samples is difficult to deal with.
- They are characterized by high speed and sample throughput, and by high stability.
- Spectral resolution is limited by stability requirements; dealing with matrices yielding line-rich spectra may be cumbersome.
- The required sample size is independent of the number of analytes.
- The instrument is compatible with excitation modes that provide transient signals of short duration, but this might require adaptation of the measuring electronics.
- The spectral range depends on the manufacturer and the type of instrument (see tabular surveys in [29, 30]). The lower wavelength limit is usually 190 nm for air-path spectrometers and 170 nm for argon- and nitrogen-flushed or vacuum instruments. The upper limit varies from 400 to 900 nm.
- Geometrical and mechanical constraints may impose limits on

    The minimum distance between spectral lines

    The maximum number of lines within a given spectral interval

    The use of higher spectral orders in addition to the first order permits acceptable compromises.

1.4.3. Monochromators

- Monochromators have a high flexibility as to line choice: in principle any wavelength for line and background measurement is accessible.
- The blaze of the grating and the spectral response of the PMT may set limitations on the useful wavelength range; however, interchangeable gratings and PMTs may provide for virtually unrestricted versatility in line choice.
- The instrument can be used for solving both research and routine problems.
- Adequate spectral resolution and optical conductance can be adopted at reasonable cost.
- Good noise characteristics can be achieved by an appropriate choice of PMT and entrance optics.
- State-of-the-art mechanical and electronic devices may provide for high-

speed slewing between spectral lines and for accurate wavelength positioning.
- The sequential mode of measurement sets lower limits to sample size and analysis time.

1.4.4. Spectrographs

- Common spectrographs provide medium-to-high spectral resolution and can easily cope with complex spectra.
- The optical conductance of spectrographs well matches the radiance of the dc arc and the response characteristics of emulsions specifically used in spectrography, but may require inconveniently long exposures with ICPs, especially at wavelengths below 250 nm.
- A photographic emulsion as detector is unexcelled as to the enormous amount of spectral information that can be covered and stored during a relatively short exposure time, generally of the order of minutes.
- This stored information remains available for visual inspection, measurements of distances on the wavelength scale, and photometric measurements.
- Therefore, photographic recording gives access to qualitative analysis of completely unknown samples, a property hitherto not yet realized in any other detector.
- The simultaneous recording of spectral lines and their adjacent background provides virtually ideal conditions for trace determinations.
- Obtaining reliable results in qualitative and quantitative analysis makes high demands upon the skill and experience of the technical staff.
- Quantitative measurements are laborious, time-consuming, restricted in precision and accuracy, and often hampered by the limited dynamic range of the photographic emulsion.
- Access to wavelengths below 230 nm and, to a lesser extent, above 650 nm is difficult.
- For fully exploiting the benefits of photographic detection in an economical way, in particular, in general survey analysis, the use of a computer-controlled microphotometer for automatic spectrum evaluation [31, 32] is indispensable; instruments that really fulfill all requirements are commercially unavailable.

1.4.5. Progress Induced by Advent of ICP

Initial ICP research and development was performed with simple, manually operated, or scanning monochromators and spectrographs. The first commercial instruments combined the ICPs with classical polychromators, up to then customarily connected to spark excitation sources for production control of metals and not designed for trace analysis. The combination of polychromators with ICPs revealed serious shortcomings for this type of application and led to these updatings:

Improvements of the stray light level.

Incorporation of devices for automatic background correction based on measurements in the close vicinity of the spectral lines.

Addition of a separate, external monochromator or an internal movable exit slit-detector assembly to increase the flexibility in line choice.

Incorporation of prealigned slits to facilitate line selection and increase flexibility.

Implementation of modern digital electronics to accommodate the large dynamic range of ICPs.

Design of vacuum or argon- or nitrogen-flushed instruments compatible with the ICP source.

Monochromators saw an evolution from the classical, manually operated or scanning type instruments to computer or microprocessor controlled slew-scan instruments, which permit rapid and often accurate access to any wavelength within the range of the instrument. Both single and double monochromators are used.

1.5. QUALITATIVE AND SEMIQUANTITATIVE EMISSION SPECTROSCOPIC ANALYSIS

Qualitative and semiquantitative emission spectroscopic analysis has found the widest application in spectrography using the dc carbon arc. The term "spectro*graphy*" implies the use of a photographic plate or film as radiation detector. The dc carbon arc permits direct excitation of solid samples, either conducting or nonconducting, and gives access to concentrations ranging from 100% down to the parts per million (μg/g) level and lower. A photographic emission permits the simultaneous recording of spectral lines and background over a broad wavelength range. For these reasons excellent dc arc methods have been developed for general survey or universal analysis of "unknown" samples [33–35]. Such methods cover up to 67 elements. Because of line coincidences and dynamic range limitations, the number of spectral lines needed for the coverage of 67 elements is a multiple of the number of elements, and may range from 400 to 600.

Since even large polychromators can hardly accommodate some 60 exit slits and PMTs, their use for general survey analysis can only be considered if sample diversity is restricted and many compromises can be tolerated. Polychromators are compatible, however, with the character of dc arc excitation, which requires integration of all spectral-line intensities over the total burning time. Obviously, monochromators cannot be used here.

The steady emission characteristics of plasma sources do not conflict with a sequential mode of measurement, whence approaches to semiquantitative analysis [36] have been reported for slew-scan systems. Improvements in wavelength positioning accuracy, speed, spectral resolution, and dynamic range can be expected to further stimulate the use of monochromators for semiquantitative analysis. Sample size and analysis time will impose severe limitations, however.

Because, on the other hand, photographic detection meets the difficulty of adequately recording the low UV region ($\lambda \leq 240$ nm) and the problems associated with emulsion calibration, limited dynamic range, and the rather cumbersome quantitative intensity measurements (cf. Section 1.4), neither of the three instruments—polychromator, monochromator, or spectrograph—ideally fulfills the requirements upon spectroscopic apparatus for general survey analysis. Therefore, if extensive and frequent general survey analyses are required in a laboratory, the combination of spectrograph and computer-controlled monochromator or polychromator, or even the combination of the three instruments, is indispensable. Setting up such analysis systems will demand considerable investments of which the instrument cost, though appreciable, will form the smallest fraction.

The procedures and expedients used for the qualitative interpretation of photographically recorded spectra are discussed in detail in classical works on emission spectroscopy, for example, [37–39].

The procedure noted here is followed. Spectra of samples, together with one or more iron spectra, juxtaposed with the sample spectra, are recorded on a plate or film. Part of the developed plate is subsequently projected with 15- or 20-fold magnification onto a chart carrying the reproduction of the corresponding part of a magnified iron spectrum (Fig. 1.4). The plate is positioned so that the projected iron spectrum exactly matches that on the chart. Wavelengths of lines in the sample spectrum can now be determined by interpolation between the known wavelengths of the iron lines stated on the chart. Often the position of the analysis lines is indicated with respect to the iron lines, thereby permitting a rapid check on the presence or absence of these lines in the sample spectrum. Ideally the indicated analysis lines and their intensity ratings should be specific to the excitation source and excitation conditions used. For dc arc procedures such data are inherent in the description of the procedures [33–35]. Data relevant to ICPs will be discussed in Chapter 7.

1.5. QUALITATIVE AND SEMIQUANTITATIVE ANALYSIS

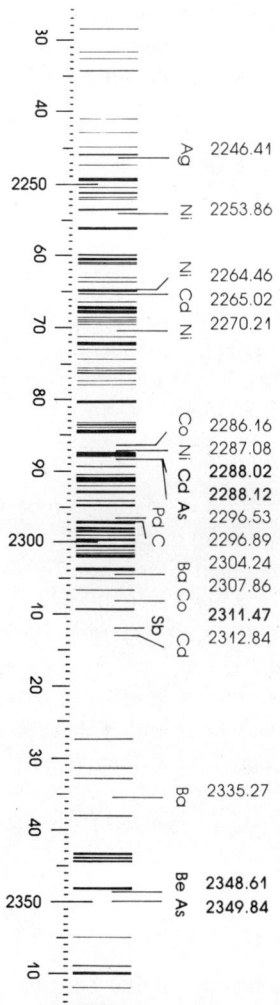

Figure 1.4. Reproduction of a part of a photographically recorded iron spectrum with indications of the positions and wavelengths of analysis lines. Such an iron spectrum may be available on charts or on a "master plate." In the former, a magnified sample spectrum is projected on the chart using a simple spectrum projector (cf. Fig. 7.20); in the latter, the sample spectrum and the master plate spectrum are projected juxtaposed using a double spectrum projector. In both methods the sample plate carries an iron spectrum to be matched with the reference iron spectrum on the chart or master plate. The spectrum shown here has been reproduced from a master plate. (Courtesy of Spex Industries, Inc., Metuchen, NJ.)

A qualitative analysis is *not* carried out by identifying *all lines* in the sample spectrum, but by establishing the presence or absence of chemical *elements* in the sample. This is accomplished by ascertaining the presence or absence of the strongest lines of the elements in the spectrum. The absence of the strongest line(s) of an element gives certainty about the absence of the element(s). In contrast, the presence of a line at the position where a strong line of an analyte will appear can prove the presence of the analyte unambiguously only if coincidences with lines of concomitants are definitely precluded. Ensuring freedom from coincidences necessitates a judicious iteration during the search of the spectrum which, in turn, requires much skill and care from the analyst, as well as the availability of at least semiquantitative data on the relevant sensitivities of the analysis and interfering lines.

1.6. QUANTITATIVE ANALYSIS: CALIBRATION AND INTERFERENCES

1.6.1. Basic Concepts

The various concepts used in quantitative emission spectroscopy are clearly defined and discussed in the IUPAC documents "Nomenclature, Symbols, Units and Their Usage in Spectrochemical Analysis" [40, 41]. These documents not only provide useful recommendations of the definition and use of terms, symbols, and units, but, owing to their narrative yet concise style, may also serve as a general introduction to the subject.

Quantitative analysis is based on an empirically established relationship between the measure (x) of the signal and the concentration (c) of the analyte in the sample. The relationship $x = g(c)$ is called "analytical calibration function," and the graph corresponding to it is termed "analytical (calibration) curve." The function or curve is determined by observations on reference samples of known concentrations frequently called "standards" or "standard samples."

The inverse function $c = f(x)$, often used in applications, is called "analytical evaluation function." The distinction between analytical evaluation and calibration functions assumes importance for multicomponent systems when the measures for the individual components are interdependent because of various interelement effects or interferences.

1.6.2. Calibration Functions

The most frequently assumed form of the calibration function is the Scheibe–Lomakin equation.[2]

$$x_A = Sc^n \qquad (1.6)$$

where ideally n is equal to unity. Departures in both directions $n < 1$ or $n > 1$ occur, the former being the more easily explained as due to self-absorption or self-reversal of spectral lines (Section 4.4.3). However, other less obvious factors may each give a contribution that makes n depart from unity: for example, data reduction using incorrect emulsion calibration constants in spec-

---

[2]In this text calibration is discussed in terms of concentration, but the discussion applies equally well to absolute amounts q.

1.6. QUANTITATIVE ANALYSIS: CALIBRATION AND INTERFERENCES

trography, the use of too large or too small a background signal (x_B) in the determination of the net line signal (x or x_A) from the gross line signal (x_{A+B}), according to

$$x_A = x_{A+B} - x_B \tag{1.7}$$

or simply the statistical uncertainty of the measurements.

In analysis, one hardly bothers about explaining departures of n from unity; the only goal is to determine the calibration (or evaluation) function in such a way that correct and/or consistent analysis results are derived from it.

If a small concentration range is covered, one assumes proportionality between x_A and c:

$$x_A = S \times c \tag{1.8}$$

The proportionality constant S now represents the sensitivity, which is generally defined as the derivative

$$S = dx_A/dc \tag{1.9}$$

Calculations usually involve the assumption of a linear relationship

$$x_A = S \times c + x_0 \tag{1.10}$$

rather than a proportionality, the intercept x_0 taking account of any systematic or incidental, but constant departures of x_A from the true net signal.

The determination of relationship Eq. (1.10) from experimental data using a least-squares fit implicitly assumes a constant standard deviation in x_A over the range covered. This may be justified, however, for a small concentration range only (cf. Section 4.2.2 and [42–46]).

If the calibration covers more than one order of magnitude in concentration, and the lower end of the range is above 50 times the detection limit, the relative standard deviation of x_A is frequently constant. Then, a least-squares fit based on the logarithmic form of Eq. (1.6) is statistically the more correct approach [42, 45]:

$$\log x_A = n \log c + \log S \tag{1.11}$$

For this reason, and because of the logarithmic response of the photographic emulsion, the use of logarithmic calibration functions is customary in spectrography. Two points should be noted here.

1. In a logarithmic calibration function it is not the slope but the intercept which is the more important parameter, since it represents a measure of the sensitivity; the slope, on the contrary, gives an indication about the linearity of the nonlogarithmic plot.
2. The linearity of a logarithmic calibration function depends to a consid-

erable extent on the correct definition of the net signal. Too large or too small a background correction can cause substantial curvature, as is illustrated in Fig. 4.26.

The large dynamic range of ICP-AES (cf. Section 4.4) generally permits the use of calibration functions as in Eqs. (1.10) and (1.11). This has become so obvious an approach that ICP spectroscopists not acquainted with other AES methods might be surprised to hear that, for example, in metal analysis using spark excitation complex calibration functions, such as polynomials, are indispensable. Useful discussions of this topic can be found in Slickers' book [47] and Scheeline's review [48].

Calibrations using equations of the type (1.6), (1.8), (1.10), or (1.11) are designated "absolute" because the analysis signal itself is taken as a measure of concentration, albeit in arbitrary relative units (volts, counts) relevant to the particular instrument only. "Absolute" calibration in this sense is feasible in flame and plasma spectroscopy, but generally meets objections in other branches of emission spectroscopy, where it is necessary to measure the analysis signal relative to a reference signal (x_R) emitted by a reference element, also called "internal standard." As reference elements one chooses either a (major) constituent present in all analysis and reference samples or elements specifically added for this purpose. Internal standards should fulfill a number of conditions, which are treated in standard texts (e.g., [49–51]). The benefits of internal standardization in ICP-AES have been recently extensively studied by Belchamber and Horlick [52], Myers and Tracy [53], and Lorber et al. [54] (cf. Section 4.2.4).

In trace analysis it may be advantageous to use the background signal as reference. This also leads to convenient expressions for the relationships between the detection limit and the instrumental parameters of the system (cf. Section 4.1.10).

1.6.3. Interferences and Approaches to Circumvent Them

The presence of concomitants in the sample can cause interferences, that is, systematic errors in the measure of the signal, so that substitution of the value of the signal into an evaluation function determined, with reference samples which do not contain these concomitants, does not yield a correct result.

An interference may be due to a particular concomitant or to the combined effect of several concomitants. A concomitant causing an interference is called an "interferent."

Interferences can arise not only from differences in chemical composition between analysis and reference samples, but also as a result of differences in physical properties such as the crystallographic structure or particle size in the case where solid samples are directly presented to the excitation source (e.g., in arc or spark AES).

The prime measure taken to avoid errors due to interference is the use of reference samples whose chemical composition and physical properties resemble those of the analysis samples as closely as possible. This is denoted as the *simulation technique*.

Other approaches for reducing errors due to interferences are the *reference element technique*, already discussed above, the *(analyte) addition technique*, the *buffer addition technique*, and the *isoformation technique*.

1.6.3.1. Addition Technique

Successive known amounts of the analyte are added to aliquots of the sample. The net measures on the samples thus obtained are plotted against the added concentrations. This plot is extrapolated to intercept the negative concentration axis. The analytical result is found from the corresponding concentration value.

Essentially one assumes a straight calibration curve and determines the sensitivity S from the increases in the net signal (Δx_1, Δx_2, Δx_3) produced by the successive increases in concentration (Δc_1, Δc_2, Δc_3). This sensitivity value is used to determine the concentration in the original sample as $c = x_A/S$, where x_A is the net signal measured in the sample without addition. It is assumed that the same value of the sensitivity applies to the original and the added amounts of analyte. This is a fair assumption for solutions, but it does not necessarily hold true for solids where differences in crystallographic structure between the sample and the added compounds may give rise to marked differences in sensitivity between the signals originating from the analyte in the sample and the added analyte.

The analyte addition technique can correct for the interferences from concomitants in that it takes their effects on the sensitivity into account, but it does require the determination of the *true* net signal in the original sample and therefore it does not circumvent in trace analysis the fundamental problem of having to subtract a correct background signal from the gross line signal (cf. Section 7.5).

Techniques using multiple additions of analytes and interferents can correct for interferences other than changes in sensitivity, in particular spectral interferences (see Sections 1.6.5 and 7.6).

1.6.3.2. Buffer Addition Technique

An additive called "spectrochemical buffer" is added to both the analysis and reference samples with the aim of making the measure of the analyte less prone to variations in interferent concentration(s). In flame spectrometry, buffers are known as "suppressors," "releasers," "protective agents," "ionization buffers," and "volatilizers" depending on the type of effect they exert [40, 41]. The principal type of buffer used in high-temperature excitation sources is the ionization buffer which is added to govern the temperature and the electron concentration and, thus, the degrees of ionization of the various elements in the plasma. Elements having a low ionization potential (Li, Na, K, Rb, or Cs) are effectively used for this purpose.

Buffers are ineffective and therefore redundant in (argon) ICPs because analytically favorable conditions can be found under which the temperature and the electron concentration are governed by mechanisms that are hardly influenced by the sample composition.

In spectrography involving the dc carbon arc ionization buffers are often indispensable, especially in "universal analysis methods," to control the conditions in the plasma (temperature, electron concentration, transport of atoms and ions) and smooth the volatilization of the sample from the lower electrode [11, 49]. Lithium salts mixed with graphite powder are frequently used [11, 49].

1.6.3.3. Isoformation Technique

Isoformation refers to pretreatment procedures intended to eliminate the effects of differences in crystallographic structure between analysis and reference samples on the analytical results. Fusion with a suitable flux (e.g., lithium metaborate) is a well-known technique in dc arc spectrography of geological materials [55–58]. Fusion also makes the results less dependent on particle size and leads to homogeneous samples for analysis.

If liquid sampling methods are applied to the analysis of solid samples, isoformation is so implicit that it is hardly ever explicitly referred to. It is, however, one of the strong points of these methods.

1.6.4. Classification of Interferences

It is convenient to distinguish between spectral and nonspectral interferences. Spectral interferences arise from the incorrect isolation of the net analysis signal from the composite radiation that passes the "spectral window" tuned to the analysis line. Spectral interferences are *additive* and connected with the intercept in Eq. (1.10). Nonspectral interferences are related to sensitivity changes (slope in Eq. (1.10) and are therefore *multiplicative*.

For illustration we assume for an analysis signal in a sample affected by multiplicative and additive interference the relationship

$$x_A^* = S^* \times c_{\text{true}} + x_0 + \Delta x_0 \tag{1.12}$$

whereas Eq. (1.10) applies to the reference samples.

1.6. QUANTITATIVE ANALYSIS: CALIBRATION AND INTERFERENCES

The concentration in the sample is then found to be

$$c_{\text{found}} = \frac{S^*}{S} \times c_{\text{true}} + \frac{\Delta x_0}{S} \quad (1.13)$$

The error thus manifests itself as a multiplicative (S^*/S) and an additive ($\Delta x_0/S$) effect. Both S^*/S and $\Delta x_0/S$ are functions of the concentration of the interferent, though not necessarily linear functions.

1.6.5. Multicomponent System with Multiple Interferences

In a system of n components we can expect that in principle the sensitivity of an analysis signal of any component also depends on the concentrations of the $n - 1$ other components, while each of the $n - 1$ other components may also contribute additively by spectral interference. Thus the signal measured at wavelength 1 (= analysis line of component 1) can be mathematically expressed as

$$x_1 = S_{11}(c_2, c_3, \ldots, c_n)c_1 + S_{12}c_2 + S_{13}c_3 + \ldots S_{1n}c_n \quad (1.14)$$

where the first index of S refers to the wavelength at which the measurement is made and the second index to the component that contributes at that wavelength. For example, S_{12} is the (partial) sensitivity of the signal due to component 2 at the wavelength of component 1 [59].

Similarly for the signals at the wavelengths 2 and n we have

$$x_2 = S_{21}c_1 + S_{22}(c_1, c_3, \ldots, c_n)c_2 + S_{23}c_3 + \ldots S_{2n}c_n \quad (1.15)$$

$$x_n = S_{n1}c_1 + S_{n2}c_2 + S_{n3}c_3 + \ldots S_{nn}(c_1, c_2, \ldots, c_{n-1})c_n \quad (1.16)$$

If one analysis line is used for each component, there exist n equations with n^2 coefficients, which can be determined, for example, by multiple additions if certain conditions are fulfilled, such as the absence of multiplicative interferences (all S_{ii} constant). Then it is possible, at least within limited ranges of concentration, to determine the corrected concentrations of all n components in a sample from the reversed set of equations having the form

$$c_i = \sum_{j=1}^{j=n} a_{ij} x_j \quad (1.17)$$

where c_i is the concentration of the ith analyte and the summation over index j extends over all components (= analytes) including analyte i.

For polychromators a variety of procedures for handling calibrations and interference corrections for multicomponent analyses has been devised. Such procedures are implemented in the manufacturer's software available with the instruments. The use of nonlinear calibration functions (e.g., polynomials) and alternative versions or extensions of the software procedures originally developed for x-ray spectrometry are not uncommon here.

The problem of interference correction in multicomponent analysis by ICP-AES has been the most extensively treated by Botto [60–63] (cf. Sections 7.6 and 7.7.6). Generalized treatments with emphasis on the mathematical solution of the problem have been published by Kowalski's group [64–66] and Fujiwara et al. [67].

1.7. SAMPLE INTRODUCTION AND EXCITATION SOURCE

1.7.1. General

Ideally one should be able to offer the sample such as it is to the excitation source. Meeting this ideal is an exception, however, rather than the rule. Sample preparation may comprise pulverization, sieving, mixing with additives, fusing, pressing, casting, and/or polishing in the case that the source "accepts" solids directly. This applies to arcs, sparks, glow discharges, and lasers. On the other hand, flames, plasma sources (ICPs, DCPs, CMPs, and MIPs), and furnaces are primarily intended for the direct analysis of liquids. The following details and differentiations should be borne in mind, however.

1.7.2. Arc

DC arcs are chiefly used for the analysis of nonconducting materials, ac arcs for metal analysis. For dc arc analysis, the samples are pulverized, mixed with graphite powder and/or additives (buffers, carriers), and tamped into a hole in a graphite rod, which, during arcing, forms the lower electrode of the arc. The sample evaporates thermally when the electrode wall is consumed so that the sample is exposed to the high-temperature arc plasma (5000–6000 K).

Although the ICP has replaced the dc arc for many purposes, the latter still is unexcelled as a source for general survey analysis where it has the advantage of the simplicity of sample preparation when solids are concerned. Well-known methods for general survey analysis have been described by Addink [35], Harvey [34], Kroonen and Vader [33], Wang [39], and Mitteldorf [68].

The dc arc is only exceptionally used for direct analysis of metals, namely in two versions of the so-called globule arc. The first version is applied to trace determinations in relatively noble metals, such as copper [47, 69, 70]. The metal is placed in a conically shaped crater of a graphite electrode. The arc first melts the metal, then the trace elements are distilled from the globule or bead formed.

The other version of globule arc technique was developed by Fassel et al. [71, 72] for the spectrographic determination of gaseous elements in metals.

The metal is placed in a specially designed carbon support electrode and arced in argon. The arc melts the metal specimen, and the molten sample dissolves the retaining wall of the electrode and forms a globule. Chemical reactions with the dissolved carbon cause oxygen, nitrogen and hydrogen to be evolved.

Increasing interest in the direct analysis of solids by ICP-AES (Part 2, Chapter 7) refocuses attention on earlier studies of selective volatilization in arcs. This is so because one of the novel ICP techniques uses a cupped graphite rod for the introduction of solids into the ICP. This approach may substantially improve the limits of detection and limits of determination if relatively volatile elements are selectively evaporated from a refractory matrix which itself produces a line-rich spectrum.

Selective volatilization has been the subject of many studies involving dc arcs. Special attention has been paid to thermochemical reactions in the electrode in general and to the so-called carrier distillation technique in particular. The following references may facilitate access to the extensive literature: Ahrens and Taylor [51], Zaidel et al. [73], Zilbershtein [74], Schroll [75, 76], Rusanov [77], Rosza [78], and Boumans [11]. For illustration Table 1.2 shows the order of volatilization of elements in the arc, while Fig. 1.5 depicts a set of volatilization curves.

Table 1.2. Order of Volatilization of Elements in the DC Carbon Arc[a]
(Reproduced with permission from L. H. Ahrens and S. R. Taylor, *Spectrochemical Analysis*, p. 82. Copyright 1961 The Benjamin/Cummings Publishing Co., Menlo Park, CA.)

		Tendency to Volatilize in Arc		
		As Oxides[b]		
As Elements	As Sulfides	Volatile	Medium Volatile	Nonvolatile
Hg > As > Cd > Zn > Sb ≥ Bi > Tl > Mn > Ag > Sn, Cu > In, Ga, Ge > Au > Fe, Co, Ni > Pt ≫ Zr, Mo, Re, Ta, W	As, Hg > Sn, Ge ≥ Cd > Sb, Pb ≥ Bi > Zn, Tl > In > Cu > Fe, Co, Ni, Mn, Ag ≫ Mo, Re	As, Hg > Cd > Pb, Bi, Tl > In, Ag, Zn > Cu, Ga > Sn > Li, Na, K, Rb, Cs >	Mn > Cr, Mo?, W?, Si, Fe, Co, Ni > Mg > Al, Ca, Ba, Sr, V >	Ti > Be, B?, Ta, Nb > Sc, La, Y and many rare earths > Zr, Hf

[a] From [51].
[b] Also sulfates, carbonates, silicates, and phosphates.

ORDER OF VOLATILIZATION IN THE D-C ARC

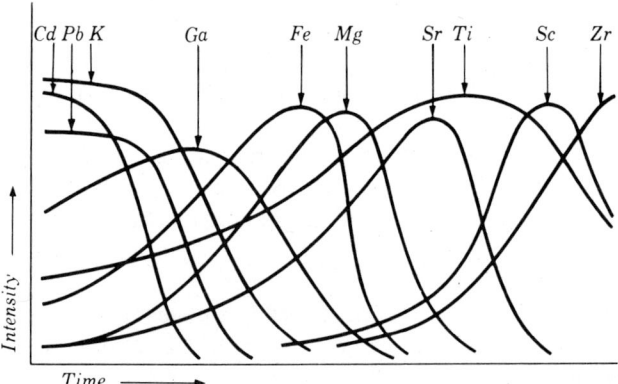

Figure 1.5. Set of somewhat idealized volatilization curves illustrating selective volatilization (fractional distillation) in the dc arc [51]. (Reprinted with permission from L. H. Ahrens and S. R. Taylor, *Spectrochemical Analysis*, p. 81. Copyright (1961), The Benjamin/Cummings Publishing Co., Menlo Park, CA.)

Arc spectroscopists have also explored "powder insufflation" and "powder sifting" techniques [77, 79–83] whereby finely ground powders of minerals or ores are either swirled by a gas stream and carried to an arc-like source or dropped into a horizontally burning arc (Figs. 1.6 and 1.7). Fine grinding of the samples (down to 10 μm) is necessary to ensure complete evaporation, which, in turn, provides maximum intensities of the analysis lines and eliminates systematic errors due to differences in crystallographic structure between analysis and reference samples [81].

Various types of arcs have also been used for the analysis of liquids. One method is known as the "Scheibe–Rivas" [84], "flat-top" [85–87], or "graphite arc" [88] technique. A droplet of solution is placed on top of a specially coated graphite rod, the solvent is evaporated, and the residue is arced. Alternative proposals are the "Gordon arc" [89] and the "vacuum cup technique" [90].

1.7.3. Spark

The various types of sparks (condensed discharges) are primarily used for the direct analysis of metals in production control. Nonconducting samples can be analyzed after grinding, mixing with graphite or metal powder, and briquetting. Useful entries to the literature are Slickers' book [47] and the reviews of Scheeline [48], Laqua [12], Walters [91], Belcher [92], and Doolan and Belcher [93].

Solution methods using spark excitation [78] have become known as the

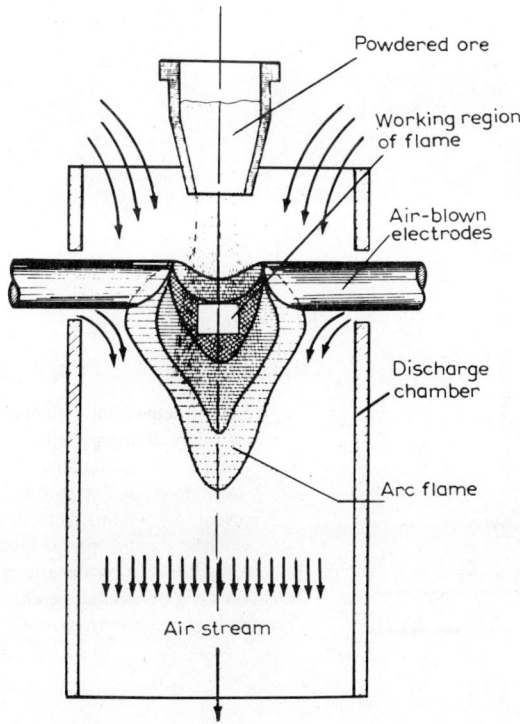

Figure 1.6. Method of introducing powdered ore into an arc in a stream of air [80]. (Reprinted with permission from A. K. Rusanov and V. G. Khitrov, "Spectrographic Analysis of Ores by Introducing the Powder into the Arc in a Stream of Air," *Spectrochim. Acta* **10**, 405 (1958). Copyright (1958), Pergamon Journals, Oxford.)

"porous cup" [94], "rotating disk" [95], "flat top" [84], "rotating platform" [96], and "plastic cup" [85] techniques. Baer and Hodge [85] assessed the five techniques as to speed, sensitivity, reproducibility, sample consumption, and overall convenience.

The flat top technique is also known as "copper spark" [86, 97] and "graphite spark" [87] method. In both versions, a drop of solution is evaporated on flat-topped electrodes, which are then subjected to spark excitation. If graphite is used, the electrodes are treated with a solution of grease to prevent seepage of the sample into the porous material.

The rotating disk ("rotrode") technique has found the widest application, inter alia, in oil [47, 98–100] and silicate [99] analysis. The availability of commercial equipment and the fact that suspended matter can be analyzed are obvious reasons for the success of the method. The technique (Fig. 1.8) employs a graphite disk as the lower electrode ("rotrode") of a spark discharge.

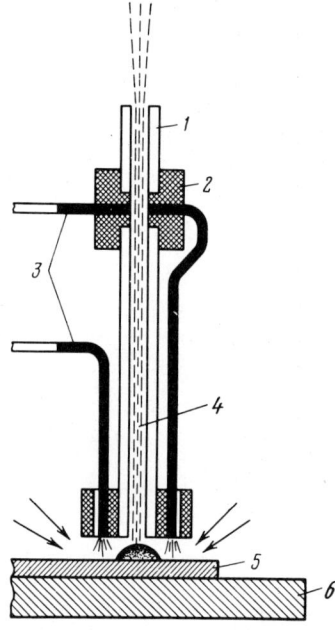

Figure 1.7. Method for introducing powdered samples into an arc discharge using a spark discharge and an air stream: *1* – suction tube, *2* – plexiglass connector, *3* – electrodes of low-power spark, *4* – air stream with powder, *5* – rotating metal disk, *6* – insulator. As the powder on the disk passes between the sparks, air streams (symbolized by the arrows) blow the powder from two sides toward the opening of the tube. (Reproduced with permission from A. K. Rusanov [77]).

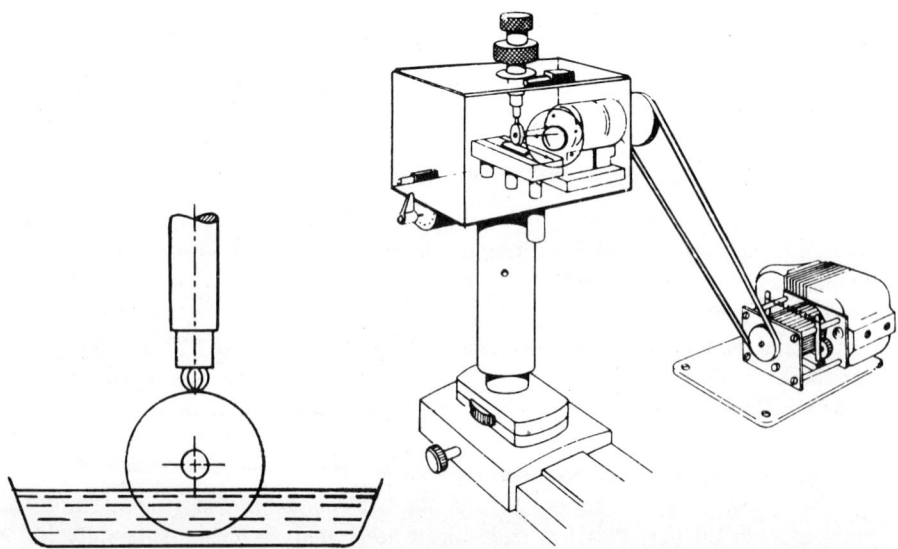

Figure 1.8. Principle of rotating disk technique for solution analysis. Left: Sample boat with sample solution, rotating electrode ("rotrode") counter electrode and spark. Right: Schematic diagram of rotrode device. (Reproduced with permission from O. Szakács [101].

1.7. SAMPLE INTRODUCTION AND EXCITATION SOURCE

The disk rotates through the sample liquid contained in a sample boat. Small amounts of sample are entrained and subsequently sparked when the relevant part of the disk passes the spark gap. Szakács [101] reviewed the literature. Using radioactive tracers, Nickel [102] investigated the transport processes in the disk and the effects of these processes on the emission.

1.7.4. Glow Discharge

The glow discharge (Fig. 1.9), as introduced by Grimm [104], is operated at reduced pressure and is used for the direct analysis of solids. The sample acts as cathode and is flat, usually disk-shaped. Metals can be directly analyzed, nonconducting materials must be mixed with, for example, copper or silver powder, and be pressed into pellets. Ablation is achieved by cathodic sputtering. Keys to the relevant literature are Laqua's survey [12], which includes a table with analytical applications, and articles by Ferreira et al. [105], Lombdahl and Sullivan [106], and Ko [107].

1.7.5. Hollow Cathode Discharge

The hollow cathode discharge (HCD) is a glow discharge with a hollow cathode (Fig. 1.10). Solid samples are introduced as the cathode itself or as disks, pel-

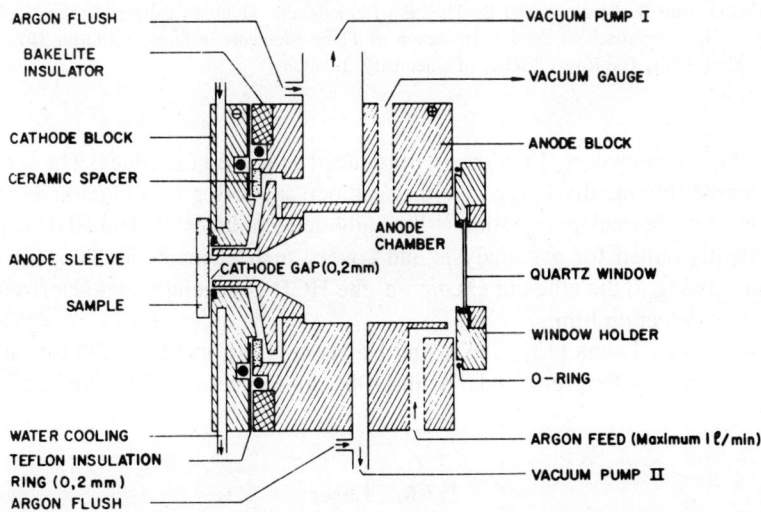

Figure 1.9. Schematic diagram of conventional front-view Grimm-type glow discharge source. The device has a cylindrical shape. The diagram shows a section through the axis. The diameter of the cathode burn spot is approximately 10 mm. (Reproduced with permission from N. P. Ferreira [103].)

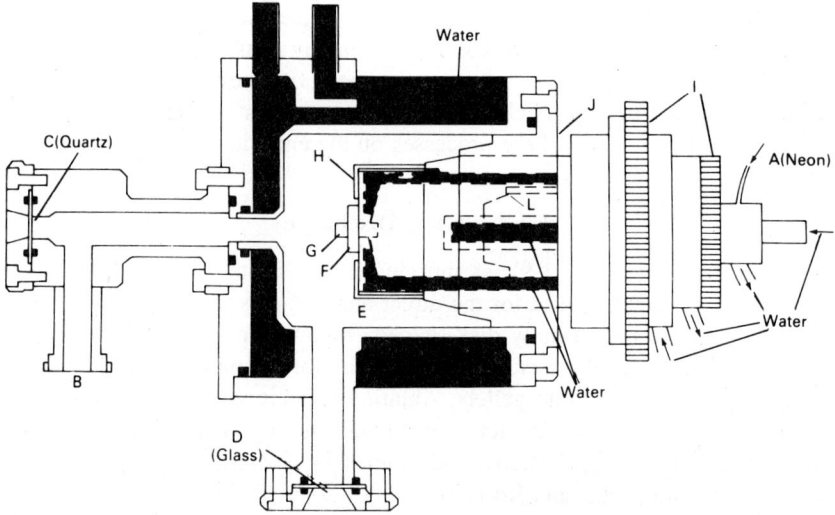

Figure 1.10. Schematic drawing of a double-chamber hollow cathode lamp. The lamp consists of a body (anode), made of stainless steel, which is evacuated and then flushed with neon through the inlet, A, and outlet, B. The sample, G, has the form of a tube with a length of about 30 mm and an inner diameter of 4 mm. The sample and holder, F, constitute the cathode, which is insulated by the glass cap, H. The cathode assembly is water cooled, E. The distance between anode and cathode can be adjusted with the wheels, I. The radiation passes through the quartz window, C, to the spectrometer. The discharge can be viewed through the window, D [108]. (Reprinted with permission from B. Berglund and B. Thelin, "Demountable Double-Chamber Hollow-Cathode Lamp: a New Approach to the Determination of Trace Elements in Steel," *Analyst* **107,** 868. Copyright (1982), The Royal Society of Chemistry, London.)

lets, chips, or powders. Liquids are introduced as solution residues. The sample is released into the discharge by either cathode sputtering ("cold cathode techniques") or thermal evaporation ("hot cathode techniques"). The HCD is also excellently suited for gas analysis and covers such elements as halogens and sulfur. Owing to the efficient excitation, the HCD is generally characterized by very low detection limits.

Reviews by Laqua [12], Slevin and Harrison [109], and Caroli [110], and a recent article by Berglund and Thelin [108], give access to the abundant literature.

1.7.6. Laser

Lasers are used for the ablation of solids and frozen liquids in atomic emission and atomic absorption spectrometry as well as in mass spectrometry. Nd-glass resonators (λ = 1060 nm), ruby lasers (λ = 694 nm), and Nd/YAG systems

1.7. SAMPLE INTRODUCTION AND EXCITATION SOURCE

($\lambda = 1060$ nm) are the most frequently employed. The method can be applied to both conducting and nonconducting samples and is particularly suited for local analysis. The laser impact produces a vapor plume that emits a spectrum. However, for most purposes one does not use this primary radiation but the spectrum obtained by additional excitation by spark, high-frequency discharge, or electron beam (Fig. 1.11).

A chapter by Laqua [111], a review "Laser Evaporation in Atomic Spectroscopy" by Dittrich and Wennrich [112], and a book by Moenke and Moenke-Blankenburg [113] lead to the relevant literature.

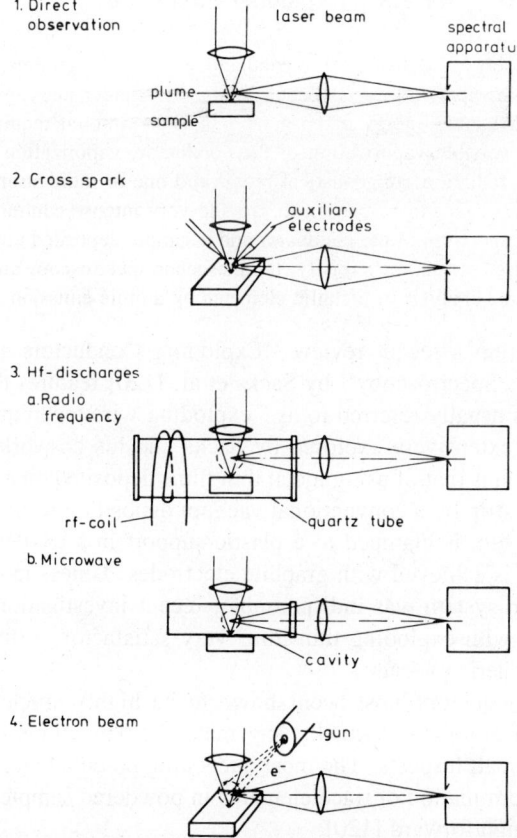

Figure 1.11. Various possibilities of atomic emission spectroscopy using laser atomizers [111]. (Reprinted with permission from K. Laqua, "Analytical Spectroscopy Using Laser Atomizers," in N. Omenetto, ed., *Analytical Laser Spectroscopy*, Ch. 2, p. 82. Copyright (1979), Wiley, New York.)

1.7.7. Furnace

The carbon furnace is the best known electrothermal atomizer in AAS. Ottaway's group has explored its potential as an emission source [114–117]. Obviously, sample introduction is the same as in AAS. Therefore, carbon furnace AES is an emission technique suitable for the analysis of small volumes of liquid samples. The method requires a monochromator of essentially higher resolving power than in AAS, while wavelength modulation is essential to achieve adequate performance. Additional possibilities offer a carbon furnace that acts concomitantly as atomizer and cathode in a low pressure discharge [118, 119].

1.7.8. Exploding Conductors

When electrical energy is delivered to a conductor at a rate much greater than its dissipation ability, rapid vaporization may occur. With a low inductance capacitive discharge circuit and total available energy much greater than the amount required for the thermodynamically reversible vaporization of the conductor, vaporization may be accompanied by intense radiation, an acoustical report and one or more strong shock waves. Experimental conditions can be chosen to provide very intense continuum radiation or a line-rich atomic spectrum of the conductor and a sample deposited on its surface. The former has been used for photochemistry, luminescence spectroscopy and laser pumping and the latter for the analysis of metallic elements by atomic emission spectroscopy.

This quote from a recent review "Exploding Conductors as Atomization Cells for Atomic Spectroscopy" by Sacks et al. [120] features the principle of a method that is usually referred to as "exploding wire technique." Its potentials have been extensively explored by Sacks and his co-workers. The most promising version is that of using metal thin films deposited on a polypropylene or polyethylene strip by a conventional vacuum deposition technique. The substrate with thin film is clamped to a plastic support in a cassette (Fig. 1.12). Electric contact is achieved with graphite electrodes. Unless in earlier studies, where a vacuum system was indispensable, recent investigations have shown [121, 122] that with exploding thin films very satisfactory results can be obtained at atmospheric pressure.

Exploding conductors have been shown to be highly efficient atomization cells capable of atomizing virtually any material. Thin films are superior to wires and foils in all respects. The most interesting potential area of application is the direct determination of trace elements in powdered samples. The method is rapid and straightforward [120].

1.7.9. Flames

The flame is the oldest established spectroscopic technique for liquid analysis and the immediate predecessor of plasma sources. It is still widely used as

1.7. SAMPLE INTRODUCTION AND EXCITATION SOURCE

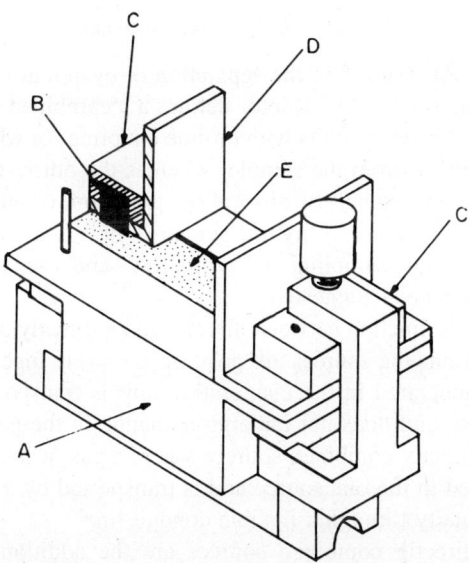

Figure 1.12. Thin film cassette for emission spectroscopy using the exploding conductor technique: *A*, plastic base, *B*, graphite electrodes, *C*, electrical contacts, *D*, shields to prevent sparkover, *E*, thin film on polymer substrate [120]. (Reprinted with permission from R. D. Sacks, J. M. Goldberg, R. J. Collins, S. Y. Suh, "Exploding Conductors as Atomization Cells and Excitation Sources for Atomic Spectroscopy," *Progr. Anal. Atom. Spectrosc.*, **5**, 120. Copyright (1982), Pergamon Journals, Oxford.)

atomizer in AAS and as excitation source for alkalis in AES. The key literature is covered in Section 1.8.

1.7.10. Plasma Sources

The desired increase in the multicapability of flame-like sources for liquid analysis was realized through the development of various types of high-temperature plasma sources, as explained in Section 1.3. ICPs, DCPs, and CMPs were initially intended primarily for the analysis of nebulized solutions, but direct analysis of solids and sample introduction using electrothermal atomizers is receiving increasing attention. Microwave induced plasmas (MIPs) have been explored for the analysis of both nebulized liquids and gases or vapors. They find application primarily as element-selective detectors in gas and liquid chromatography.

Plasma sources in general and the ICP in particular constitute the subject of this book. Aspects directly related to sample introduction are covered in Chapters 2, 3, and 6 and in Part 2, Chapters 1–8.

1.7.11. Combined Sources

Reviewing "New approaches to the separation of evaporation and atomization in atomic spectrometry" [123], Kántor defines a "combined source" as a system of two high-temperature units with on-line coupling, of which one produces a gas, vapor or aerosol from the sample, whereas the other serves for the generation of spectroscopic signals by absorption, emission, or fluorescence. Terms such as "combined source," "hybrid source," or "two-stage source," and expressions as "separate sampling and excitation" and "separate volatilization and detection" have been suggested.

Kántor makes a distinction between directly and indirectly combined sources. In direct combinations the sources are positioned closely together or overlap in part. The vapor generated in the evaporation unit is transported by diffusion, convection, or both and does not undergo a change in the gaseous state. This contrasts with indirectly combined sources where a gas or highly dispersed dry aerosol is generated in the one source and is transported by a carrier gas to the second source, usually through a flexible conduit line.

Examples of directly combined sources are the additional electrothermal heating of graphite rods and cups or metal wires introduced into flames, and laser ablation in a graphite furnace. Combinations of spark, laser, or graphite furnace with an ICP are examples of indirectly combined sources.

Kántor's review shows that a large variety of combinations has been explored. On the whole, such combinations provide many possibilities of which those applied in ICP–AES will be discussed in Chapter 6 and Part 2, Chapters 7 and 12.

1.8. GUIDE TO THE LITERATURE ON EMISSION SPECTROSCOPIC ANALYSIS

1.8.1. Introduction

Section 1.8 points to the principal sources of information on AES as available in the form of books, chapters, reviews, and journals. Details on review articles specifically dealing with the ICP are given in Section 3.3. Wavelength tables and related works are covered in Sections 7.4.3 and 7.4.4.

In contrast to the custom adopted throughout this book, the works mentioned in this literature survey are referred to by quoting the name of the author, followed by the year of publication. These references are listed in Section 1.8.7. The subsequent sections list abstract compilations, annual and biennial reviews, and journals relevant to atomic emission spectroscopy.

1.8. GUIDE TO THE LITERATURE ON EMISSION SPECTROSCOPIC ANALYSIS 35

1.8.2. Encyclopedical Information

Although not up to date, Clark's *Encyclopedia of Spectroscopy* (1960) contains compact and still useful information on many items of emission spectroscopy. An outstanding survey of emission spectroscopy by Laqua (1980) can be recommended to readers understanding German. This article, which excels in both compactness and coverage of the subject, includes tables with (1) detection limits of various spectroscopic methods, (2) analytical ranges of methods using medium voltage sparks, and (3) characterizations of analytical applications involving hollow cathode discharges, glow discharges, and microwave induced plasmas.

Useful sources of information with an encyclopedical character are the IUPAC recommendations "Nomenclature, Symbols, Units and Their Usage in Spectrochemical Analysis" (IUPAC, 1972–1982).

1.8.3. Classical Treatises on the Emission Spectroscopic Methodology

The most recent works dealing with the principles of analytical emission spectroscopy are the textbooks by Slavin (1971), Grove (1971/1972), Schrenk (1975), Grove (1978), Rusanov (1978), and a chapter by Kántor (1975).

"Methods for Emission Spectrochemical Analysis," published by the American Society for Testing and Materials (ASTM, 1971), should also be mentioned here.

Grove's *Analytical Emission Spectroscopy* (1971/1972) encompasses chapters on historical developments (Grove), origins of atomic spectra (Devlin), prism spectrographs and spectrometers (Faust), gratings and grating spectrometers (Barnes and Jarrell), spectroradiometric principles (Betz and Johnson), excitation of spectra (Boumans), flame spectrometry (Vickers and Winefordner), qualitative and semiquantitative analysis (Wang, Cave, and Coakley), and quantitative analysis (Rosza).

Grove's *Applied Atomic Spectroscopy* (1978) comprises chapters on photographic photometry (Anderson), laser emission excitation spectroscopy (Scott and Strasheim), electrode material and design for emission spectroscopy (Mellichamp), behavior of refractory materials in a dc arc plasma: new approaches for spectrochemical analysis of trace impurities in refractory matrices (Avni), preparation and evaluation of spectrochemical standards (Gillieson), precious metals (Jaeger), petroleum industry analytical applications of atomic spectroscopy (Buell), analytical emission spectroscopy in biomedical research (Niedermeier), and application of spectroscopy to toxicology and clinical chemistry (Berman).

Schrenk's book (1975) covers AES, AAS, and AFS, while Slavin's work (1971) specifically deals with AES. The latter book is appended with a list of sources of information covering the great majority of the books on AES and related subjects up to 1970; therefore, only a few of these will be mentioned below.

The emphasis in the works mentioned is on the methodology of classical emission spectroscopy using flames, arcs, or sparks as excitation sources. Although this methodology is also the basis of modern "plasma emission spectrochemistry," it is hardly treated in the present book; therefore, the various classical treatises remain indispensable as works of reference.

Among the classical works we further list Mika and Török (1974), Török Mika, and Gegus (1978), Zilbershtein (1971/1977), Zaidel, Kaliteevskii, Lipis, and Chaika (1963), Ahrens and Taylor (1963), Sawyer (1963), Seith, Ruthardt and Rollwagen (1970), and Mannkopf and Friede (1975).

Not specifically devoted to emission spectroscopy is Winefordner's book *Trace Analysis*, which covers many items of interest for emission spectroscopists.

1.8.4. Recent Works on Emission Spectroscopic Methodology

Three recent works deserve special attention:

1. Slickers (1980): a practical book with emphasis on routine industrial applications involving sparks, glow discharges and, to a lesser extent, plasma sources.
2. Thompson and Walsh (1983): also a practical book that discusses the methodology (80 pages) and applications (170 pages) of ICP-AES. Ten chapters cover introduction, analytical characteristics, instrumentation, silicate rock analysis, multielement applications in applied geochemistry, gas phase sample injection, discrete sample injection, methods for solid samples, water analysis, the analysis of environmental materials, and ICP-AES now and in the future.
3. Laqua's chapter "Analytical Spectroscopy using Laser Atomizers" in Omenetto's book *Analytical Laser Spectroscopy* (1979).

1.8.5. Works Dealing with Fundamentals of Excitation Sources

Boumans' book (1966) covers the theory of excitation, ionization, and emission for plasmas in LTE with special reference to the dc arc. Part of this theory is also found in Boumans' chapter, "Excitation of Spectra," in Grove (1972). This chapter covers the principles of excitation sources such as arcs, sparks, discharges at reduced pressure, plasma sources, and lasers.

A comprehensive treatment of spark excitation has been given by Mika and Török (1974). Recent progress has been reviewed by Scheeline (1984).

Alkemade and Herrmann's book (1979) deals with the fundamentals of flame

1.8. GUIDE TO THE LITERATURE ON EMISSION SPECTROSCOPIC ANALYSIS

spectroscopy and includes many topics relevant to plasma spectrochemistry. This holds a fortiori for the comprehensive treatment: *Metal Vapours in Flames* by Alkemade, Hollander, Snelleman, and Zeegers (1982) and the chapter "Excitation and De-excitation Processes in Flames" by Alkemade and Zeegers in *Spectrochemical Methods of Analysis* edited by Winefordner (1971).

Other books on fundamentals to be mentioned in this context are the works by Griem (1964), Unsoeld (1955), and Dresvin (1972/1977).

1.8.6. Books on Flame Spectrometry

The methodology of plasma spectrochemistry has a large tangent plane with that of flame spectrometry; therefore, works on flame spectrometry often contain information useful for plasma spectrochemists. In addition to the works referred to in Section 1.8.5 we mention here the works of Herrmann and Alkemade (1963), Mavrodineanu and Boiteux (1965), Dean (1960), Pungor (1967), Dean and Rains (1969, 1971, 1975), and Mavrodineanu (1970).

1.8.7. Bibliography of Sections 1.8.2 to 1.8.6

Ahrens, L. H., and S. R. Taylor (1961). *Spectrochemical Analysis*. Addison-Wesley, Reading, MA.

Alkemade, C. Th. J., and R. Herrmann (1979). *Fundamentals of Flame Spectroscopy*. Adam Hilger, Bristol.

Alkemade, C. Th. J., Tj. Hollander, W. Snelleman, and P. J. Th. Zeegers (1982). *Metal Vapours in Flames*. Pergamon Press, Oxford.

ASTM (1971). *Methods for Emission Spectrochemical Analysis*. American Society for Testing and Materials, Philadelphia, PA.

Boumans, P. W. J. M. (1966). *Theory of Spectrochemical Excitation*. Adam Hilger, London/Plenum Press, New York.

Clark, G. L., ed. (1960). *The Encyclopedia of Spectroscopy*. Reinhold, New York.

Dean, J. A. (1960). *Flame Photometry*. McGraw-Hill, New York.

Dean, J. A., and T. C. Rains, ed. (1969, 1971, 1975). *Flame Emission and Atomic Absorption Spectrometry*, Vols. 1, 2, 3. Marcel Dekker, New York.

Dresvin, S. V., ed. (1977). *Physics and Technology of Low-Temperature Plasmas*. Atomizdat, Moscow (1972); Eng. Transl.: Iowa State University Press, Ames.

Griem, H. (1964). *Plasma Spectroscopy*. McGraw-Hill, New York.

Grove, E. L., ed. (1971, 1972). *Analytical Emission Spectroscopy*, Vol. 1, Pt. 1 and 2. Marcel Dekker, New York.

Grove, E. L., ed. (1978). *Applied Atomic Spectroscopy*, Vols. 1 and 2. Plenum, New York.

Herrmann R., and C. Th. J. Alkemade (1963). *Chemical Analysis by Flame Photometry*. Wiley, New York.

I.U.P.A.C., Analytical Chemistry Division, Commission on Spectrochemical and Optical Procedures for Analysis, Nomenclature, Symbols, Units and Their Usage in Spectrochemical Analysis:
1. General Atomic Emission Spectroscopy, *Pure Appl. Chem.* **30**, 653 (1972); *Spectrochim. Acta* **33B**, 219 (1978).
2. Data Interpretation, *Pure Appl. Chem.* **30**, 99 (1976); *Spectrochim. Acta* **33B**, 241 (1978).
3. Data Interpretation, *Pure Appl. Chem.* **30**, 105 (1976); *Spectrochim. Acta* **33B**, 247 (1978).
4. Radiation Sources, *Pure Appl. Chem.* **53**, 1913 (1981); *Spectrochim. Acta* **41B**, 507 (1986).

Kántor, T. (1975). "Emission Spectroscopy," in G. Svehla, ed., *Wilson and Wilson's Comprehensive Analytical Chemistry*, Vol. V, Ch. 1. Elsevier, Amsterdam/New York.

Laqua, K. (1980). "Emissionsspektroskopie," in *Ullmanns Encyklopaedie der technischen Chemie*, 4th ed., Vol. 5, Verlag Chemie, Weinheim, F.R.G., p. 441.

Mannkopf, R., and G. Friede (1975). *Grundlagen und Methoden der chemischen Emissionsspektralanalyse*. Verlag Chemie, Weinheim, F.R.G.

Mavrodineanu, R., and H. Boiteux (1965). *Flame Spectroscopy*. Wiley, New York.

Mavrodineanu, R., ed. (1970). *Analytical Flame Spectroscopy*, MacMillan, London.

Mika, J., and T. Török (1974). *Analytical Emission Spectroscopy, Fundamentals*. Butterworths, London.

Omenetto, N., ed. (1979). *Analytical Laser Spectroscopy*. Wiley, New York.

Pungor, E. (1967). *Flame Photometry Theory*. Van Nostrand, New York.

Rusanov, A. K. (1978). *Fundamentals of the Quantitative Spectrographic Analysis of Ores and Minerals* (in Russian), Nedra, Moscow.

Sawyer, R. A. (1963). *Experimental Spectroscopy*, 3rd ed. Dover, New York.

Scheeline, A. (1984). *Progr. Anal. Atomic Spectrosc.* Vol. 7, Pergamon Press, Oxford/New York, p. 21.

Schrenk, W. G. (1975). *Analytical Atomic Spectroscopy*. Plenum, New York.

Seith, W., K. Ruthardt, and W. Rollwagen (1970). *Chemische Spektralanalyse*. Springer Verlag, Berlin/Heidelberg/New York.

Slavin, M. (1971). *Emission Spectrochemical Analysis*. Wiley, New York.

Slickers, K. (1980). *Automatic Emission Spectroscopy*. Bruehl Druck & Pressehaus Giessen, Giessen, F.R.G.

Thompson, M. and J. N. Walsh (1983). *A Handbook of Inductively Coupled Plasma Spectrometry*. Blackie, Glasgow/Chapman and Hall, New York.

Török, T., J. Mika, and E. Gegus (1978). *Emission Spectrochemical Analysis*. Adam Hilger, Bristol.

Unsoeld, A. (1955). *Physik der Sternatmosphaeren*. Springer Verlag, Berlin Goettingen Heidelberg.

1.8. GUIDE TO THE LITERATURE ON EMISSION SPECTROSCOPIC ANALYSIS

Winefordner, J. D., ed. (1971). *Spectrochemical Methods of Analysis.* Wiley, New York.

Winefordner, J. D., ed. (1976). *Trace Analysis, Spectroscopic Methods for Elements.* Wiley, New York.

Zaidel, A. N., N. I. Kaliteevskii, L. V. Lipis, and M. P. Chaika (1963). *Emission Spectrum Analysis of Atomic Materials*, State Publishing House of Physicomathematical Literature, Moscow (1960); Eng. Transl.: U.S. Atomic Energy Commission, AEC-tr-5745, Washington, DC.

Zilbershtein, Kh. I., ed. (1977). *Spectrochemical Analysis of Pure Substances.* Khimia Press, Leningrad (1971); Eng. Transl.: Adam Hilger, Bristol.

1.8.8. Abstracts, Compilations, and Regular Reviews

Meggers, W. F., and B. F. Scribner (1941, 1947, 1954, 1959). *Index to the Literature on Spectrochemical Analysis* (4 Vols.). American Society for Testing and Materials, Philadelphia, PA.

van Someren, H. S. E. (1938–1973). *Spectrochemical Abstracts* (19 Vols.). Hilger, London.

Analytical Abstracts. The Royal Society of Chemistry, London.

Annual Reports on Analytical Atomic Spectroscopy (from 1971). The Royal Society of Chemistry, London.

Emission Spectrometry, Fundamental Reviews in: *Analytical Chemistry.* April, even years. American Chemical Society, Washington, DC.

Barnes, R. M., ed. *ICP Information Newsletter* (from 1975). University of Massachusetts, Department of Chemistry, Amherst.

Atomic Absorption and Emission Spectrometry Abstracts. PRM Science and Technology Agency Ltd., London.

1.8.9. Journals

Analusis. Société de Chimie Industrielle et Société Francaise de Chimie, Paris.

Analytica Chimica Acta. Elsevier, Amsterdam/New York.

Analytical Chemistry. American Chemical Society, Washington, DC.

Applied Spectroscopy. Society for Applied Spectroscopy, Frederick, MD.

Atomic Spectroscopy. Perkin-Elmer, Ridgefield, CT.

Fresenius Zeitschrift fuer Analytische Chemie. Springer Verlag, Berlin/Goettingen/Heidelberg.

ICP Information Newsletter. University of Massachusetts, Department of Chemistry, Amherst.

Industrial Laboratory (USSR): English Translation of Zavodskaya Laboratoriya. Consultants Bureau, New York.

Journal of Analytical Atomic Spectrometry. The Royal Society of Chemistry, London.

Journal of Analytical Chemistry of the USSR: English Translation of Zhurnal Analiticheskoi Khimii. Consultants Bureau, New York.
Progress in Analytical (Atomic) Spectroscopy. Pergamon Press, Oxford/New York.
Spectrochimica Acta, Part B: Atomic Spectroscopy. Pergamon Press, Oxford/New York.
Talanta. Pergamon Press, Oxford/New York.
The Analyst. The Royal Society of Chemistry, London.

REFERENCES

Note: Section 1.8.7 contains a list of general literature: books, abstracts, compilations and reviews; only a few of those references are also included in the list below.

1. A. Montaser and V. A. Fassel, *Anal. Chem.* **48,** 1492 (1976).
2. R. H. Wendt and V. A. Fassel, *Anal. Chem.* **38,** 337 (1966).
3. S. Greenfield, P. B. Smith, A. E. Breeze, and N. M. D. Chilton, *Anal. Chim. Acta* **41,** 385 (1968).
4. M. E. Britske, V. M. Borisov, and Yu. S. Sukah, *Ind. Lab.* **33,** 301 (1967).
5. C. Bordonali and M. A. Biancifiori, *Met. It.*, No. 8, 631 (1967).
6. C. Bordonali and M. A. Biancifiori, *Proc. 14th Coll. Spectr. Int.*, Debrecen 1967, Vol. 3, Hilger, London (1968), p. 1153.
7. W. B. Barnett, V. A. Fassel, and R. N. Kniseley, *Spectrochim. Acta* **23B,** 643 (1968).
8. C. Veillon and M. Margoshes, *Spectrochim. Acta* **23B,** 503 (1968).
9. M. H. Abdallah, R. Diemiaszonek, J. Jarosz, J. M. Mermet, J. Robin, and C. Trassy, *Anal. Chim. Acta* **84,** 271 (1976).
10. L. de Galan, G. R. Kornblum, and M. T. C. de Loos-Vollebregt, in K. Fuwa, ed., *Recent Advances in Analytical Spectroscopy.* Pergamon Press, Oxford/New York (1982), p. 33.
11. P. W. J. M. Boumans, "Excitation of Spectra," in E. L. Grove, ed., *Analytical Emission Spectroscopy*, Vol. 1, Pt. 2, Ch. 6, Dekker, New York (1972), p. 155.
12. K. Laqua, "Emissionspektroskopie," in *Ullmans Encyklopaedie der technischen Chemie*, 4th ed., Vol. 5, Verlag Chemie GmbH, Weinheim (1980), p. 441.
13. C. E. Moore, *Ionization Potentials and Ionization Limits Derived from the Analysis of Optical Spectra*, NSRDS-NBS-34. Superintendent of Documents, Washington, DC (1970).
14. R. C. Weast, ed., *Handbook of Chemistry and Physics*, 65th ed. CRC Press, Boca Raton, FL (1984).
15. F. Malamand, B. Daigne, and F. Girard, *Spectrochim. Acta* **33B,** 463 (1978).
16. F. Malamand, B. Daigne, and M. Armand, *Spectrochim. Acta* **35B,** 243 (1980).
17. S. K. Hughes and R. C. Fry, *Appl. Spectrosc.* **35,** 493 (1981).

18. J. D. Winefordner, J. J. Fitzgerald, and N. Omenetto, *Appl. Spectrosc.* **29,** 369 (1975).
19. P. W. J. M. Boumans, *Philips Tech. Rev.* **34,** 305 (1974).
20. M. D. Amos and J. B. Willis, *Spectrochim. Acta* **22,** 1325, 2128 (1966).
21. J. B. Willis, "Atomic Absorption Spectrometry," in R. Mavrodineanu, ed., *Analytical Flame Spectroscopy*, Ch. 10, MacMillan, London (1970), p. 525.
22. V. A. Fassel and D. W. Golightly, *Anal. Chem.* **39,** 466 (1967).
23. E. E. Pickett and S. R. Koirtyohann, *Spectrochim. Acta* **23B,** 235, 673 (1968); **24B,** 325 (1969).
24. E. E. Pickett and S. R. Koirtyohann, *Anal. Chem.* **41,** (14), 28A (1969).
25. G. D. Christian and F. J. Feldman, *Appl. Spectrosc.* **25,** 660 (1971).
26. P. W. J. M. Boumans and F. J. de Boer, *Spectrochim. Acta* **27B,** 391 (1972).
27. A. Danielsson and P. Lindblom, *Appl. Spectrosc.* **30,** 151 (1976).
28. A. Danielsson, E. Söderman, P. Lindblom, G. Sundkvist, and T. Berggren, 21st CSI/8th ICAS, Cambridge 1979, Keynote Lectures, Heyden, London (1979), p. 1.
29. *Annual Reports on Analytical Atomic Spectroscopy*, Vol. 1. The Society for Analytical Chemistry, London (1971).
30. J. A. C. Broekaert, *Spectrochim. Acta* **36B,** 563, 931 (1981); **37B,** 69, 359 (1982).
31. A. W. Witmer, J. A. J. Jansen, G. H. van Gool, and G. Brouwer, *Philips Tech. Rev.* **34,** 322 (1974).
32. K. Zimmer, Gy. Heltai, and K. Florian, *Progr. Anal. Atom. Spectrosc.* **5,** 341 (1982).
33. J. Kroonen and D. Vader, *Line Interference in Emission Spectrographic Analysis*. Elsevier, Amsterdam (1963).
34. C. E. Harvey, *Semiquantitative Spectrochemistry*. Applied Research Laboratories, Glendale, CA (1964).
35. N. W. H. Addink, *DC Arc Analysis*. MacMillan, London (1971).
36. P. D. P. Taylor, J. de Donder, and R. Dams, *Spectrochim. Acta* **41B,** No. 8 (1986).
37. M. Slavin, *Emission Spectrochemical Analysis*. Wiley, New York (1971).
38. W. G. Schrenk, *Analytical Atomic Spectroscopy*. Plenum, New York (1975).
39. M. S. Wang, W. T. Cave, and W. S. Coakley, "Qualitative and Semiquantitative Analysis," in E. L. Grove, ed., *Analytical Emission Spectroscopy*, Vol. 1, Pt. 2, Ch. 8. Dekker, New York (1972), p. 395.
40. IUPAC, "Nomenclature, Symbols, Units and Their Usage in Spectrochemical Analysis," Parts I–III, V, *Pure Appl. Chem.* **30,** 653 (1972); **45,** 99 (1976); **53,** 1913 (1981).
41. IUPAC, "Nomenclature, Symbols, Units and their Usage in Spectrochemical Analysis," Parts I–III, V, Reprint of [40]; *Spectrochim. Acta* **33B,** 219, 241, 247 (1978); **41B,** 507 (1986).

42. F. J. M. J. Maessen and H. Balke, *Spectrochim. Acta* **37B**, 37 (1982).
43. J. Agterdenbos, *Anal. Chim. Acta* **108**, 315 (1979).
44. R. Klockenkaemper and H. Bubert, *Spectrochim. Acta* **37B**, 127 (1982).
45. H. Bubert and R. Klockenkaemper, *Spectrochim. Acta* **38B**, 1087 (1983).
46. H. Bubert, R. Klockenkaemper, and H. Waechter, *Spectrochim. Acta* **39B**, 1465 (1984).
47. K. Slickers, *Automatic Emission Spectroscopy*. Bruehl Druck & Pressehaus Giessen, Giessen, F. R. G.
48. A. Scheeline, *Progr. Anal. Atom. Spectrosc.* **7**, 21 (1984).
49. P. W. J. M. Boumans, *Theory of Spectrochemical Excitation*. Adam Hilger, London/Plenum, New York (1966).
50. W. R. Barnett, V. A. Fassel, and R. N. Kniseley, *Spectrochim. Acta* **25B**, 139 (1970).
51. L. H. Ahrens and S. R. Taylor, *Spectrochemical Analysis*. Addison-Wesley, Reading, MA (1961).
52. R. M. Belchamber and G. Horlick, *Spectrochim. Acta* **37B**, 1037 (1982).
53. S. A. Myers and D. H. Tracy, *Spectrochim. Acta* **38B**, 1227 (1983).
54. A. Lorber, Z. Goldbart, and M. Eldan, *Anal. Chem.* **56**, 43 (1984).
55. F. J. M. J. Maessen and P. W. J. M. Boumans, *Spectrochim. Acta* **23B**, 739 (1968).
56. P. W. J. M. Boumans and F. J. M. J. Maessen, *Spectrochim. Acta* **24B**, 585 (1969).
57. P. W. J. M. Boumans and F. J. M. J. Maessen, *Spectrochim. Acta* **24B**, 611 (1969).
58. F. J. M. J. Maessen, J. W. Elgersma, and P. W. J. M. Boumans, *Spectrochim. Acta* **31B**, 179 (1976).
59. P. W. J. M. Boumans, *Spectrochim. Acta* **31B**, 147 (1976).
60. R. I. Botto, *Developments in Atomic Plasma Spectrochemical Analysis* (R. M. Barnes, ed.), Heyden, London/Philadelphia (1981), p. 141.
61. R. I. Botto, *Anal. Chem.* **54**, 1654 (1982).
62. R. I. Botto, *Spectrochim. Acta.* **38B**, 129 (1983).
63. R. I. Botto, *Spectrochim. Acta.* **39B**, 95 (1984).
64. B. E. H. Saxberg and B. R. Kowalski, *Anal. Chem.* **51**, 1031 (1979).
65. J. H. Kalivas and B. R. Kowalski, *Anal. Chem.* **53**, 2207 (1981).
66. C. Jochum, P. Jochum, and B. R. Kowalski, *Anal. Chem.* **53**, 85 (1981).
67. K. Fujiwara, J. M. McHard, S. J. Foulk, S. Bayer, and J. D. Winefordner, *Can. J. Spectrosc.* **25**, 18 (1980).
68. A. J. Mitteldorf, *The Spex Speaker* **2** (2) (1957).
69. G. Maassen, Proc. 7th Coll. Spectrosc. Int., Liege 1958. *Rev. Universelle Mines* **15**, 300 (1959).
70. W. E. Publicover, *Anal. Chem.* **37**, 1680 (1965); **38**, 220 (1966).

REFERENCES

71. V. A. Fassel and R. W. Tabeling, *Spectrochim. Acta* **8**, 210 (1956).
72. V. A. Fassel and J. W. Goetzinger, *Spectrochim. Acta* **21**, 289 (1965).
73. A. N. Zaidel, N. I. Kaliteevskii, L. V. Lipis, and M. P. Chaika, *Emission Spectrum Analysis of Atomic Materials*, State Publishing House of Physicomathematical Literature, Moscow (1960); Eng. Transl.: U.S. Atomic Energy Commission, AEC-tr-5745, Washington, DC (1963).
74. Kh. I. Zilbershtein, ed., *Spectrochemical Analysis of Pure Substances*. Hilger, Bristol (1977).
75. E. Schroll, *Z. Anal. Chem.* **198**, 40 (1963).
76. E. Schroll, *Proc. 14th Coll. Spectr. Int., Debrecen 1967*, Vol. 1. Hilger, London (1968), p. 397.
77. A. K. Rusanov, *Fundamentals of the Quantitative Spectrographic Analysis of Ores and Minerals* (in Russian), Nedra, Moscow (1978).
78. J. T. Rosza, "Quantitative Analysis," in E. L. Grove, ed., *Analytical Atomic Spectroscopy*, Vol. 1, Pt. 2, Ch. 9. Dekker, New York (1972), p. 451.
79. A. K. Rusanov and N. T. Batova, *J. Anal. Chem. USSR* (Eng. Transl. **20**, 387, 607, 725 (1965).
80. A. K. Rusanov and V. G. Khitrov, *Spectrochim. Acta* **10**, 405 (1958).
81. I. G. Yudelevich and A. S. Cherevko, *Spectrochim. Acta* **31B**, 93 (1976).
82. Z. Radwan, B. Strzyzewska, and J. Minczewski, *Appl. Spectrosc.* **17**, 1 (1963).
83. Ya. D. Raikhbaum, V. D. Malykh, and M. A. Luzhnova, *Ind. Lab.* (USSR) (Eng. Transl.) **29**, 721 (1963).
84. G. Scheibe and A. Rivas, *Angew. Chem.* **49**, 443 (1936).
85. K. Baer and E. S. Hodge, *Appl. Spectrosc.* **14**, 141 (1960).
86. M. Fred, N. H. Nachtrieb, and F. S. Tomkins, *J. Opt. Soc. Am.* **37**, 279 (1947).
87. J. M. Morris and F. X. Pink, *Symp. Spectrochemical Analysis for Trace Elements*, Am. Soc. Test. Mat., Spec. Tech. Publ. No. 221 (1957), p. 39.
88. N. Krasnobaeva, Z. Zadgorska, and N. Nedjalkova, *Spectrochim. Acta* **33B**, 655 (1978).
89. W. A. Gordon and G. B. Chapman, *Spectrochim. Acta* **25B**, 123 (1970).
90. T. H. Zink, *Appl. Spectrosc.* **13**, 94 (1959).
91. J. P. Walters, *Science* **198**, 787 (1977).
92. C. B. Belcher, *CRC Crit. Rev. Anal. Chem.* **7**, 121 (1978).
93. K. J. Doolan and C. B. Belcher, *Progr. Anal. Atom. Spectrosc.* **3**, 125 (1980).
94. C. Feldman, *Anal. Chem.* **21**, 1041 (1949).
95. M. Pierucci and L. Barbanti-Silva, *Nuovocimento* **17**, 275 (1940).
96. J. T. Rosza and L. E. Zeeb, *Petrol. Processing* **8**, 1708 (1953).
97. T. Nakajima and H. Kawaguchi, *Spectrochim. Acta* **18**, 1479 (1962).
98. J. P. Pagliasotti and F. W. Porsche, *Anal. Chem.* **23**, 189 (1951).
99. N. H. Suhr and C. O. Ingamells, *Anal. Chem.* **38**, 730 (1966).
100. J. P. Pagliasotti, *Anal. Chem.* **28**, 1774 (1956).

101. O. Szakács, *Ann. Univ. Sci. Budapest, R. Eötvös, Sec. Chim.* **7**, 55 (1965).
102. H. Nickel, *Z. Anal. Chem.* **245**, 250 (1969).
103. N. P. Ferreira, Ph.D. thesis, University of Stellenbosch, Stellenbosch, South Africa (1980).
104. W. Grimm, *Naturwiss.* **54**, 586 (1967); *Spectrochim. Acta* **23B**, 443 (1968).
105. N. P. Ferreira, J. A. Strauss, and H. G. C. Human, *Spectrochim. Acta* **38B**, 899 (1983).
106. G. S.Lomdahl and J. V. Sullivan, *Spectrochim. Acta* **39B**, 1395 (1984).
107. J. B. Ko, *Spectrochim. Acta* **39B**, 1405 (1984).
108. B. Berglund and B. Thelin, *Analyst* **107**, 867 (1982).
109. P. J. Slevin and W. W. Harrison, *Appl. Spectrosc. Rev.* **10**, 202 (1976).
110. S. Caroli, *Progr. Anal. Atom. Spectrosc.* **6**, 253 (1983).
111. K. Laqua, "Analytical Spectroscopy using Laser Atomizers," in N. Omenetto, ed., *Analytical Laser Spectroscopy*, Ch. 2. Wiley, New York (1979), p. 47.
112. K. Dittrich and R. Wennrich, *Progr. Anal. Atom. Spectrosc.* **7**, 139 (1984).
113. H. Moenke and L. Moenke-Blankenburg, *Laser Micro-Spectrochemical Analysis*. Crane, Russak and Co., New York (1979).
114. J. M. Ottaway, R. C. Hutton, D. Littlejohn, and F. Shaw, *Wiss. Z. Karl-Marx-Univ. Leipzig, Math. Naturwiss. R.* **28**, 357 (1979).
115. J. M. Ottaway, L. Bezur, and J. Marshall, *Analyst* **105**, 1130 (1980).
116. J. Marshall, D. Littlejohn, J. M. Ottaway, J. M. Harnly, N. J. Miller-Ihli, and T. C. O'Haver, *Analyst* **108**, 178 (1983).
117. L. Bezur, J. Marshall, J. M. Ottaway, and R. Fakhrul-Aldeen, *Analyst* **108**, 553 (1983).
118. H. Falk, E. Hoffmann, and Ch. Luedke, *Spectrochim. Acta* **36B**, 767 (1981).
119. H. Falk, E. Hoffman, C. Ludke, J. M. Ottaway, and S. K. Giri, *Analyst* **108**, 1459 (1983).
120. R. D. Sacks, J. M. Goldberg, R. J. Collins, and S. Y. Suh, *Progr. Anal. Atom. Spectrosc.* **5**, 111 (1982).
121. R. D. Sacks and D. V. Duchane, *Anal. Chem.* **50**, 1757 (1978).
122. E. M. A. Clark, Ph.D. thesis, University of Michigan, Ann Arbor (1979).
123. T. Kántor, *Spectrochim. Acta* **38B**, 1483 (1983).

CHAPTER

2

PLASMA SOURCES OTHER THAN INDUCTIVELY COUPLED PLASMAS

P. W. J. M. BOUMANS

Philips Research Laboratories
Eindhoven, The Netherlands

2.1. Introduction
2.2. DC Plasmas
2.3. Capacitively Coupled Microwave Plasmas
2.4. Microwave-Induced Discharges
References

2.1. INTRODUCTION

The incentive for the development of plasma sources was the need for methods that permit direct multielement analysis of solutions. As has been set out in Section 1.3, an emission spectroscopic method using an excitation source with a temperature of at least 5000 K is in principle suited for this purpose. The main difficulty has been the realization of such source configurations that a nebulized solution could be introduced with high efficiency into the plasma without disturbing the plasma stability and markedly affecting the thermodynamic conditions of the plasma. Evidently efficient sample introduction, high source stability, and minimum interference of the sample with the conditions in the source are prerequisites for attaining high detection power, high precision, and high accuracy.

The efforts have concentrated on three classes of electrically generated, atmospheric pressure plasmas: (1) dc plasmas, differentiated in current-carrying and current-free plasmas; (2) radio frequency (rf) plasmas with either capacitive or inductive coupling, and operated at frequencies between 1 and 100 MHz; and (3) microwave plasmas either capacitively coupled or induced, and operated at frequencies above 1 GHz, usually at 2.45 GHz. Discharges with rf or microwave excitation operating at *reduced* pressure have been also investigated as excitation sources for AES or as primary radiation sources for AAS or AFS. This topic is beyond the scope of this book. Useful introductions to this subject

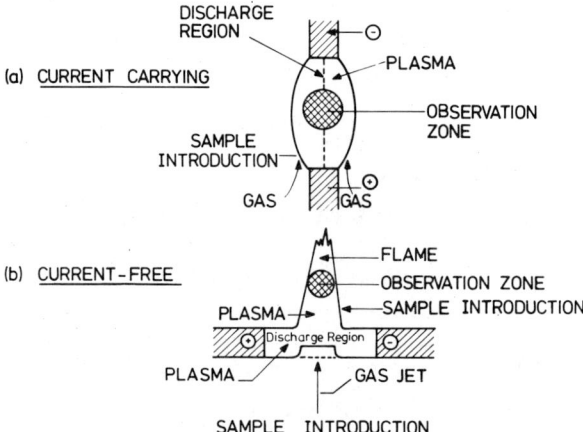

Figure 2.1. Schematic representation of the distinction between current-carrying and current-free dc plasmas [6]. (Reprinted with permission from H. G. C. Human and R. H. Scott, "Electrical Flames," in J. W. Robinson, ed., *Handbook of Spectroscopy*, Vol. 1, Ch. XIV, p. 817. Copyright (1974) CRC Press, Boca Raton, FL.)

and further references are given by Boumans [1], Barnes [2], Tschöpel [3], and Zander and Hieftje [4]. An interesting recent development is an ICP operated at reduced pressure [5] which permits, for example, the excitation of halogens from organic halogen compounds.

2.2. DC PLASMAS

DC plasmas (DCP) are discharges in which a gas is heated by a dc current passing between two electrodes. If spectroscopic observations are made in the current-carrying portion, the configuration is referred to as a "current-carrying DCP" (Fig. 2.1a). The principal types in this category are the gas-stabilized arcs and the disk- or wall-stabilized arcs.

In a "current-free DCP" a gas stream forces a portion of the plasma generated in an arc discharge to flow out of the discharge gap in such a way as to form a flame-like plume in which spectroscopic observations are made (Fig. 2.1b). Plasmas of this type have been designated "transferred plasmas," "plasma jets," or "plasmatrons."

Devices in the two categories have many features in common, but there are marked differences in the properties of the plasmas. One should note in particular that in a current-carrying DCP the sample must be introduced into the discharge region; consequently the excitation conditions are affected by easily

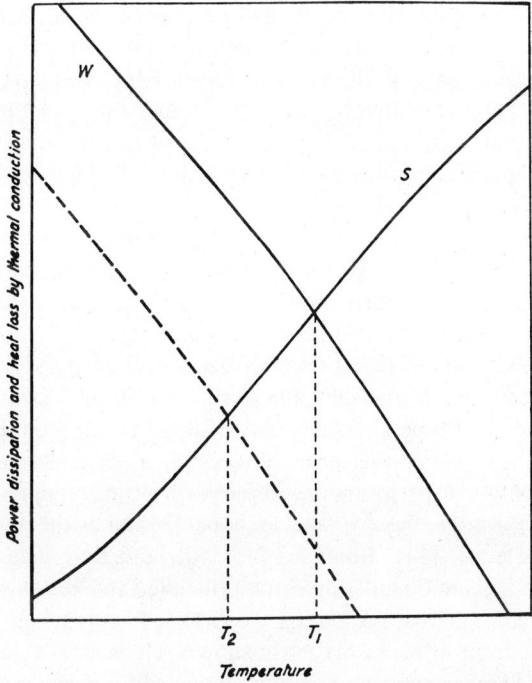

Figure 2.2. Schematic representation of the power dissipation (W) and the thermal conductivity (S) in a dc carbon arc. The arc takes up the temperature (T_1) at which the power supplied equals the energy loss by thermal conduction. The addition of a substance of low ionization potential produces a marked change in the intrinsic electrical conductivity of the gas, whereas it leaves the intrinsic thermal conductivity unaffected. The repercussion of the altered dependence of power dissipation on temperature (broken curve) causes the temperature to drop to the value T_2, where again the power supplied is balanced by the energy loss by thermal conduction. (cf. Boumans [7].)

ionizable elements (EIE) in the sample in a way similar to that in the classical dc carbon arc [1, 7].

Briefly, the EIE disturbs the power balance in that it enhances the electrical conductivity of the gas so that the power input decreases. The power loss, however, remains the same because the thermal conductivity is not altered by the EIE. The result is a decrease in arc temperature to such an extent that again power input and power loss balance each other. This is schematically illustrated in Fig. 2.2. If the concentration of the EIE is steadily increased, the temperature will first decrease sharply, but then gradually approach an approximately constant level at which the electron concentration starts to increase with a further increase in the concentration of the EIE in the plasma [7].

In a current-free DCP, the sample can be introduced in either the current-carrying zone or the current-free plume. In the former case the plasma temperature will sharply respond to the introduction of EIEs. In the latter, however, the influence of EIEs is relatively small: the temperature is not affected but the electron concentration is increased and therefore ionization suppression will occur. This is comparable to what happens in flames. An increase in the electron concentration also modifies the radial electric field, which in turn has repercussions on the radial transport of ions and electrons (ambipolar diffusion), as has been shown by Schirrmeister [8, 9] for a plasma jet after Kranz (see below). Since no power is dissipated in a current-free plasma jet, the entry of the solvent of an aerosol may have a marked cooling effect.

The first applications of DCPs to spectrochemical analysis were reported in 1959 independently by Margoshes and Scribner [10] and Korolev and Vainshtein [11]. Since that time a large variety of designs has been described. Historical developments, operating principles, physical characterizations, and assessments of analytical performance are covered by the chapters of Boumans [1], Keliher [12], Butler et al. [6], and Tschöpel [3] and by the reviews of Fassel [13], Greenfield et al. [14], Boumans [15, 16], and Keirs and Vickers [17]. Discussion here aims at elucidating some principles and sketching some major developments rather than at presenting a complete historical survey.

The simplest form of a plasma jet is shown schematically in Fig. 2.3. An arc burns in a chamber between a carbon anode and a perforated carbon cathode. A stream of gas introduced into the chamber ejects the plasma through the cathode orifice to form a jet. For solution analysis, Margoshes and Scribner [10] introduced a ring-shaped anode through which aerosol from a direct neb-

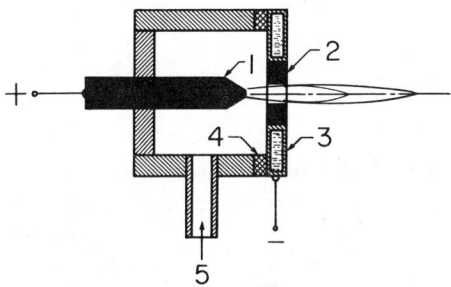

Figure 2.3. Schematic diagram of a plasma jet with a carbon rod as the anode and a carbon ring as the cathode (Korolev–Vainshtein configuration): *1*, carbon anode; *2*, carbon cathode; *3*, cooling water; *4*, isolation; *5*, gas inlet. (Reprinted with permission from E. Kranz, "Aufbau und Eigenschaften eines verunreinigungsfreien Plasmabrenners fuer spektroskopische Zwecke," in R. Ritschl and G. Holdt, eds., *Emissionsspektroskopie*, p. 161. Copyright (1964), Akademie-Verlag, Berlin, G.D.R.)

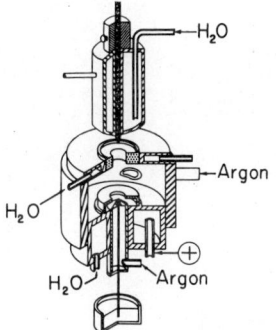

Figure 2.4. Cross-sectional view of Margoshes and Scribner's plasma-jet assembly. (From Margoshes and Scribner [20].)

ulizer was fed. The approach was not very successful because (a) the cathode spot was not fixed in space with the result that an unstable and rotationally asymmetric jet was obtained, and (b) effective mixing of the aerosol with the plasma was not achieved. Attempts to improve the stability of the jet and the efficiency of sample injection led to various modifications of the original design, (e.g., [19, 20]). One of these is shown in Fig. 2.4. This configuration no longer is a plasma jet but a gas-stabilized arc with additional wall stabilization by an electrically floating graphite ring.

The gas stream cools the peripheral regions of the arc, makes them less conductive, and constricts the arc column. Consequently the power input per unit volume increases. This effect is referred to as "thermal pinch." If the current density is high enough, the thermal pinch causes a magnetohydrodynamic pinch by the self-induced magnetic field. The latter effect is qualitatively understood by considering that parallel conductors through which currents flow in the same direction are attracted to each other by their self-induced magnetic fields. The charged particles in the channel of an arc discharge behave similarly.

Although a satisfactory plasma stability could be achieved when the device shown in Fig. 2.4 was operated without aerosol, difficulties were encountered when aerosols were introduced. The plasma had the appearance of a cone having the anode ring as a basis. It was thus expected that the aerosol could be injected through the open bottom of the cone into the plasma. However, high-speed motion pictures revealed that the anode ring was seized in a single spot which was rapidly spinning over the ring [21]. The aerosol flow was therefore blown from the outside to the discharge channel. The cooling effect of this impinging aerosol flow causes an arc to contract, to raise its current density and to change its position in order to balance the additional heat losses at the point where the aerosol flow contacts the periphery of the arc. These effects, which hamper the effective introduction of aerosols into arc plasmas, have been discussed in detail by Kranz [22, 23], who indicated desolvation of the aerosol

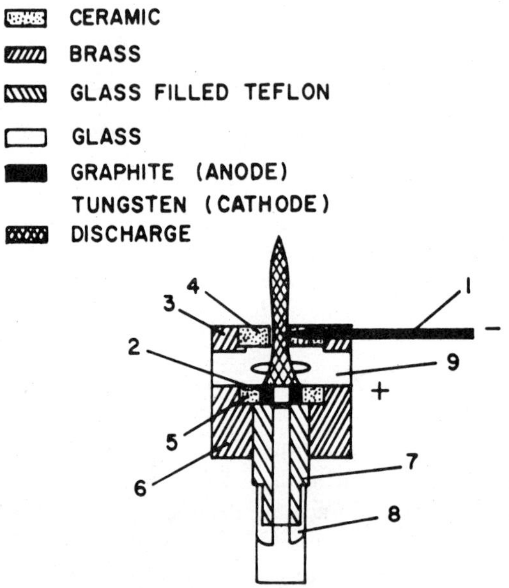

CERAMIC
BRASS
GLASS FILLED TEFLON
GLASS
GRAPHITE (ANODE)
TUNGSTEN (CATHODE)
DISCHARGE

Figure 2.5. Cross section of the dc arc plasma jet explored by Rippetoe and Vickers [24]. (Reprinted with permission from W. E. Rippetoe and T. J. Vickers, "Rotating Arc Plasma Jet for Emission Spectroscopy," *Anal. Chem.* **47**, 2082 (1975). Copyright (1975) American Chemical Society, Washington, DC.)

prior to feeding it to the plasma and turbulent mixing of the aerosol with the plasma as means to overcome the problems. Further research aimed at improving the Margoshes–Scribner design followed various pathways.

One of them is the approach of Rippetoe and Vickers [24] who devised a "rotating arc," schematically depicted in Fig. 2.5. The rotation of the arc was achieved by an argon stream introduced tangentially into the arc chamber. The reasoning was that if the arc rotation rate was sufficiently rapid as compared to the time required for an aerosol particle to cross the interelectrode space, then the aerosol particles would be constrained to travel in a channel efficiently heated by the surrounding high-temperature arc column. The rotation rate was found to increase linearly with the flow rate of the tangential gas stream and a rotation frequency of 500 Hz was achieved with 13 L/min of argon. The gas carrying the aerosol entered the discharge chamber from below. Expansion of the volatilized aerosol and carrier gas offered resistance to the rotation of the arc column and, hence, the rate of rotation was found to decrease as the flow rate of the carrier gas increased. To maintain rapid rotation it was found to be desirable to desolvate the aerosol and to use a nebulizer which could be operated with a low carrier gas flow. An ultrasonic nebulizer was then used.

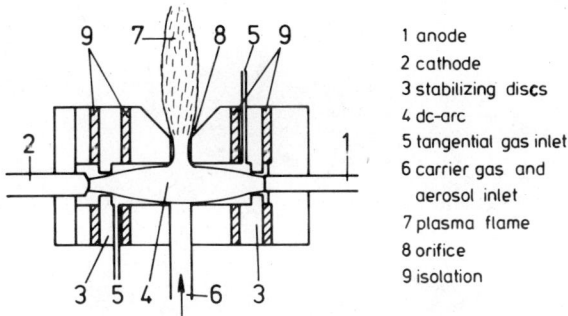

1 anode
2 cathode
3 stabilizing discs
4 dc-arc
5 tangential gas inlet
6 carrier gas and aerosol inlet
7 plasma flame
8 orifice
9 isolation

Figure 2.6. Schematic diagram of the Kranz dc plasma jet [3]. (Reproduced with permission from P. Tschoepel, "Plasma Excitation in Spectrochemical Analysis," in G. Svehla, ed., *Comprehensive Analytical Chemistry*, Vol. 9, Ch. 3, p. 264. Copyright (1979), Elsevier Science Publishers, Amsterdam.)

A few detection limits were reported, but these were not impressive. Atomization efficiency was studied using the interference effect of phosphate on calcium as criterion. A 32% depression was found of the Ca emission for a molar ratio of P/Ca of 1 when the arc rotation frequency was 300 Hz. This interference was eliminated when the rotation frequency was increased to 600 Hz, but desolvation of the aerosol was required to achieve this.

Earlier Kranz [18, 25, 26] exploited turbulent mixing of the aerosol and the plasma in a current-free DCP, developed from a wall-stabilized arc. The design, schematically shown in Fig. 2.6, closely resembles a "tandem Gerdien plasma jet" described by McGinn [27]. Kranz used the device chiefly for physical investigations, but also gave some analytical results. Schirrmeister [8, 9] studied the effect of EIEs in a modified Kranz plasma jet. On the whole, however, extensive and convincing results that demonstrate the analytical performance for real samples were never reported.

Riemann [28-30], Holdt and Hoffmann [31, 32], and Marinkovic et al. [33-35] explored disk-stabilized arcs for solution analysis. Disk stabilization is a type of wall stabilization introduced by Maecker [36] for research in high-temperature physics. The basic arrangement (without device for aerosol introduction) is shown schematically in Fig. 2.7. This configuration is also referred to as a "cascade arc."

Marinković et al. [33-35] investigated, inter alia, the effects of adding EIEs as buffers. Depending on the buffer concentration, the arc could then be operated as either a high-temperature or a low-temperature arc. The authors concluded that "high flexibility of the arc might be considered its weakness, for it requires careful standardization of the operations and special precautionary measures for avoiding matrix effects. On the other hand, this is also its merit

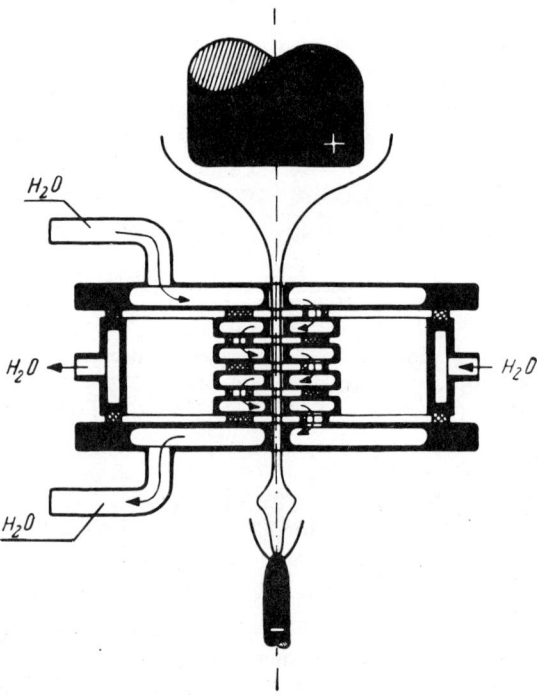

Figure 2.7. Illustration of the principle of the wall-stabilized cascade arc. (Reprinted with permission from H. Maecker, "Ein zylindrischer Bogen fuer hohe Leistungen," *Z. Naturforsch.* **11a**, 458 (1956). Copyright (1956), Verlag der Zeitschrift fuer Naturforschung, Mainz, F.R.G.)

since it can be successfully used for solving many different analytical problems." Obviously, as happens with many of these devices, their proper operation requires considerable skill, experience, and faith.

Using the configuration shown in Fig. 2.8, Rippetoe et al. [37] studied the effect of KCl on the analytical utility of this DCP. With solutions containing 2 M KCl (16% salt w/v) they obtained a "completely diffuse discharge mode." Among the benefits of this "seeded arc" they mention improved aerosol penetration into the plasma, as demonstrated by elimination of the depressing effect of phosphate on calcium emission. Some detection limits, which compared favorably with those of flame AES, were reported, but on the whole the analytical significance of this approach was not encouraging, in particular because "constant spraying of KCl complicates the experimental arrangement."

Considering the developments sketched here, as well as alternative approaches treated in the reviews referred to above, one is inclined to conclude that, in spite of several promising efforts, the problem of effective sample in-

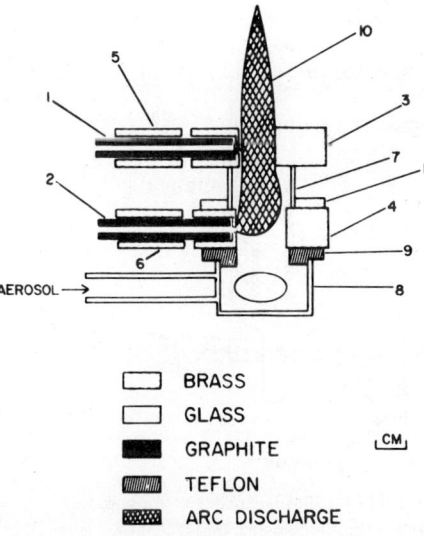

Figure 2.8. Cross-sectional view of the dc arc plasma explored by Rippetoe et al. [37]. (Reprinted with permission from W. E. Rippetoe, E. R. Johnson, and T. J. Vickers, "Characterization of the Plume of a Direct Current Plasma Arc for Emission Spectrometric Analysis," *Anal. Chem.* **47,** 437 (1975). Copyright (1975), American Chemical Society, Washington, DC.)

troduction into DCPs while maintaining the plasma stability could not be adequately overcome. This, in fact, is a common conclusion of the various reviews covering the literature up to about 1977. Since that time the situation has changed, at least in the respect that a commercial DCP spectrometer was developed which found acceptance as an analytical tool for analyzing a variety of sample types. This DCP followed a line of development not yet discussed.

Valente and Schrenk [38] enclosed two electrodes in chambers of the type shown in Fig. 2.9. The two chambers were first aligned vertically to strike an arc by extending the cathode until it touched the anode upon which the cathode was withdrawn to its proper position inside the chamber. After ignition of the arc, one of the electrode units was rotated about the center of the discharge to a position so that the angle formed between the axes of both jets was approximately 30°. The plasma then appeared approximately as an inverted "V." Direct spraying of sample into the plasma immediately quenched the discharge, while introduction of wet aerosol from an indirect nebulizer by means of the tangential gas flow resulted in extremely poor detection limits. A desolvated aerosol [39], however, could be successfully and reproducibly introduced with the tangential gas. Valente and Schrenk measured the emission intensities of a Ca I and a Ca II line and the background as a function of the various parameters.

Fig. 2.10 shows spatial mappings of the emission distribution from which the location of regions with maximum signal-to-background could be derived. In all cases, the best region was found at the base of the anode jet, and this region was common within a small range of displacements for the elements

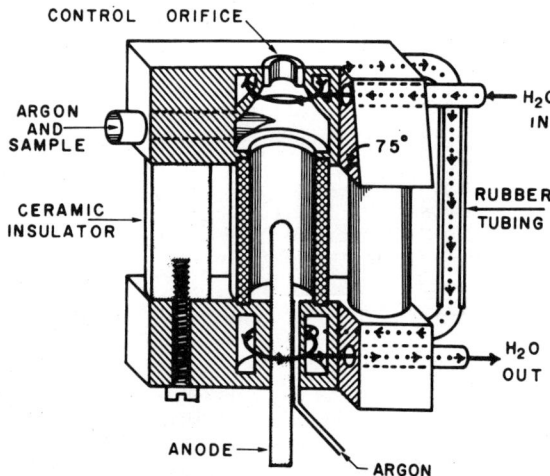

Figure 2.9. Cut-away drawing of the anode chamber of the plasma arc device proposed by Valente and Schrenk [38]. (Reprinted with permission from S. E. Valente and W. G. Schrenk, *Appl. Spectrosc.* **24**, 197 (1970). Copyright (1970), Society for Applied Spectroscopy, Frederick, MD.)

studied. The careful selection of an observation zone with low background and high analyte emission intensities led to unusually good detection limits. This feature is paramount in the following development.

The inverted "V" configuration of the Valente–Schrenk plasma jet was adopted in a commercial DCP [40] which superseded an earlier design described by Elliott [41] and was further characterized by Merchant and Veillon [42]. In the inverted V-shape DCP, the aerosol was introduced from a separate nozzle, as shown in Fig. 2.11, which depicts the eventual configuration with three electrodes (inverted "Y"), two graphite anodes, and a tungsten cathode, each flushed with an argon jet. The introduction of the third electrode led to improved stability, while the existence of a spatially very limited zone with analytically useful excitation characteristics (low background, efficient excitation) was maintained.

Characteristics of the two-electrode version, including a helium-doped DCP [43], were studied by various authors [44–46], and a critical comparison of the two- and three-electrode versions was made by Johnson et al. [47] who evaluated the capabilities of the complete commercial system, encompassing the DCP and an echelle spectrometer, for the determination of 18 common elements in water.

An investigation of Decker [48] revealed, as advantages of the three-electrode DCP, low detection limits, high precision, good stability, and three orders of magnitude dynamic range, while the disadvantages were identified as the

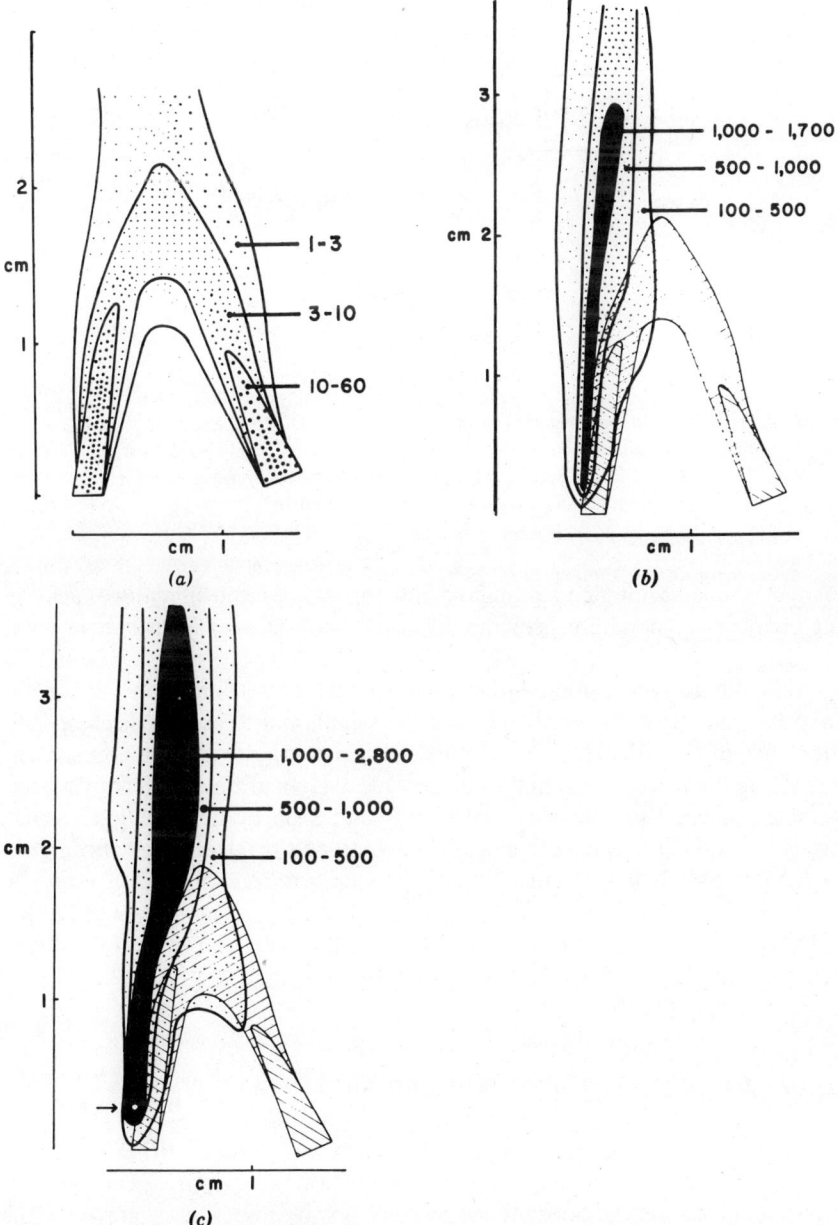

Figure 2.10. Spatial mappings of emission distributions in the Valente–Schrenk DCP: (*a*) intensity distribution of the continuous background; (*b*) intensity distribution of Ca I 422.7 nm; (*c*) intensity distribution of Ca II 393.3 nm. The numbers in the figures refer to relative intensities [38]. (Reprinted with permission from S. E. Valente and W. G. Schrenk, *Appl. Spectrosc.* **24,** 200 (1970). Copyright (1970), Society for Applied Spectroscopy, Frederick, MD.)

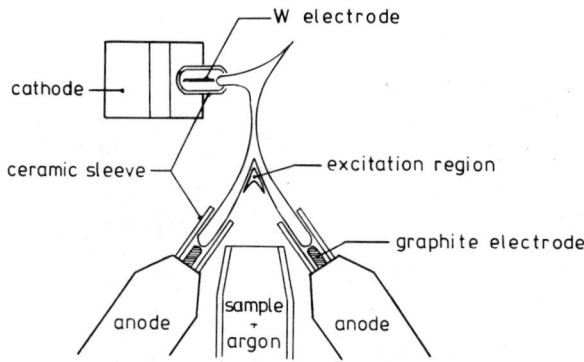

Figure 2.11. Schematic representation of the three-electrode DCP (inverted "Y" configuration) [48]. (Reprinted with permission from R. J. Decker, "Some Analytical Characteristics of a Three Electrode D.C. Argon Plasma Source for Optical Emission Spectrometry," *Spectrochim. Acta* **35B**, 21 (1980). Copyright (1980), Pergamon Journals, Oxford.)

extremely small excitation region favorable for analysis, the difficulty of achieving compromise conditions for multielement analysis, and interferences from alkalis.

Although analytical observations are made in a current-free zone, it is clear from the configuration (Fig. 2.11) that the sample must pass the discharge region and, thus, will influence the power dissipation. However, the excitation and ionization mechanisms in the current-free analytical zone have not yet been definitely cleared up. Various interesting studies on these excitation mechanisms, including the enhancements of analyte signals by alkalis, have been published [49–54]. It is clear that observations cannot be explained in terms of simple LTE ([7]) considerations, and that mechanisms and models proposed for ICPs may be, at least partly, also applicable to the analytical zone of the three-electrode DCP [52, 55].

2.3. CAPACITIVELY COUPLED MICROWAVE PLASMAS

The principle of the capacitively coupled microwave plasma (CMP) is elucidated in Fig. 2.12. A magnetron, energized by a power supply, generates microwaves (2.45 GHz), which are led by coaxial waveguides to a hollow coaxial electrode. This electrode is provided, at its upper end, with orifices and a tip, as shown in the insert in Fig. 2.12. The electrode forms a capacitance against earth and transfers the microwave power to a gas streaming through the electrode, once this gas has been made electrically conductive by external means such as Tesla sparks. The working gas, usually argon or nitrogen, serves at the

2.3. CAPACITIVELY COUPLED MICROWAVE PLASMAS

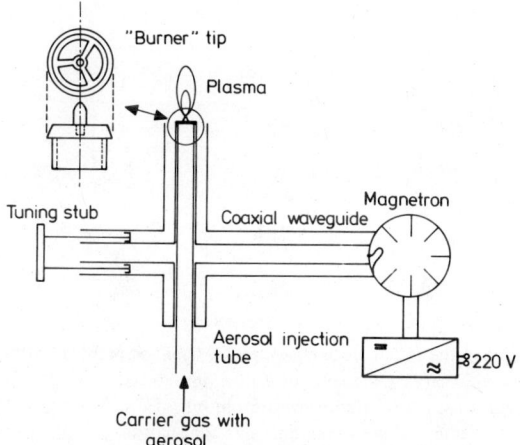

Figure 2.12. Schematic representation of a typical CMP with nebulizer. (Reproduced with permission from Spectrochimica Acta, Part B [56].)

same time as nebulizer gas and plasma gas. Thus the plasma is built up from the gas that generates the aerosol in a nebulizer and carries this aerosol to the electrode (see Fig. 2.12). The system is provided with a tuning stub for impedance matching of the cavity to obtain maximum power transfer to the plasma. The power input to the magnetron can be controlled by the power supply unit and is generally in a range between 0.5 and 3 kW. The gas flow rate is of the order of a few liters per minute.

The plasma has the appearance of a flame (up to 30 cm in height) and comprises three distinct zones (Fig. 2.13): the inner cone, the outer cone, and the tail flame. The spatial structure of the plasma is markedly affected if alkali metals are introduced.

Capacitively coupled plasmas (CCP) have been generated at various frequencies in both the rf and microwave regions. CCPs were first described in 1941 by Critescu and Grigorovici [57–59]. Analytical applications were for the first time reported by Stolov [60] and Badarau et al. [61] in the mid 1950s. Several types of CCP were explored in the beginning of the 1960s [62–68]. The configurations finally developed for analytical work used microwave excitation and were derived from the initial work of Cobine and Wilbur [69]. The further developments centered about the work of Kessler and his co-workers [70–74] in Germany and the work of Murayama and Yamamoto [75–79] and Goto [80] in Japan. The various efforts resulted in CMP equipment commercialized by (1) Hitachi, Tokyo, Japan, designated "ultra high frequency (UHF) plasma," (2) Applied Research Laboratories, Ecublens–Lausanne, Switzerland, and (3) Erbe–Elektromedizin, Tuebingen, Federal Republic of Germany.

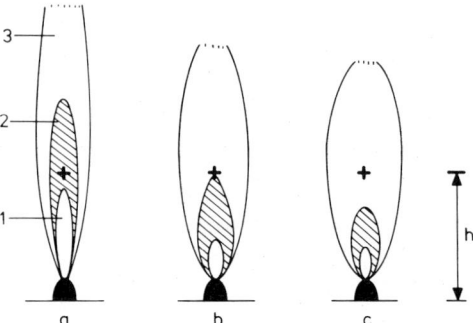

Figure 2.13. Variation of the visual appearance of a nitrogen-supported CMP when a pure aqueous solution or solutions with varying amounts of Cs are introduced: (*a*) pure aqueous solution. (*b*) solution with 2 mg/mL of Cs. (*c*) solution with 5 mg/mL of Cs. *1*, inner cone, characterized by strong emission of molecular bands (region unusable for analysis); *2*, outer cone (best region for analysis); *3*, tail flame (region hardly usable for analysis). h = observation height; + = optical axis of monochromator. (Reproduced with permission from Spectrochimica Acta, Part B [56].)

CMPs have never met widespread interest because their analytical performance cannot compete with that of ICPs, as has been established in comparative investigations of Boumans et al. [56], Larson and Fassel [81], and Burman and Boström [82]. In spite of this inferiority, CMPs have been successfully applied by Govindaraju et al. [83] to silicate rock analysis and by Dahmen [84] to trace analysis of high-purity chemicals.

Reviews of the literature usually cover both CMPs and MIPs. Excellent historical accounts, including discussions of the instrumentation and the analytical characteristics, have been published by Greenfield et al. [85] and Tschoepel [3]. A review by Zander and Hieftje [4], in which the MIP is emphasized, covers the physical properties and excitation characteristics, the instrumentation, and the analytical properties of various types of discharge energized by microwaves. The 1981–1983 literature on MIPs and CMPs is covered in Dahmen's reviews [86–88], which have their annual follow-up.

2.4. MICROWAVE-INDUCED DISCHARGES

In contrast to CMPs, MIPs are electrodeless discharges generated in a glass or quartz capillary tube having an inner diameter on the order of a few millimeters. For coupling microwave power into a gas streaming through the capillary, the latter must be placed in a resonant cavity or another plasma supporting structure. This is schematically illustrated in Fig. 2.14. for both the atmospheric pressure and reduced pressure MIP.

2.4. MICROWAVE-INDUCED DISCHARGES

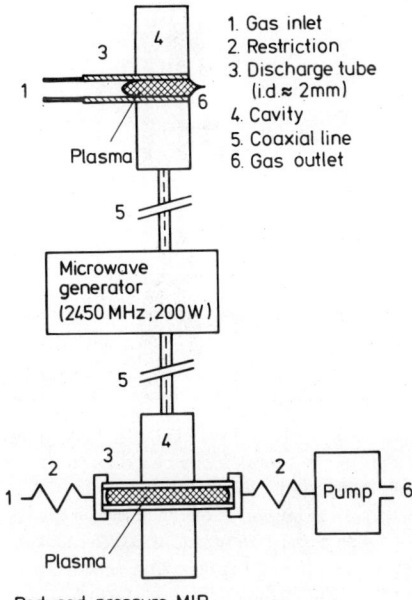

Figure 2.14. Schematic drawing of an MIP at atmospheric and reduced pressure. The plasma is generated in a discharge tube placed in a resonant cavity which is coupled by a coaxial line to a microwave generator. (Reprinted with permission from P. W. J. M. Boumans, "Inductively Coupled Plasma—Atomic Emission Spectroscopy: Its Present and Future Position in Analytical Chemistry," *Fresenius Z. Anal. Chem.* **299**, 357 (1979). Copyright (1979), Springer-Verlag, Berlin, F.R.G.)

A resonant cavity is a hollow metal container having a shape and size which allow a standing electromagnetic wave to be established within it or along it [4]. Because the standing wave is at microwave frequencies, the cavity dimensions will be on the order of several centimeters. Both cylindrical and rectangular cavities are in use [3, 4]. To generate the standing wave, microwave energy is sent into the cavity by means of a circuit loop, circuit short, or an antenna, which, in turn, is connected to the microwave power supply via a coaxial cable or waveguide. The discharge tube is usually placed in the cavity along an axis which is parallel to the line of electric field oscillation. Various transverse electric (TE) and transverse magnetic (TM) modes of oscillation enable such a configuration [3, 4].

Whether an MIP can be operated at atmospheric pressure or will require a vacuum depends on the cavity and the type of gas. Most cavities permit the generation of atmospheric pressure MIPs in argon, but necessitate the use of a vacuum system for operation of an MIP in helium. An exception is the TM_{010} cavity proposed by Beenakker [90–92], with which MIPs in both argon and helium can be produced at atmospheric pressure. For this reason the "Beenakker cavity" (Fig. 2.15), with some modifications or adaptations, has gained considerable popularity in analytical chemistry.

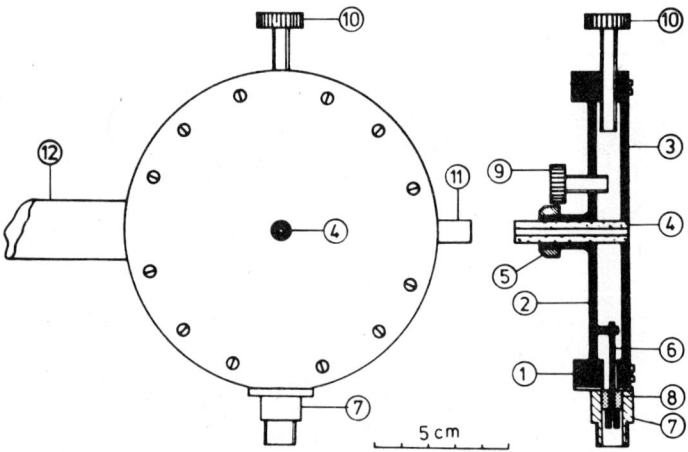

Figure 2.15. Schematic drawing of the original T_{010} cavity ("Beenakker cavity"). Left: front view. Right: side view. The cavity has the shape of a pillbox and consists of a cylindrical wall, *1*, with a fixed bottom, *2*, and a removable lid, *3*. Other items: *4*, discharge tube, *5*, holder of discharge tube, *6*, coupling loop, *7*, connector, *8*, vacuum sealing kit, *9, 10*, tuning screws, *11*, hole for viewing or air cooling, *12*, support [90]. (Reprinted with permission from C. I. M. Beenakker, "A Cavity for Microwave-Induced Plasmas Operated in Helium and Argon at Atmospheric Pressure," *Spectrochim. Acta* **31B**, 485 (1976). Copyright (1976), Pergamon Journals, Oxford.)

MIPs are commonly operated with relatively small microwave generators (2450 MHz) capable of delivering a maximum power of 200 W to the plasma. The gas flow is on the order of 1 L/min or less. A low-power plasma of this type is strongly affected by molecular species and has only a limited tolerance for foreign substances in general; therefore, MIPs have been really successfully applied chiefly in those instances where small amounts of sample per unit time are fed into the discharge, that is, in gas chromatography and microsample analysis using electrothermal evaporation of liquid or solid samples. Solution analysis using the MIP in combination with a nebulizer has been also extensively studied, with moderate success; however, the detection limits are appreciably worse than those attainable in a conventional ICP, interferences of various type are rather pronounced, and the MIP is unable to sustain dissolved solid concentrations of, say, 10 mg/mL, as would be required for its general analytical applicability. Consequently, for applications in which solutions are continuously fed to a nebulizer and subsequently introduced as aerosols into the plasma, the MIP lags too far behind plasma sources, such as the conventional ICP and the three-electrode DCP, to be competitive. However, the recent development of moderate-power (500 W) MIPs appears to be more promising for the analysis of nebulized solutions [93, 94].

A basic property of the Beenakker cavity is that it permits the generation of a plasma in helium at atmospheric pressure. This possibility has greatly prompted the application of the MIP as an element-selective detector in gas chromatography (GC) because nonmetals (C, N, O, F, S, Cl, and Br) can be readily excited under these conditions. Initially only wavelengths between 190 and 850 nm were used [95], but recently the near-infrared region (700–1200 nm) has been more fully explored and found to bring advantages [96, 97]. On the whole, it is evident from various reviews [3, 4, 98–101] that the MIP has become an important analytical tool as GC detector. The current literature shows that developments have not yet come to a halt, while the area of applications is being ever more extended (see, e.g., [102–104] and further publications in the

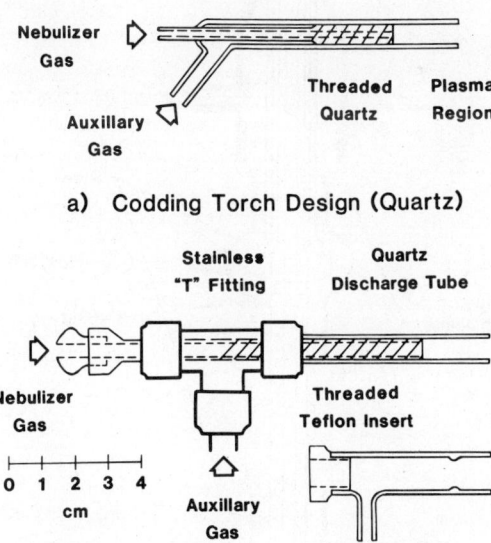

Figure 2.16. Torch designs with tangential gas flow for plasma centering within the discharge tube: (*a*) Codding torch design with externally threaded discharge tube placed concentrically in a plasma containment tube [119]; (*b*) design as modified by Haas and Caruso [93] for easy replacement of threaded discharge tube [93]. (Reprinted with permission from D. L. Haas and J. A. Caruso, "Characterization of a Moderate-Power Microwave-Induced Plasma for Direct Solution Nebulization of Metal Ions," *Anal. Chem.* **56**, 2018 (1984). Copyright (1984), American Chemical Society, Washington, DC.)

62 PLASMA SOURCES OTHER THAN INDUCTIVELY COUPLED PLASMAS

Proceedings of the 1982 and 1984 Winter Conferences on Plasma Spectrochemistry [105, 106] and journals referred to in Section 1.8).

The MIP is also used as an excitation source for microsamples that are separately evaporated and atomized. The relatively low cost of the equipment and the economy of gas consumption, the multielement capability, and the coverage of a wide range of elements including nonmetals make the MIP a most useful device in applications where overloading of the plasma with sample is not likely

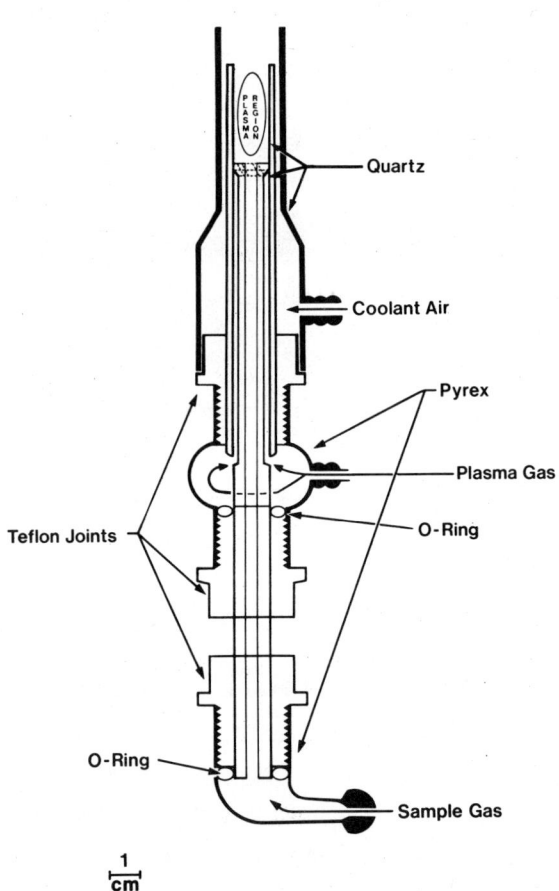

Figure 2.17. Air cooled, tangential flow, demountable torch ("slotted torch") for operating an MIP capable of maintaining stable plasmas of air, nitrogen, helium, and argon at powers up to 500 W. [94]. (Reprinted with permission from K. G. Michlewicz, J. J. Urh, and J. W. Carnahan, "A Microwave Induced Plasma System for the Maintenance of Moderate Power Plasmas of Helium, Argon, Nitrogen and Air," *Spectrochim. Acta* **40B**, 495 (1985). Copyright (1985), Pergamon Journals, Oxford.)

2.4. MICROWAVE-INDUCED DISCHARGES

to occur, thus, in combinations with electrothermal atomizers. This might also apply to the recently described microwave-induced nitrogen discharge at atmospheric pressure (MINDAP) [107–109].

In this brief survey of the MIP, we have purposely omitted recounting historical developments and discussing instrumental details and analytical results because the subject has been most adequately covered in an excellent chapter by Tschoepel [3] and in various reviews [4, 98–100]. The biennial reviews in *Analytical Chemistry* (e.g., [110]), the Annual Reports "ARAAS" [111], and Dahmen's annual bibliography [86–88] provide further keys to the extensive literature. From recent work we still mention the following:

Abdallah et al. [112] assessed the analytical performance of an atmospheric pressure helium microwave plasma produced by a surfatron [113, 114] but experienced problems with sample introduction. Excitation temperatures and other characteristics of microwave plasmas generated with a surfatron have been studied [115, 116].

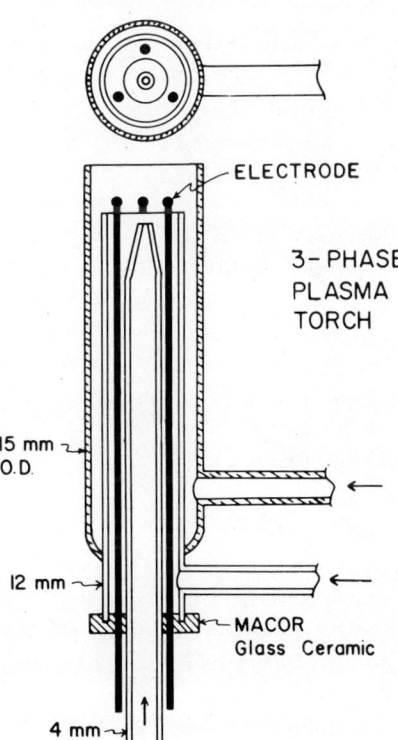

Figure 2.18. Cross section of a new torch design for a dc plasma. The torch is similar in size to those of typical ICP torches (Chapter 3). However, the torch contains three thoriated tungsten electrodes to which power is supplied for generating a dc discharge [120]. (Reprinted from R. A. Masters and E. H. Piepmeier, "Sample Entraining 3-Electrode Argon Plasma Source for Atomic Emission Spectroscopy," *Spectrochim. Acta* **40B**, 86 (1985). Copyright (1985), Pergamon Journals, Oxford.)

Kollotzek et al. [117, 118] obtained different forms of an Ar–MIP in a TM_{010} cavity by using special mountings for the discharge tube. They describe a stable three-filament Ar–MIP for the excitation of dry argon and a toroidal Ar–MIP with aerosol passing through the center of the plasma, analogous to the ICP.

Bollo-Kamara and Codding [119] have realized a toroidal MIP using a unique discharge tube with a tangential plasma support gas flow, which was further developed by Goode et al. [103] and Haas and Caruso [93] (Fig. 2.16).

The latter authors aimed in particular at better plasma containment at higher power input. Generally the use of an auxiliary ("coolant") gas and more sophisticated tube configurations ("torches") appear to be upgrading the MIP to a type of micro-ICP (Figs. 2.16 and 2.17) [93, 94, 103]. Curiously DCPs, too, seem to be more and more tailored to resemble ICPs (Fig. 2.18) [120].

REFERENCES

1. P. W. J. M. Boumans, "Excitation of Spectra," in E. L. Grove, ed., *Analytical Emission Spectroscopy*. Vol. 1, Pt. 2, Ch. 6, Dekker, New York (1972), p. 155.
2. R. M. Barnes, *CRC Crit. Rev. Anal. Chem.* **7**, 203 (1978).
3. P. Tschoepel, "Plasma Excitation in Spectrochemical Analysis," in G. Svehla, ed., *Comprehensive Analytical Chemistry*. Vol. 9, Ch. 3, Elsevier, Amsterdam (1979), p. 173.
4. A. T. Zander and G. M. Hieftje, *Appl. Spectrosc.* **35**, 357 (1981).
5. C. J. Seliskar and D. K. Warner, *Appl. Spectrosc.* **39**, 181 (1985).
6. L. R. P. Butler, H. G. C. Human, and R. H. Scott, "Electrical Flames," in J. W. Robinson, ed., *Handbook of Spectroscopy*. Vol. 1, CRC Press, Cleveland (1974), p. 816.
7. P. W. J. M. Boumans, *Theory of Spectrochemical Excitation*. Adam Hilger, London/Plenum, New York (1966).
8. H. Schirrmeister, *Spectrochim. Acta* **23B**, 709 (1968).
9. H. Schirrmeister, *Spectrochim. Acta* **24B**, 1 (1969).
10. M. Margoshes and B. F. Scribner, *Spectrochim. Acta* **15**, 138 (1959).
11. V. V. Korolev and E. E. Vainshtein, *J. Anal. Chem. USSR* **14**, 731 (1959).
12. P. N. Keliher, "Flame and Plasma Emission Analysis," in T. Kuwana, ed., *Physical Methods in Modern Chemical Analysis*. Plenum, New York (1978).
13. V. A. Fassel, *Proc. 16th Coll. Spectr. Int., Heidelberg 1971*, Plenary Lectures and reports. Adam Hilger, London (1972), p. 63.
14. S. Greenfield, H. McD. McGeachin, and P. B. Smith, *Talanta* **23**, 1 (1976).
15. P. W. J. M. Boumans, *Fresenius Z. Anal. Chem.* **299**, 337 (1979).

16. P. W. J. M. Boumans, *Fresenius Z. Anal. Chem.* **279,** 1 (1976).
17. C. D. Keirs and T. J. Vickers, *Appl. Spectrosc.* **31,** 273 (1977).
18. E. Kranz, in *Emissionsspektroskopie*, p. 160. Akademie-Verlag, Berlin (1964).
19. L. E. Owen, *Appl. Spectrosc.* **15,** 150 (1961).
20. M. Margoshes and B. F. Scribner, *J. Res. Nat. Bur. Std. A.* **67A,** 561 (1963).
21. M. Margoshes, Private communication.
22. E. Kranz, *Proc. 15th Coll. Spectr. Int., Madrid 1969*, Vol. 4. Iberica, Tarragona, 34-Madrid-7 (1971), p. 95.
23. E. Kranz, *Spectrochim. Acta* **27B,** 327 (1972).
24. W. E. Rippetoe and T. J. Vickers, *Anal. Chem.* **47,** 2082 (1975).
25. E. Kranz, *Proc. 12th Coll. Spectr. Int., Exeter 1965*. Adam Hilger, London (1965), p. 574.
26. E. Kranz, *Proc. 14th Coll. Spectr. Int., Debrecen 1967*, Vol. 2. Adam Hilger, London (1968), p. 697.
27. J. H. McGinn, *Proc. 5th Int. Conf. Ionization Phenomena Gases, Munich 1961*, (1962), p. 967.
28. M. Riemann, *Z. Physik* **179,** 38 (1964).
29. M. Riemann, in *Emissionsspektroskopie*, Akademie-Verlag Berlin (1964), p. 173.
30. M. Riemann, *Z. Anal. Chem.* **215,** 407 (1966).
31. G. Holdt and E. Hoffmann, *Z. Anal. Chem.* **225,** 114 (1967).
32. E. Hoffmann and G. Holdt, *Can. Spectrosc.* **12,** 10 (1967).
33. M. D. Marinkovic and A. M. Antic-Jovanovic, *Analyst* **92,** 645 (1967).
34. M. Marinkovic and B. Dimitrijevic, *Spectrochim. Acta* **23B,** 257 (1968).
35. M. Kliska and M. Marinkovic, *Spectrochim. Acta* **25B,** 545 (1970).
36. H. Maecker, *Z. Naturforsch.* **11a,** 457 (1956).
37. W. E. Rippetoe, E. R. Johnson, and T. J. Vickers, *Anal. Chem.* **47,** 436 (1975).
38. S. E. Valente and W. G. Schrenk, *Appl. Spectrosc.* **24,** 197 (1970).
39. C. Veillon and M. Margoshes, *Spectrochim. Acta* **23B,** 553 (1968).
40. P. N. Keliher, *Res. Developm.* **27,** (6), 26 (1976).
41. W. G. Elliott, *Amer. Lab.* **2,** (3), 67 (1970).
42. P. Merchant, Jr. and C. Veillon, *Anal. Chim. Acta* **70,** 17 (1974).
43. G. N. Coleman, W. P. Braun, and A. M. Allen, *Appl. Spectrosc.* **34,** 24 (1980).
44. D. W. Golightly and J. L. Harris, *Appl. Spectrosc.* **29,** 233 (1975).
45. R. K. Skogerboe, I. T. Urasa, and G. N. Coleman, *Appl. Spectrosc.* **30,** 500 (1976).
46. R. K. Skogerboe and I. T. Urasa, *Appl. Spectrosc.* **32,** 527 (1978).
47. G. W. Johnson, H. E. Taylor, and R. K. Skogerboe, *Spectrochim. Acta* **34B,** 197 (1979).
48. R. J. Decker, *Spectrochim. Acta* **35B,** 19 (1980).
49. D.-L. Eastwood, M. S. Hendrick, and G. Sogliero, *Spectrochim. Acta* **35B,** 421 (1980).

50. D.-L. Eastwood, M. S. Hendrick, and M. H. Miller, *Spectrochim. Acta* **37B**, 293 (1982).
51. G. W. Johnson, H. E. Taylor, and R. K. Skogerboe, *Appl. Spectrosc.* **34**, 19 (1980).
52. M. H. Miller, D.-L. Eastwood, and M. S. Hendrick, *Spectrochim. Acta.* **39B**, 13 (1984).
53. R. R. Williams and G. N. Coleman, *Spectrochim. Acta.* **38B**, 1171 (1983).
54. R. R. Williams and G. N. Coleman, *Appl. Spectrosc.* **35**, 312 (1981).
55. M. Miller, E. Keating, D. Eastwood, and M. S. Hendrick, *Spectrochim. Acta* **40B**, 593 (1985).
56. P. W. J. M. Boumans, F. J. de Boer, F. J. Dahmen, H. Hoelzel, and A. Meyer, *Spectrochim. Acta* **30B**, 449 (1975).
57. G. D. Critescu and R. Grigorovici, *Bull. Soc. Roum. Phys.* **42**, 37 (1941).
58. G. D. Critescu and R. Grigorovici, *Opt. Spektrosk.* **6**, 129 (1959).
59. G. D. Critescu and R. Grigorovici, *Naturwissenschaften* **29**, 571 (1941).
60. A. L. Stolov, *Uch. Zap. Kazan. Gos. Univ.* **116**, 118 (1956).
61. E. Badarau, M. Giurgea, G. H. Giurgea, and A. T. H. Trutia, *Proc. 6th Coll. Spectr. Int., Amsterdam 1956. Spectrochim. Acta* **11**, 441 (1957).
62. H. Dunken, W. Mikkeleit, and W. Kniesche, *Acta. Chim. Acad. Sci. Hung.* **33**, 67 (1962).
63. W. Tappe and J. van Calker, *Z. Anal. Chem.* **198**, 13 (1968).
64. V. Trunecek, *Z. Chem.* **4**, 358 (1964).
65. H. Dunken, G. Pforr, and W. Mikkeleit, *Z. Chem.* **4**, 237 (1964).
66. G. Pforr and K. Langner, *Z. Chem.* **5**, 115 (1965).
67. H. Dunken and G. Pforr, *Z. Phys. Chem.* **230**, 48 (1965).
68. R. Mavrodineanu and R. C. Hughes, *Spectrochim. Acta* **19**, 1309 (1963).
69. J. D. Cobine and D. A. Wilbur, *J. Appl. Phys.* **22**, 835 (1951).
70. U. Jecht and W. Kessler, *Z. Anal. Chem.* **198**, 133 (1964).
71. U. Jecht and H. Kessler, *Z. Physik* **178**, 133 (1964).
72. W. Kessler and F. Gebhardt, *Glastech. Ber.* **40**, 194 (1967).
73. W. Kessler, *Glastech. Ber.* **44**, 479 (1971).
74. F. Gebhardt and H. Horn, *Glastech. Ber.* **44**, 483 (1971).
75. M. Yamamoto and S. Murayama, *Spectrochim. Acta* **23A**, 773 (1967).
76. S. Murayama, *Spectrochim. Acta* **25B**, 191 (1970).
77. S. Murayama, *J. Appl. Phys.* **39**, 5478 (1968).
78. S. Murayama, *Bunko Kenkyu* **19**, 237 (1970).
79. S. Murayama, H. Matsumo, and M. Yamamoto, *Spectrochim. Acta* **23B**, 513 (1968).
80. H. Goto, K. Hirokawa, and M. Suzuki, *Z. Anal. Chem.* **225**, 130 (1967).
81. G. F. Larson and V. A. Fassel, *Anal. Chem.* **48**, 1325 (1976).
82. I.-O. Burman and K. Boström, *Anal. Chem.* **51**, 516 (1979).

REFERENCES

83. K. Govindaraju, G. Mevelle, and C. Chouard, *Anal. Chem.* **48**, 1325 (1976).
84. J. Dahmen, *ICP Information Newslett.* **6**, 576 (1981).
85. S. Greenfield, H. McD. McGeachin, and P. B. Smit, *Talanta* **22**, 553 (1975).
86. J. Dahmen, *ICP Information Newslett.* **7**, 441 (1982).
87. J. Dahmen, *ICP Information Newslett.* **9**, 81 (1983).
88. J. Dahmen, *ICP Information Newslett.* **10**, 1 (1984).
89. P. W. J. M. Boumans, *Mikrochim. Acta* **1978 I**, 399.
90. C. I. M. Beenakker, *Spectrochim. Acta* **31B**, 483 (1976).
91. C. I. M. Beenakker, *Spectrochim. Acta* **32B**, 173 (1977).
92. C. I. M. Beenakker, P. W. J. M. Boumans, and P. J. Rommers, *Philips Tech. Rev.* **39**, 65 (1980).
93. D. L. Haas and J. A. Caruso, *Anal. Chem.* **56**, 2014 (1984).
94. K. G. Michlewicz, J. J. Urh, and J. W. Carnahan, *Spectrochim. Acta* **40B**, 493 (1985).
95. K. Tanabe, H. Haraguchi, and K. Fuwa, *Spectrochim. Acta* **36B**, 119 (1981).
96. J. E. Freeman and G. M. Hieftje, *Spectrochim. Acta* **40B**, 475 (1985).
97. J. E. Freeman and G. M. Hieftje, *Spectrochim. Acta* **40B**, 653 (1985).
98. J. W. Carnahan, K. J. Mulligan, and J. A. Caruso, *Anal. Chim. Acta* **130**, 227 (1981).
99. T. H. Risby and Y. Talmi, *CRC Crit. Rev. Anal. Chem.* **14**, 231 (1983).
100. S. R. Goode and K. W. Baughman, *Appl. Spectrosc.* **38**, 755 (1984).
101. J. P. Matousek, B. J. Orr, and M. Selby, *Progr. Anal. Atom. Spectrosc.* **7**, 275 (1984).
102. N. Rait, D. W. Golightly, and C. J. Massoni, *Spectrochim. Acta* **39B**, 931 (1984).
103. S. R. Goode, B. Chambers, and N. P. Buddin, *Spectrochim. Acta* **40B**, 329 (1985).
104. G. W. Jansen, F. A. Huf, and H. J. de Jong, *Spectrochim. Acta* **40B**, 307 (1985).
105. R. M. Barnes, ed., "Plasma Spectrochemistry," Proc. 1982 Winter Conf. Plasma Spectrochemistry, Orlando, *Spectrochim. Acta* **38B** (1/2), (1983) 1.
106. R. M. Barnes, ed., "Plasma Spectrochemistry," Proc. 1982 Winter Conf. Plasma Spectrochemistry, San Diego, *Spectrochim. Acta* **38B** (1/2), (1983) 1.
107. R. D. Deutsch and G. M. Hieftje, *Appl. Spectrosc.* **39**, 214 (1985).
108. R. D. Deutsch and G. M. Hieftje, *Appl. Spectrosc.* **39**, 19 (1985).
109. R. D. Deutsch and G. M. Hieftje, *Anal. Chem.* **56**, 1923 (1984).
110. W. J. Boyko, P. N. Keliher, J. M. Patterson III, and J. W. Hershey, *Anal. Chem.* **56**, 133R (1984).
111. *Annual Reports on Analytical Atomic Spectroscopy*, Vol. 1 (1971). The Society for Analytical Chemistry, London.
112. M. H. Abdallah, S. Coulombe, and J. M. Mermet, *Spectrochim. Acta* **37B**, 583 (1982).
113. M. Moisan, R. Pantel, J. Hubert, E. Bloyet, P. Leprince, J. Marec, and A. Ricard, *J. Microwave Power* **14**, 57 (1979).

114. J. Hubert, M. Moisan, and A. Ricard, *Spectrochim. Acta* **34B**, 1 (1979).
115. M. H. Abdallah and J. M. Mermet, *Spectrochim. Acta* **37B**, 391 (1982).
116. P. S. Moussounda, P. Ranson, and J. M. Mermet, *Spectrochim. Acta* **40B**, 641 (1985).
117. D. Kollotzek, P. Tschoepel, and G. Toelg, *Spectrochim. Acta* **37B**, 91 (1982).
118. D. Kollotzek, P. Tschoepel and G. Toelg, *Spectrochim. Acta* **39B**, 625 (1984).
119. A. Bollo-Kamara and E. G. Codding, *Spectrochim. Acta* **36B**, 973 (1981).
120. R. A. Masters and E. H. Piepmeier, *Spectrochim. Acta* **40B**, 85 (1985).

CHAPTER 3

INDUCTIVELY COUPLED PLASMAS

P. W. J. M. BOUMANS

Philips Research Laboratories
Eindhoven, The Netherlands

3.1. Historical
3.2. Principles of ICP Generation
    3.2.1. Basic Setup
    3.2.2. ICP Configuration and Appearance
3.3 Performance of ICP–AES: An Outline
    3.3.1. Performance Characteristics
    3.3.2. Sample Types
    3.3.3. Sample Size
    3.3.4. Sample Preparation
    3.3.5. Elements Covered
    3.3.6. Multielement Capability and Selectivity
    3.3.7. Detection Limits
    3.3.8. Selectivity and Limits of Determination
    3.3.9. Precision
    3.3.10. Accuracy
    3.3.11. Dynamic Range—Analytical Range
    3.3.12. Synopsis
3.4. ICP Atomic Fluorescence Spectrometry (ICP–AFS)
3.5. ICP Mass Spectrometry (ICP–MS)
3.6. Review of Reviews
References

3.1. HISTORICAL

Reviews including discussions of the history of ICP–AES [1–8] give credit to Babat [9, 10] as the individual who first succeeded in sustaining induction heated plasmas at atmospheric pressure. Babat generated his plasmas in a closed system and used power inputs as high as 30–50 kW, having in mind industrial applications of rf discharges.

    Stabilization of an inductively heated plasma operated at atmospheric pressure in gases flowing through an open-ended tube was achieved in the early

1960s. Reed [11-13] was the first to describe such a discharge, which he designed with the primary aim of crystal growing and not for spectrochemical analysis, although he made reference to this possibility.

Fassel [6, 7] and Greenfield [5] recount their independent starting of analytical studies of ICPs in 1962, the first results of which were communicated in 1964 [14] and 1965 [15]. Both authors had recognized that Reed's "plasma torch" offered, in principle, unique potentials as a high-temperature atomization-excitation source, free of contamination from electrode vapors. Major efforts by the groups of Fassel [15-18] and Greenfield [14, 19-21] during the 1960s established the viability of the principle as a useful basis for spectroanalytical measurements. This work and that of other authors exploring the potentialities of ICPs for spectrochemical analysis in the initial period [23-34] is discussed in various reviews (see Section 3.6).

One criterion of the progress in the early years is the improvement in the detection limits. From Barnes' tabular survey of the various reuslts [2], it is evident that the paper of Dickinson and Fassel [16], published in 1969, heralded a new era in ICP-AES. Those authors succeeded in bringing down the detection limits to the 0.1-10 ng/mL range for many elements, which meant an improvement by two or more orders of magnitude compared to the results achieved previously. Therefore, this report stands as a major landmark paper in ICP progress [26].

Dickinson and Fassel had recognized the concept of Reed [11, 12] and Greenfield [20] to exploit the skin effect of high-frequency currents and to also configure their ICP arrangement in such a way that a toroidal instead of an ellipsoid ("tear drop") discharge was formed. Thus, the sample could be effectively introduced through the central channel of the toroid (cf. Section 3.2). Since the aerosol was generated with an ultrasonic nebulizer and desolvated prior to its introduction into the plasma, the high efficiency of ultrasonic nebulization further contributed to the success of the efforts.

By 1971, the major advantages of ICPs as excitation sources for AES had been documented and were summarized by Fassel [17] as follows:

(a) effective injection of the sample into the hot portion of the plasma; (b) relatively long-residence time of the sample in the plasma; (c) higher temperature than combustion flames; (d) continuous temperature gradient from 9000 K to room temperature, allowing greater latitude in selecting optimal temperature; (e) free atoms may be generated in the hottest zones of the plasma and then observed in lower temperature zones where background emission is lower; (f) chemical environment may be manipulated, within limits; and (g) no electrode contamination.

The results reported by Dickinson and Fassel were verified independently in 1972 by Souilliart and Robin [35] and Boumans and de Boer [36] using different ICP instruments and operating conditions. The potentials of low-power argon

ICPs as excitation sources for simultaneous multielement analysis using "compromise operating conditions" were recognized and firmly established [36–39]; the exceptionally high sensitivity of ionic lines in the spectra emitted from argon ICPs was noted and exploited [35, 38, 40], and such performance characteristics as low-interference levels under judiciously chosen operating conditions, large dynamic range, and fair stability were further documented by various groups of authors who followed Fassel's line of exploring and developing *low-power argon* ICPs. Although a variety of experimental facilities was used, the results obtained in different laboratories showed a marked degree of congruence. These results, along with those of the early efforts, form the basis of present-day ICP-AES.

Greenfield and his co-workers followed a different line and became the champions of the *high-power nitrogen–argon* ("nitrogen cooled") ICPs, of which higher sensitivity, smaller liability to interferences, and better precision were postulated to be major advantages in comparison to low-power argon ICPs [3, 41]. The past has seen many—often confusing—arguments in favor of either low-power argon or high-power nitrogen–argon ICPs, but it is only since rather recently that unambiguous experimental data are beginning to clear up the complex picture. This topic will be further considered in Sections 4.8 and 4.9.

The confusion about the advantages and disadvantages of the two categories of ICPs should not detract from the fact that Greenfield et al. [41] could state in 1975 that they had been using their high-power nitrogen–argon ICP system on a daily routine basis for practical analysis of a large variety of samples during the previous four years. Although various applications of low-power argon ICPs had been reported by 1975 [37, 42–45], the era of widespread use of these ICPs for practical applications was just beginning with the advent of commercial instruments.

Instrument manufacturers had become gradually aware of the potentialities of ICP-AES and thus 1974 saw the first modern commercial ICP instrument [46, 47]. Rather explosive developments were to follow, as may be deduced from the tabular surveys of commercial equipment in the *Annual Reports on Analytical Atomic Spectroscopy* (ARAAS) [48]. These developments will be disussed in the appropriate chapters of this work. Because of the rapid aging of detailed data on commercial equipment, we have purposely omitted surveys of such equipment, and refer the reader for up-to-date information to instrument columns in journals such as *Spectrochimica Acta (Part B)*, *Analytical Chemistry*, and *Applied Spectroscopy*, the latest volume of ARAAS and the program of the Pittsburgh Conference on Analytical Chemistry and Applied Spectroscopy.

Concluding this brief historical survey we wish to point to the common phases through which an analytical method passes during its life [49, 50]:

1. Conception.

2. Experimental verification.
3. Instrument development.
4. Principles and mechanisms.
5. Standardized procedures.
6. Senescence.

Assessing the position of ICP–AES anno 1981 in the perspective of the above phases, Barnes [51] concluded that ICP instruments have developed from laboratory curiosities into something usable by nonspecialists, while the introduction of new ICP systems bears witness to the continuation of the development phase. Concurrently, steps four and five occur. It is predicted that with the resources available at present, the ICP will probably become one of the best characterized excitation sources of emission spectroscopy, while the ever-widening range of applications, extending toward standardized procedures, indicates the maturing of the technique as an accepted approach in competition and cooperation with other methods. According to Barnes, ICP–AES will probably not reach in the 1980s the stage in which it will be gradually replaced by other methods with greater speed, economy, convenience, sensitivity, and selectivity.

3.2. PRINCIPLES OF ICP GENERATION

3.2.1. Basic Setup

Figure 3.1 is a schematic drawing of an assembly of three concentric tubes, most frequently made of silica, used for operating an ICP. The assembly of tubes, called "torch," is set up in a water-cooled coil of an rf generator. The torch consists of three tubes here designated unambiguously "outer tube," "intermediate tube," and "inner tube." Flowing gases are introduced into the torch, the rf field is switched on and the gas in the coil region is made electrically conductive by Tesla sparks. Then a plasma will be formed, provided the magnetic field strength is high enough and the gas streams follow a particular, rotationally symmetrical pattern. This inductively coupled plasma thus formed, the ICP, is maintained by inductive heating of the flowing gas in a way similar to the inductive heating of a metallic cylinder placed in the induction coil; that is, the rf currents flowing in the coil generate oscillating magnetic fields with lines of force axially oriented inside the coil. These induced magnetic fields generate in turn high-frequency, annular electric currents in the conductor, which is then heated as a result of its ohmic resistance. If the conductor is a flowing gas, an insulating confinement tube must surround the gas to prevent it from extending to the coil, which would lead to short-circuiting. On the other hand, the gas flow should be made so that a thin sheath of cold gas separates the

3.2. PRINCIPLES OF ICP GENERATION

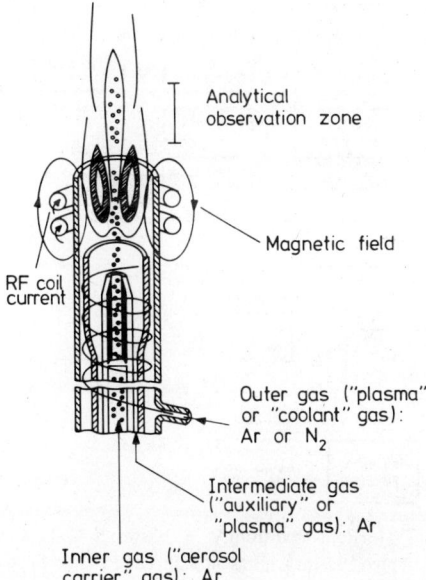

Figure 3.1. Schematic drawing of a toroidal ICP. The figure shows three concentric tubes (the "torch") placed in a water-cooled induction coil of an RF generator. The rf current through the coil induces a magnetic field, which, in turn, produces an RF current in the flowing, conducting gas. Two or three gas flows are used. The ranges of the gas flow rates, the power input to the ICP, and the frequency used for generating typical, conventional argon or nitrogen-argon ICPs are given in Table 3.1. (Reprinted with permission from P. W. J. M. Boumans, "Inductively Coupled Plasma—Atomic Emission Spectroscopy: Its Present and Future Position in Analytical Chemistry", *Fresenius Z. Anal. Chem.* **299,** 341 (1979). Copyright (1979), Springer-Verlag, Berlin, F.R.G.)

plasma from the outer confinement tube in order to prevent the latter from melting. The thermal isolation of the plasma can be achieved by Reed's vortex stabilization technique [11, 12] using a tangentially introduced gas flow, as shown in Fig. 3.1. For that reason, this outer gas flow was originally termed "coolant gas flow," as it was thought that the plasma itself was sustained by the "plasma gas," introduced through the space between the outer and intermediate tubes. Using argon for both the outer and intermediate gas, it was found, however, that an ICP could be sustained with the outer gas only, which showed to play various roles, namely those of maintaining the plasma, stabilizing its position, and thermally isolating the plasma from the outer confinement tube. Therefore, a change in terminology has been proposed: "coolant gas" would become "plasma gas" and "plasma gas" would become "auxiliary gas." Although these terms are rational, a more confusing proposal could not have been made to bring historical developments into line with new insights. To avoid this confusion, in the present text we shall use the unambiguous terms: "outer gas" and intermediate gas." It should be noted that in ICPs with nitrogen, or air, as outer gas and argon as intermediate and inner gas, the nitrogen or air serves primarily as a coolant, so the term "nitrogen-cooled ICP" still appears appropriate.

The gas flowing through the inner tube may be called "inner gas," but the terms "carrier gas" or "aerosol carrier gas" have become common, because this gas is used to carry the sample to the plasma.

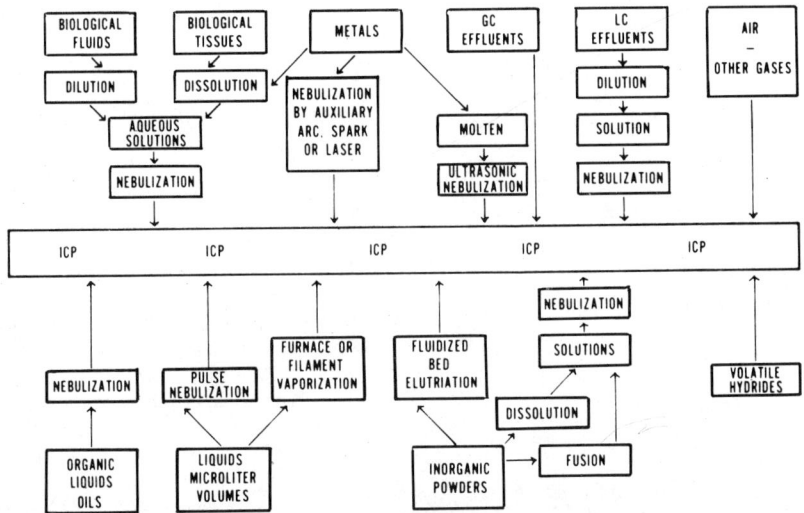

Figure 3.2. Modes of sample injection into ICPs [54]. (Reprinted from V. A. Fassel, in K. Fuwa, ed., "Analytical Spectroscopy with Inductively Coupled Plasmas—Present Status and Future Prospects," *Recent Advances in Analytical Spectrometry*. Copyright (1982), Pergamon Journals, Oxford, p. 2.)

In spectrochemical applications of ICPs, argon is the most frequently used as carrier and intermediate gas, while either argon or nitrogen (sometimes oxygen or air) is used as outer gas. More recently, ICPs with nitrogen as carrier gas or ICPs running completely on nitrogen or air have been investigated (see Section 4.9). Also, an ICP operated with helium has been studied [53].

A schematic diagram representing modes of sample introduction, as given by Fassel [7, 8, 54], is shown in Fig. 3.2. The majority of the samples at present analyzed by ICP-AES are offered to the ICP instrument in the form of liquids which are converted into an aerosol by pneumatic or ultrasonic nebulization. A fraction of the aerosol is eventually carried to the ICP. The interest in other sampling techniques, such as nebulization by auxiliary arc, spark, or laser, by furnace or filament vaporization, or by hydride generation is increasing, however.

Sample introduction in general, solid sampling techniques, and fundamentals of aerosol generation are specifically covered in Chapter 6 and Part 2, Chapters 7 and 8.

3.2.2. ICP Configuration and Appearance

The torch dictates the flow patterns of the two or three gas streams in the ICP and is the most critical component of the ICP assembly. The various consid-

3.2. PRINCIPLES OF ICP GENERATION

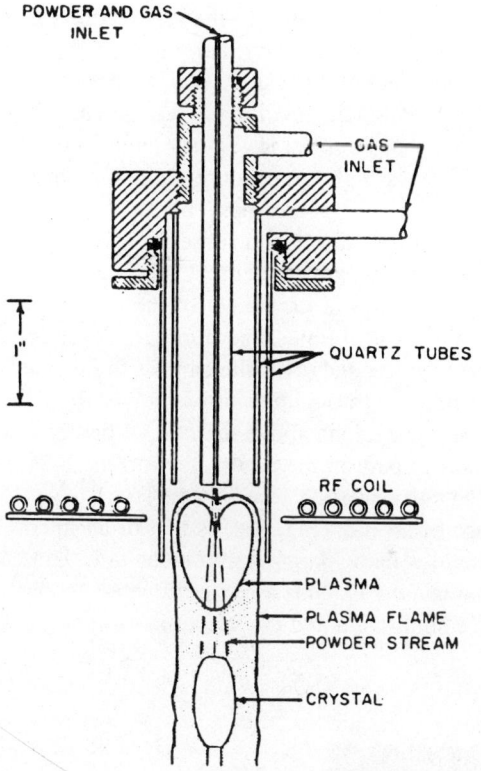

Figure 3.3. One of Reed's original configurations for generating an ICP for crystal growth [12]. (Reprinted with permission from T. B. Reed, *J. Appl. Phys.* **32**, 2534 (1961). Copyright (1961), American Institute of Physics, New York.)

erations that underlie the design of the present torches used for analytical ICPs have been gradually developed from Reed's original concepts [11, 12] and the early work of Greenfield and Fassel (see Section 3.1). A glance at Reed's torch configuration (Fig. 3.3) immediately reveals all essential features still found in the present-day torches for spectrochemical analysis. Reed introduced the intermediate gas tangentially to produce a low-pressure area at the center of the torch and cause recirculation of the plasma giving a continuous supply of ions in the coil region. The outer gas flow too was introduced tangentially to provide axial stabilization.

Greenfield et al. [20] had initially recognized the necessity to form a toroidal discharge. This toroidal configuration, which was more or less evident from Reed's work (Fig. 3.3), was explicitly featured in a patent of Greenfield et al. [20] as being essential to the effective introduction of aerosols into an ICP for spectrochemical analysis. A toroidal shape implies that the axial zones of the plasma are relatively cool compared to the surrounding zones. This phenomenon is related to the skin effect of the rf current and to aerodynamic factors. RF inductive heating provides the possibility to generate a plasma into which a stream of gas with a relatively small cross section can "bore" a hole without disturbing the stability of the plasma. Thus a major problem of introducing an aerosol effectively into a high-temperature plasma, with its inherent high viscosity [55, 56], is elegantly overcome. The aerosol is injected via a "weak spot" in the bottom of the plasma and subsequently carried through a tunnel to the plasmas zones located above the coil. Under the conditions normally used in analysis, the temperature in the tunnel of the ICP is sufficiently high (i.e., 4000–5000 K) and the transit time of aerosol particles sufficiently long (i.e., of the order of a few milliseconds) for the sample to become volatilized and atomized during its passage through the coil region of the ICP, so atomic spectra can be produced in the regions above the coil. Whether complete atomization is achieved depends in part on the power input to the ICP and on the flow rate and velocity of the carrier gas.

The appearance of an ICP (Fig. 3.4) is that of an intensely luminous, nontransparent core and a flamelike, less luminous tail. The core fills the region inside the coil and usually extends a few millimeters below and above the coil. It emits an intense continuum and the spectrum of hydrogen and neutral argon,

Figure 3.4. Side-on photograph of a conventional, 50-MHz ICP produced with the torch shown in Fig. 3.8.

3.2. PRINCIPLES OF ICP GENERATION

Figure 3.5. End-on photograph of the ICP shown in Fig. 3.4.

and various band spectra (in particular, if aerosols are introduced). Above the coil, the core becomes conical and fades into a still bright but slightly transparent zone. In argon ICPs the transition region between the core and this second zone is the best region for analytical observations, since it yields the highest SBR combined with low background flicker noise; so this region, which is located between 10 and 20 mm above the coil in low-power argon ICPs, yields the best detection limits. The precise value of the optimum observation height must be established empirically in combination with the optimization of the two principal ICP parameters, power input to the plasma and carrier gas flow rate (see Sections 4.7–4.9).

The third region of an ICP is the tail flame, which is hardly visible when pure water is nebulized but assumes typical colors as observed in flames when aerosols of metals are injected into the plasma. Such colors are also seen in the tunnel of the toroid when the metal concentration in the solution is made sufficiently high. The coloring of the tunnel can be seen by both side-on and end-on observations. The latter way of observing reveals an ICP as a luminous ring (Fig. 3.5) with a dark hole when pure water is nebulized. Side-on observation is the most common in ICP–AES, although some work with end-on observation

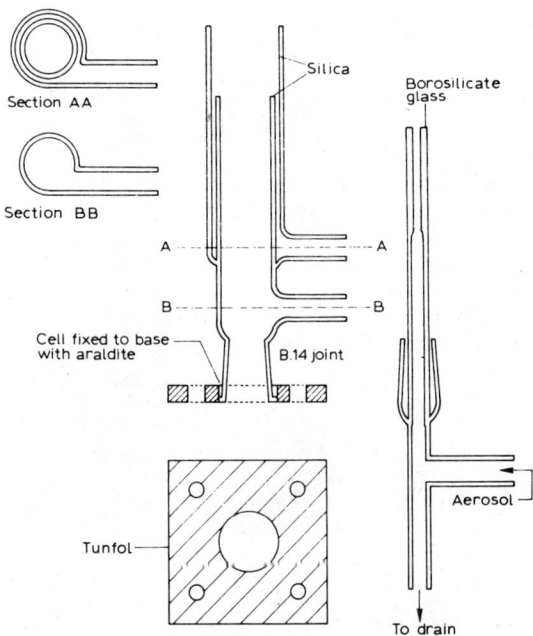

Figure 3.6. Schematic drawing of the "Greenfield torch." The internal diameter of the outer tube is approximately 23 mm [41]. (Reprinted with permission from S. Greenfield, I. Ll. Jones, H. McD. McGeachin, and P. B. Smith, *Anal. Chim. Acta* **74**, 234 (1975). Copyright (1975), Elsevier Science Publishers, Amsterdam.)

has been reported [57–61]. More recently, the possible advantages of "top-down" versus "side-on" viewing have again come into the limelight [62].

Broadly speaking, two main families of ICPs have initially resulted from Greenfield and Fassel's work, high-power nitrogen–argon (or "nitrogen-cooled") ICPs and low-power argon ICPs. A fairly large torch (Fig. 3.6) is employed for operating high-power nitrogen–argon ICPs, in which nitrogen is the outer gas and argon the intermediate and carrier gas. The torches for low-power argon ICPs are narrower (Figs. 3.7 and 3.8), while in general only two argon flows (outer and carrier) are used with aqueous solutions containing inorganic matter only, and three argon flows with organic liquids and aqueous solutions containing substantial amounts of organic matter. The ranges of the operating parameters of conventional "analytical ICPs" are stated in Table 3.1.

Most of the work of Greenfield et al. in search for designing an optimum torch was conducted on a trial-and-error basis and has been summarized in a paper [41] and a review [3]. The torch configuration considered the most satisfactory in terms of stability and ease of operation is that shown in Fig. 3.6. This configuration is commonly referred to as the "Greenfield torch." The bore

3.2. PRINCIPLES OF ICP GENERATION

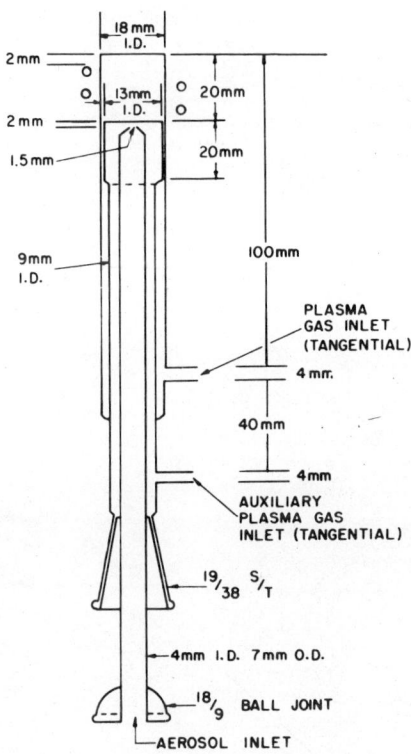

Figure 3.7. Schematic drawing of the "Fassel torch" [37, 63]. (Reprinted with permission from A. Montaser and V. A. Fassel, "Inductively Coupled Plasmas as Atomization Cells for Atomic Fluorescence Spectrometry," *Anal. Chem.* **48**, 1492 (1976). Copyright (1976), American Chemical Society, Washington, DC.)

of the injector tube was said to be the only critical dimension. Small changes in any other dimensions could be compensated by changing the gas flows. The concentricity of the work coil and the tubes of the torch seemed to be very important for stability, however.

The plasma tube arrangements used by Fassel et al. [15, 18] and Veillon and Margoshes [65] in the initial years did not permit the formation of the toroidal plasma. By choosing the appropriate torch configuration, Dickinson and Fassel [16] adopted the concept of Reed and Greenfield to "punch" a hole in the plasma by the carrier gas. They then achieved detection limits far better than those published previously. The torch used in Fassel's group was further developed [37, 39] and received the final form depicted in Fig. 3.7, which is often referred to as the "Fassel torch."

The "classical" torches shown in Figs. 3.6–3.8, as well as modifications of them as reviewed by Barnes [2], cover most of the needs for routine analytical work if used in connection with the rf equipment and under the operating conditions (power, types of gases, and gas flows) recommended by their designers. Adaptations may be required if different rf apparatus and/or operating conditions

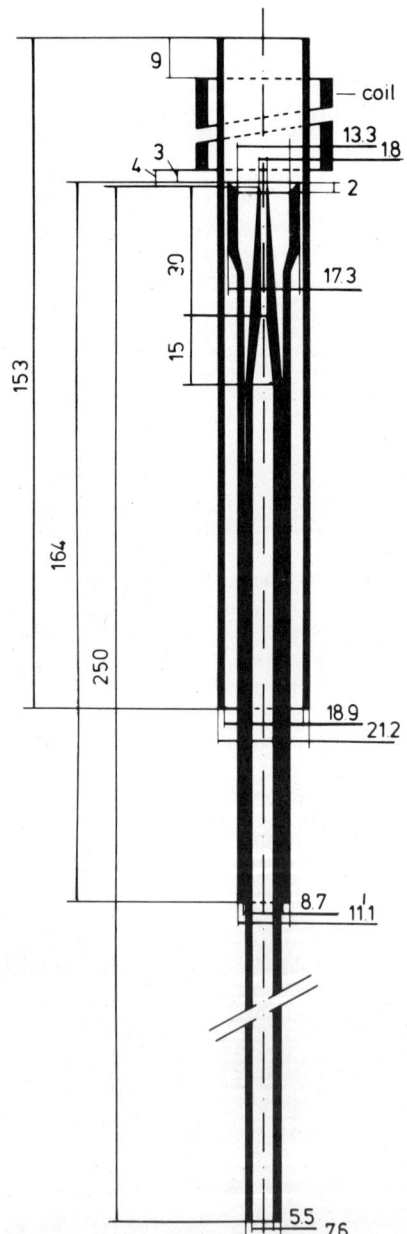

Figure 3.8. Geometry of a streamlined torch used for generating a 50-MHz ICP. Dimensions in mm. (Reproduced with permission from *Spectrochimica Acta, Part B* [64].)

Table 3.1. Ranges of Operating Conditions for Conventional Argon and Nitrogen–Argon ICPs

Parameter	Argon ICP	Nitrogen–Argon ICP
Frequency (MHz)	27–50	7–27
Power (kW)	1–2	3–7
Outer gas flow (L/min)	15–20 (Ar)	20–50 (N_2)
Intermediate gas flow (L/min)	0–1 (Ar)	10–20 (Ar)
Carrier gas flow (L/min)	0.5–1 (Ar)	1–2 (Ar)
Observation height above coil (mm)	12–18	4–10
Sample uptake rate (mL/min)	1–2	1–2

are employed. For adequately judging the literature on torch development, the following must be recognized.

During the period between 1964 and 1975, ICP research was explorative and involved the use of a variety of rf generators designed for other purposes. The primary aim of this research was to establish the effects of essential parameters, including the torch configuration, on the analytical characteristics of the ICP to optimize the analytical performance. The relevant literature reflects the explorative nature of this research and covers a diversity of torch configurations and operating conditions. This literature is summarized in the reviews of Greenfield et al. [3] and Barnes [2], which include tabular surveys of the rf generator characteristics, torch configurations, and operating parameters. For popular arrangements also see [66].

The various developments have led to a few standardized torches of the "conventional type," at present used in commercial equipment. These torches, in turn, serve as starting points in the development of a second generation of ICPs that are being developed with the primary goal of reducing the operating cost (lower argon consumption or replacement of argon by nitrogen or air) and the generator cost (lower power) without losing the excellent analytical performance achieved with the classical torches. This topic is discussed in Chapter 5.

3.3. PERFORMANCE OF ICP-AES: AN OUTLINE

3.3.1. Performance Characteristics

The performance of an analysis method can be assessed by considering the following characteristics:

- Sample types covered
- Sample preparation required

- Sample size required
- Elements covered
- Simultaneous multielement capability
- Limits of detection
- Precision
- Selectivity
- Limits of determination
- Dynamic range
- Analytical range
- Accuracy (multiplicative and additive interferences)
- Speed
- Overall ease and convenience
- Skill required from the operator
- Cost (investment, operating cost)

If an accurate and unambiguous rating or appraisal of these characteristics could be given in tabular form, the present book would find its end in that table. The fact that there are 18 chapters more to come in this book and Part 2 might warn the reader that at least "some" differentiations and nuances cannot be bypassed. This is logic, because the characteristics are partly interdependent; in particular, there will always exist a balance between the analytical capabilities, on the one hand, and such factors as instrument cost, speed, ease of operation, and skill required, on the other hand. Therefore, we shall limit this outline to a few general statements, chiefly as an introduction to the detailed discussion in the chapters that follow.

3.3.2. Sample Types

ICP–AES is principally intended for the determination of chemical elements dissolved or suspended in liquids. Both aqueous and organic liquids can be directly analyzed. These samples are nebulized and introduced into the ICP as aerosols (Chapter 6 and Part 2, Chapter 8). Handling of organic solvents is more difficult than that of aqueous solutions (Section 4.7.8 and Part 2, Chapter 6). High solids contents in aqueous solutions may give clogging problems in the nebulizer and the carrier gas tube of the torch (Chapter 6). The direct analysis of solids requires special devices and techniques (Part 2, Chapter 7).

3.3.3. Sample Size

Analyses of liquids are commonly based on steady-state signals and require a few milliliters of sample for a simultaneous multielement analysis. Depending on the number of analytes, larger amounts of sample are generally needed for sequential analyses. Special sample introduction techniques combined with appropriate electronics for signal measurement and processing permit analyses of microliter size samples (cf. Chapter 6).

3.3.4. Sample Preparation

Sample preparation techniques are specifically linked with the fields of application and escape an assessment in general terms in the scope of this outline. The topic is discussed in detail in the chapters on applications (Part 2, Chapters 1-6).

3.3.5. Elements Covered

In principle, all metallic elements are accessible. The coverage of alkalis at low concentrations is difficult. A vacuum or purged spectrometer is desired or indispensable for the determination of boron, carbon, phosphorus, nitrogen, and sulphur. Hydride generation may greatly improve coverage of hydride-forming elements.

Typically precluded under common conditions are the halogens as inorganic halides (except for iodine) and argon, oxygen and hydrogen. A new reduced-pressure ICP torch [67] appears to offer interesting prospects for halogen determinations.

3.3.6. Multielement Capability and Selectivity

The simultaneous multielement capability is high: not only is this a typical feature of any AES method, but it culminates in ICP–AES in that the majority of the elements can be virtually optimally excited under one and the same set of ICP operating conditions. Which conditions will be optimum depends on the sample type, in particular the solvent (cf. Section 4.7).

It is not so much the ICP but the compromise achieved in the spectrometer that limits the multielement capability (cf. Section 1.4 and Chapter 8). This compromise is dictated by the sample types and results from the lack of selectivity inherent in AES methods. Favorable excitation characteristics, such as those found in ICPs, lead to the emission of both the desired analysis lines and undesirable lines of concomitants. This situation creates the problem of spectral interferences and the need for careful line selection and, therefore, the demand for spectrometer flexibility. The sample types and analytes then dictate the degree of flexibility that will be required in any concrete situation.

Spectrometer flexibility and *simultaneous* multielement capability are contradictory requirements, but the greater the spectrometer's flexibility the larger its multielement capability in *sequential* analysis (cf. Sections 1.4, 4.5, 4.6 and

Chapter 7). Unless a user makes extravagantly high demands upon the number of analytes (say, more than 50) and the variety of sample types to be covered, it will be possible to find among the commercially available simultaneous or sequential spectrometers that instrument which answers the user's needs as to multielement capability. The combination of a simultaneous and a sequential instrument provides for virtually unlimited simultaneous and sequential multielement capability, but it is costly.

3.3.7. Detection Limits

Detection limits appear to constitute the easiest assessable performance characteristic of ICP–AES. Referring to the periodic table reproduced in Fig. 4.13 might settle this point. Here too, however, differentiation is required because the detection limits depend on

- The ICP (frequency, gases, torch)
- The nebulizer (pneumatic vs. ultrasonic)
- The spectrometer (bandwidth, optical conductance, detector, atmosphere in the optical path)
- The use of special techniques (e.g., hydride generation)
- The sample type

Therefore, the periodic table of Fig. 4.13 can only give an impression of the detection limits of ICP–AES. Margins amounting to factors of 1/10 or 10 should be at least allowed for. This is detailed in Section 4.1.

3.3.8. Selectivity and Limits of Determination

For an analysis line lying on a smooth, structureless background, it can be shown that the relative standard deviation (RSD) in the net line signal is 10% when the concentration equals five times the detection limit, c_L (see Section 4.2.2). This concentration is designated "limit of determination," c_D.

The ideal $c_D = 5c_L$ is not reached if the analysis line experiences overlap from interfering lines. This may often happen as a consequence of the lack of selectivity inherent in ICP–AES. Then c_D may easily exceed c_L by one or two orders of magnitude. The reason is that an AES method does not permit a "look under the spectral line." The magnitude of interfering signals other than smooth background has to be determined by indirect means, and this introduces an additional uncertainty in the eventual net line signal. It is, in particular, with major elements which emit line-rich spectra that large departures from the ideal $c_D = 5c_L$ may occur, while at the same time c_L tends to be raised. Increasing the selectivity by either chemical separation of the elements or improving the

3.3.9. Precision

At concentrations of a few hundred times the detection limits, the RSD of a single concentration measurement ranges from 0.5 to a few percent. The precise value of the RSD depends on the equipment, the precautions, the analytes, and the samples. At the concentration level mentioned above, it is the flicker noise of the nebulizer that dictates the RSD (cf. Section 4.2 and Chapter 6).

3.3.10. Accuracy

As happens with most instrumental analysis methods, the accuracy of ICP–AES depends on the extent to which the standards match the analysis samples. However, ICP–AES is known for its exceptionally low multiplicative interferences. This low interference level can be achieved by optimizing the ICP parameters (cf. Section 4.7). Thus it is possible to reach, for a large variety of samples, an accuracy of 10% with simple standards, for example, pure aqueous solutions. However, it should not be a priori assumed that ICP–AES is by definition free from multiplicative interferences (cf. Chapter 6 and Part 2, Chpater 8). In trace analysis, spectral interferences may severely endanger accuracy (cf. Section 4.3).

An accuracy far better than 10% is often needed and can be reached, but this requires greater efforts (see Part 2, Chapters 1–6). Then, a level of accuracy limited by the precision only can be attained.

3.3.11. Dynamic Range—Analytical Range

No ICP feature has been so firmly established as the large dynamic range of four to even six orders of magnitude (cf. Section 4.4). Thus the analytical range extends from traces to major constituents.

3.3.12. Synopsis

Table 3.2 summarizes the essentials of ICP–AES performance.

3.4. ICP ATOMIC FLUORESCENCE SPECTROMETRY (ICP-AFS)

In an exhaustive tutorial review, Omenetto and Winefordner [68] discussed the basic principles and applications of AFS, while Ullman [69] extensively re-

Table 3.2. Synopsis of ICP–AES Performance

Sample types:	Liquids directly
	Solids after dissolution
	Solids directly with special techniques
Sample size:	A few mL
	A few μL with special techniques
Element coverage:	At least 70 elements
	Typically precluded: inorganic F, Cl,and Br, and Ar, O, H
Detection limits:	0.1 to 100 ng/mL, depending on the analyte and the sample
Multielement capability:	Simultaneous: high
	Sequential: virtually unlimited
Precision:	0.5 to 2% RSD
Accuracy:	Reasonable, with little effort
	Accuracy = precision, with efforts and precautions
Dynamic range:	4 to 6 orders of magnitude
Analytical range:	Trace, minor, and major constituents
Interferences:	Multiplicative: low
	Additive (spectral): severest limitation of the method

viewed multielement AFS systems, especially with regard to future developments. Winefordner [70] condensed past, present, and future of AFS in tabular form with brief explanations. He answered the question "Why use atomic fluorescence spectrometry?" as follows:

1. High sensitivity and detection power.
2. Large linear dynamic range.
3. Spectral selectivity.
4. Minimal chemical interferences.
5. Cost and simplicity of use of systems.

An AFS system for single-element analysis consists of a radiation source, an atomizer, a spectral isolation device, a detector with appropriate electronics, and optics to transfer exciting radiation from source to atomizer and fluorescence radiation from atomizer to spectral isolator. The latter may be a dispersive monochromator, an interference filter, or a resonance monochromator. The usual

3.4. ICP ATOMIC FLUORESCENCE SPECTROMETRY (ICP-AFS)

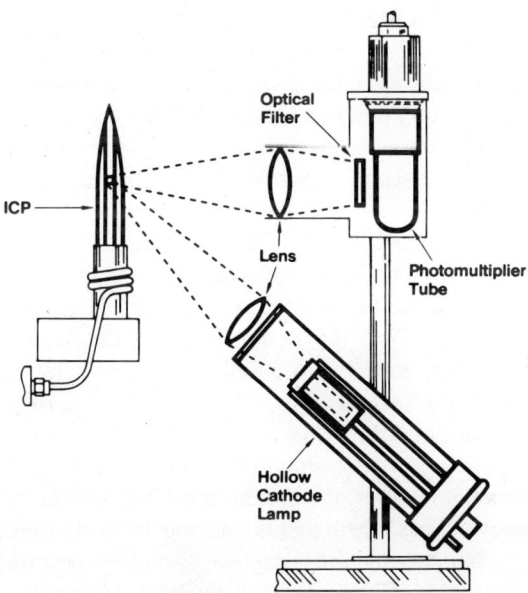

Figure 3.9. Schematic diagram of instrumental setup for ICP–AFS using hollow cathode lamp excitation [72]. (Reprinted from *American Laboratory*, volume 14, number 3, page 168, 1982. Copyright 1982 by International Scientific Communications, Inc.)

detector is a photomultiplier tube (PMT). Commonly, the exciting radiation is pulsed or chopped while the fluorescence radiation is synchronously detected in order to eliminate nonfluorescence signals emitted from the atomizer.

Montaser and Fassel [63] were the first to demonstrate the feasibility of an ICP as atomizer for AFS. Later, Demers and Allemand [71] explored an ICP as atomizer using pulsed hollow cathode lamps (HCL) as sources. This work has led to a commercial instrument which is further developed and updated [72–75]. The instrument comprises up to 12 element modules (Fig. 3.9) placed around the ICP. The method has been shown to be essentially free from scatter interferences, spectral interferences, and chemical interferences. In particular, the freedom from spectral interferences, resulting in baseline stability, is an important advantage over ICP–AES, while the large dynamic range and freedom from light scattering are advantages over flame AAS.

Initially acceptable detection limits were achieved only for nonrefractory elements. The reason is that the AFS measurements must be made in an ICP region with low background, thus high above the load coil—this is required because high background saturates the PMT or at least increases the noise—then it becomes difficult to maintain the refractory elements as free atoms. To overcome this problem, a small amount of propane is added to the carrier gas [73], which results in the reduction of metal oxides in much the same way as in a fuel-rich flame. Obviously this approach also complicates analysis and makes the choice of compromise conditions for multielement analysis more difficult. Further improvement of the detection power is achieved by the use of boosted HCLs and ultrasonic instead of pneumatic nebulization [75]. Numerical values of ICP-AFS detection limits are stated in Section 4.1.9. The literature provides additional performance data of the HCL-excited AFS system [73-75].

Laser-excited fluorescence in flames, furnaces, and ICPs is another approach to AFS [68, 70, 76]. Omenetto and co-workers recently discussed the parameters in general [77] and the fluorescence characteristics and detection limits of 14 elements in particular [78], Al, B, Ba, Ga, Mo, Pb, Si, Sn, Ti, Tl, U, V, Y, and Zr. As atomizer, an ICP was mainly used. The detection limits, found to be in a range between 0.4 and 20 ng/mL, are better than or competitive with the best detection limits of ICP-AES (cf. Section 4.1.9). Omenetto et al. pointed out the high-sensitivity, low-detection limits and extremely high spectral selectivity as the paramount features of a system consisting of a pulsed dye laser and an ICP. As disadvantages, they mentioned the by definition single-element character of the techniques and the complexity and high cost of the equipment.

Long and Winefordner [79] evaluated AFS detection limits with one ICP as an excitation source and a second one as atomization cell. This ICP-ICP-AFS setup allowed an improvement of the limits of detection by one to two orders of magnitude compared to previous ICP-ICP-AFS work [80]. The detection limits attained in the new setup are of the same magnitude as those achieved with HCL-ICP-AFS (see Section 4.1.9).

Also, a new mode for operation of the ICP was discussed. Through torch position and flow rate adjustments, a thin "pencil plasma" was formed, extending 20-30 cm above the torch. With the use of these operating conditions, propane can be added to the argon nebulizer gas to aid the dissociation of refractory oxides. Essential is the use of a high nebulizer gas flow rate of the order of 4 L/min.

Pioneers of ICP-AES will remember their experience with such pencil plasmas at high-carrier gas flow rates. In emission, disastrous interference effects were found as the carrier gas flow rate essentially exceeded about 1 L/min (see, e.g., [38, 52, 64, 81-83] and Section 4.7). Therefore, plasma configurations obtained with carrier gas flow rates of the order of various liters per minute were rejected from further consideration.

Kornblum and de Galan [84, 85] compared the physical characteristics and some interference effects in a "low-flow ICP" and a "high-flow ICP," the carrier gas flow rates being 1.36 and 4.5 L/min respectively. They concluded that the "low-flow ICP" was analytically a more useful source because higher line intensities, especially of ionic lines, were reached while interferences were significantly reduced.

This illustrates that conditions which have to be rejected in the one approach may deserve reconsideration in the other. However, a study of interferences was not yet covered by Long and Winefordner [79] in their exploration of the pencil plasma for ICP-ICP-AFS.

3.5. ICP MASS SPECTROSCOPY (ICP-MS)

Conventional inorganic mass spectrometry is an established method for elemental analysis. It is commonly designated "spark source mass spectrometry" (SSMS), although it uses not only an rf spark but also a triggered dc arc or a laser for the production of ions, that is, as ion source. Basic characteristics are the direct analysis of solids, the high detection power and selectivity, the multielement capability and wide element coverage, and the capability to obtain isotope abundance information. However, solutions cannot be directly analyzed.

Ahearn's books [86, 87], Elser's chapter [88], and the reviews by Adams [89] and Bacon and Ure [90] may serve as entries to the literature.

It is logic that the success of the ICP as an excitation source for the direct AES analysis of liquids would also induce investigations of its capabilities as an ion source for mass spectrometry (MS). Research in this field has attracted more attention from ICP workers than from mass spectrometrists so that ICP-MS appears to be an extension of ICP-AES rather than an area of MS.

Gray [91] was the first to show the feasibility of plasma source MS using a dc capillary arc source and a quadrupole mass filter. A small fraction of the plasma gas was extracted through a pinhole sampling orifice into a differentially pumped vacuum system containing an electrostatic ion lens. The first explorations of an ICP as ion source are described by Gray et al. [92, 93] and Houk et al. [95-96]. The potentials of a microwave induced plasma (MIP) as ion source have also been investigated [97, 98]. Two commercial ICP mass spectrometers have become available [99, 100]. A compilation of brief articles in the ICP Information Newsletter [101], including a chronological survey of ICP-MS publications, provides access to the literature.

Figure 3.10 shows a schematic diagram of an ICP mass spectrometer, while Fig. 3.11 illustrates the principle of the plasma sampling interface. The tail flame of the ICP is brought into contact with a water-cooled cone carrying a small molybdenum disk with a hole through which ions are extracted.

Two modes for ion extraction have been explored [93, 102, 103]:

1. The boundary layer sampling method.
2. The continuum sampling method.

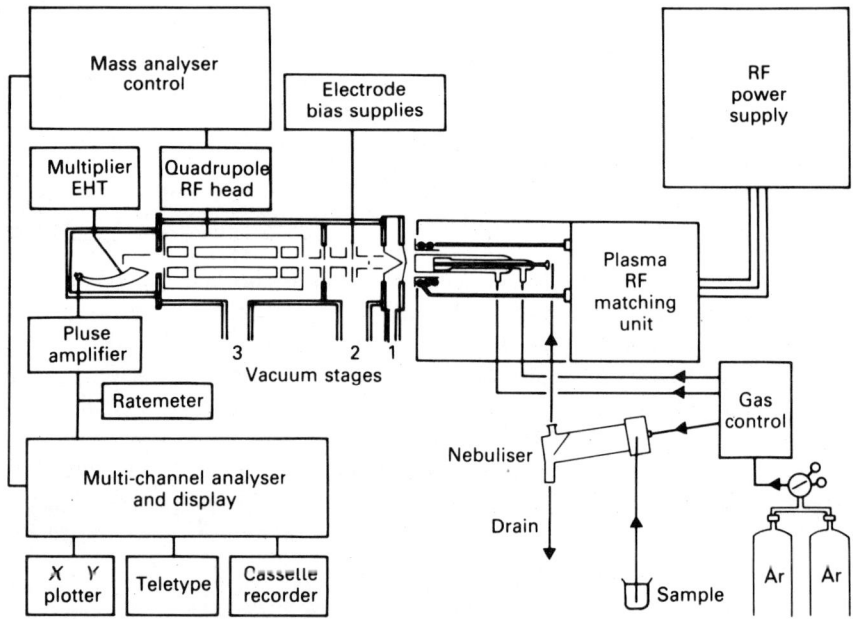

Figure 3.10. Schematic diagram of an ICP mass spectrometer using continuum mode sampling [102]. (Reprinted with permission from A. R. Date and A. L. Gray, *Analyst* **108,** 160 (1983). Copyright (1983), The Royal Society of Chemistry, London.)

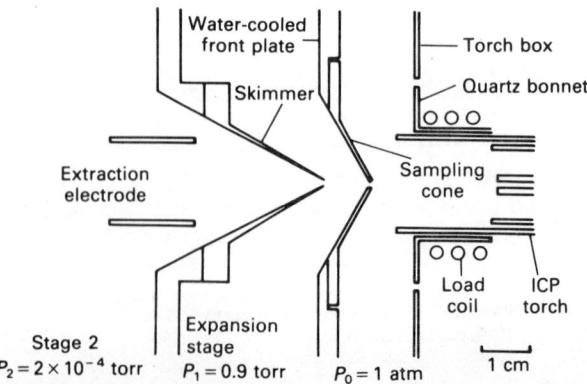

Figure 3.11. Schematic diagram of plasma sampling interface in ICP mass spectrometer using continuum mode sampling [102]. (Reprinted with permission from A. R. Date and A. L. Gray, *Analyst* **108,** 160 (1983). Copyright (1983), The Royal Society of Chemistry, London.)

3.5. ICP MASS SPECTROSCOPY (ICP-MS)

In the boundary layer sampling mode, the ions are not directly extracted from the bulk of the plasma but from the cool boundary layer that forms on the cold metal probe and the aperture. This happens with small apertures having a diameter of the order of 70 μm. Ions entering the boundary layer from the plasma may thus be modified in this layer by ion-molecule reactions, recombinations, or clustering. Although very low detection limits (Section 4.1.9) were reported, the full advantages of the ICP, in particular with respect to matrix tolerance and dynamic range, were not achieved. The technique is confined to analyses of simple solutions, requires aerosol desolvation, and limits the total salt concentration to less than 10 μg/mL in order to prevent blockage of the aperture.

In the continuum sampling mode a larger extraction aperture is used to induce continuum flow from the bulk plasma. However, this results in a "pinch" discharge in the aperture mouth caused by compression of the free electron population of the plasma, which, in turn, gives rise to photon background interference and raises the ion energy leading to a degradation in mass spectrometer resolution. The "pinch" discharge may be avoided by adjustment of the expansion rate through the sampling aperture. Thus the use of a 0.4 mm diameter aperture has become feasible. Although the detection limits in the continuum sampling mode (Section 4.1.9) are not as good as in the boundary layer method, the overall analytical capabilities are far more promising.

It appears too early, however, for a definitive assessment of ICP-MS and for a detailed comparison with ICP-AES; therefore, we shall limit ourselves to the following statements.

1. The performance of ICP-MS depends on the ICP parameters (power, gas flows), the sampling position in the ICP, the diameter of the sampling aperture, and the matrix concentration. Much knowledge about the dependence of the performance on these parameters has still to be compiled.

2. The background response is generally low, but blank problems arise at atomic mass units (amu) where hydrogen and oxygen (16–19 amu) or argon and hydrogen (40–41 amu) contribute.

3. The spectra are simple, although peaks of doubly charged ions and molecular ions do appear; the problem of spectral interferences (line overlap) is far less severe than in ICP-AES; the inherently far greater selectivity doubtless is a vital advantage over ICP-AES; a penalty of the simplicity of the spectra is that for some elements it may be difficult to find even a single analysis line.

4. At low mass numbers, the detection limits (Section 4.1.9) are comparable to those achieved with ICP-AES using a high resolution spectrometer; at high mass numbers the balance is tipped in favor of ICP-MS, which then yields detection limits that are at least one order of magnitude better.

5. The possibility to measure isotope ratios can be exploited not only in the fields where such information is needed (nuclear industry, mining, geology), but also for isotope dilution analysis.

Table 3.3. Chronological Survey of Reviews

Authors	Year	Class[1]	Note	Ref.
Keirs–Vickers	1971	R	a	105
Jordan	1971	R	b	106
Fassel	1972	TR		17
Fassel–Kniseley	1974	TR		37
Boumans	1974	TR		107
Butler–Human–Scott	1974	R		66
Greenfield	1975	V	c	5
Greenfield–McGeachin–Smith	1976	R	d	3
Greenfield–McGeachin–Smith	1976	R	e	108
Greenfield–McGeachin–Smith	1976	R	f	109
Boumans	1976	V	g	110
Sharp	1976	TR	h	111
Fassel	1977	V	i	112
Fassel	1977	TR		113
Boumans	1977	TR		114
Fassel	1978	TR		1
Barnes	1978	R		2
Barnes	1978	TR		115
Keliher	1978	TR		139
Fassel	1979	V	j	7
Robin	1979	TR	k	116
Jarosz–Mermet–Mouton–Robin–Trassy	1979	TR	l	117
Boumans	1979	TR		52
Tschoepel	1979	TR	h	4
Haas–Fassel	1979	TR	m	118
Haas–Fassel–Grabau–Kniseley–Sutherland	1979	TR	m	119
Greenfield	1980	V		120
Browner	1980	TR	n	121
Laqua	1980	TR	o	122
Barnes	1981	V		51
Zander–Hieftje	1981	R	p	123
Carnahan–Mulligan–Caruso	1981	R	p	124
Dahmen	1981	R	f	125
Robin	1982	TR	k	126
Barnes	1982	V		127
Dahmen	1982	R	f	128
Fassel	1982	TR		54
Fassel	1982	V		8
de Galan–Kornblum–de Loos Vollebregt	1982	TR	q	129
Winefordner	1982	TR	r	70
Dahmen	1983	R	f	130
Barnes	1983	V		131
Risby–Talmi	1983	R	p	132

Table 3.3. (*Continued*)

Authors	Year	Class[1]	Note	Ref.
Thompson–Walsh	1983	TR	s	133
Hieftje	1983	TR	t	134
Goode–Baughman	1984	TR	p	135
Matousek–Orr–Selby	1984	R	p	136
Dahmen	1984	R	f	137

[a] Direct current plasmas (DCP).
[b] "Physical" ICPs.
[c] Emphasis on historical developments.
[d] ICPs.
[e] DCPs.
[f] Microwave plasmas.
[g] In German.
[h] Covers high-frequency electrodeless plasmas.
[i] Emphasis on competition between ICP-AES and ETA-AAS.
[j] Award address with emphasis on historical developments.
[k] Emphasis on fundamentals.
[l] In French.
[m] Emphasis on analysis of biological samples.
[n] Concerns aerosol production in atomic spectroscopy.
[o] Covers plasma sources in encyclopedic review of AES (in German).
[p] Microwave-induced plasmas (MIP).
[q] Covers ICP-AES in a review of "Automatic atomic spectrometry."
[r] Covers ICP in a review on AFS.
[s] Book on ICP spectrometry.
[t] "Micro ICPs" (cf. Chapter 5).
[1] R = "review," TR = "tutorial review," V = "view."

It can be expected that ICP-MS will rapidly develop until a stage has been reached where both the strengths and the weaknesses of the method have been fully recognized and an assessment in terms of cost/performance ratio has shown its true position among other analysis methods.

3.6. REVIEW OF REVIEWS

Although, in many respects, reviews tend to age rapidly, they maintain considerable value because they permit the reconstruction of historical developments, provide details of these developments, and show the views of leading scientists on these developments.

Table 3.3 presents a survey of reviews, chapters, books, and reports with a review character. They are classified into three categories:

(R) Reviews that cover the literature almost completely and include tabulations with specifications of instruments or devices, working conditions, and figures of merit.

(TR) Reviews of tutorial character with often extensive, though not necessarily exhaustive coverage of the literature.

(V) Reviews of tutorial character with limited literature coverage and emphasis on the reviewer's "view" and/or a particular aspect of ICP–AES and/or related topics.

Where it was felt useful, footnotes have been added to provide further information.

Clearly, no classification is entirely free from ambiguity and bias, but it is hoped that the distinctions made here will serve the reader.

Brief status reports of national correspondents, such as those which annually appear in the *ICP Information Newsletter* [104], should be also noted. They have not been included, however, in the present "review of reviews."

REFERENCES

1. V. A. Fassel, *Science* **202**, 183 (1978).
2. R. M. Barnes, *CRC Crit. Rev. Anal. Chem.* **7**, 203 (1978).
3. S. Greenfield, H. McD. McGeachin, and P. B. Smith, *Talanta* **23**, 1 (1976).
4. P. Tschoepel, "Plasma Excitation in Spectrochemical Analysis," in G. Svehla, ed., *Comprehensive Analytical Chemistry*, Vol. 9, Ch. 3, Elsevier, Amsterdam (1979), p. 173.
5. S. Greenfield, *ICP Information Newslett.* **1**, 3 (1975).
6. V. A. Fassel, *ICP Information Newslett.* **1**, 267 (1976).
7. V. A. Fassel, *Anal. Chem.* **51**, 1290A (1979).
8. V. A. Fassel, *ICP Information Newslett.* **8**, 69 (1982).
9. G. I. Babat, *Vestn. Elektroprom.*, (1942), No. 2, p. 1; No. 3, p. 2.
10. G. I. Babat, *J. Inst. Elect. Eng.* **94**, 27 (1947).
11. T. B. Reed, *J. Appl. Phys.* **32**, 821 (1961).
12. T. B. Reed, *J. Appl. Phys.* **32**, 2534 (1961).
13. T. B. Reed, *Int. Sci. Technol.* **6**, 42 (1962).
14. S. Greenfield, I. Ll. Jones, and C. T. Berry, *Analyst* **89**, 713 (1964).
15. R. H. Wendt and V. A. Fassel, *Anal. Chem.* **37**, 920 (1965).
16. G. W. Dickinson and V. A. Fassel, *Anal. Chem.* **41**, 1021 (1969).
17. V. A. Fassel, *Proc. 16th Coll. Spectr. Int.*, Heidelberg 1971, Plenary lectures and reports, Adam Hilger, London (1972), p. 63.
18. R. H. Wendt and V. A. Fassel, *Anal. Chem.* **38**, 337 (1966).

19. S. Greenfield, C. T. Berry, and L. G. Bunch, *Spectroscopy with a High Frequency Plasma Torch*, Radyne International, Wokingham, England (1965).
20. S. Greenfield, I. L. W. Jones, and C. T. Berry, U.S. Patent 3,467,471, September 16, 1969.
21. S. Greenfield, I. L. W. Jones, C. T. Berry, and L. G. Bunch, *Proc. Soc. Anal. Chem.* **2,** 111 (1965).
22. S. Greenfield, P. B. Smith, A. E. Breeze, and N. M. D. Chilton, *Anal. Chim. Acta* **41,** 385 (1968).
23. H. C. Hoare and R. A. Mostyn, *Anal. Chem.* **39,** 1153 (1967).
24. M. E. Britske, V. M. Borisov, and Yu. S. Sukah, *Ind. Lab.* **33,** 301 (1967).
25. J. M. Mermet and J. Robin, *Proc. 14th Coll. Spectr. Int.*, *Debrecen 1967*, Vol. 2, Hilger, London (1968), p. 715.
26. R. M. Barnes, *Emission Spectroscopy*. Dowden, Hutchinson & Ross, Stroudsburg, PA (1976).
27. H. Dunken and G. Pforr, *Z. Phys. Chem.* **230,** 48 (1965).
28. G. Pforr, *Proc. 14th Coll. Spectrosc. Int.*, *Debrecen 1967*, Vol. 2, Hilger, London (1968), p. 687.
29. C. Bordonali and M. A. Biancifiori, *Met. It.*, No. 8, 631 (1967).
30. C. Bordonali and M. A. Biancifiori, *Proc. 14th Coll. Spectr. Int.*, *Debrecen 1967*, Vol. 3, Hilger, London (1968), p. 1153.
31. V. M. Gold'farb and V. K. Goikhman, *J. Appl. Spectrosc. USSR* **8,** 119 (1968).
32. I. Kleinmann and V. Svoboda, *Anal. Chem.* **41,** 1029 (1969).
33. G. H. Morrison and Y. Talmi, *Anal. Chem.* **42,** 809 (1970).
34. D. Truitt and J. W. Robinson, *Anal. Chim. Acta* **49,** 401 (1970); 51, 61 (1970).
35. J. C. Souilliart and J. Robin, *Analusis* **1,** 427 (1972).
36. P. W. J. M. Boumans and F. J. de Boer, *Spectrochim. Acta* **27B,** 391 (1972).
37. V. A. Fassel and R. N. Kniseley, *Anal. Chem.* **46,** 1110A, 1155A (1974).
38. P. W. J. M. Boumans and F. J. de Boer, *Spectrochim. Acta* **30B,** 309 (1975).
39. R. H. Scott, V. A. Fassel, R. N. Kniseley, and R. N. Nixon, *Anal. Chem.* **46,** 75 (1974).
40. J. M. Mermet, *C. R. Acad. Sci. Ser. B* **281,** 273 (1975).
41. S. Greenfield, I. Ll. Jones, H. McD. McGeachin, and P. B. Smith, *Anal. Chim. Acta* **74,** 225 (1975).
42. C. C. Butler, R. N. Kniseley, and V. A. Fassel, *Anal. Chem.* **47,** 825 (1975).
43. P. W. J. M. Boumans and F. J. de Boer, *Proc. Anal. Div. Chem. Soc.* **12,** 140 (1975).
44. R. N. Kniseley, V. A. Fassel, and C. C. Butler, *Clin. Chem.* **19,** 801 (1973).
45. D. E. Nixon, V. A. Fassel, and R. N. Kniseley, *Anal. Chem.* **46,** 210 (1974).
46. A. L. Davison, J. R. Bethune, and R. M. Ajhar, *24th Pittsburgh Conf. Anal. Chem. and Appl. Spectrosc.*, Abstr. Paper No. 30 (1973).
47. J. L. Jones, R. L. Dahlquist, J. W. Knoll, and R. H. Hoyt, *24th Pittsburgh Conf. Anal. Chem. and Appl. Spectrosc.*, Abstr. Paper No. 147 (1974).

48. *Annual Reports on Analytical Atomic Spectroscopy*, Vol. 1 (1971). The Society for Analytical Chemistry, London.
49. H. A. Laitinen, *Anal. Chem.* **45**, 2305 (1980).
50. R. M. Barnes, *ICP Information Newslett.* **2**, 62 (1976).
51. R. M. Barnes, *Trends Anal. Chem. (TrAC)* **1**, 51 (1981).
52. P. W. J. M. Boumans, *Fresenius Z. Anal. Chem.* **299**, 337 (1979).
53. J. P. Robin, J. M. Mermet, M. H. Abdallah, A. Batal, and C. Trassy in K. Fuwa, ed., *Recent Advances in Analytical Spectroscopy*. Pergamon Press, Oxford (1982), p. 75.
54. V. A. Fassel in K. Fuwa, ed., *Recent Advances in Analytical Spectroscopy*, Pergamon Press, Oxford/New York (1982), p. 1.
55. E. Kranz, *Proc. 15th Coll. Spectr. Int., Madrid 1969*, Vol. 4. Iberica, Tarragona, 34-Madrid-7 (1971), p. 95.
56. E. Kranz, *Spectrochim. Acta* **27B**, 327 (1972).
57. M. E. Britske, J. S. Sukach, and L. N. Filimov, *Zh. Prikl. Spektrosk.* **25**, 5 (1976).
58. K. I. Zil'bershtein, *ICP Information Newslett.* **5**, 508 (1980).
59. F. E. Lichte and S. R. Koirtyohann, *ICP Information Newslett.* **2**, 192 (1976).
60. C. T. Apel, D. V. Duchane, B. A. Palmer, T.M. Bieniewski, H. V. Pena, L. E. Cox, D. L. Gallimore, K. Vincent, M. Lopez, J. V. Kline, and D. W. Steinhaus in R. M. Barnes, ed., *Developments in Atomic Plasma Spectrochemical Analysis*. Heyden, London/Philadelphia (1981), p. 383.
61. J. Robin and C. Trassy, *C. R. Acad. Sc.* (Paris) **281**, 345 (1975).
62. L. M. Faires, T. M. Bieniewski, C. T. Apel, and T.M. Niemczyk, *Appl. Spectrosc.* **39**, 5 (1985).
63. A. Montaser and V. A. Fassel, *Anal. Chem.* **48**, 1490 (1976).
64. P. W. J. M. Boumans and M. Ch. Lux-Steiner, *Spectrochim. Acta* **36B**, 97 (1982).
65. C. Veillon and M. Margoshes, *Spectrochim. Acta* **23B**, 503 (1968).
66. L. R. P. Butler, H. G. C. Human, and R. H. Scott, "Electrical flames," in J. W. Robinson, ed., *Handbook of Spectroscopy*, Vol. 1. CRC Press, Cleveland (1974), p. 816.
67. C. J. Seliskar and D. K. Warner, *Appl. Spectrosc.* **39**, 181 (1985).
68. N. Omenetto and J. D. Winefordner, *Progr. Anal. Atom. Spectrosc.* **2**, 1 (1979).
69. A. H. Ullman, *Progr. Anal. Atom. Spectrosc.* **3**, 87 (1980).
70. J. D. Winefordner in K. Fuwa, ed., *Recent Advances in Analytical Spectroscopy*. Pergamon Press, Oxford (1982), p. 151.
71. D. R. Demers and C. D. Allemand, *Anal. Chem.* **53**, 1915 (1980).
72. D. R. Demers, D. A. Busch, and C. D. Allemand, *Am. Lab.* **14** (3), 167 (1982).
73. D. R. Demers, *Spectrochim. Acta* **40B**, 93 (1985).
74. R. L. Lancione and D. M. Drew, *Spectrochim. Acta* **40B**, 107 (1985).
75. E. B. M. Jansen and R. D. Demers, *Analyst* **110**, 541 (1985).

REFERENCES

76. M. Omenetto (ed.), *Analytical Laser Spectroscopy*. Wiley, New York (1979).
77. N. Omenetto and H. G. C. Human, *Spectrochim. Acta* **39B**, 1333 (1984).
78. H. G. C. Human, N. Omenetto, P. Cavalli, and G. Rossi, *Spectrochim. Acta* **39B**, 1345 (1984).
79. G. L. Long and J. D. Winefordner, *Appl. Spectrosc.* **38**, 563 (1984).
80. M. A. Kosinski, H. Uchida, and J. D. Winefordner, *Anal. Chem.* **55**, 688 (1983).
81. G. F. Larson, V. A. Fassel, R. H. Scott, and R. N. Kniseley, *Anal. Chem.* **47**, 238 (1975).
82. P. W. J. M. Boumans, L. C. Bastings, F. J. de Boer, and L. W. J. van Kollenburg, *Fresenius Z. Anal. Chem.* **291**, 10 (1978).
83. D. J. Kalnicky, V. A. Fassel, and R. N. Kniseley, *Appl. Spectrosc.* **31**, 137 (1977).
84. G. R. Kornblum and L. de Galan, *Spectrochim. Acta* **32B**, 71 (1977).
85. G. R. Kornblum and L. de Galan, *Spectrochim. Acta* **32B**, 455 (1977).
86. A. J. Ahearn, ed., *Mass Spectrometric Analysis of Solids*. Elsevier, Amsterdam/New York (1966).
87. A. J. Ahearn, ed., *Trace Analysis by Mass Spectrometry*. Academic Press, New York (1972).
88. R. C. Elser, "Spark Source Mass Spectrometry," in J. D. Winefordner, ed., *Trace Analysis: Spectrometric Methods for Elements*, Ch. 10. Wiley, New York (1976), p. 383.
89. F. Adams, *Spectrochim. Acta* **38B**, 1379 (1983).
90. J. R. Bacon and A. M. Ure, *Analyst* **109**, 1229 (1984).
91. A. L. Gray, *Analyst* **100**, 289 (1975).
92. A. L. Gray and A. R. Date, in D. Price and J. F. J. Todd, eds., *Dynamic Mass Spectrometry*, Vol. 6, Ch. 20. Heyden, London (1981), p. 252.
93. A. R. Date and A. L. Gray, *Analyst* **106**, 1255 (1981).
94. R. S. Houk, V. A. Fassel, G. D. Flesch, H. J. Svec, A. L. Gray, and C. E. Taylor, *Anal. Chem.* **52**, 2283 (1980).
95. R. S. Houk, V. A. Fassel, and H. J. Svec, in D. Price and J. F. J. Todd, eds., *Dynamic Mass Spectrometry*, Vol. 6, Ch. 19. Heyden, London (1981), p. 234.
96. R. S. Houk, H. J. Svec, and V. A. Fassel, *Appl. Spectrosc.* **35**, 380 (1981).
97. D. J. Douglas and J. B. French, *Anal. Chem.* **53**, 37 (1981).
98. D. J. Douglas, E. S. K. Quan, and R. G. Smith, *Spectrochim. Acta* **38B**, 39 (1983).
99. D. J. Douglas, *ICP Information Newslett.* **10**, 196 (1984).
100. J. E. Cantle, *ICP Information Newslett.* **10**, 206 (1984).
101. R. M. Barnes, ed., *ICP Information Newslett.* **10**, 191–212 (1984).
102. A. R. Date and A. L. Gray, *Analyst* **108**, 159 (1983).
103. A. L. Gray and A. R. Date, *Analyst* **108**, 1033 (1983).
104. R. M. Barnes, ed., *ICP Information Newsletter*, University of Massachusetts, Department of Chemistry, Amherst, MA (from 1975).

105. C. D. Keirs and T. J. Vickers, *Appl. Spectrosc.* **31,** 273 (1977).
106. G. R. Jordan, *Rev. Phys. Technol.* **2,** 128 (1971).
107. P. W. J. M. Boumans, *Philips Tech. Rev.* **34,** 305 (1974).
108. S. Greenfield, H. McD. McGeachin, and P. B. Smith, *Talanta* **23,** 1 (1976).
109. S. Greenfield, H. McD. McGeachin, and P. B. Smit, *Talanta* **22,** 553 (1975).
110. P. W. J. M. Boumans, *Fresenius Z. Anal. Chem.* **279,** 1 (1976).
111. B. L. Sharp in *Selected Annual Reviews of the Analytical Sciences,* Vol. 4. The Chemical Society, London (1976), p. 37.
112. V. A. Fassel, *Spec. Tech. Publ. 618.* American Society for Testing and Materials, Philadelphia, PA (1977), p. 22.
113. V. A. Fassel, *Pure Appl. Chem.* **49,** 1533 (1977).
114. P. W. J. M. Boumans, *ICP Information Newslett.* **3,** 71 (1977).
115. R. M. Barnes in *Applications of Inductively Coupled Plasma Emission Spectroscopy.* The Franklin Institute Press, Philadelphia, PA (1978).
116. J. Robin, *Analusis* **6,** 89 (1978); *ICP Information Newslett.* **4,** 495 (1979).
117. J. Jarosz, J. M. Mermet, J. L. Mouton, J. Robin, and C. Trassy, "Excitation spectrographique, Plasmas," *Techniques de l'Ingenieur,* P2719- 1, 7 (1979).
118. W. J. Haas and V. A. Fassel, "Inductively Coupled Plasma Atomic Emission Spectrometry," in *Elemental Analysis of Biological Materials,* Ch. 9. Int. Atomic Energy Agency, Tech. Rept. 197, Vienna, Austria (1979), p. 167.
119. W. J. Haas, Jr., V. A. Fassel, F. Grabau IV, R. N. Kniseley, and W. L. Sutherland, "Simultaneous Determination of Trace Elements in Urine by Inductively Coupled Plasma—Atomic Emission Spectrometry," in *Ultratrace Metal Analysis in Science and Environment,* Ch. 8. American Chemical Society, Washington, DC (1979), p. 91.
120. S. Greenfield, *Analyst* **105,** 1032 (1980).
121. R. F. Browner, *Anal. Chem.* **56,** 786A, 875A (1984).
122. K. Laqua, "Emissionsspektroskopie," in *Ullmanns Encyklopaedie der technischen Chemie,* 4th ed., Vol. 5. Verlag Chemie GmbH, Weinheim (1980), p. 441.
123. A. T. Zander and G. M. Hieftje, *Appl. Spectrosc.* **35,** 357 (1981).
124. J. W. Carnahan, K. J. Mulligan, and J. A. Caruso, *Anal. Chim. Acta* **130,** 227 (1981).
125. J. Dahmen, *ICP Information Newslett.* **6,** 576 (1981).
126. J. P. Robin, *Progr. Anal. Atom. Spectrosc.* **5,** 79 (1982).
127. R. M. Barnes, *Phil. Trans. Roy. Soc. London A* **305,** 499 (1982).
128. J. Dahmen, *ICP Information Newslett.* **7,** 441 (1982).
129. L. de Galan, G. R. Kornblum, and M. T. C. de Loos-Vollebregt, in K. Fuwa, ed., *Recent Advances in Analytical Spectroscopy.* Pergamon Press, Oxford/New York (1982), p. 33.
130. J. Dahmen, *ICP Information Newslett.* **9,** 81 (1983).

131. R. M. Barnes, *ICP Information Newslett.* **9,** 419 (1983).
132. T. H. Risby and Y. Talmi, *CRC Crit. Rev. Anal. Chem.* **14,** 231 (1983).
133. M. Thompson and J. N. Walsh, *A Handbook of Inductively Coupled Plasma Spectrometry.* Blackie, Glasgow/Chapman and Hall, New York (1983).
134. G. M. Hieftje, *Spectrochim. Acta* **38B,** 1465 (1983).
135. S. R. Goode and K. W. Baughman, *Appl. Spectrosc.* **38,** 755 (1984).
136. J. P. Matousek, B. J. Orr, and M. Selby, *Progr. Anal. Atom. Spectrosc.* **7,** 275 (1984).
137. J. Dahmen, *ICP Information Newslett.* **10,** 1 (1984).
138. P. N. Keliher, "Flame and Plasma Emission Analysis," in T. Kuwana, ed., *Physical Methods in Modern Chemical Analysis.* Plenum, New York (1978).

CHAPTER

4

BASIC CONCEPTS AND CHARACTERISTICS OF ICP–AES

P. W. J. M. BOUMANS

Philips Research Laboratories
5600 JA Eindhoven, The Netherlands

4.1. Detection Limits
    4.1.1. The Concept of Detection Limit
    4.1.2. The Statistical Interpretation of Detection Limits and Associated Figures of Merit
    4.1.3. Relationships between Detection Limits and Instrumental Parameters and the Significance of Signal-to-Background Ratio, Relative Standard Deviation of the Background Signal, and Signal-to-Noise Ratio
        4.1.3.1. The SBR–RSD Approach
    4.1.4. Relative Standard Deviation of the Background Signal and Detection Limit in ICP–AES Using a Photomultiplier as Detector
        4.1.4.1. Definitions
        4.1.4.2. Experimental Results
        4.1.4.3. Theoretical Relationships
        4.1.4.4. Experimental Values of Background RSD
        4.1.4.5. Detection Limits in Dependence on Background RSD and SBR
    4.1.5. Relative Standard Deviation of the Background Signal in Spectrography
    4.1.6. Detection Limit in the Presence of a Blank
    4.1.7. Numerical Values of Detection Limits in Low-Power Argon ICPs
        4.1.7.1. General Considerations
        4.1.7.2. Detection Limits of 67 Elements for Air-Path Spectrometers
        4.1.7.3. Detection Limits of Elements with Prominent Lines in the Vacuum Ultraviolet
        4.1.7.4. Analysis Lines in the Infrared Region
        4.1.7.5. Detection Limits of Alkali Metals

BASIC CONCEPTS AND CHARACTERISTICS

        4.1.7.6. Detection Limits Obtained with Hydride Generation Techniques
        4.1.7.7. Detection Limits and Spectral Resolution
        4.1.7.8. Effect of Sample Matrix on the Detection Limits and the Limits of Determination
   4.1.8. Detection Limits in ICPs Operated in Molecular Gases
   4.1.9. Detection Limits for ICP-AFS and ICP-MS
   4.1.10. Comprehensive Theoretical Treatment of the Dependence of Detection Limits on the Parameters of the Spectroscopic Apparatus and the Detector

4.2. Precision
   4.2.1. Relative Standard Deviation
   4.2.2. Dependence of the RSD of the Net Analyte Signal on the Analyte Concentration—Limit of Determination
   4.2.3. Noise Power Spectra—Effect of Signal Integration Period
   4.2.4. Precision of ICP-AES—Internal Standardization

4.3. Accuracy

4.4. Dynamic Range
   4.4.1. Calibration Function Curvature
   4.4.2. Dynamic Range—Analytical Range
   4.4.3. Self-Absorption

4.5. Multielement Capability

4.6. Line Selection and Spectral Interferences: An Outline

4.7. Optimization of ICP Operating Conditions
   4.7.1. Definition of the Problem
   4.7.2. Norm Temperature: A Useful Concept for Classifying Spectral Lines and Understanding Their Behavior
   4.7.3. Parameters To Be Optimized
   4.7.4. Trends Observed in Argon ICPs and Compromise Conditions for Inorganic Samples in Aqueous Solutions
   4.7.5. Rules for the Selection of Compromise Conditions in Argon ICPs for Analysis of Inorganic Samples in Aqueous Solutions
   4.7.6. Selection of Compromise Conditions for Aqueous Solutions Containing both Inorganic and Organic Matter (Argon ICPs)
   4.7.7. Some Trends Observed in Argon ICPs Fed with MIBK as Organic Solvent
   4.7.8. Feeding Organic Solvents into Argon ICPs: Significance of Solvent Load
   4.7.9. Other Views on and Approaches to Optimization of the ICP Parameters: Simplex Optimization

4.8. The Eternal Controversy: High Power Versus Low Power, Nitrogen-Argon Versus Argon ICPs

4.9. New Systematic Studies of ICPs in which Argon Is Partly or Wholly Replaced by Molecular Gases

References

4.1. DETECTION LIMITS

4.1.1. The Concept of Detection Limit

A detection limit is a statistical figure of merit that appraises an analytical method for the smallest detectable concentration (c_L) or absolute amount (q_L) of an element. For convenience, the topic will be discussed here in terms of the *relative* detection limit c_L, but the considerations can be easily extended to the absolute detection limit q_L.[1]

Conventionally, the detection limit is a concentration associated with the smallest signal (x_L) that can be distinguished with a predetermined chance from the random fluctuations of the background. This definition is usually interpreted as follows.

The smallest signal x_L is equalized to z times the standard deviation (σ_B) of the background signal:

$$x_L = z\,\sigma_B \tag{4.1}$$

The analytical evaluation function is assumed to be a straight line through the origin,

$$c = \frac{x_A}{S} \tag{4.2}$$

where c is concentration, x_A the net analyte signal, and S sensitivity ($= dx_A/dc$).

The detection limit is then computed by substituting x_L from Eq. (4.1) into (4.2):

$$c_L = \frac{z\,\sigma_B}{S} \tag{4.3}$$

A vast amount of literature documents that the problem is not as simple as the previous statements would suggest. Access to this literature and the concepts developed therein is facilitated if we distinguish between the following aspects of detection and detection limits:

1. The statistical interpretation of detection limits and some associated figures of merit.

---

[1] In compliance with IUPAC recommendations [1], we shall designate c_L as detection *limit*, not as detection *power*, which would refer to $1/c_L$ if Kaiser's proposal [2] were followed. The term *detection power* will be used here only for comparative appraisals of methods as a whole. Detection power then refers to the statement of a *set* of detection limits for the analytes covered by the method (cf. Section 7.7.7.5).

2. Theoretical and experimental relationships between detection limits and the instrumental parameters of an analysis method.
3. Prescriptions for measuring detection limits.
4. Signal-to-noise ratio and sources and types of noise.
5. Optimization of ICP systems to achieve the lowest possible detection limits.

Issues 2–5 are closely linked, whereas issue 1 occupies a more or less separate, though basic position.

4.1.2. The Statistical Interpretation of Detection Limits and Associated Figures of Merit

Concepts involved in the statistical interpretation of detection limits and associated figures of merit have been basically developed and/or reviewed by Kaiser [2–6], Kaiser and Specker [7], Nalimov [8], Ehrlich [9, 10], Ehrlich and Mai [11], Curry [12], Svoboda and Gerbatsch [13], Koch and Koch–Dedic [14], Winefordner [15], Zilbershtein [16], Zimmer [17], and Long and Winefordner [18].

In a tutorial review, Boumans [19] discussed the concepts of limit of detection, c_L, limit of identification (= limit of guarantee for purity), c_I, and limit of determination, c_D, from an elementary, statistical point of view. The interrelation between c_L and c_I is treated in terms of the "α error" and "β error."

The α error is elucidated in Fig. 4.1 which shows the probability distribution of a net signal x_A (or x) about the mean $\mu = 0$; this is in fact the distribution of the background measure, since x_A is defined as the difference of the gross signal x_{A+B} and the background signal x_B,

$$x_A = x_{A+B} - x_B \qquad (4.4)$$

and the mean of x_A is by definition equalized to zero ($\mu = 0$). As Fig. 4.1 shows, a critical value x_L is adopted such that we run a definite risk of taking an erroneous decision, that is, mistaking a random fluctuation of the background for a true analyte signal. This risk is defined in terms of the probability α. The magnitude of α depends on x_L and the probability distribution function. Usually one assumes a normal distribution and expresses x_L as a multiple (z) of the standard deviation (σ_B) of the background signal (Eq. (4.1)). If $z = 3$, then $\alpha = 0.0013$; if $z = 2$, then $\alpha = 0.0228$. Actually, the risks associated with these values of z are larger because experimental distributions are not necessarily normal and an estimate s_B, derived from a finite number of measurements, replaces the true standard deviation σ_B. For this reason a value of $z = 3$ has been

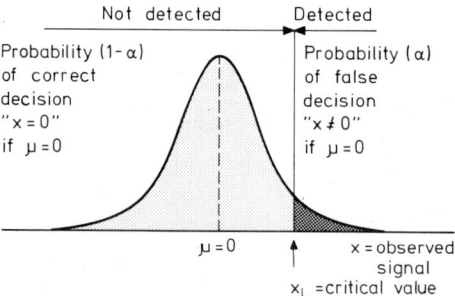

Figure 4.1. Schematic representation of the "α error." Probability distribution of a net line signal about the mean $\mu = 0$, which in fact is the distribution of the background measure. If a critical value x_L is adopted for the decision "detected" or "not detected," there is a risk α of taking a false decision. (Reproduced with permission from *Spectrochimica Acta, Part B* [19].)

favored in (arc) emission spectrography, and this value is at present recommended by the IUPAC [1]. However, flame spectrometrists have adopted $z = 2$, and this custom has proliferated in ICP-AES. As long as one is not immediately concerned with an exact statistical interpretation of an experimental result, the value of z is not important, provided that it is stated.

When it comes to an interpretation, it is crucial to recognize not only the existence of the α error but also that of the β error. The latter is explained in Fig. 4.2, which shows a probability distribution of net signals with a true mean $\mu = \mu_I$. This represents a distribution of observations associated with a concentration c_I. Let us arbitrarily fix a critical value x_L for rejecting those observations of x that we shall consider as "not detected." If the area at the lefthand

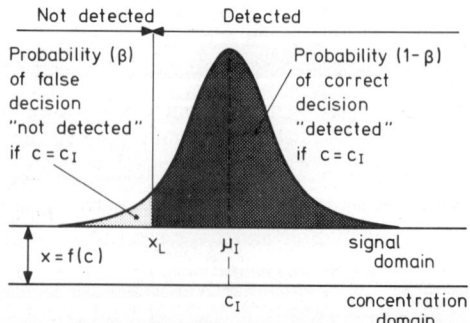

Figure 4.2. Schematic representation of the "β error." Probability distribution of a net line signal about the mean $\mu = \mu_I$. If a critical value x_L is adopted for the decision "detected" or "not detected," there is a risk β of taking a false decision. (Reproduced with permission from *Spectrochimica Acta, Part B* [19].)

4.1. DETECTION LIMITS

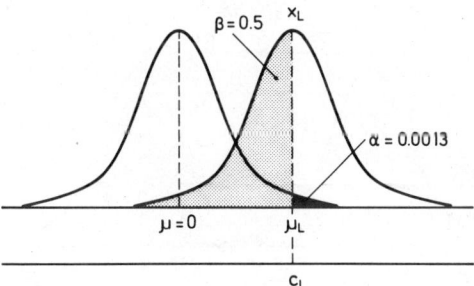

Figure 4.3. Illustration of the fact that there is a 50% chance of falsely not detecting an analyte if its concentration c equals the detection limit, that is, $c = c_L$. (Reproduced with permission from *Spectrochimica Acta, Part B* [19].)

side of x_L is equal to β, then there is a probability β that a true concentration c_I will remain undetected.

Synthesizing Figs. 4.1 and 4.2, such that $c_I = c_L$, we obtain Fig. 4.3 which illustrates the fact that there is a 50% chance ($\beta = 0.5$) of falsely not detecting an analyte if its concentration c equals the detection limit. Most frequently this is not the meaning that "customers" of analytical chemists attach to the concept of detection limit, as they are apt to interpret it as a *maximum* concentration that may escape attention with a definite risk (β). Therefore, Kaiser [2, 4, 6] introduced the concept of "limit of guarantee for purity,"[2] c_I, that is the largest concentration that may be left undetected with a risk β, or the smallest concentration that we can detect with a probability $(1 - \beta)$ if it is present in the sample. Concentration c_I is defined so that $x_L = 3\sigma_B$ and $\alpha = \beta$. This situation is depicted in Fig. 4.4 for two normal distributions having both a standard deviation σ. Since now $\alpha = \beta = 0.0013$, the following criteria will hold:

1. By definition no analysis can ever yield a result $c < c_L$.
2. If we find a result $c \geq c_L$, there is a chance $\alpha \leq 0.0013$ that the result originates from random fluctuations of the background.
3. Concentrations $c \geq c_I$, if present, will escape attention with a risk $\beta \leq 0.0013$; in other words, such concentrations will be identified with a probability $1 - \beta \geq 0.9987$.

It follows from the above that the signal x_I associated with c_I is given by

$$x_I = x_L + 3\sigma = 6\sigma \tag{4.5}$$

---

[2] The term *limit of identification* is preferred by this author [19] since the concept plays a part not only in the analysis of high-purity materials but in any field of analytical chemistry where a constituent must be identified with sufficient certainty.

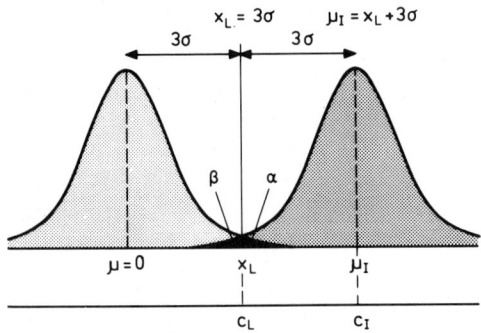

Figure 4.4. Combination of the probability distributions about $\mu = 0$ (Fig. 4.1) and $\mu = \mu_I$ (Fig. 4.2), where the critical level has been chosen so that $x_L = 3\sigma$ and μ_I has been chosen so that $\alpha = \beta$. (Reproduced with permission from *Spectrochimica Acta, Part B* [19].)

or generally by

$$x_I = z_{U_B} + z_{U_A} \tag{4.6}$$

if the background and net signal do not have the same standard deviation in the vicinity of the detection limit and the α and β risks are generally defined in terms of parameter z.

The last quantity to be considered in this elementary treatment is the "number of measurements." Two different numbers must be distinguished.

1. The number (N) of measurements used in the determination of the standard deviation. This number should be sufficiently large, at least equal to 20, to permit a reliable estimate of σ_B to be made. If this condition cannot be fulfilled, one must consider that the calculated standard deviation is an estimate (s_B) with an uncertainty that can be taken into account by replacing z in Eqs. (4.1), (4.2), and (4.3) with Student's t.

2. The number (n) of measurements underlying the determination of the means of the background (x_B) and gross line (x_{A+B}) signals when a sample is tested for the presence of the analyte. Let us denote these numbers by n_B and n_{A+B} respectively. Statistical tests involving a comparison of the means x_B and x_{A+B} use the standard deviation (s_m) of the difference of the means as a statistic:

$$s_m^2 = \frac{s_{A+B}^2}{n_{A+B}} + \frac{s_B^2}{n_B} \tag{4.7}$$

If

$$s_{A+B} \approx s_B \tag{4.8}$$

in the vicinity of the detection limit, then

$$s_m = s_B \left(\frac{1}{n_{A+B}} + \frac{1}{n_B}\right)^{1/2} \tag{4.9}$$

We thus find the critical level

$$x_L = t s_m = t s_B \left(\frac{1}{n_{A+B}} + \frac{1}{n_B}\right)^{1/2} \tag{4.10}$$

where t is Student's t, which, in turn, is a function of α and the number of degrees of freedom $(N - 1)$ underlying the determination of s_B. Two special cases are of interest:

$$x_L = t s_B \sqrt{2} \tag{4.11}$$

which applies if $n_{A+B} = n_B = 1$, that is, when a single measurement on a sample is compared with a single measurement on a blank; and

$$x_L = t s_B \tag{4.12}$$

which applies if $n_{A+B} = 1$ and $n_B \geq 20$, that is, when a single measurement on a sample is compared with a well-known blank. However, $n_{A+B} = n_B = 2$ leads to the same value of x_L.

The values to be adopted for N, n, and t (or z) constitute part of the prescriptions that define a complete analytical procedure. It is only within the scope of such a complete analytical procedure that limits of detection and limits of identification have an unambiguous and realistic meaning.

Frequently the literature produces detection limits that are not associated with a complete analytical procedure, but only characterize a set of exploratory experimental conditions which have interesting potentials for the further development of a new, complete analytical method. Then the precise interpretation of the detection limit is hardly relevant, so the choice of the critical level $x_L = 2\sigma_B$ can be justified in comparisons with results obtained on a similar basis. The step towards an interpretation within the scope of a complete procedure, however, will have to involve the multiplication of the "exploratory detection limits" with some factor, in the most pessimistic case, the factor $3\sqrt{2}$ if one proceeds from a "2σ detection limit" to a "$6\sigma\sqrt{2}$ limit of identification." Usually this will not be the only factor to be sacrificed if one steps over from idealized, exploratory laboratory conditions to real sample analysis!

We must eventually note that even if the critical level is fixed, detection limits have an inherently large uncertainty of, say, a factor of 2 to 3, because the standard deviation of the background may fluctuate from one experiment to another and also because the sensitivity has a statistical uncertainty that may be large in the vicinity of the detection limit. Finally, the RSD of a concentration determination at the detection limit is theoretically 50% if $x_L = 2\sigma_B\sqrt{2}$ and 33% if $x_L = 3\sigma_B\sqrt{2}$ (see Section 4.2.2).

4.1.3. Relationships between Detection Limits and Instrumental Parameters and the Significance of Signal-to-Background Ratio, Relative Standard Deviation of the Background Signal, and Signal-to-Noise Ratio

To achieve the lowest possible detection limits one must aim (Eq. (4.3)) at maximizing the sensitivity (S) and minimizing the standard deviation (σ_B) of the background signal. An effective approach requires precise knowledge of the contributions from the separate parts of the system to S and σ_B, that is, from the excitation source, the optics, the detector, and the electronics, so that the greatest emphasis can be put on the optimization of those parameters that have the largest influence on S and σ_B.

Various groups of authors have published theoretical approaches to express S and σ_B in terms of instrumental parameters in order to define optimum conditions for detecting low concentrations or absolute amounts of analytes [20–38]. The theory developed by Laqua and his associates covering the parameters of the spectrometer and the detector will be briefly discussed in Section 4.1.10. Here we shall primarily consider some concepts of the approach using the SBR and the RSD of the background signal, and we shall discuss some experimental findings that may provide an insight into the problems that are treated comprehensively in elaborate theories.

It should be noted that the RSD is identical to the reciprocal of the SNR, and that very often "signal" in SNR can be identified with background signal. This recognition may clear up [39] some apparent gaps between approaches in terms of SNRs elaborated by authors such as Alkemade, Winefordner, and their co-workers [40–44] and approaches in terms of SBR-RSD originating from the "Dortmund-School" of Kaiser, Laqua, Hagenah, and their associates [3, 11, 20, 21, 23–27, 45–47]. The present treatment links up mainly with the SBR-RSD approach.

Epstein and Winefordner recently reviewed the usefulness of signal-to-noise treatment in analytical spectrometry [48]. McGeorge and Salin published interesting theoretical and practical results for rapid scan ICP-AES [49] and for ICP-AES using photodiode arrays [50] in terms of SNR.

The SBR-RSD approach roots in theories of detection limits for photographic detection where the critical level corresponding to x_L in Eq. (4.1) is the difference (ΔS) in the blackenings of line plus background (S_{A+B}) and background (S_B), which, in view of the logarithmic response of a photographic emulsion, translates into a difference of two logarithmic intensities or the logarithm of the ratio of the intensities of line plus background and background:

$$\underline{\Delta Y}_{A+B,B} = \log \underline{I}_{A+B}/I_B \qquad (4.13)$$

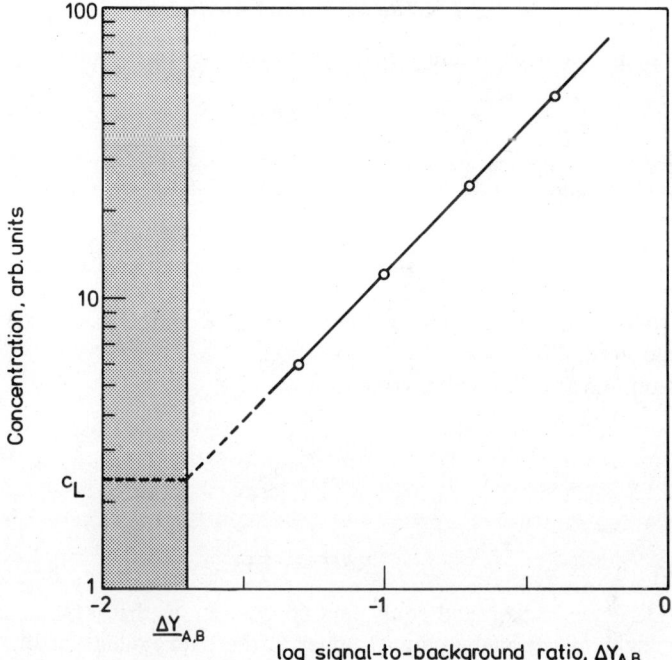

Figure 4.5. Double logarithmic plot of the analytical function $c = f(I_A/I_B)$: $\log c = \eta \Delta Y_{A,B} + K$. The figure illustrates how the detection limit c_L is found by extrapolating the analytical curve to the critical level ("noise level"), defined here in terms of the logarithm of the smallest detectable SBR $\underline{\Delta Y}_{A,B}$. In the example $\underline{\Delta Y}_{A,B} = -1.70$ ($I_A/I_B = 0.02$), as derived from $\underline{\Delta Y}_{A,B} = 3\sqrt{2}\, \sigma_{Y(B)}$ using the conversion formula

$$\Delta Y_{A,B} = \log(10^{\Delta Y_{A+B,B}} - 1)$$

where $\sigma_{Y(B)}$ is the standard deviation of the log background intensity. In the example $\sigma_{Y(B)}$ is about 0.002, which corresponds to an RSD of $2.3 \times 0.002 \approx 0.005$ in the intensity. If the latter is converted to the critical value of I_L/I_B using the factor $3\sqrt{2}$, one consistently finds a value of 0.02 (cf. Section 4.1.5 and e.g., [28, 45]).

The underlining symbolizes that the critical level is a lower limit.

A detection limit is found if the log *net* line to background ratio ($\underline{\Delta Y}_{A,B}$), derived from $\underline{\Delta Y}_{A+B,B}$ is associated with an analytical function of log concentration vs. log line-to-background ratio (Fig. 4.5). The critical level $\underline{\Delta Y}_{A+B,B}$ is, similarly to x_L in Eq. (4.1), expressed as a multiple of a standard deviation which now is the standard deviation of a logarithmic intensity and, therefore, proportional to a *relative* standard deviation of the (background) intensity.

The approach using line-to-background ratio (or SBR) and RSD of the background signal has turned out to be most useful not only because it fits the logarithmic response of photographic detection so well, but even more so because it introduces the background as a natural reference level [20].

4.1.3.1. The SBR–RSD Approach

By introducing the background signal x_B into Eq. (4.3) we obtain

$$c_L = \frac{z\sigma_B}{x_B} \frac{x_B}{S} \qquad (4.14)$$

which converts into

$$c_L = \frac{z\sigma_B}{x_B} \frac{x_B}{x_0} c_0 \qquad (4.15)$$

if we introduce the signal-to-background ratio, SBR = x_0/x_B associated with a concentration c_0.

Equation (4.15) can now be written as

$$c_L = z(\text{RSD})_B \left(\frac{c_0}{\text{SBR}}\right) \qquad (4.16)$$

or as

$$c_L = z(\text{RSD})_B \cdot \text{BEC} \qquad (4.17)$$

where BEC is the background equivalent concentration, that is, the concentration that yields a net analyte signal equal to the background signal. In Eqs. (4.16) and (4.17) the RSD must be expressed as a fraction; if it is expressed as a percentage, the right-hand sides of the equations must be multiplied by 0.01.

From Eq. (4.16) one finds that the SBR at the detection limit equals $z(\text{RSD})_B$. Thus if $z = 2$ and $(\text{RSD})_B = 0.01$, the SBR at the detection limit is 1/50.

Writing the detection limit as a product of an RSD and a reciprocal SBR (or a BEC) greatly facilitates a theoretical or experimental analysis of the influences of the instrumental parameters on the detection limits, as will be illustrated in Section 4.4.

4.1.4. Relative Standard Deviation of the Background Signal and Detection Limit in ICP–AES Using a Photomultiplier as Detector

4.1.4.1. Definitions

The RSD of the background signal is usually derived from a series of, say, ten consecutive integrations of 10 or 15 s. Ambiguity may arise in defining the background signal x_B.

Direct-Current versus Alternating-Current Measurements. In dc measurements the dark current signal is often included in the background and gross line signals because it cancels in taking the difference x_A (Eq. (4.1)).

This approach leads to correct values of the detection limit, also when Eq.

(4.16) is used, provided that the dark current signal is consistently included in *both* $(RSD)_B$ and the SBR. However, the numerical values of $(RSD)_B$ and SBR determined in this way are not equivalent to those found in ac measurements using lock-in amplification, where the dark current signal is electronically eliminated. This may give rise to ambiguity when $(RSD)_B$ and SBR values obtained in either mode of measurement are mutually compared: the dc mode will tend to yield lower $(RSD)_B$ and SBR values than the ac mode.

To avoid this ambiguity and to permit significant comparisons of systems, the "ICP Detection Limits Committee" [51] recommended that reported values of $(RSD)_B$ and SBR be corrected for the dark current signal in dc measurements, or that at least the dark current signals be separately stated so that the results can be eventually reduced to the same denominator as results obtained with ac measurements. Evidently it is misleading to include the dark current signal in the definition of $(RSD)_B$ and not in that of SBR.

"Blank" Measurement versus "Off-Peak Background" Measurement. Basically the background signal can be taken as either

1. The signal in the spectral window of the analysis line ($\lambda_a \pm \Delta\lambda/2$, where $\Delta\lambda$ = spectral bandwidth) while a "blank" solution is aspirated, or
2. The signal in a spectral window ($\lambda_a' \pm \Delta\lambda/2$) adjacent to the analysis line, again while a "blank" solution is aspirated ("off-peak setting").

With pure aqueous solutions, differences between (1) and (2) may result from

a. A residual analyte impurity (e.g., Si, Ca)
b. Plasma emission other than continuous radiation, in particular bands (e.g., OH, NH, N_2) and lines (Ar, H)

A contribution of type (b) to the signal in the spectral window of the analysis line is inherent in the system and must, therefore, be included in the background measurement. A contribution of type (a), on the contrary, depends on the incidental impurity of the chemicals used and would ideally have to be precluded from the results. This can be achieved by the off-peak measuring mode, which, however, will introduce ambiguity with respect to the effect of contributions of type (b).

This difficulty will show up in amplified form if solutions with a matrix instead of pure aqueous solutions are taken as the basis for a determination of detection limits. For example, with an iron salt as matrix in water, the iron spectrum will essentially contribute to the background in the spectral windows of the analysis lines, and *these* contributions, and *not* those found at the off-peak settings, constitute the relevant background spectrum for the iron matrix.

Therefore, the "ICP Detection Limits Committee" [51] considered a background measurement in the spectral window of the analysis line using a blank solution of the matrix as being the best alternative, but recommended to supplement the results by a wavelength scan in the vicinity of the analysis line.

In agreement with the above, we shall define the background signal in the following discussion as the "blank signal" in the spectral window of the analysis line and preclude the dark current contribution from this background signal.

Reagent blanks frequently impose essential limitations on the detection power of a method. The main problems lie in the chemical rather than the spectroscopic domain, for a discussion of which we refer the reader to the literature, for example, [52–56].

4.1.4.2. Experimental Results

In the initial landmark ICP papers of Fassel et al. [57, 58] and Greenfield et al. [59], only "bare" detection limits were reported without additional information about $(RSD)_B$ and SBR. The first data of this type appeared in an article of Boumans and de Boer [60], who reported $(RSD)_B$ values between 1 and 3% for analysis lines between 250 and 670 nm. Using an improved experimental setup, this group of authors [61–64] later reported the $(RSD)_B$ to fluctuate about a value of 1% at wavelengths between 260 and 650 nm, where a sufficiently high background radiant flux reached the detector to make the source flicker noise the predominant noise contribution. Similar findings were explicitly reported by other authors for a low-power argon ICP [65] and an ICP with reduced argon consumption [66]. The result of an approximately constant $(RSD)_B$, that is, a standard deviation proportional to the background signal, appears to apply generally provided that particular conditions are fulfilled. Recently Boumans et al. [67] making a more refined analysis of the $(RSD)_B$, reported a definite trend with wavelength (Fig. 4.6), and indicated by which factors the $(RSD)_B$ is controlled (Section 4.1.4.3).

For a high-power nitrogen–argon ICP, Greenfield [68] reported the background noise to be proportional to the background signal, but also noted that a square root dependence on the background signal had been found for his experimental facilities, which comprised a somewhat outdated polychromator system using classical integration of dc signals. In a subsequent comprehensive theoretical treatment, Greenfield and Thorburn Burns [69] defined the background signal as a *gross* background signal, thus including the dark current contribution. Those authors showed that both the dark current and the PMT gain may play an essential part in the dependence of the background noise on the gross background signal. In the light of the results of Boumans et al. [67] (Section 4.1.4.3) one may ask to what extent the complex treatment of the problem by Greenfield and Thorburn Burns [69] specifically roots in their experience with a particular spectrometer, and whether much of the complication would not have been avoided if state-of-the-art facilities had been at their disposal.

4.1.4.3. *Theoretical Relationships*

A theoretical treatment requires the background signal x_B to be accurately defined. Here we shall define it as the anode current[3] drawn from the PMT when a background radiant flux $\Phi_{B,\lambda}\,d\lambda$ strikes the photocathode. Flux $\Phi_{B,\lambda}\,d\lambda$ is the flux in the spectral window of an analysis line that originates from the ICP when no analyte is present.

We assume that there are three contributions to the background noise: source flicker noise, shot or photon noise, and detector noise (see, e.g., [70–74] and references given therein for the origin of some basic formulas used below). The noise contributions are expressed as standard deviations (σ), and it is assumed that the total variance (σ^2) is the sum of the variances of the components; in other words, the different types of noise are uncorrelated (see in particular [42] for the treatment of correlations). This implies that the electronics are assumed to contribute negligible noise.

Source Flicker Noise. A radiant flux $\Phi_{B,\lambda}\,d\lambda$ (W) produces a photocathodic current i_Φ^c (A):

$$i_{\Phi,B}^c = S_\lambda \Phi_{B,\lambda}\,d\lambda \qquad (4.18)$$

where S_λ (A/W) is the spectral sensitivity of the PMT.

The corresponding anode current is

$$i_{\Phi,B}^a = g S_\lambda \Phi_{B,\lambda}\,d\lambda \qquad (4.19)$$

where g is the PMT gain ($= i_\Phi^a/i_\Phi^c$).

The source flicker noise, $\sigma_{F,B}^a$ (A), is proportional to $i_{\Phi,B}^a$

$$\sigma_{F,B}^a = \alpha_B i_{\Phi,B}^a \qquad (4.20)$$

where α_B is the proportionality coefficient for the background flicker noise.

Shot Noise. The shot noise, σ_{Sh}^c (A), of the photocathodic current due to the incident flux is

$$\sigma_{Sh}^c = (2 e i_{\Phi,B}^c)^{1/2}\,\Delta f^{1/2} \qquad (4.21)$$

where Δf (Hz) is the effective noise bandwidth and e the electronic charge ($= 1.6 \times 10^{-19}$ C). In terms of anode current,

---

[3] Obviously, an equivalent alternative is the voltage over the anode resistance.

$$\sigma_{Sh}^a = ga(2ei_{\Phi,B}^c)^{1/2} \Delta f^{1/2} \tag{4.22}$$

where a (no unit) is a factor that accounts for the fluctuations in gain due to statistical fluctuations of secondary electron emission ($1.1 < a < 1.4$).

Detector Noise. Two contributions must be considered:

1. Dark current noise, which, in terms of the shot noise σ_D^a (A) of the anode dark current, can be expressed as

$$\sigma_D^a = ga(2ei_D^c)^{1/2} \Delta f^{1/2} \tag{4.23}$$

where i_D^c is the cathode dark current (A).

2. Johnson or thermal noise, σ_T^a (A), in the anode load resistor, given by the Nyquist formula

$$\sigma_T^a = \left(\frac{4kT}{R_a}\right)^{1/2} \Delta f^{1/2} \tag{4.24}$$

where k is the Boltzmann constant ($= 1.38 \times 10^{-23}$ J/K), T the absolute temperature (K), and R_a the anode load resistance (Ω).

Total Noise. From Eqs. (4.20), (4.21), (4.23), and (4.24) we obtain for the total background noise, σ_B:

$$\sigma_B^2 = \alpha_B^2 (i_{\Phi,B}^a)^2 + g(a^2\, 2e\, \Delta f)\, i_{\Phi,B}^a + [(g^2 a^2\, 2ei_D^c + 4kT/R_a)\, \Delta f] \tag{4.25}$$

or

$$\sigma_B^2 = \alpha_B^2 x_B^2 + g\beta x_B + \gamma \tag{4.26}$$

where $\beta = a^2 2e\, \Delta f$, γ is the function in brackets, and x_B has replaced $i_{D,B}^a$.

For the RSD we then have

$$(RSD)_B = \left(\alpha_B^2 + \frac{g\beta}{x_B} + \frac{\gamma}{x_B^2}\right)^{1/2} \tag{4.27}$$

It can be easily shown that the Johnson noise term in Eq. (4.25) is at least two orders of magnitude smaller than the dark current term if $R_a = 1$ MΩ, $g = 10^6$, and the anode dark current (gi_D^c) is smaller than 100 nA. This is the most common situation in practice. Then Eq. (4.27) can be written in the form

$$(RSD)_B = \left(\alpha_B^2 + \frac{a^2 2e\Delta f}{S_\lambda \Phi_{B,\lambda} d\lambda} + \frac{a^2 2ei_D^c \Delta f}{(S_\lambda \Phi_{B,\lambda} d\lambda)^2}\right)^{1/2} \tag{4.28}$$

Expression (4.28) is useful for discussing the differences in $(RSD)_B$ found with different PMTs at constant background radiant flux. Equations (4.26) and (4.27)

4.1. DETECTION LIMITS

are convenient for discussing the dependence of the $(RSD)_B$ on the background signal measured with a given PMT.

Equation (4.28) does not contain the PMT gain because it was assumed that the anode dark current arises entirely from electrons thermally emitted from the cathode. Deviations from this ideal situation have been reported, however. For example, Greenfield and Thorburn Burns [22] classified the PMTs in their spectrometer into three groups: those with a dark current proportional to the gain (i.e., tubes with a dark current due to thermally emitted electrons from the cathode); those with a dark current independent of gain (i.e., with a dark current attributed to anode leakage); and hybrids of these two.

Various other departures from the ideal situation described by the above-stated formulas may occur (see, e.g., [70–74] for details).

4.1.4.4. Experimental Values of Background RSD

Figure 4.6 shows the dependence of $(RSD)_B$ and x_B on wavelength as reported by Boumans et al. [67] for their ICP system with two types of entrance optics,

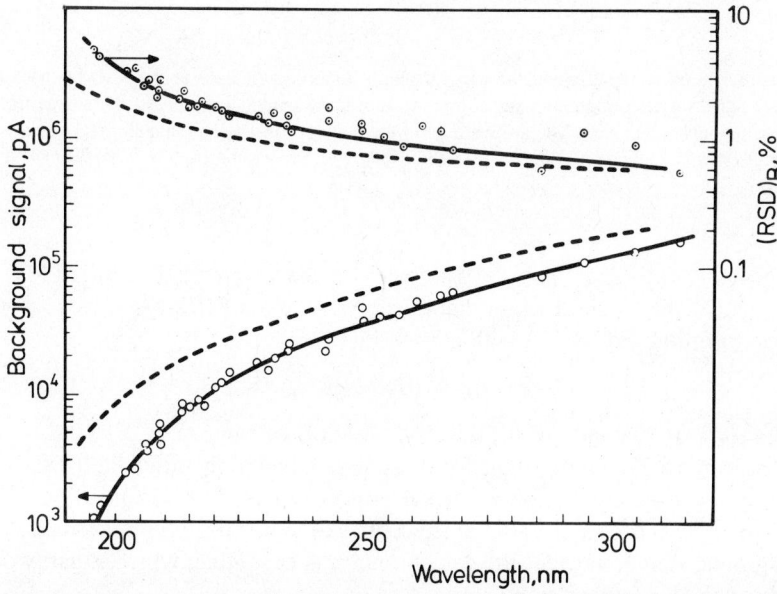

Figure 4.6. Experimental plots of the background signal (x_B) and the RSD of x_B versus wavelength for the experimental facilities used in [67]. The continuous and broken curves refer to different entrance optics in front of the monochromator. (Reproduced with permission from *Spectrochimica Acta, Part B* [67].)

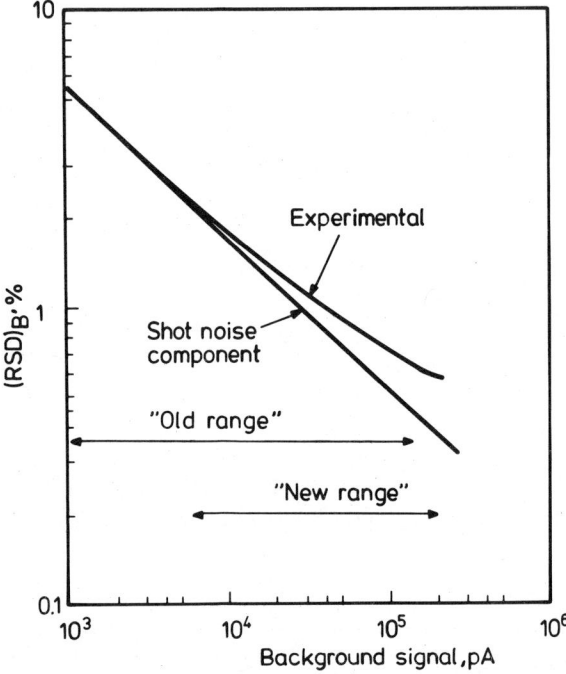

Figure 4.7. Curves representing the experimentally found dependence of the RSD of the background signal on the background signal and the shot noise component alone if the source flicker noise component were zero. The arrows refer to the ranges of the background signal covered with different entrance optics (cf. Fig. 4.6). (Reproduced with permission from *Spectrochimica Acta, Part B* [67].)

designated "old" and "new."[4] The measurements were made with a PMT of type EMI 9789QA and lock-in amplification. A plot of $(RSD)_B$ versus x_B was shown to follow Eq. (4.27) with $\gamma = 0$:

$$(RSD)_B = (\alpha_B^2 + g\beta/x_B)^{1/2} \qquad (4.29)$$

where $\alpha_B = 0.005$ and $g\beta = 2.88$, if x_B is expressed in pA.

Equation (4.29) implies that for those experimental facilities the $(RSD)_B$ is built up from a constant source flicker noise component (corresponding to an RSD of about 0.5%) and a shot noise contribution inversely proportional to the background signal, whereas the detector noise is negligible, which is partly due to the applied ac modulation.

This result is graphically depicted in Fig. 4.7 which shows the total experi-

---

[4]The main difference between the two is the transmittance in the low wavelength region, which in the "old" entrance optics was lower owing to degraded optical components.

mentally found $(RSD)_B$ and the shot noise component alone as functions of x_B. Evidently, the system with the "old optics" is shot-noise limited when x_B is smaller than about 40 nA, that is, at wavelengths below 250 nm. Under such conditions the $(RSD)_B$ can be substantially reduced if the radiant flux is increased, which in the system under discussion, could be achieved by changing from the "old" to the "new" optics. This is indicated by the arrows that mark the ranges of x_B in Fig. 4.7.

Results consistent with those for the EMI 9789QA PMT were reported for other types of PMT [67], again with $\alpha_B = 0.005$, but with values of $g\beta$ depending on the PMT gain.

Fundamentally, the shot noise term $(g\beta/x_B)$ is independent of PMT gain, because x_B is defined in terms of the output signal of the PMT and thus is proportional to the PMT gain. This is readily seen from Eq. (4.28).

Equation (4.29) was found to hold not only when pure aqueous solutions were fed into the ICP, but also when solutions with substantial amounts of matrices such as calcium chloride, sodium chloride, and a mixture of nickel and cobalt nitrates were introduced (Fig. 4.8). In the latter case, however, de-

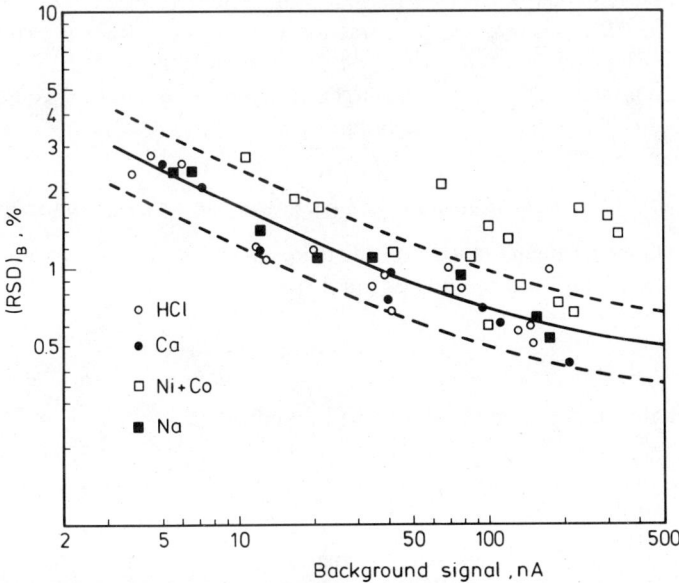

Figure 4.8. RSD of the background signal for a 0.1 M HCl solution and for solutions with various matrices, viz. 10 mg/mL NaCl, 27.7 mg/mL $CaCl_2$, and 5 mg/mL of each of the hexahydrates of Ni and Co. The continuous curve has been derived from the measurements shown in Fig. 4.6. The broken curves represent the RSD values that differ by a factor of $\sqrt{2}$ from those on the continuous curve ("error margins"). The departures of part of the results for the Ni–Co matrix are attributed to line coincidences. (Reproduced with permission from *Spectrochimica Acta, Part B* [67].)

partures did occur at wavelengths where lines of the matrix appeared in the spectral windows of the analysis lines.

The validity of the rather simple Eq. (4.29) for the system under discussion is partly due to the choice of the PMT and partly due to the use of ac modulation, which not only eliminates the dark current signal, but also reduces the dark current noise substantially if this is flicker noise [42]. With dc measurements, such as are commonly used in polychromators, the dark current signal and dark current noise may give rise to more complex relationships (cf. Eq. (4.28) and the results reported for the polychromator in [69]). From the results produced by Winge et al. [65] for their polychromator system, one may conclude, however, that less complication tends to arise with state-of-the-art polychromators, even when operated in the dc mode.

Exploring high-resolution ICP spectroscopy, Boumans and Vrakking [75] applied Eq. (4.27) to a 1.5-m echelle monochromator with predisperser in parallel slit arrangement. Signals were detected with an EMI 9789 QA PMT and measured by photon counting. The authors postulated the validity of counting statistics (Poisson distribution) and, therefore, assumed the shot noise component of $(RSD)_B$ to be equal to the square root of the background signal. Since the dark current was negligible, $(RSD)_B$ was described by Eq. (4.29) with $g\beta = 1$ if the RSD is expressed as a fraction ($g\beta = 10^4$ if RSD is expressed in percent). The flicker noise was slightly less than 0.005 (or 0.5%) for pure aqueous solutions and somewhat larger (0.008) for a solution containing 1 mg/mL of each Ni and Co as the nitrates.

4.1.4.5. Detection Limit in Dependence on $(RSD)_B$ and SBR

Under such experimental conditions that Eq. (4.29) is valid, the expression for the detection limit becomes (cf. Eq. (4.17))

$$c_L = z(\alpha_B^2 + g\beta/x_B)^{1/2} \frac{c_0}{\text{SBR}} \quad (4.30)$$

Accordingly, the detection limit can be minimized by

1. Minimizing the source flicker noise (α_B), that is, by maximizing the source stability.
2. Maximizing the background signal by
   a. The maximization of the background radiant flux to the detector.
   b. Choosing a PMT with maximum spectral sensitivity (and low dark current).
3. Maximizing the SBR.

4.1. DETECTION LIMITS

Table 4.1. Effect of ICP Power (P) on the Background Signal (x_B), the RSD of this Background Signal ($(RSD)_B$), the Signal-to-Background Ratio (SBR) of the Net Line Signal, and the Detection Limit (c_L) for Some Spectral Lines for which the Power Has a Relatively Small Influence on the SBR [67]

Spectral Line (nm)		P (kW)	x_B (a.u.)	$(RSD)_B$ (%)	SBR	c_L (ng/mL)
Zn II	202.548	1.15	14.9	1.46	9.1	1.0
		1.40	32.8	1.04	7.0	0.9
		1.65	63.5	0.81	4.9	1.0
As I	193.696	1.15	7.8	1.98	16.5	18
		1.40	16.3	1.40	10.5	20
		1.65	33.3	1.03	6.1	25
Se I	196.026	1.15	9.2	1.83	11.9	23
		1.40	19.5	1.30	8.2	24
		1.65	39.9	0.96	4.8	30

Maximizing the background radiant flux and maximizing the SBR tend to be contradictory conditions. Frequently a change in the parameter setting of the ICP system produces an increase in x_B at the cost of the SBR, or conversely, if the SBR increases, a lower background must be tolerated. We illustrate this with an example in which the ICP power is varied [67].

Table 4.1 shows the effect of the power on x_B, SBR, $(RSD)_B$, and c_L for three lines for which a relatively small effect on the SBR can be expected. The effects of increased power on the SBR and $(RSD)_B$ are seen to balance each other for the Zn line, whereas the adverse effect on the SBR predominates for the other two lines, so the detection limits increase.

Often ICP optimization is performed with the SBR as the only criterion. Although this approach has proved to provide interesting results, it is justified only if $(RSD)_B$ is constant. A prerequisite for this is that shot noise in the background signal is negligible over the entire range of background signals covered in the optimization experiment. This happens only when the background radiant flux per unit of wavelength interval impinging on the detector is relatively high. Unfortunately this condition conflicts with the requirements for a high practical resolving power. Therefore, if a spectrometer of relatively high resolving power is used, consideration of both the SBR and the background RSD is indispensable in optimization experiments. This also implies that the slit width should be included among the parameters varied in the optimization.

A comprehensive theoretical treatment of slit width optimization has been given by Laqua and his associates (cf. Section 4.1.10). Boumans and Vrakking [75] described an experimental approach to slit width optimization for the 1.5-m echelle monochromator referred to above. As an example, Figs. 4.9 and

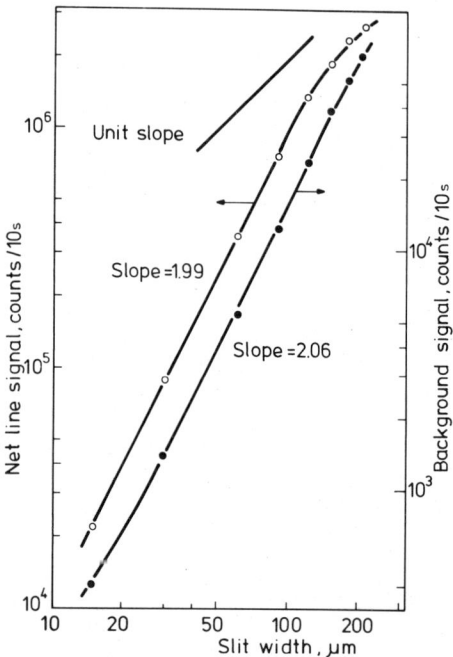

Figure 4.9. Double logarithmic plots of net line and background signals versus slit width for Mn II 257.610 nm. Mn concentration: 400 ng/mL. (Reproduced with permission from *Spectrochimica Acta, Part B* [75].)

4.10 show some results from this work (cf. Section 4.1.4.4). The figures are for equal entrance and exit slit widths (s).

Figure 4.9 shows double logarithmic plots of net line and background signal vs. s for Mn II 257.610 nm. The slope of the background plot is somewhat larger than the theoretically expected value of 2. The corresponding slope for the net line signal, on the contrary, is substantially higher than the expected value of 1, the latter value being approached only at slit widths beyond the range accessible. Consequently the SBR changes little, as shown in the upper diagram of Fig. 4.10. However, since the background intensity varies steeply with slit width, the RSD of the background signal changes substantially because the shot noise contribution sharply diminishes when the slits are widened. This is illustrated by the RSD plot in the lower diagram of Fig. 4.10. The combination of the SBR and RSD curves depict the dependence of the detection limit on slit width. The flat minimum of this curve indicates that the choice of slit width is not very critical between 150 and 240 μm where the effect of a decrease in SBR is balanced by the effect of a decrease in background RSD. Should the picture be further followed towards higher slit widths, then the SBR alone will

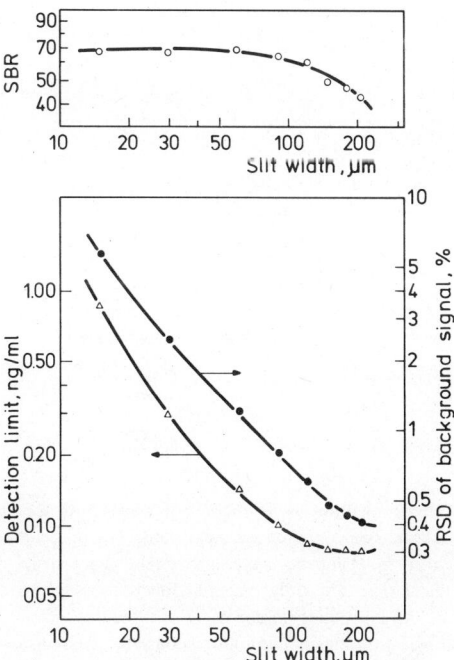

Figure 4.10. Double logarithmic plots of SBR, $(RSD)_B$, and c_L versus slit width for Mn II 257.610 nm. Mn concentration for SBR plot: 400 ng/mL. (Reproduced with permission from *Spectrochimica Acta, Part B* [75].)

govern the detection limit, the RSD of the background being dominated by constant source flicker noise. This is the region in which common monochromators providing medium resolving power operate. Whether, for a given instrument, the background RSD will then be dominated by flicker noise depends on the actual background radiant flux reaching the detector and the latter's spectral response [67].

Extending slit width optimization for the echelle monochromator to the entire wavelength range between 190 and 500 nm, Boumans and Vrakking [75] arrived via various intermediate steps at the picture shown in Fig. 4.11, which reveals the dependence of "the" detection limit on wavelength and slit width for the relevant instrument. This picture does not reflect individual characteristics of the spectral lines other than their wavelengths. This is so because (1) all lines were found to behave virtually similarly as to the dependence of their log intensity on log slit width, (2) the dependence of the log background signal on log slit width was identical at all wavelengths, and (3) the background RSD proved to be a unique function of the background signal (Eq. (4.29)), independent of wavelength.

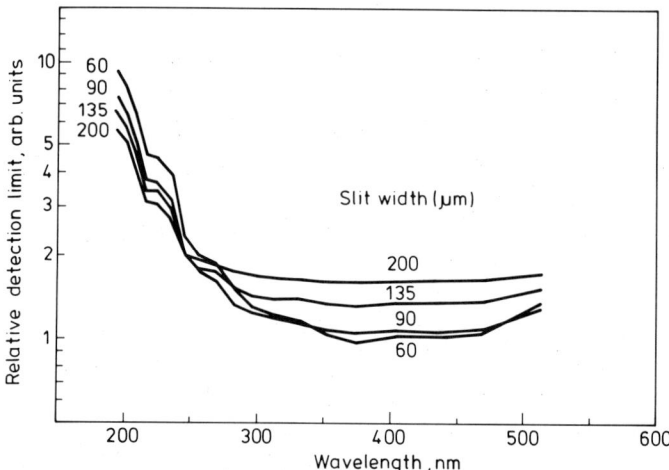

Figure 4.11. Relative detection limit as a function of wavelength for different slit widths. The results are for pure water and ICP compromise conditions. The diagram expresses that the choice of the optimum slit width is independent of the individual characteristics of spectral lines, in particular their absolute intensities. The only essential functions involved are the dependencies of relative line and background intensities on slit width and the dependence of the noise (expressed in terms of background RSD) on the background response (expressed in counts). The relative detection limit has been arbitrarily assigned a value of 1 at the plateau of curves for the 60 and 90 μm slit widths. In this plateau the intrinsic detection power of the source is fully used. The worsening towards lower (and higher) wavelengths indicates the increasing influence of shot noise due to the light gathering properties of the spectrometer and the response characteristics of the detector. (Reproduced with permission from *Spectrochimica Acta, Part B* [75].)

Figure 4.11 depicts the extent to which the intrinsic detection power of the pertinent 50-MHz ICP at different wavelengths can actually be exploited with the high-resolution echelle monochromator. The broad minimum between 300 and 500 nm results from the fact that the background response is so high that shot noise is entirely negligible at all slit widths considered, making the system completely dominated by flicker noise with a characteristic RSD of approximately 0.5%. Thus that slit width can be used at which the maximum SBR is practically reached, that is, 60 μm. For this slit width the background response below 300 nm reaches a level where shot is no longer negligible and contributes at an increasing rate when one travels to lower wavelengths. This adverse effect on the detection limit can then be partly compensated for by using wider slits, which, in turn, demands a sacrifice from the SBR. Consequently the achievable detection power at wavelengths below 300 nm lags behind the value that would be intrinsically reached if a higher background response were available. This "lag" amounts to a factor of about 5 at 200 nm.

4.1.5. Relative Standard Deviation of the Background Signal in Spectrography

In spectrography, part of the spectrum emitted by the excitation source is recorded on a photographic emulsion (plate or film). Ideally a single spectrogram may then provide the information about the presence or absence of chemical elements in the sample. Under these conditions, in which a smooth background without structure, as seen on the wavelength scale, is assumed, the detection power is completely limited by the detector noise as dictated by the "graininess" of the emulsion. The fluctuations due to this graininess can be quantified in terms of the standard deviation ($\sigma_{S(G)}$) of the blackening (S) as measured by scanning smooth background in a microphotometer using a rectangular measuring area, for example, 10 mm long and 10 μm wide.

To illustrate some important features of photographic detection we shall use here an empirical relationship between $\sigma_{S(G)}$ and S established by Kaiser [3],

$$\sigma_{S(G)} = \sigma_0 (S + 0.5) \tag{4.31}$$

where the coefficient σ_0 depends on the type of emulsion and is inversely proportional to the square root of the measured area [76].

Standard deviation $\sigma_{S(G)}$ can be expressed in terms of the standard deviation (σ_I) of the corresponding relative exposure, for convenience hereafter called "intensity" (I). To that end we express the emulsion calibration function in terms of the photographic parameter (P) [1, 28, 77, 78]:

$$Y = Y_0 + P/\gamma \tag{4.32}$$

where Y is log I, Y_0 an emulsion constant related to the response of the emulsion ("speed"), γ the slope of the straight calibration curve P versus Y, and P the photographic parameter defined by

$$P = S - \kappa D \tag{4.33}$$

or in terms of the transmittance (T)

$$P = -\log T + \kappa \log (1 - T) \tag{4.34}$$

where κ is the transformation constant used to extend the linear portion of the common Hurter and Driffield curve $S = f(Y)$ down to low blackenings (see [79], however).

From Eqs. (4.32)–(4.35), it can be derived that [3, 21, 27, 28]

$$\frac{\sigma_I}{I} = \frac{2.303}{\gamma} \sigma_S \left(1 + \frac{\kappa}{10^S - 1}\right) \tag{4.35}$$

where σ_S is the total standard deviation of S including contributions such as the photometer error (see caption of Fig. 4.12).

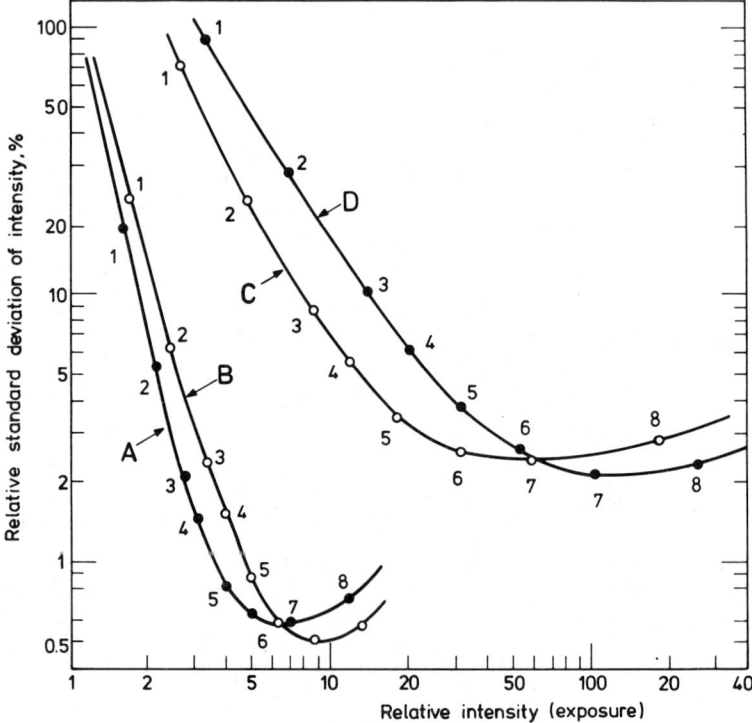

Figure 4.12. Curves representing the RSD of the intensity in photographic detection according to Eq. (4.35) with

$$\sigma_S = \sqrt{\sigma_{S(G)}^2 + \sigma_{S(P)}^2}$$

This equation covers the error due to the graininess $\sigma_{S(G)}$ (Eq. (4.31)) and a photometric error $\sigma_{S(P)}$, which is crucial only at high blackenings. For $\sigma_{S(P)}$ a value was taken equivalent to a standard deviation of 0.001 in the transmittance.

The curves are presented as functions of the relative intensity using the calibration functions: A, $\kappa = 0.6$, $\gamma = 2.1$; B, $\kappa = 0.8$, $\gamma = 2.8$; C, $\kappa = 0.5$, $\gamma = 1.0$; D, $\kappa = 0.8$, $\gamma = 1.3$.

The points inserted in the curves provide a key to the blackenings corresponding to the intensities, as follows: 1, $S = 0.003$; 2, $S = 0.01$; 3, $S = 0.03$; 4, $S = 0.05$; 5, $S = 0.1$; 6, $S = 0.2$; 7, $S = 0.4$; 8, $S = 0.8$. Curves A and B are representative of hard emulsions (e.g., Kodak SA1), and curves C and D pertain to soft emulsions (e.g., the former Ilford N30 and N40 emulsions).

The value of σ_0 has been equalized to 0.0043 for curves A and B so that $\sigma_{S(G)} = 0.003$ for $S \approx 0.2$, in agreement with experimental results [28]. For curves C and D, σ_0 was arbitrarily taken twice as large.

The value of Y_0 has been arbitrarily adjusted for all curves so as to make the relative intensity for a transmittance of 0.999 equal to 1.

4.1. DETECTION LIMITS

By combining Eqs. (4.31) and (4.33) σ_I/I can be written as a function of S. If appropriate values for σ_0 and Y_0 are chosen (see caption of Fig. 4.12), the relationships between σ_I/I and S, T, Y, or I can be calculated for any pair of (κ, γ) values. Fig. 4.12 depicts a few examples for realistic values of κ and γ. The curves are shown here as functions of I, the correspondence with S being given in the figure caption, which further explains the curves.

The diagram, which is essentially similar to a corresponding figure published by Kaiser [3] in 1947, is seen to reveal the following vital features of photographic detection if one recognizes that σ_I/I must be identified with $(RSD)_B$ in Eq. (4.16) or (4.17):

1. For optimum detection of spectral lines an emulsion must be exposed at least to such an extent that a blackening of the order of 0.15 to 0.2 is reached for the background; then σ_I/I will approach a minimum.
2. Emulsions having a fine grain and consequently high gamma ("hard" emulsions) will yield the lowest values of σ_I/I.
3. σ_I/I will become disproportionately large if exposures leading to too low blackenings are chosen.

For example, if emulsion type D is used, an available intensity $I = 7$ will yield a blackening of only 0.01 ($T = 0.977$), which leads to $(RSD)_B = 28\%$. An increase in intensity by a factor of 8 would be necessary to make $S \approx 0.2$ and $(RSD)_B \approx 2.5\%$, that is, to bring $(RSD)_B$ close to its minimum value for this "soft" emulsion.

For a hard emulsion of type B, however, an "initial" blackening of 0.01 ($I = 2.5$) corresponds to $(RSD)_B = 6\%$, while a 2.6-fold increase in intensity would be needed to bring the blackening and $(RSD)_B$ to their optimum values of 0.2 and 0.5% respectively. Note that the intensity scales for the different emulsion types in Fig. 4.12 are incomparable, since the value of Y_0 has been arbitrarily chosen (see caption).

These examples teach that large losses in detection power will result from insufficient exposures. Achieving optimum exposures for the background over a broad wavelength range may be difficult, because the intensity in the source, the response of the emulsion, and the calibration constants all vary with wavelength. Thus the use of step filters and/or exposures with different exposure times may be required to cover a large wavelength range.

In argon ICPs the absolute intensity of the background decreases rapidly with wavelength below 250 nm [80] and so does the response of photographic emulsions suitable for spectrography. To the author's experience, the region below about 240 nm is hardly accessible with a hard emulsion if the exposure time should be kept at a reasonable level.

It is rational in spectrography to take z in Eqs. (4.16) and (4.17) equal to $3\sqrt{2}$, because the detection of a spectral line in a single spectrum on a smooth background involves the comparison of a line-plus-background and a background measurement, which leads to factor $\sqrt{2}$. The factor 3 has been always customary in spectrography for statistical reasons (Section 4.2), and is further justified by the fact that it leads to photometrically measured detection limits which closely link up with the results of visual observations using a spectrum projector.

If the background is not ideally smooth, but shows molecular band structure with analysis lines located on rotational lines, then a single spectrum is no longer sufficient to ascertain the presence or absence of those analysis lines. Comparisons with "blank" spectra are now necessary. This complicates the problem in that the fluctuations of the blank and the higher background caused by the blank have to be taken into account (see Section 4.1.6). A comparison of a sample spectrum with a series of blank spectra further involves that the case of the "well-known blank" with $z = 3$ is approached (Section 4.1.2). The visual determination of detection limits from a series of spectra of samples with stepwise decreasing analyte concentration(s) closely links up with the latter situation and tends to provide detection limits on a "3σ basis" (cf. Section 4.1.7.2).

As Fig. 4.12 shows, the numerical value of $(RSD)_B$ in spectrography approaches 0.5% in the case of a hard emulsion, and this is about the best value achievable. With $z = 3\sqrt{2}$, this translates into an SBR of about 1/50 at the detection limit.

Further access to the abundant literature on detection limits involving photographic detection may be obtained by consulting [9, 10, 13, 17, 21, 27, 28, 45]. Applications of photographic detection to ICP–AES have been described by Broekaert et al. [27], Brenner et al. [81], Newland and Mostyn [82, 83], Watson et al. [84], and Boumans et al. [85–87].

4.1.6. Detection Limit in the Presence of a Blank

Blanks adversely affect the detection limits because they introduce additional noise and contribute additional "background." The term "blank" may be taken in its widest sense so as to cover both analyte reagent blanks and any signals (lines of molecular bands, interfering lines of concomitants) that appear in the spectral windows of the analysis lines and manifest themselves as structural maxima above a smooth background when a scan of the spectrum is made.

The ratio of the detection limit (c'_L) in the presence of the blank to the ideal detection limit (c_L) in the absence of the blank has been shown by Laqua [28] to be

$$\frac{c'_L}{c_L} = \frac{(\text{RSD})_{Bl+B}}{(\text{RSD})_B} \left[1 + \frac{x_{Bl}}{x_B}\right] \qquad (4.36)$$

for the case that a double logarithmic plot of a calibration curve has unit slope. In Eq. (4.36), $(\text{RSD})_{Bl+B}$, and $(\text{RSD})_B$ are the RSDs of the blank plus background and background respectively, and x_{Bl} and x_B are the net blank and background signal respectively.

Boumans and Bosveld [85] derived an expression which in its generalized form reads

$$\frac{c'_L}{c_L} = (\text{RSD})_{Bl+B} \left[\frac{1}{(\text{RSD})_B} + z\frac{c_{Bleq}}{c_L}\right] \qquad (4.37)$$

In this equation the ratio of the "blank equivalent concentration" (c_{Bleq}) and the ideal detection limit has replaced the ratio of the blank to background signal in Eq. (4.36). The two expressions are based on the same assumptions and can be mutually converted.

4.1.7. Numerical Values of Detection Limits in Low-Power Argon ICPs

4.1.7.1. General Considerations

Methods such as ICP–AES that went through a stormy development have seen the publication of large varieties of detection limits for equally large varieties of explorative experimental conditions. Now ICP–AES has matured, we may make an assessment of the state-of-the art detection limits in the light of our present knowledge.

Looking through historical reviews of detection limits in ICP–AES [88, 89] we see that in the initial years (1964–1968) the majority of the detection limits published for nitrogen–argon and argon ICPs were in a range between 10 ng/ml and 10 µg/mL [57, 59, 90–95]. A substantial progress resulted when the toroidal configuration exploited by Greenfield et al. [96] was experimentally also realized in the low-power argon ICP explored by Dickinson and Fassel [58] in conjunction with an ultrasonic nebulizer (USN). This step brought the range of the detection limits down to 0.01–10 ng/mL for many elements. These results were substantiated or improved by Souilliart and Robin [97] and Boumans and de Boer [60] using other experimental facilities.

The work of these groups also brought the recognition of the "ionic line advantage" of argon ICPs [61, 97, 98], that is, the fact that ionic lines are far more sensitive compared to atomic lines than would be derived from measurements of excitation temperatures and LTE considerations (see, however, Part 2, Chapter 11). The introduction of pneumatic nebulizers [99], which were developed in view of their greater convenience and lower cost, meant a small step backwards as to the achievable detection power, and this was convincingly proved by Olson et al. [100] to be attributable to the lower efficiency of pneumatic nebulizers, which lacks a factor of 3 to 10 behind that of USNs.

At present, pneumatic nebulizers are the most frequently used and, therefore, the assessments in the following sections emphasize results obtained with pneumatic nebulization.[5]

4.1.7.2. Detection Limits of 67 Elements for Air-Path Spectrometers

It has been established [61, 101–103] that in low-power argon ICPs single-element optimization of detection limits hardly repays the associated inconvenience: a few sets of compromise operating conditions generally suffice to cover a broad range of sample materials. Such compromise conditions are different, for example, for aqueous or organic solutions, but can always be chosen so that the detection limits are well equilibrated among the elements (except for the alkalis) and that at the same time a good balance is achieved between the overall detection power and multiplicative interferences (see Section 4.7 for details). Thus the detection limits published by the groups of Fassel and Boumans, for example, after 1972 refer to compromise conditions.

Often, in comparisons of detection limits, one uses the results published by Winge et al. [65] as "standards." The main reasons for this are the comprehensiveness of the data and the uniformity of the detection limits which were stripped of incidental noise characteristics in that they were based on SBRs and a single value of 1% for the background RSD (Eq. (4.16) and Section 4.1.4). Boumans [104, 105] found excellent correlations between the detection limits of Winge et al. and those obtained with his experimental facilities and, therefore, used Winge's data for converting the sensitivities in the "Tables of Spec-

---

[5]Numerical values of detection limits will be given in units of volume (μg/mL or ng/mL). The conversion into units of mass per unit of mass of dissolved sample is effected by th formula

$$c_{m,a} = c_{v,a}/c_{v,s}$$

where $c_{m,a}$ = analyte concentration (mass/mass)

$c_{v,a}$ = analyte concentration (mass/volume)

$c_{v,s}$ = sample concentration (mass/volume)

Evidently consistent units must be used for all three concentrations. A numerical coefficient must be included if the units are inconsistent; for example, for a very common conversion the formula will read:

$$c_{m,a} \; (\mu g/g) = 0.1 \; c_{v,a} \; (ng/mL)/c_{v,s} \; (\% \; m/v)$$

where % m/v is the dissolved solid concentration in percent mass per unit of volume (1% m/v = 10 mg/mL).

It is further recommended to use the units ppm and ppb (parts per 10^6 and 10^9 respectively only for μg/g and ng/g, but not for μg/mL and ng/mL.

tral-Line Intensities'' [106] for the National Bureau of Standards (NBS) copper arc into sensitivities appropriate to argon ICPs. This has resulted in the publication of "Line Coincidence Tables for ICP-AES" [104, 107]. These tables include a listing of detection limits, which, aside from some amendations and supplementations, are essentially the Winge detection limits.

As the following data show, the detection limits of Winge et al. are comprehensive as to the number of lines covered, but at the same time represent upper limits and thus tend to underestimate the performance of ICP-AES using conventional argon ICPs. This feature, too, has made them attractive as figures of merit in comparisons!

Table 4.2 lists detection limits as reported by four groups of authors using pneumatic nebulization. All data are for $(RSD)_B = 1\%$ and have been recalculated for $z = 2$ (Eq. 4.1) for the sake of mutual consistency and consistency with other tables and figures in this chapter. In principle, the best three or four lines according to Winge's tabulation [65] were chosen, with some exceptions. For a few elements more lines have been listed to ease comparisons. The footnotes summarize the experimental conditions and the references.

We limit discussion to the following remarks:

1. The data reflect differences in source characteristics (27 vs. 50 MHz ICP, nebulizers, choice of operating conditions) and differences in spectral bandwidth.

2. A breakdown of the ratio of detection limits obtained under different conditions has been recently described by Boumans and Vrakking [108] for their detection limits obtained at high spectral resolution with a 50-MHz ICP and those of Winge et al. [65], determined at medium resolution using a 27-MHz ICP (cf. Section 4.1.7.7). The lower detection limits achieved under the former conditions were attributed partly to a difference between the sources, as follows: a factor of 2 to 4 for ionic lines and a factor of 1 to 2 for atomic lines. These factors are also involved in the data in columns II and III of Table 4.2, which are for comparable spectral bandwidths.

3. Wohlers' results in column IV have been derived from the background equivalent concentrations listed in his comprehensive experimental "ICP-AES Wavelength Table" [110]. Wohlers' detection limits are usually somewhat lower than Winge's, although the bandwidth of the monochromator was poorer. Therefore, Wohlers' ICP operating conditions tend to be more favorable. In the low UV, however, Wohlers' detection limits grow worse, which must be due to the monochromator (type of grating, blaze, and stray light characteristics).

4. Comparison of the three sets of detection limits obtained under compromise conditions (columns II to IV) shows that the differences are most frequently less than a factor of 3 and hardly ever exceed a factor of 4.

5. Fernando's results obtained with single-element optimization and an

Table 4.2. Detection Limits in ng/mL ("2σ Basis") as Reported by Four Groups of Authors
(The results are for conventional argon ICPs and pneumatic nebulization. Background RSD = 1%. The experimental conditions are further specified in the footnotes.)

I Spectral Line (nm)		II Winge[a]	III Boumans[b]	IV Wohlers[c]	V Fernando[d]
Ag I	328.068	5		3	2
Ag I	338.289	9		5.5	3
Ag II	243.779	80		160	
Ag II	224.641	90	50		
Al I	309.271	15	15 a	7	
Al I	308.215	15	15 a	13	
Al I	396.152	20		8	20
Al I	226.916	20	10	25	
As I	193.696	35	13	45	60
As I	197.197	50	12	50	
As I	228.812	55	20	40	60
As I	234.984	90	40	65	
Au I	242.795	11	1.8	6	7
Au I	267.595	20	3.5	9.5	8.5
Au I	197.819	25	4.0		
Au I	197.754			20	
Au II	208.209	30		17	
B I	249.773	3	1.1	2	2
B I	249.678	4	1.6	4.5	
B I	208.959	6.5	2	4	12
B I	208.893	8	3.5	8	
Ba II	455.403	0.85		0.2	1.4
Ba II	493.409	1.5		0.4	
Ba II	233.527	2.5		2.5	1.6
Ba II	230.424	2.5	0.9	3.5	
Be II	313.042	0.18	0.05 a	0.2	0.055
Be I	234.861	0.2	0.07	0.2	0.18
Be II	313.107	0.5	0.1 a	0.4	
Be I	249.473	2.5		7.5	
Bi I	223.061	25	8	15	40
Bi I	306.772	50	11 a	20	200
Bi I	222.825	55	20		
Bi I	206.170	55	35	45	
C I	193.091	30			
C I	247.856	120			

Table 4.2. (*Continued*)

I Spectral Line (nm)		II Winge[a]	III Boumans[b]	IV Wohlers[c]	V Fernando[d]
Ca II	393.366	0.13		0.2	0.07
Ca II	317.933	6.5	2 a	9	
Ca II	315.887		4 a	15	
Ca I	422.673	6.5		6	2.4
Cd II	214.438	1.7	0.6	1.4	1.4
Cd I	228.802	1.8	0.7	1.4	2
Cd II	226.502	2.5	0.6	2	
Ce II	413.765	30		25	
Ce II	413.380	35		20	
Ce II	418.660	35		20	
Co II	238.892	4		4	2.5
Co II	228.616	4.5	1	3.5	
Co II	237.862	6.5		7	
Co II	230.786	6.5	3	6.5	
Cr II	205.552	4	0.8	3	1.7
Cr II	206.149	4.5	1	3.5	
Cr II	267.716	4.5		3.5	
Cr II	283.563	4.5	0.9 a	3	
Cu I	324.754	3.5		1.6	2.5
Cu I	224.700	5		6	3
Cu I	219.958	6.5	5	11	
Cu I	327.396	6.5		3	
Dy II	353.170	6.5			
Dy II	364.540	15			
Dy II	340.780	18			
Er II	337.271	6.5			
Er II	349.910	11			
Er II	323.058	12	4		
Eu II	381.967	1.8			
Eu II	412.970	3			
Eu II	420.505	3			
Fe II	238.204	3		1.6	1.5
Fe II	239.562	3.5		2.4	
Fe II	259.940	4		1.2	1.5
Ga I	294.364	30	15 a	15	18
Ga I	417.206	45		18	15
Ga I	287.424	50	20 a	25	

Table 4.2. (*Continued*)

I Spectral Line (nm)		II Winge[a]	III Boumans[b]	IV Wohlers[c]	V Fernando[d]
Gd II	342.247	9			
Gd II	336.223	13			
Gd II	335.862	14			
Gd II	310.050	15	4 a		
Ge I	209.426	25	7.5	17	
Ge I	265.118	30	8	12	11
Ge I	206.866	40		24	
Hf II	277.336	10		8.5	2.6
Hf II	273.876	11		17	
Hf II	264.141	12		9	
Hf II	232.247	12		8.5	
Hg II	194.227	17		55	
Hg I	253.652	40		17	25
Ho II	345.600	4			
Ho II	339.898	8.5			
Ho II	389.102	11			
In I	230.606	40	10	20	95
In I	325.609	80		30	
In I	303.936	100	30	45	
Ir II	224.268	18	3	11	7
Ir II	212.681	20	4	14	
Ir I	205.222	40	10	40	
La II	379.478	6.5			2
La II	333.749	6.5			
La II	408.672	6.5			
Li I	670.784			0.2	
Li I	610.362			5.5	8
Li I	460.286	550		170	35
Li I	323.263	700	650 a		
Lu II	261.542	0.65			
Lu II	291.139	4	1.4 a		
Lu II	219.544	5.5			
Lu II	307.760	6	1.8 a		
Mg II	279.553	0.1		0.2	
Mg II	280.273	0.2		0.2	
Mg I	285.213	1.1	0.45 a	1.2	
Mg II	279.079	20		24	

Table 4.2. (*Continued*)

I Spectral Line (nm)		II Winge[a]	III Boumans[b]	IV Wohlers[c]	V Fernando[d]
Mn II	257.610	0.95	0.15	0.40	0.45
Mn II	259.373	1.1		0.40	
Mn II	260.569	1.4		0.80	
Mo II	202.030	5.5	1	10	
Mo II	203.844	8	3		
Mo II	204.598	8	3		
Mo II	281.615	9.5	3 a	4	11
Mo II	287.151	18	6 a		
Mo I	379.825			6.5	7
Mo I	386.411			7.5	
Na I	588.995	20		6	1.4
Na I	589.592	45		13	
Na I	330.237	1200		1500	700
Nb II	309.418	25[e]	3 a	6	2.5
Nb II	316.340	25[e]	2.5 a	8	
Nb II	313.079	35[e]	3.5 a	7	
Nb II	269.706	45[e]		11	
Nd II	401.225	35			
Nd II	430.358	50			
Nd II	406.109	65			
Ni II	221.647	6.5	1.4	5	
Ni II	232.003	10	4	7	
Ni II	231.604	10		5	2.5
Ni II	216.556	11	3	10	
Os II	225.585	0.25[f]		7	0.5
Os II	228.226	0.4[f]		12	
Os II	233.680	0.8[f]		24	
P I	213.618	50	10	70	11
P I	214.914	50	18	80	100
P I	253.565	180	90		
Pb II	220.353	30	8	11	20
Pb I	216.999	60	20		
Pb I	261.418	85	30	30	
Pb I	283.306	95	50	30	
Pd I	340.458	30		14	5.5
Pd I	363.470	35		25	
Pd II	229.651	45		60	14
Pd I	324.270	50		30	

Table 4.2. (*Continued*)

I Spectral Line (nm)		II Winge[a]	III Boumans[b]	IV Wohlers[c]	V Fernando[d]
Pr II	390.844	25			
Pr II	414.311	25			
Pr II	417.939	25			
Pt II	214.423	20		20	25
Pt II	203.646	35		70	
Pt I	265.945	55		30	15
Pt I	224.552	55		55	
Re II	197.313	4			
Re II	197.248			20	
Re II	221.426	4		6	15
Re II	227.525	4		5.5	10
Rh II	233.477	30		35	20
Rh II	249.077	40		30	
Rh II	343.489	40		20	8
Rh II	252.053	50		45	
Ru II	240.272	20		12	6
Ru II	245.650	20		15	
Ru II	267.876	20		12	
Sb I	206.833	20	5	12	
Sb I	217.581	30	7.5	16	30
Sb I	231.147	40		20	35
Sc II	361.384	1			0.7
Sc II	363.075	1.4			
Sc II	364.279	1.8			
Se I	196.026	50	10	110	30
Se I	203.985	75	20	180	50
Se I	206.279	200			
Si I	251.611	8		6.5	8
Si I	212.412	11		18	50
Si I	288.158	18	10 a	12	
Sm II	359.260	30			
Sm II	442.434	35			
Sm II	360.949	40			
Sn I	189.980	17		180	
Sn I	235.484	65	30	85	
Sn I	242.949	65	20	85	
Sn I	283.999	75		75	10

Table 4.2. (*Continued*)

I Spectral Line (nm)		II Winge[a]	III Boumans[b]	IV Wohlers[c]	V Fernando[d]
Sr II	407.771	0.3		0.2	0.045
Sr II	421.552	0.5		0.4	
Sr II	216.596	5.5		11	5
Sr II	346.446	15		15	
Ta II	226.230	17			15
Ta II	226.162			22	
Ta II	226.142			22	
Ta II	240.063	20		11	6.5
Ta II	268.517	20		13	
Ta II	233.198	20		22	
Ta II	233.219			22	
Ta II	263.558	20		11	
Ta II	267.590	30		22	
Tb II	350.917	15			
Tb II	384.873	35			
Tb II	367.635	40			
Te I	214.281	25	30	55	60
Te I	238.578	120		190	200
Te I	225.902	120			
Te I	214.725	140			
Th II	283.730	45	6.5 a	11	9.5
Th II	283.231	45	8.0 a	16	
Th II	274.716	55		20	
Th II	401.913	55		15	5.5
Ti II	334.941	2.5		1.6	0.55
Ti II	336.121	3.5		2	
Ti II	337.280	4.5		2.5	13
Ti II	334.904	5			
Ti II	308.802	5	1.1 a	3.5	
Ti II	307.864	5.5	1.6 a	3.5	
Tl II	190.864	25		140	
Tl I	276.787	80		65	
Tl I	351.924	130		120	170
Tl I	377.572	150		120	100
Tl I	535.046			100	
Tm II	346.220	3.5			
Tm II	384.802	5.5			
Tm II	342.508	6.5			

Table 4.2. (*Continued*)

I Spectral Line (nm)		II Winge[a]	III Boumans[b]	IV Wohlers[c]	V Fernando[d]
U II	385.958	170		80	30
U II	367.007	200		130	
U II	409.014	200		130	
U II	393.203	200		160	
U II	294.192	300	60 a		
V II	309.311	3.5	0.9 a	2	
V II	310.230	4.5	1.8 a	3	11
V II	292.402	5	2 a	3.5	
V II	311.071	6	1.4 a	3.5	
V II	289.332	6.5	3.5 a	7.5	
V II	268.796	6.5		4	
V II	311.838	7.5	1.8 a	4.5	
W II	207.911	20		40	
W II	224.875	30		30	
W II	218.936	30		70	
W II	209.475	30		50	
W II	239.709	35		20	7
Y II	371.030	2.5			0.2
Y II	324.228	3			0.35
Y II	360.073	3			
Yb II	328.937	1.2			
Yb II	369.419	2			
Yb II	289.138	5.5	1.2 a		
Zn I	213.856	1.2	0.35	1.6	2
Zn II	202.548	2.5	0.45	6.5	
Zn II	206.200	4		7.5	8
Zn I	334.502	90		95	
Zr II	343.823	4.5		2	
Zr II	339.198	5		1.6	0.6
Zr II	257.139	6.5		5	2.4
Zr II	349.621	6.5		2.6	

[a] Winge et al. [65]: 27-MHz ICP, compromise conditions, 1.1-kW power; 1-m monochromator, 1180 grooves/mm grating, 250-nm blaze, 20-μm slits; theoretical spectral bandwidth = 17 pm.

[b] Boumans et al. [67] and unpublished results obtained under the same conditions: 50-MHz ICP, compromise conditions, 1.15-kW power; 1-m monochromator, 2400 grooves/mm grating, 240-nm blaze, 20-μm entrance slit, 30-μm exit slit; theoretical spectral bandwidth = 12 pm.

Results labeled "a" were obtained with the same ICP at 1.4-kW power and a 1.5-m

4.1. DETECTION LIMITS

echelle monochromator with crossed dispersion (column V) are usually of the same magnitude as those in columns II to IV, with some notable exceptions, however.

6. On the whole, Table 4.2 provides an idea of the detection limits of ICP-AES as obtained with conventional argon ICPs, pneumatic nebulization, and monochromators. Since the values listed are based on a 1% background RSD, actual detection limits may be up to a factor of 2 better or worse depending on the noise characteristics of the equipment.

7. Recent work by Boumans and Vrakking [307] substantially extends the data in Table 4.2 in that it contains (a) a tabulation of the detection limits of about 350 most prominent lines of 64 elements measured at high spectral resolution, and (b) the breakdown of these detection limits and the corresponding values of both Winge et al. [65] and Wohlers [110] into the factors contributed by the spectral resolution, the noise, and the source characteristics. An improved version of the approach discussed in Section 4.1.7.7 was used for this breakdown as noted at the end of that section.

Table 4.3 shows additional sets of detection limits reported in the literature. Winge's results [65] have been included for comparison. Owing to lack of data a smaller number of lines was covered.

We note the following items:

1. The data in column III illustrate the magnitude of actual detection limits achieved with polychromators using pneumatic nebulization. Generally these data do not differ widely from those of Winge et al.
2. The results in columns IV and V reveal an improvement in the detection

Table 4.2. (*Continued*)

echelle monochromator with 210 μm slits; practical spectral bandwidth: 11–15 pm at the wavelengths considered (cf. [108, 109]). These detection limits will improve by a factor of 2 and 1.5 at least for atomic and ionic lines respectively when the power is reduced to 1.15 kW (cf. [307]).

^{c}Wohlers [110]: 27-MHz ICP, compromise conditions, 1.1-kW power; 0.5-m monochromator, 1180 grooves/mm grating, 300-nm blaze, 15-μm slits; theoretical spectral bandwidth: 24 pm.

^{d}Fernando [111]: 27-MHz ICP, single-element conditions without rigorous optimization; 0.75-m echelle monochromator with crossed dispersion; 79 grooves/mm echelle, 63°26' blaze angle, 50-μm entrance slit; 100-μm exit slit; theoretical spectral bandwidth: 6 pm at 200 nm, 9 pm at 300 nm, 12 pm at 400 nm and 15 pm at 500 nm.

^{e}Result probably erroneous owing to incorrect Nb concentration [87, 108].

^{f}Detection limit more than a factor of 10 too low probably owing to the use of the volatile OsO_4 resulting in enhanced nebulizer efficiency [307].

Table 4.3. Detection Limits in ng/mL ("2σ basis") for Conventional Argon ICPs Operated Under Compromise Conditions with Pneumatic (PN) or Ultrasonic (USN) Nebulizers
(The experimental conditions are further specified in the footnotes.)

Spectral line (nm)	PN Winge[a]	PN Polychr[b]	USN Polychr/Monochr[c]	USN Floyd[d]	PN High Resoln[e]
Ag I 328.068	5	3.5	0.5		
Al I 309.271	15				2.5
Al I 308.215	15	20	0.2–2	1.5	4
Al I 396.152	20	4.5	0.2		
As I 193.696	35	40	2–5		7
As I 197.197	50	40	5	7	25
As I 228.812	55		6		
B I 249.773	3		0.1		0.7
B I 249.678	4	4	2.5		1.3
B I 208.959	6.5				2
Ba II 455.403	0.85	0.15	0.01–0.06	0.3	
Ba II 493.409	1.5				0.25
Be II 313.042	0.18	0.3	0.03	0.03	0.015
Be I 234.861	0.2		0.003		
Bi I 223.061	25	10	10	3	5
Bi I 306.772	50				16
Bi I 222.825	55				15
Ca II 393.366	0.13	0.07			
Ca II 317.933	6.5	10			0.8
Ca II 315.887	6.5	4	1.5		1.3
Cd II 214.438	1.7			0.6	0.2
Cd I 228.802	1.8		0.2		
Cd II 226.502	2.5	3	0.1–0.5		0.35
Ce II 413.765	30			3	
Ce II 418.660	35		0.4		
Co II 238.892	4	3	0.1		1
Co II 228.616	4.5		0.3–0.5	1.5	
Co II 237.862	6.5		0.6		
Cr II 205.552	4		0.9		
Cr II 267.716	4.5	3	0.1–0.5	0.7	
Cr II 283.563	4.5		0.2		0.3
Cu I 324.754	3.5	1.4	0.04–0.5	0.3	
Cu I 327.396	6.5		0.06		

Table 4.3. (*Continued*)

Spectral line (nm)	PN Winge[a]	PN Polychr[b]	USN Polychr/Monochr[c]	USN Floyd[d]	PN High Resoln[e]
Dy II 353.170	6.5			0.7	
Er II 337.271	6.5			0.7	
Er II 323.058	11				0.8
Eu II 420.505	3			0.3	
Fe II 238.204	3			0.3	
Fe II 239.562	3.5				0.6
Fe II 259.940	4	1.4	0.1–0.5		
Ga I 294.364	30			4	4
Ga I 417.206	45		0.6		
Ga I 287.424	50				7
Ge I 209.426	25				10
Ge I 265.118	30		0.5		3
Ge I 206.866	40				15
Gd II 342.247	9			3	
Gd II 310.050	15				0.9
Hf II 277.336	10			1.5	
Hf II 282.022	12				0.8
Hg I 253.652	40	14	3		
Ho II 389.102	11			0.6	
In I 230.606	40			15	15
Ir II 224.268	18			0.4	
La II 408.672	6.5		0.1		
Li I 670.784			0.02		
Lu II 261.542	0.65			0.15	
Lu II 291.139	4				0.7
Lu II 307.760	6				0.7
Mg II 279.553	0.1	0.3	0.003	0.03	0.01
Mg I 285.213	1.1		0.02		
Mg II 279.079	20	20	2		
Mn II 257.610	0.95	0.25	0.01–0.05	0.07	0.08
Mn II 259.373	1.1		0.03		
Mn II 260.569	1.4		0.03		0.15
Mo II 202.030	5.5			4	2
Mo II 281.615	9.5		0.3		0.9

Table 4.3. (*Continued*)

Spectral line (nm)	PN Winge[a]	PN Polychr[b]	USN Polychr/Monochr[c]	USN Floyd[d]	PN High Resoln[e]
Mo II 287.151	18	9	0.6–1.5		1.8
Mo I 386.411			0.3–3		
Na I 588.995	20		0.2		
Na I 330.237	1200	1000			
Nb II 309.418	25		0.2		1.2
Nb II 316.340	25			0.7	0.8
Nb II 313.079	35				1.2
Nd II 401.225	35		0.3		2
Nd II 430.358	50			2	
Ni II 231.604	10	8	0.3–0.7	1.5	
P I 213.618	50	40	15		
Pb II 220.353	30	25	1–8	15	7
Pb I 283.306	95		2		15
Pd I 340.458	30			0.7	
Pr II 390.844	25			0.2	
Pt I 265.945	55		0.9		
Re II 221.426	4			3	
Rh II 233.477	30			7	
Rh II 249.077	40				7
Ru II 240.272	20			5	
Sb I 206.833	20	35	2.5		
Sb I 217.581	30			20	
Sc II 361.384	1			0.07	
Se I 196.026	50	35	1–3	50	
Si I 288.158	18	10			3
Sm II 359.260	30			8	
Sm II 442.434	35				4
Sn I 235.484	65			7	15
Sn I 242.949	65				10
Sn I 283.999	75				10
Sr II 407.771	0.3	0.05	0.003–0.02	0.007	
Ta II 226.230	17			10	

Table 4.3. (*Continued*)

Spectral line (nm)	PN Winge[a]	PN Polychr[b]	USN Polychr/Monochr[c]	USN Floyd[d]	PN High Resoln[e]
Tb II 350.917	15		0.1	0.7	
Te I 214.281	25	35	0.7	3	
Te I 238.578	120	80			
Th II 283.730	45				1.6
Th II 283.231	45			7	2.5
Th II 401.913	55	8	0.9		
Ti II 334.941	2.5		0.03	0.3	0.2
Ti II 308.802	5				0.3
Ti II 307.864	5.5				0.4
Tl II 190.864	25		3		
Tl I 351.924	130	50	15	5	
Tl I 377.572	150		11		
Tm II 346.220	3.5			2	
U II 385.958	170		1.5	7	
V II 309.311	3.5		0.06		0.3
V II 310.230	4.5				0.3
V II 292.402	5	1	0.1–0.3		0.5
V II 311.071	6		0.1		0.4
V II 289.332	6.5				1.1
V II 311.838	7.5	2	0.1		0.5
W II 207.911	20			7	
Y II 371.030	2.5	0.5	0.05	0.05	0.15
Y II 324.228	3				0.2
Yb II 328.937	1.2		0.02	0.07	
Yb II 289.138	5.5				0.3
Zn I 213.856	1.2	2.5	0.1–0.5	0.15	
Zn II 202.548	2.5	2			
Zr II 343.823	4.5		0.06	0.7	

[a] Winge et al. [65]: 27-MHz ICP, monochromator, medium resolution, background RSD = 1%.
[b] Winge et al. [112] and/or Taylor and Floyd [113]: 27 MHz ICP, polychromators, medium resolution, background RSD as measured. The geometric mean has been listed when the literature values differed.
[c] (1) Boumans and de Boer [61] and Boumans [114]: 50 MHz ICP, monochromator, medium resolution, background RSD = 1%; (2) Olson et al. [100] and Taylor and Floyd [113]: 27 MHz ICP, polychromators, medium resolution,, background RSD as measured; and (3) Montaser et al. [115]: 27-MHz ICP, monochromator, medium resolution, background RSD as measured.
[d] Floyd et al. [116, 117]: 27-MHz ICP, monochromator, medium resolution, background RSD as measured.
[e] Boumans and Vrakking [75, 108]: 50-MHz ICP, monochromator, high resolution, background RSD as measured.

limits by one-half to over one order of magnitude achievable by replacing pneumatic by ultrasonic nebulization.

3. The combination of a 50-MHz ICP with pneumatic nebulization and a high resolution spectrometer (column VI) is seen to yield results that compete well with those obtained with a 27-MHz ICP with ultrasonic nebulization and a medium resolution monochromator (column V). The effects of the spectral resolution will be further discussed in Section 4.1.7.7.

Finally, one should keep in mind that the detection limits in Tables 4.2 and 4.3 are for pure aqueous solutions. The presence of concomitants generally worsens the detection limits as will be discussed in Section 4.1.7.8.

The data in Tables 4.2 and 4.3 are supplemented by Figs. 4.13–4.15, which are all based on the results of Winge et al. [65]. Figure 4.13 shows a "three-dimensional" periodic table which not only indicates the detection limits attainable with the best lines, but also reveals for each element the number of prominent lines that yield a detection limit within a factor of three from that of the best line. The more such lines are available, the greater the chance that a free line with a reasonable detection limit can still be found if the matrix yields a line-rich spectrum. Then it may often happen that the best prominent lines cannot be used as a result of line coincidences (cf. Section 4.1.7.8 and Chapter 7).

Figure 4.14 supplements Fig. 4.13 and shows the distributions of the prominent lines that give detection limits below 10 and 50 ng/mL respectively. In particular, transition elements and lanthanides are seen to have lines providing detection limits better than 10 or 50 ng/mL. For these elements it will be relatively easy to reach a detection power of this magnitude, even if the most prominent lines are unusable as a result of line coincidences. More critical is the situation for most elements of the 3a–4a and 5a–6a groups, the platinum metals, and Hg, Nd, and U, for which the ≤ 10 ng/mL level is mostly not accessible, and also the number of lines giving access to the ≤ 50 ng/mL level is very small.

The histogram in Fig. 4.15 shows the distribution over the wavelength region of those prominent lines that yield a detection limit not worse than three times the minimum detection limit. This figure marks the wavelength regions of paramount importance for trace analysis involving the various groups of elements and reveals the high density of prominent lines in the 190–250-nm wavelength region, as pointed out by Winge et al. [65].

4.1.7.3. Detection Limits of Elements with Prominent Lines in the Vacuum Ultraviolet

Strong oxygen absorption bands below 200 nm may interfere with the use of prominent lines located in that region (cf. Section 7.3.4). Down to about 190

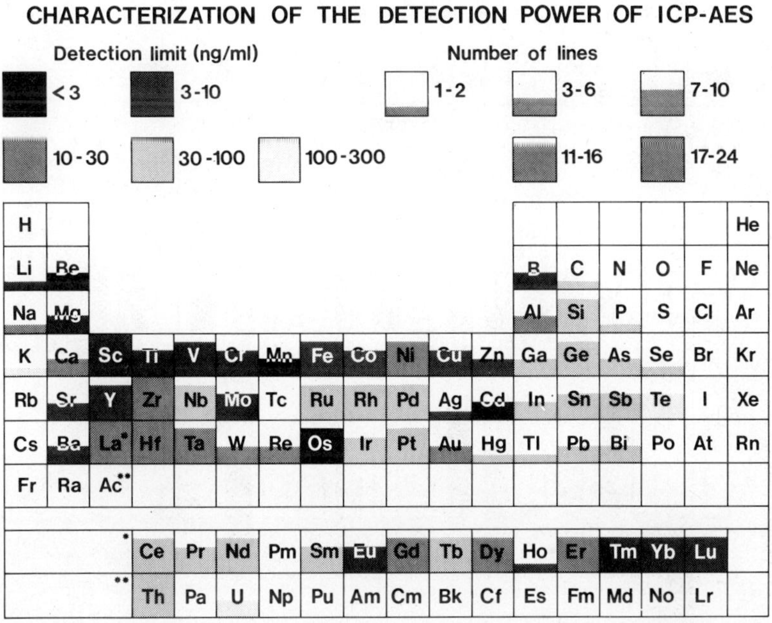

4.1. DETECTION LIMITS

CHARACTERIZATION OF THE DETECTION POWER OF ICP-AES

Detection limit (ng/ml): <3, 3-10, 10-30, 30-100, 100-300

Number of lines: 1-2, 3-6, 7-10, 11-16, 17-24

Figure 4.13. Periodic Table characterizing the detection power of ICP-AES for argon ICPs operated with a pneumatic nebulizer. The degree of the shading indicates the range of the detection limits of the most prominent lines of each element. The area of the shading is representative of the number of most prominent lines having a detection limit in the relevant range. The diagram is based on the data published by Winge et al. [65] with the extensions by Boumans [104, 105]. The detection limits are on a "2σ basis."

nm, this interference is still at a tolerable level so that reasonable detection limits with the relevant prominent lines can be attained, even with air-path spectrometers (Tables 4.2 and 4.3), although shot noise will tend to limit the detection power to an increasing extent the lower the wavelength [67]. Below 190 nm the oxygen must be removed from the light path by purging it with a gas optically transparent in this region. This is achieved with atmospheric pressure, nitrogen-purged or vacuum argon-purged spectrometers, combined with a purged light path between ICP and spectrometer. A variety of systems and devices, including extended torches with side arms, is offered by the instrument manufacturers.

Access to the vacuum ultraviolet, VUV (<200 nm) is important for the determination of halogens and B, C, P, S, Se, As, Sn, Hg, N, and O. Kirkbright et al. [118, 119] were the first to report detection limits of S, P, I, Hg, As, and Se using lines down to 177 nm. More recently various studies for argon- or nitrogen-purged system [120-125] and vacuum systems [126-129] have been published. These publications include discussions of spectral interferences.

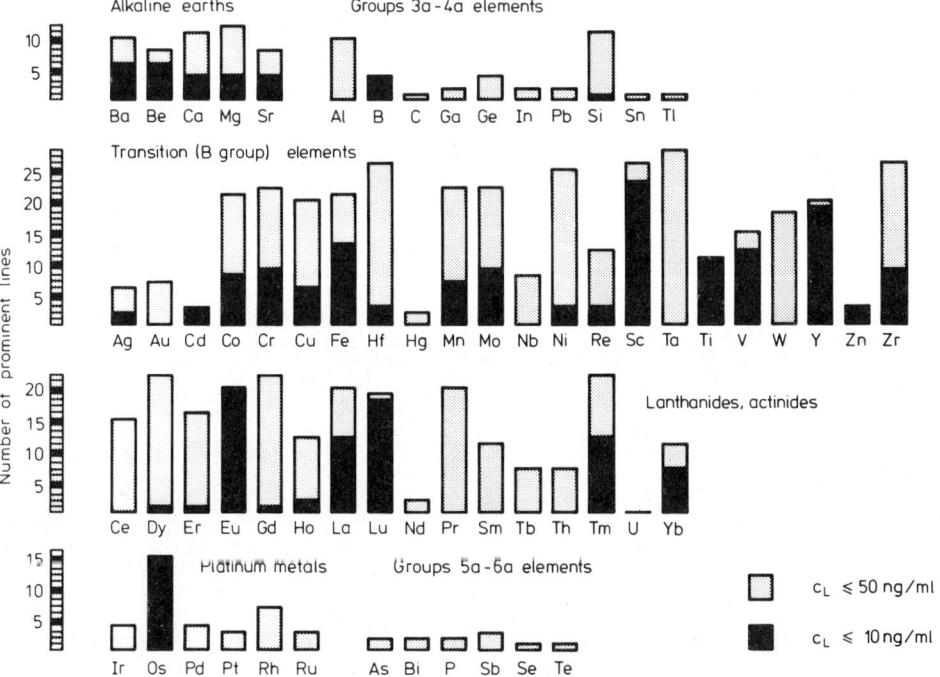

Figure 4.14. Distributions of prominent lines according to groups of elements and wavelengths. Only those of the 896 prominent lines [104, 105] have been considered whose detection limits are up to a factor of 3 from the detection limit attainable with the best line of the element. The alkalis have been left out of consideration. (Reproduced with permission from *Spectrochimica Acta, Part B* [105].)

Further data may be found in the application data sheets of instrument manufacturers offering vacuum or purged spectrometers. For illustration, Table 4.4 lists some detection limits and/or background equivalent concentrations reported in recent papers.

Heine et al. [129] investigated the spectral region between 120 and 185 nm and identified a number of promising analytical lines for oxygen, nitrogen, carbon, bromine, sulfur, and chlorine using an extended torch with a side arm purged with helium. The wavelengths of the lines along with their relative intensities are reported.

Butler-Sobel and Cass [128] used an extended torch to eliminate air entrainment into the plasma in order to determine nitrogen at the N I 174.270 nm line in solutions in both the inorganic and organic form. A detection limit of 30 μg/mL was reported.

For the determination of nitrogen, Broekaert and Zeeman [130] used an extended torch, an optical path purged with argon, and a small (0.3 m) vacuum

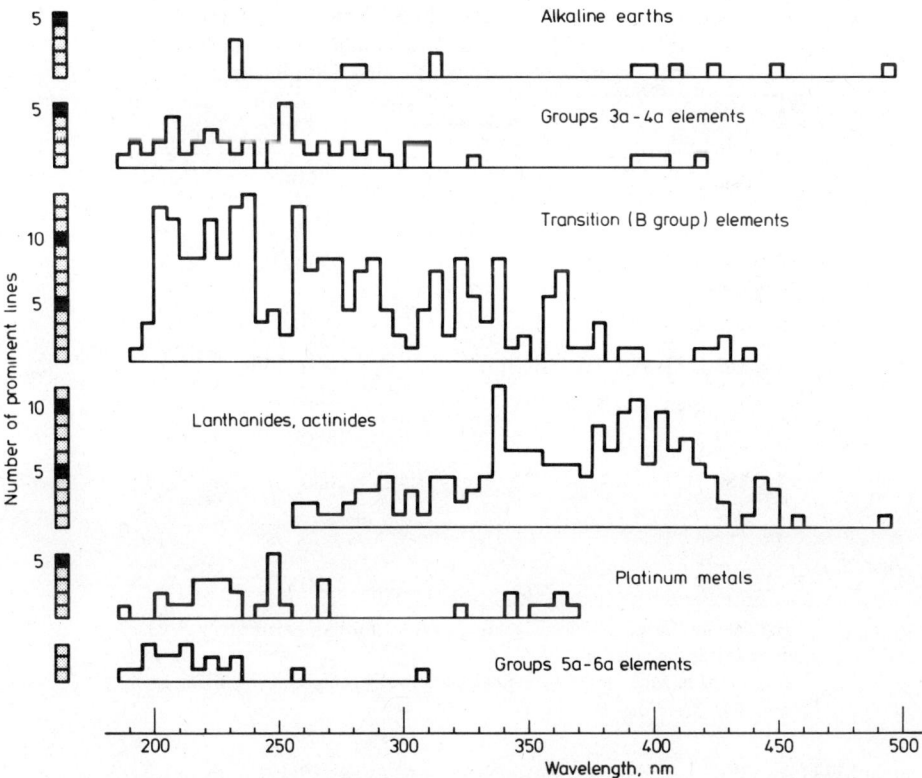

Figure 4.15. Distribution of prominent lines over the elements. The figure shows the numbers of prominent lines that yield detection limits up to 10 and 50 ng/mL in pure aqueous solution with spectroscopic equipment of medium spectral resolution. (Reproduced with permission from *Spectrochimica Acta, Part B* [105].)

monochromator. They compared various VUV lines and showed that the detection limit is dictated by the blank, despite their using an extended torch and purging the optical path.

Nakahara [125] studied the determination of phosphorus for the VUV lines 185.94, 178.29, and 177.50 nm and the UV lines 213.62, 214.91, and 253.57 nm using a purged optical path and a 0.5-m purged monochromator. The effects of the argon flow in the monochromator and spectral interferences from Al, Co, Cr, Cu, Fe, Mn, Mo and Ni were covered. Major spectral interferences from Fe and Cu could be circumvented by the use of P I 178.29 nm. For the elimination of some spectral interferences from minor elements, a spectrometer of higher resolving power than the one used (reciprocal dispersion: 1.6 nm/mm) was recommended.

Additional information on the exploration of the VUV region can be found

Table 4.4. Detection Limits and Background Equivalent Concentrations (BEC) of As, B, I, P, S and Sn as Determined with Vacuum or Nitrogen-Purged Equipment[a]

Analysis Line (nm)	Detection Limit (ng/mL) [126]	BEC (ng/mL) [126]	BEC (ng/mL) [123, 124]
As I 189.0 42	13	3900	
B I 182.6 41	6	800	180
I 182.5 91			380
I I 178.2 76	16 (5)[b]	1500 (40)[b]	
I 183.0 38	80 (3)[b]	5000 (160)[b]	
P I 177.4 99	8	120	
I 178.2 87	18	2300	900
S I 180.7 31	24	2600	2000 #
I 182.0 34			3200 #
I 182.6 24			5200 #
Sn II 189.9 89	18	2300	

[a] The data are for pure aqueous solutions except for those marked "#," which refer to xylene as solvent.
[b] The results in parentheses were obtained in the presence of an oxidizing agent.

in [131] for a DCP. Wavelengths and spectroscopic data for lines of all elements in the VUV have been tabulated by Kelly [132] and Kelly and Palumbo [133]. Kelly's table [132] also contains separate lists of the most intense lines according to both element and wavelength.

4.1.7.4. Analysis Lines in the Infrared Region

Fry and his associates made an extensive study of ICP excited near infrared atomic spectra of O, N, F, Cl, Br, S and C; [134] includes references to all preceding work.

4.1.7.5. Detection Limits of Alkali Metals

The determination of alkalis by ICP–AES meets two difficulties:

1. The compromise conditions favorable for the majority of the elements are unsuited for the alkalis.
2. The most prominent lines of Li, K, Rb, and Cs are located at near in-

4.1. DETECTION LIMITS

Table 4.5. Detection Limits of Alkali Metals ("2σ Basis")

		Detection Limit (μg/mL)			
		h [135][b]			h [65][c]
Spectral Line (nm)[a]		1 mm	5 mm	10 mm	15 mm
Li I	670.7 84	0.00007	0.001	0.002	
I	460.2 86				0.6
I	323.2 63				0.7
Na I	588.9 95	0.006	0.006	0.01	0.03
I	330.2 37				1.2
I	330.2 98				3
K I	766.0 23	0.03	0.05	0.1	
I	404.7 21				30
Rb I	780.0 23	0.15	0.3	0.35	
I	420.1 85				25
Cs I	894.3 59	0.9	1.5	2.5	
II	452.6 73				30
I	455.5 31				100

[a] h = observation height above the coil and refers to the center of a 5–6-mm high zone.
[b] ICP conditions: 800-W power, 1 L/min carrier argon flow, 14-L/min outer argon flow, cross-flow pneumatic nebulizer.
Spectrometer: 25-cm Ebert, 1200 grooves/mm grating, blaze 600 nm, PD triptych detector.
[c] ICP conditions: 1100-W power, 1-L/min carrier argon flow, 20-L/min outer argon flow, cross-flow pneumatic nebulizer.
Spectrometer: 1-m focal length, 1180 grooves/mm grating, blaze 250 nm, PMT detector (not specified).

frared wavelengths (Table 4.5) which may intervene with the range commonly accessible with spectroscopic instruments primarily designed for work in the UV-visible region.

In polychromators such measures as using the zero-th order reflection from the grating in combination with a filter or an additional grating may give access to the Na, Li, and K lines.

Boumans [135] made a brief investigation of alkali determinations using a "triptych" of silicon PDs with separate outputs [136] and electronics of the type described by Bubert and Hagenah [26, 47, 137, 138]. The PDs replaced the exit slit of the monochromator.

Table 4.5 states the detection limits attained at three observation heights in the ICP. It should be noted that an observation height of 10 mm was considered desirable if the interference of K on the other alkalis had to be minimized.

For comparison, Table 4.5 includes the detection limits measured by Winge et al. [65] with a medium size monochromator at far less sensitive analysis lines under normal compromise conditions.

4.1.7.6. Detection Limits Obtained with Hydride Generation Techniques

The detection limits of As, Bi, Ge, Pb, Sb, Se, and Te can be considerably improved if the common sample introduction by nebulization is replaced by a hydride generation technique (cf. Chapter 6). Hydride generation is ever more applied in analytical atomic spectrometry: AAS, AES, and AFS. Nakahara recently published an excellent review [139] from which a table with detection limits is reproduced here as Table 4.6. The data permit a comparison of the results obtainable with the various methods and provide an idea about the benefits of hydride generation compared to direct introduction via nebulization. Two points should be stressed, however. First, Nakahara's remark that "there is much confusion over the definition of the detection limit, so users of such data [as shown in Table 4.6] should always check the definition in the original paper." Second, one must realize that "detection limits" is only one aspect of hydride generation; the precise experimental arrangement and execution, as well as interferences in the solution during hydride formation constitute other crucial aspects. They are well covered in Nakahara's survey, which reviews more than 300 papers.

4.1.7.7. Detection Limits and Spectral Resolution

Table 4.3 included detection limits obtained at high spectral resolution (column VI). These values are generally a factor of 3 to 15 better than those of Winge et al. (column II). This improvement is only partly due to the resolution, however, as was recently shown by a precise breakdown of the factor by which the detection limits differ [108]. It was demonstrated that for ionic lines a factor of 2 to 4 and for atomic lines a factor of 1 to 2 is attributable to a difference between the sources, a factor of 1.3 to 4.0, depending on the physical line width, is due to the difference in spectral resolution, and a factor of 1 to 2 stems from the difference in background RSDs.

In general the detection limits $c_L(I)$ and $c_L(II)$ obtained with methods I and II will differ by a factor $F(I/II)$:

$$F(I/II) = c_L(I)/c_L(II) \qquad (4.38)$$

where for convenience we assume method I to yield the poorer detection limit, so that $F > 1$. Factor F can be written as

$$F(I/II) = F_{\text{noise}} \times F_{\text{width}} \times F_{\text{source}} \qquad (4.39)$$

Table 4.6. Comparison of Detection Limits (ng/mL) for Analytical Atomic Spectrometric Techniques Using Conventional Solution Nebulization or Hydride Generation. [139] (Reprinted with permission from T. Nakahara, *Progr. Anal. Atom. Spectrom.* 6, 189. Copyright (1983), Pergamon Journals, Oxford.)

	AAS		AES				AFS	
			Solution Nebulization[a]	Hydride Generation				
Element	Solution Nebulization[a]	Hydride Generation	ICP	ICP	MIP[b]	APAN[c]	Solution Nebulization[a]	Hydride Generation
As	630	0.8[d]	40	0.02[g]	0.35	0.2	100	0.1[n]
Bi	44	0.2[d]	50	0.3[h]		0.1	5	0.005[o]
Ge	20	3.8[e]	150	0.2[i]	0.15	20	100	
Pb	17	0.6[f]	8	1[j]		5	10	
Sb	60	0.5[d]	200	0.08[h]	0.5	1	50	0.1[n]
Se	230	1.8[d]	30	0.03[k]	1.25	5	40	0.06[n]
Sn	150	0.5[d]	300	0.05[l]	2	20	50	1.2[p]
Te	44	1.5[d]	80	0.7[m]		50	5	0.08[n]

[a] W. B. Robbins and J. A. Caruso, *Anal. Chem.* **51**, 889A (1979).
[b] F. L. Fricke, W. B. Robbins, and J. A. Caruso, *J. Assoc. Off. Anal. Chem.* **61**, 1118 (1978).
[c] A. P. D'Silva, G. W. Rice, and V. A. Fassel, *Appl. Spectrosc.* **34**, 578 (1980).
[d] K. C. Thompson and D. R. Thomerson, *Analyst* **99**, 595 (1974).
[e] J. R. Castillo, J. Lanaja, and J. Aznarez, *Analyst* **107**, 89 (1982).
[f] P. N. Vijan and G. R. Wood, *At. Absorption Newslett.* **13**, 33 (1974).
[g] (1) R. C. Fry, M. B. Denton, D. L. Windsor, and S. J. Northway, *Appl. Spectrosc.* **33**, 399 (1979). (2) P. D. Goulden, D. H. J. Anthony, and K. D. Austen, *Anal. Chem.* **53**, 2027 (1981). (3) See h.
[h] M. H. Hahn, K. A. Wolnik, F. L. Fricke, and J. A. Caruso, *Anal. Chem.* **54**, 1048 (1982).
[i] M. A. Eckhoff, J. P. McCarthy, and J. A. Caruso, *Anal. Chem.* **54**, 165 (1982).
[j] M. Ikeda, J. Nishibe, S. Hamada, and R. Tsujino, *Anal. Chim. Acta* **125**, 109 (1981).
[k] See g (2).
[l] T. Nakahara, *Appl. Spectrosc.* **37**, 539 (1983).
[m] K. A. Wolnik, F. L. Fricke, M. H. Hahn, and J. A. Caruso, *Anal. Chem.* **53**, 1030 (1981).
[n] K. C. Thompson, *Analyst* **100**, 307 (1975).
[o] S. Kobayashi, T. Nakahara, and S. Musha, *Talanta* **26**, 951 (1979).
[p] K. Tsujii and K. Kuga, *Anal. Chim. Acta* **101**, 199 (1978).

where F_{source} = SBR(II)/SBR(I) for the radiation as emitted from source I and source II, respectively

F_{width} = SBR(II)/SBR(I) in the case that one and the same source is observed with spectrometer II and spectrometer I respectively; the factor accounts for the modification of the original SBRs by the spectroscopic apparatuses as a result of the difference in practical resolving power

F_{noise} = (background RSD for method I)/(background RSD for method II); this factor is made up of contributions from both the sources and the detectors

Equation (4.39) follows straightforwardly from the definition of the detection limit according to Eq. (4.16) and is the basis of Laqua's theoretical treatment of the dependence of detection limits on the parameters of the spectroscopic apparatus and the detector (Section 4.1.10).

Factor $F(\text{I/II})$ depends in two ways on the spectral resolution of the spectrometers involved, directly via factor F_{width} and indirectly via factor F_{noise}. The latter dependence is related to the radiant flux and the contribution from shot noise to the total noise, as has been discussed in detail in the preceding sections (see in particular Section 4.1.4.5).

The value of F_{width} depends on the spectral bandwidths and is directly connected with the effective line widths, as was shown by Boumans and Vrakking [108] by comparing SBRs of the same lines at "high resolution" (HR) and "medium resolution" (MR), respectively. Experimentally HR and MR conditions were realized by adjusting the slits widths of a 1.5 m echelle monochromator at 60 and 210 μm respectively. It was shown that F_{width} is linearly related to the ratio of the effective line widths ($\Delta\lambda_{\text{eff}}$),

$$F_{\text{width}}(\text{MR/HR}) = b \frac{(\Delta\lambda_{\text{eff}})_{\text{MR}}}{(\Delta\lambda_{\text{eff}})_{\text{HR}}} + a \qquad (4.40)$$

where $b \approx 0.9$ and $a \approx 0.2$, so that, to a first approximation, Eq. (4.40) expresses a proportionality.

The effective line width (Section 7.7.2) can be expressed in the physical line width ($\Delta\lambda_{\text{phys}}$) and the spectral bandwidth ($\Delta\lambda_{\text{instr}}$) by

$$\Delta\lambda_{\text{eff}}^2 = \Delta\lambda_{\text{phys}}^2 + \Delta\lambda_{\text{instr}}^2 \qquad (4.41)$$

If the physical line width can be neglected with respect to the spectral bandwidth, F_{width} will approach the ratio of the spectral bandwidths. In other words, SBRs will then increase nearly inversely proportionally to the spectral bandwidths. This is the maximum SBR profit obtainable by increasing the spectral resolution. Boumans and Vrakking showed that this maximum profit can be realized for many spectral lines when the spectral bandwidth is decreased from 16 to 4 pm. However, there are also many exceptions, in particular lines of elements with low atomic mass, which show substantial Doppler broadening, and lines that are apparently broadened by hyperfine structure (HFS). For those lines the contribution from the true or apparent physical line width can no longer be ignored: increasing the resolution then yields a smaller SBR profit than would be expected from the change in spectral bandwidth.

As an example Table 4.7 shows results for six spectral lines considered in

Table 4.7. Effect of Spectral Bandwidth on the SBR of Spectral Lines with Different Physical Widths: The SBR Improves Less than Inversely Proportionally to the Bandwidth if the Physical Width Essentially Contributes to the Effective Line Width [108] (cf. [307])

Spectral line (nm)		HR		MR		r^a	f^b
		$\Delta\lambda_{\text{eff}}$	$\Delta\lambda_{\text{instr}}$	$\Delta\lambda_{\text{eff}}$	$\Delta\lambda_{\text{instr}}$		
Be II	313.042	7.9	4.4	15.0	13.2	1.90	1.85
Er II	323.058	5.3	5.2	15.3	15.3	2.85	3.00
Hf II	282.022	4.0	4.0	11.9	11.9	3.00	2.90
Lu II	291.139	12.7	4.6	15.6	13.5	1.25	1.40
Nb II	309.418	15.7	4.7	19.5	13.8	1.24	1.45
Nb II	319.498	6.1	5.5	16.0	15.9	2.60	2.85

$^a r = (\Delta\lambda_{\text{eff}})_{\text{MR}}/(\Delta\lambda_{\text{eff}})_{\text{HR}}$.
$^b f = (\text{SBR})_{\text{HR}}/(\text{SBR})_{\text{MR}} = F_{\text{width}}$.

[108]. The data are experimental values obtained with the 1.5-m echelle monochromator referred to above. The table shows the effective lines widths and spectral bandwidths at HR and MR, the ratio of the effective line widths and F_{width}. The latter is seen to lag behind the ratio of the bandwidths (i.e., 3.0) for the Doppler broadened Be line and the Lu 291.139- and Nb 309.418-nm lines, both broadened by HFS.

Three conclusions should be stated:

1. In comparing detection limits obtained with different apparatuses, only a precise breakdown of the ratios can put the results into the right perspectives.
2. SBRs will increase inversely proportionally to the spectral bandwidth until the contribution of the physical line width to the effective line width can no longer be ignored.
3. Detection limits will decrease inversely proportionally to the SBRs only if the background radiant flux remains high enough to keep shot noise at a negligible level.

On the whole, when the spectral bandwidth is decreased from about 15 to 4 pm, detection limits will improve by a factor of maximally 3 to 4. This applies to both analysis lines located on smooth background and lines suffering partial line overlap. However, in the latter situation a far more important advantage is obtained from a decrease in spectral bandwidth, namely, an increase in selectivity and, therefore, a vast improvement of the limit of determination and the "true" limit of detection, as will be discussed in Section 7.7.7 (cf. [306]).

As mentioned in Section 4.1.7.2 (item 7), Boumans and Vrakking [307] recently reported an improved approach to the breakdown of detection limits. The principle was the same as discussed above, but the use of exact values of physical line widths, determined in the same experiment [308], permitted a more rigorous treatment of the experimental data. The approach was applied to the detection limits of about 350 prominent lines of 64 elements, and covered results obtained at high spectral resolution as well as those of Winge et al. [65] and Wohlers [110].

4.1.7.8. Effect of Sample Matrix on the Detection Limits and the Limits of Determination

In general the detection limits in aqueous solutions with a matrix (e.g., salts, acids, or albumine) or in organic solvents will not be the same as in pure aqueous solution. Possible causes of differences include multiplicative interferences, increased flicker noise, and spectral interference.

Multiplicative Interference (cf. Section 1.6). The matrix may change the sensitivity of the net analyte signal so that the SBR is modified. The effect may be positive or negative. For aqueous solutions at least this effect is relatively small, that is, less than 10–20%, if the ICP is operated under compromise conditions (see Section 4.7). Although a sensitivity change of this magnitude is not negligible in analysis, it is hardly relevant for a detection limit. This also implies that detection limits cannot be essentially improved by additives (e.g., excess alkalis) as happens in some types of dc and microwave plasmas.

Increased Flicker Noise. The matrix may adversely affect the stability of the nebulization and/or aerosol transport conditions resulting in a higher flicker noise contribution (α_B) to the background RSD (cf. Section 4.1.4.3–4.1.4.5).

Spectral Interference (cf. Section 1.6). Spectral interferences may drastically deteriorate the detection limits. These interferences are discussed in Section 4.6 and Chapter 7. The prime effect of spectral interference is a decrease in SBR as a result of enhanced background due to recombination continua, stray light, line wings, or direct line overlap. The background enhancement also has a compensatory effect in that the shot noise contribution to the background RSD ($g\beta/x_B$ in Eq. 4.27, 4.29, or 4.30) is reduced. However, in the case of line overlap the flicker noise (α_B) tends to be higher (Section 4.1.4.4 and [67, 75, 306]).

On the whole it follows from Eq. (4.30) that a detection limit (c_L) always increases with an increase in the background signal (x_B). If α_B is constant, c_L increases proportionally to x_B in the worst case ($g\beta/x_B \ll \alpha_B$) and proportionally to $\sqrt{x_B}$ in the best case ($\alpha_B \ll g\beta/x_B$).

Finally the following should be stressed: since the type and magnitude of spectral interferences are specific for the sample type, the effect of the sample on the detection limits cannot be predicted in general terms. Only detailed experimental data can reveal the situation for each analysis line and analyte separately. All that can be said in general is that samples whose major and minor constituents emit simple spectra (e.g., the alkalis, see Table 7.1) will have little effect on the detection limits. Samples that emit line-rich spectra, on the contrary, may substantially worsen the detection limits. If in the latter case, serious line overlap occurs, then the numerical value of a conventionally determined detection limit may even represent a useless figure of merit because the interference may enhance the limit of determination (Section 4.2.2) to such an extent that the link between the conventional limit of detection and the limit of determination is disrupted (Section 7.7.7).

For details on detection limits attainable in real samples refer to the chapters dealing with applications (Part 2, Chapters 1–6).

4.1.8. Detection Limits in ICPs Operated in Molecular Gases

Table 4.9 provides a survey of detection limits reported by various groups of authors. The preceding table, Table 4.8, serves as a key and summarizes the most crucial experimental conditions. To ease comparisons, Table 4.9 includes three sets of detection limits obtained with argon ICPs:

1. The detection limits of Winge et al. [65] for pneumatic nebulization (PN): column II.
2. Results of Montaser et al. [115] using ultrasonic nebulization (USN): column V.
3. Results of Meyer and Thompson [142] for PN: column IX.

The Winge detection limits again serve as overall standards of comparison. The Montaser and Meyer detection limits for argon ICPs are useful as references for the other results reported by these groups.

The most striking feature of the data in Table 4.9 is that an ICP with a molecular gas as outer gas yields acceptable detection limits only if a Greenfield torch is used (see results of Moore and Broekaert in columns III and IV).

Montaser's detection limits could approach Winge's results only because USN was employed. From the comparison of Montaser's results in columns VI, VII and VIII with those of the reference argon ICP in column V, it is evident that only for a few atomic lines in the wavelength region between 340 and 390 nm did a nitrogen–argon ICP configuration yield better or similar detection limits, provided that low power (1 kW) and a high carrier gas flow (2 L/min) were used. This point is further commented on in Section 4.9.

Table 4.8. Essential Operating Characteristics Underlying the Determination of the Detection Limits Listed in Table 4.9[a]

Clmn	Torch	Neb	Power (kW)	Outer Gas	Intm Gas	Carrier Gas	F_c (L/min)	Reference
II	F	PN	1.1	Ar	Ar	Ar	1	Winge et al. [65]
III	G	PN	4.7	N_2	Ar	Ar	1.5	Moore et al. [140]
IV	G	PN	3.0	N_2	Ar	Ar	0.9	Broekaert et al. [141]
V	F	USN	1.0	Ar	Ar	Ar	1	Montaser et al. [115]
VI	F	USN	1.0	N_2	Ar	Ar	2	Montaser et al. [115]
VII	F	USN	1.7	N_2	Ar	N_2	1	Montaser et al. [115]
VIII	F	USN	3.0	N_2	Ar	Ar	2	Montaser et al. [115]
IX	S-F	PN	1.3?	Ar	Ar	Ar	0.3–0.7	Meyer–Thompson [142]
X	S-F	PN	1.5–2	Air	Air	Ar	0.3–0.7	Meyer–Thompson [142]
XI	S-F	PN	1.5–2	Air	Air	Air	0.3–0.7	Meyer–Thompson [142]

[a]Key: Clmn = column number in Table 4.9, Neb = nebulization Intm = intermediate, F_c = carrier gas flow, F = Fassel torch (Fig. 3.7), G = Greenfield torch (Fig. 3.6), S-F = torch of special design, outer diameter approximately equal to that of the Fassel torch; PN = pneumatic nebulization, USN = ultrasonic nebulization.

Table 4.9. Detection Limits (ng/mL) in Argon ICPs and in ICPs in which Argon Has Been Partly or Wholly Replaced by Nitrogen or Air ("2σ Basis"). The Essential Operating Characteristics and the References are Stated in Table 4.8

Spectral line (nm)		WIN [65]	MOO [140]	BRO [141]	MONTASER [115]				MEYER [142]		
I	a	II	III	IV	V	VI	VII	VIII	IX	X	XI
Ag I	328.0	5	20[b]								
Al I	308.2	15	55[b]								
Al I	396.1	20							40	80	1,400
Al I	236.7								700	40,000	40,000
As I	193.6	35			4.5	240	230	35			
As I	197.1	50	95								
Au I	267.5	20	11*								
B I	249.7	3	3.5	0.9							
Ba II	455.4	0.8							0.8	8	6,500
Ba II	493.4	1.5							1.2	8	4,500
Be II	313.0	0.1							1.8	35	13,000
Be I	234.8	0.2							4	80	13,000
Ca II	393.3	0.1							0.2	0.5	1,700
Ca I	422.6	6.5							11	5	850
Cd II	214.4	1.7							60	40,000	450,000
Cd I	228.8	1.8								4,000	
Cd II	226.5	2.5		8	0.55	35	60	1.5	75	25,000	200,000
Cd I	470.9								5,000	80,000	10,000

Table 4.9. (Continued)

Spectral line (nm)		WIN [65] II	MOO [140] III	BRO [141] IV	MONTASER [115]						MEYER [142]	
I	a				V	VI	VII	VIII	IX	X	XI	
Co II	238.8	4	6	10								
Co II	228.6	4.5			0.45	25	20	2.0				
Co II	237.8	6.5			0.6	40	30	2.5				
Co I	345.3				2.5	1.7	4.5	15				
Cr II	205.5	4			1	65	75	4				
Cr II	267.7	4.5			0.45	13	35	2.5				
Cr II	283.5	4.5	3	8	0.2	6.5	5.5	0.6				
Cr I	357.8	15			1.3	0.3	2.5	55				
Cu I	324.7	3.5	5[b]	17								
Fe II	259.9	4	4	2.5								
Ga I	294.3	30							250	1,800	5,000	
Ga I	417.2	45							120	500	10,000	
K I	766.4								270	650	1,500	
K I	769.9								600	2,500	3,000	
La II	333.7	6.5							30	600	14,000	
La II	494.9	6.5							10	400	25,000	
La II	442.9	6.5							30	1,100	25,000	
Li I	670.7								110	110	650	
Li I	610.3								80	550	4,500	

Mg II	279.5	0.1	1.7	0.5					0.45	250	400,000
Mg I	285.2	1.1							4.0	410	71,000
Mg I	383.8	20							55	500	6,500
Mn II	257.6	0.9	0.65						2.5	270	75,000
Mn I	402.0			1					650	100	100
Mn I	280.1	14							150	370	1,700
Mn I	279.4	8							100	200	1,300
Mo I	386.4				2.5	0.3	5.0	16			
Mo II	277.5	17			1	20	40	3.5			
Na I	588.9	20							45	40	600
Na I	589.5	45							17	50	2,000
Nb II	309.4	25	8						75	750	6,500
Nb II	269.7	45							190	4,500	17,000
Ni II	231.6	10	11		0.75	110	100	9			
Ni II	227.0	17		85							
Ni I	341.4	30			2	1.5	2.5	10			
Ni I	351.5	30			2	1.2	4.0	40			
P I	214.9	50	85		8	1200	2000				
Pb II	220.3	30	65					45			
Pb I	405.7	180		260							
Pd I	363.4	35							130	120	650
Pd II	229.6	45							1,500	10,000	4,500
Pd II	248.8	70							1,200	5,500	3,000
Pt I	265.9	55	40[b]								

Table 4.9. (Continued)

Spectral line (nm)		WIN [65]	MOO [140]	BRO [141]	MONTASER [115]				MEYER [142]		
I	[a]	II	III	IV	V	VI	VII	VIII	IX	X	XI
Rb I	780.0								950	9,500	7,000
Rb I	794.7								130	15,000	75,000
S I	182.0		150								
Se I	196.0	50			2.5						
Si I	251.6	8	17			250	300	55			
Sn I	189.9	17	45								
Ta II	240.0	20	20								
Ti II	334.9	2.5	7.5								
Ti II	337.2	4.5							11	75	4,000
Ti II	338.3								13	75	4,500
Tl II	190.8	25			2.5	900	650	50			
Tl I	276.7	80		240							
Tl I	377.5	150			11	2.5	12	220			
V II	292.4	5			0.35	1.5	5	6			
Zn I	213.8	1.2	9.5	19	0.25	12	14	1.1	45	450	7,000
Zn II	202.5	2.5							300		
Zn II	206.2	4							200		

[a]Wavelength truncated.
[b]With optimization for each line individually: Al I: 5.5; Ag I: 2.0; Au I: 2.5; Cu I: 0.3; Pt I: 11.

4.1. DETECTION LIMITS

The results of Meyer and Thompson (columns IX to XI) are disappointing in all respects. The reasons are not clear; also the argon ICP did not yield satisfactory results. The larger spectral bandwidth of about 40 pm, compared to Winge's 17 pm, can only partly explain the discrepancies. Meyer and Thompson also reported results for oxygen–argon and oxygen ICPs, but these are generally still worse than those for the air–argon and air ICPs.

By contrast, Ohls and Sommer [143], using a Greenfield torch, achieved substantial improvements in the detection limits for various spectral lines, in particular at wavelengths above about 340 nm, when nitrogen or air as outer gas was replaced by oxygen. This is primarily connected with changes in background structure due to molecular band emission, as will be further discussed in Section 7.3.6 (cf. [306]).

Table 4.9 does not include detection limits reported by Ohls' group; these are covered in Part 2, Chapter 1.

4.1.9. Detection Limits for ICP-AFS and ICP-MS

Table 4.10 lists state-of-the-art detection limits for ICP-AFS using either a tunable dye laser [144, 145] or hollow cathode lamps [146, 147] as excitation

Table 4.10. Detection Limits (ng/mL) for ICP-AFS[a]

Element	Tunable Dye Laser [144, 145]	Hollow Cathode Lamp [146, 147]	Element	[144, 145]	[146, 147]
Ag	—	2	Na	—	0.7
Al	0.3	13 p	Ni	—	5
B	3	300 p	P	—	13000
Ba	0.5	300	Pb	0.7	70
Be	—	0.5 p	S	—	700
Ca	—	0.3	Si	0.7	200
Cd	—	0.5 cv	Sn	2	200 p
Cr	—	4 p	Ta	—	1300 p
Cu	—	2	Ti	0.7	300 p
Fe	—	7	Tl	5	—
Ga	0.7	—	V	2	200 p
Ge	—	130 p	W	—	1300 p
Hg	—	0.2 cv	Y	0.4	—
K	—	1.3	Yb	—	13 p
Mg	—	0.4	Zn	—	0.4
Mn	—	2	Zr	2	—
Mo	4	130 p			

[a] The data have been normalized to 2σ in order to link up with those in the previous tables.
Key: p, Propane introduced into the plasma. cv, Cold vapor technique.

Table 4.11. Detection Limits (ng/mL) for ICP Mass Spectrometry[a]

Element	Single Ion	Multielement	Element	Single Ion	Multielement
Ag	0.03	0.3	Nb	0.03[b]	—
As	0.04	1.0	Nd	0.03[b]	—
Au	0.06	2.1	Pb	0.05	1.5
B	0.4	5.7	Pr	0.02[b]	—
Ba	—	0.3	Sb	—	0.5
Cd	0.06	0.6	Sc	—	0.4
Ce	0.02[a]	—	Se	0.75	—
Co	0.05	0.3	Si	10[b]	—
Cr	0.06	0.7	Sn	0.06	1.2
Ga	—	0.3	Sr	0.04[b]	0.3
Ge	0.02	1.1	Te	0.08	5.8
Hg	0.02	0.4	Th	0.02	0.2
La	0.02[b]	0.3	Ti	0.1[b]	0.5
Li	0.1	3.6	U	0.03	0.1
Mg	0.1	1.2	V	—	0.3
Mn	0.1	1.1	W	0.05	0.3
Mo	0.04	0.7	Zn	0.2	2.4
Na	—	1.1	Zr	0.06[b]	0.5

[a] Results quoted from Gray [149] unless otherwise stated. The data have been normalized to 2σ in order to link up with those in the previous tables.
[b] From Douglas and Houk [150].
[c] From Gray [151].

sources (cf. Section 3.4). The data are for aqueous solutions introduced with a pneumatic nebulizer. As in ICP-AES, the detection limits can be improved by an order of magnitude by the use of ultrasonic nebulization [148].

Table 4.11 shows detection limits for ICP-MS using the continuum sampling mode [149-150] (cf. Section 3.5). Single ion detection limits are compared to multielement values under worst case conditions. Results reported for the boundary layer method have not been included since this method is of little practical use [151].

4.1.10. Comprehensive Theoretical Treatment of the Dependence of Detection Limits on the Parameters of the Spectroscopic Apparatus and the Detector

A theoretical description of the dependence of detection limits on the parameters of the spectroscopic apparatus and the detector has been given by Laqua et al. [22-25, 28, 152]. Essentially this treatment follows the SBR-RSD approach discussed in Sections 4.1.3-4.1.5, but is far more comprehensive.

4.1. DETECTION LIMITS

The basis of Laqua's theory is found in Kaiser's initial treatment of the calibration function and the detection limit [20]. Intensities are normalized to the background so that the detection limit is found from the intersection of the calibration curve and the "noise level" (cf. Fig. 4.5).

Laqua [28] summarizes the theory of the calibration curve as follows. If the background (I_u) is taken as the reference signal, the equation for the calibration curve reads

$$(I/I_u) = (c/c_u)^n \qquad (4.42)$$

where I is the net line signal, c the concentration, n the slope of the logarithmic calibration function log I versus log c, and c_u the calibration constant, that is, the background equivalent concentration if $n = 1$, which may be assumed to apply in the vicinity of the detection limit.

For c_u the following expression can be derived:

$$c_u = \frac{(dB_u/d\lambda)\Delta\lambda_L}{B_0} \frac{\sqrt{1 + \hat{s}_e^2 + \hat{R}^2 + \hat{R}_z^2}}{\hat{R}} \times \frac{\sqrt{\pi/\ln 16} \cdot \hat{s}_a/s_{\text{eff}}}{\int \exp(-y^2 \ln 16)\, dy} \qquad (4.43)$$

where $dB_u/d\lambda$ is the spectral radiance of the background, $B_0 = \int_{-\infty}^{+\infty} B_{0,\lambda}\, d\lambda$, that is, the integrated radiance of the analysis line at concentration $c = 1$, and $\Delta\lambda_L$ the physical width of the analysis line (cf. Section 7.7.2). The integral is often taken between $-\Delta\lambda_L/2$ and $+\Delta\lambda_L/2$.

For a further discussion of Eq. (4.43), it is convenient to write the equation in concise form,

$$c_u = c_{u,\infty} \times A_1 \times A_2 \qquad (4.44)$$

where $c_{u,\infty}$, A_1, and A_2 represent the three factors in Eq. (4.43).

Factor $c_{u,\infty}$ is the background equivalent concentration for "infinite" spectral resolution. This factor covers the influence of the excitation source stripped from apparatus functions. $c_{u,\infty}$ is related to the signal-to-background ratio $(\text{SBR})_\infty$ for "infinite" spectral resolution by

$$c_{u,\infty} = c_0/(\text{SBR})_\infty \qquad (4.45)$$

Factor A_1 describes the influence of the spectroscopic apparatus in terms of "reduced" quantities as defined in [22, 28]. Factor A_1 decreases with increasing resolving power and approaches unity for apparatus of high resolving power.

Factor A_2 describes the influences of the profile of the analysis line and the "effective width" of the measuring slit, which may be the width of the densitometer slit or the exit slit of a monochromator or polychromator. The expression for factor A_2 in Eq. (4.43) underlies the assumption of a Gaussian line profile.

Combining Eqs. (4.42) and (4.44), assuming $n = 1$ and rearranging one obtains for the calibration function:

$$c = c_{u,\infty} \times A_1 \times A_2 \times (I/I_u) \qquad (4.46)$$

and for the detection limit:

$$c_L = c_{u,\infty} \times A_1 \times A_2 \times (I_L/I_u) \qquad (4.47)$$

where (I_L/I_u) is the "noise level," that is, the smallest SBR that can be distinguished from the background. This noise level can be expressed as a multiple, for example $2\sqrt{2}$, of the RSD of the background, $(RSD)_B$.

Theoretical expressions have been derived for A_1, A_2, and (I_L/I_u) for both photographic and photoelectric detection [22–25, 28, 152]. The theory covers the effects of the spectral resolution as dictated by the theoretical resolving power and the widths of the entrance and exit slits of the spectrometer (or the width of the measuring slit in a densitometer) and the effect of the line profile. Conditions with a constant noise level, a noise level dictated by photon noise, and a noise level determined by statistical effects in the PMT are considered. The papers include a description of an experimental study [24] using an array of exit slits composed of "sandwiches" of glass fibers, each sandwich being connected with a separate PMT. This arrangement permitted simultaneous measurements at maximally 11 adjacent positions within a 0.005 nm spectral region.

The series of publications was concluded [152] with a fundamental comparison of photographic and photoelectric detection for the determination of low concentrations in the case of unlimited sample consumption. The basic conclusion was that the photographic emulsion can be distinctly better than a photoelectric detector provided the necessary measuring time is not too long. The relevant paper also contains detailed numerical data for two photographic emulsions, eight types of PMTs and five types of spectrometers. Formulas are given that allow extension of the results to other emulsions and PMTs.

4.2. PRECISION

4.2.1. Relative Standard Deviation

According to the IUPAC recommendations [1], precision represents "the random uncertainty in the value for the measure, x, or the corresponding uncertainty in the estimate of concentration, c, or quantity, q." This random uncertainty is conveniently expressed in terms of standard deviation or RSD. The use of the term "coefficient of variation" instead of "relative standard deviation" is discouraged by the IUPAC. It will not be used in this book.

We do not embark here upon a statistical treatment of precision and, thus, we omit discussing concepts such as confidence limits, regression, and analysis of variance and covariance. For treatments of these topics refer to textbooks on statistical analysis (e.g., [8, 153]) and papers on spectrochemical analysis specifically dealing with statistical calculations (e.g., [154–158]).

Usually one measures the RSD of an analyte signal and assumes that this value also applies to the concentration measurement. This is true only if the random error in the analytical curve is negligible.

It should be further noted that the standard deviation σ_A of a net analyte signal (x_A) is related to the standard deviations, σ_{A+B} and σ_B, the gross signal (x_{a+B}), and background signal (x_B) by

$$\sigma_A^2 = \sigma_{A+B}^2 + \sigma_B^2 \qquad (4.48)$$

if x_{A+B} and x_B are uncorrelated, as happens when a blank and a sample are measured sequentially. The corresponding RSDs are then related by

$$(RSD)_A^2 = \left(1 + \frac{x_B}{x_A}\right)^2 (RSD)_{A+B}^2 + \left(\frac{x_B}{x_A}\right)^2 (RSD)_B^2 \qquad (4.49)$$

Equation (4.49) shows that $(RSD)_A$ depends not only on $(RSD)_{A+B}$ and $(RSD)_B$, but also on the SBR $(= x_A/x_B)$. This dependence can be expressed in terms of the ratio (c/c_L) of the concentration in the sample to the detection limit.

4.2.2. Dependence of the RSD of the Net Analyte Signal on the Analyte Concentration—Limit of Determination

Again we shall assume that x_{A+B} and x_B are uncorrelated so that Eq. (4.48) applies. Let us further assume the following noise contributions to σ_{A+B} (cf. Section 4.1.4): flicker noise in the net and background signals ($\alpha_A x_A$ and $\alpha_B x_B$), shot noise in the total signal ($g\beta[x_A + x_B]$), and detector noise (γ), as follows:

$$\sigma_{A+B}^2 = \alpha_A^2 x_A^2 + \alpha_B^2 x_B^2 + g\beta(x_A + x_B) + \gamma \qquad (4.50)$$

Analogously for σ_B:

$$\sigma_B^2 = \alpha_B^2 x_B^2 + g\beta x_B + \gamma \qquad (4.51)$$

Therefore

$$\sigma_A^2 = \alpha_A^2 x_A^2 + g\beta x_A + 2\sigma_B^2 \qquad (4.52)$$

For the RSD we thus obtain

$$(RSD)_A = \left(\alpha_A^2 + \frac{g\beta}{x_A} + \frac{2\sigma_B^2}{x_A^2}\right)^{1/2} \qquad (4.53)$$

which can be converted into

$$(RSD)_A = \left[\alpha_A^2 + \frac{g\beta}{x_A} + \frac{1}{2^2(c/c_L)^2}\right]^{1/2} \qquad (4.54)$$

if we recall the analytical evaluation function (Eq. (4.2)) and the Eq. (4.3) for the detection limit with $z = 2\sqrt{2}$. The factor $\sqrt{2}$ must be included because in the present context we have assumed that a single measurement of the blank and not the mean of a series of blank measurements is involved. Therefore,

$$c/c_L = \frac{x_A}{2\sigma_B\sqrt{2}} \qquad (4.55)$$

If concentration c approaches the detection limit c_L, the first two terms in brackets in Eq. (4.54) can be neglected with respect to the third term[6], so that

$$(\text{RSD})_A = \frac{1}{2(c/c_L)} \qquad (4.56)$$

Accordingly, at the detection limit the RSD of the net analyte signal will be always 0.5 or 50% if $z = 2\sqrt{2}$ and 0.33 or 33% if $z = 3\sqrt{2}$. Often one considers a concentration equal to five times the detection limit as the "limit of determination" (c_d). In fact c_d is defined as the concentration at which $(\text{RSD})_A = 10\%$. It follows from Eq. (4.56) that this condition is fulfilled if the concentration equals five times the detection limit, whence

$$c_d = 5c_L \qquad (4.57)$$

is commonly accepted. However, this is justified only if the analysis line is located on smooth background. In the case that the line experiences line overlap, c_d may become appreciably larger than $5c_L$, as will be explained in Section 7.7.7.

At high analyte concentrations Eq. (4.54) reduces to

$$(\text{RSD})_A = \alpha_A \qquad (4.58)$$

For intermediate concentrations ($x_A \approx x_B$) the shot noise term $g\beta/Sc$ may play a role, but only if the shot noise in the background is not negligible with respect to the background flicker noise and the detector noise.

Figure 4.16 shows curves of $(\text{RSD})_A$ versus c/c_L for some values of α_A in the case where the shot noise term can be neglected. If considered as a function of c/c_L, relative standard deviation $(\text{RSD})_A$ is independent of $(\text{RSD})_B$. However, different concentration scales will apply for different values of $(\text{RSD})_B$, as is illustrated in the lower part of Fig. 4.16. In ICP-AES, α_A and α_B are likely to

---

[6]Actually the second term in brackets in Eq. (4.54) is not entirely negligible in the vicinity of the detection limit. However, a term of this type has been neglected in defining the detection limit, since there no distinction was made between the shot noise in the pure blank signal and that in the blank plus an analyte signal corresponding to the detection limit. If there had been, the shot noise term in brackets would cancel so that Eq. (4.56) would result by merely neglecting the flicker noise term α_A^2. Although the question is of academic interest only, it should be noted.

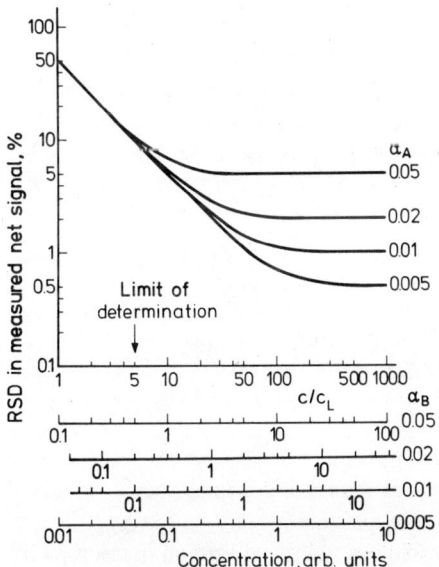

Figure 4.16. Dependence of the RSD of a measured net analyte signal on the ratio of the *analyte* concentration (c) to the detection limit (c_L) for four values of the analyte flicker noise parameter (α_A). The lower part of the figure shows the different concentration scales that apply for different values of the *background* flicker noise parameter (β). In ICP-AES it can be often assumed that $\alpha_A \approx \alpha_B$ so that a particular concentration scale corresponds to a particular curve. (Reprinted with permission from P. W. J. M. Boumans, Inductively Coupled Plasma-Atomic Emission Spectroscopy: Its Present and Future Position in Analytical Chemistry, *Fresenius Z. Anal. Chem.* **299**, 345 (1979). Copyright (1979), Springer-Verlag, Berlin.)

be of the same magnitude. Then, if $(RSD)_B \approx \alpha_B$, a single parameter $\alpha \approx \alpha_A \approx \alpha_B$ will dictate the dependence of $(RSD)_A$ on concentration. In an earlier treatment of the problem, discussion was limited to this particular case [159, 160].

Theoretical and experimental plots of the type shown in Fig. 4.16 have been produced and discussed by various authors (e.g., [16, 154, 161, 162]). As an example, Fig. 4.17 reproduces an experimental plot from Maessen and Balke's article [154]. These authors established for an ICP-polychromator system the following criteria:

1. A useful analytical concentration range covering four to five orders of magnitude.
2. A constant standard deviation of the net analyte signal over the lower 2 to 2.5 orders of magnitude of the analytical range.
3. A constant RSD of the net analyte signal over the upper 2 to 2.5 orders of magnitude of the range.

These findings agree with theoretical expectations (Eqs. (4.56 and 4.58)).

In the light of (1)–(3) Maessen and Balke critically examined the calculation of calibration curves using either linear or logarithmic scales. The former is statistically preferable if the standard deviation of the net analyte signal is constant, the latter if the RSD is constant, so that in fact two scales would be needed

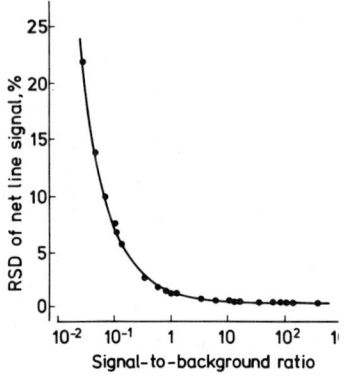

Figure 4.17. Experimental plot of the RSD of the net analyte signal of Cd II 226.502 nm as a function of the corresponding SBR [154]. (Reprinted with permission from F. J. M. J. Maessen and H. Balke, "A Critical Examination of the Analytical Significance of Extended Linear Working Ranges in Inductively Coupled Argon Plasma Emission Spectrometry," *Spectrochim. Acta* **37B**, 41 (1982). Copyright (1982), Pergamon Journals, Oxford.)

to cover the range. However, using a logarithmic scale over the entire range was shown to yield definitely better results than using a linear scale. Maessen and Balke eventually concluded that "for maintaining a distinct level of precision over the entire analytical range twice as many calibration measurements are required than would be the case if the error variances of ICP-AES were of the same order of magnitude over the entire linear range." With a view to Eq. (4.52) it is clear that a nonconstant variance over the entire range is not a specific characteristic of ICP-AES, but is inherent in any AES method because flicker noise (α_A) cannot be eliminated and, even if this contribution could be made negligibly small (as in x-ray fluorescence spectrometry), the fundamentally inevitable shot noise will make the variance depend on the magnitude of the net analyte signal. For a statistical treatment of calibration curves with either constant or variable standard deviation (including constant RSD) refer to Bubert and Klockenkaemper [157, 158, 163].

4.2.3. Noise Power Spectra—Effect of Signal Integration Period

A fundamental understanding of the noise characteristics of the ICP requires the knowledge of the noise power spectra of emission signals as obtained by fast Fourier transform (FTT) digital techniques. A noise power spectrum is a plot of noise power per unit bandwidth (W/Hz) as a function of frequency (Hz) or as a plot of rms current per square root bandwidth (A/V Hz) as a function of frequency. Noise components at their respective frequencies are displayed in a noise spectrum; therefore, the spectrum can be used to identify noise sources and predict the noise origins. This knowledge can be used for an intelligent choice of the measuring conditions such as signal integration periods or possible modulation frequencies. Knowledge of noise power spectra may further provide sound rationales for future system improvement, and the noise power can provide a monitor of the effectiveness of such improvements [164].

Noise power spectra have proven useful in the characterization of noise encountered in several analytical systems [165–167]. Specific studies have been reported for both flame emission [168–170] and atomic absorption [171] spectrometry, ICP-AES [172], and, more recently, for a microwave-induced nitrogen discharge at atmospheric pressure [173]. A tutorial elementary treatment of noise and associated concepts can be found in a chapter by Cova and Longoni [174]. A review and tutorial discussion of noise and SNR in analytical spectroscopy has been given by Alkemade et al. [42, 44] and Boutilier et al. [43], while Alkemade's group has reported [175] the results of an experimental study concerning spatial and spectral cross-correlation in the noise of the background and sodium spectrum emitted from a sheathed acetylene–air flame. Although this study does not deal with an ICP, the approach may be of interest for future investigations of ICPs. Finally, we should note here an excellent review by Epstein and Winefordner [48]: "Summary of the Usefulness of Signal-to-Noise Treatment in Analytical Spectrometry."

In noise analysis [42–44, 48, 172, 176] the observed noise is broken down into three major types:

1. White noise.
2. Low frequency (or excess noise), usually referred to as $1/f$ noise.
3. High frequency proportional noise (or whistle noise).

White noise is a type of noise having a constant amplitude at all frequencies and as such forms the baseline for measurements of other noise components above this level. White noise is due to completely random variations in the analytical system.

Low frequency noise contributions are components that occur at the low frequency end of the power spectrum. In some cases, they follow a $1/f$ function. The principal cause for such noise is the slow drift in the instrumental components. High frequency proportional noise occurs at given frequencies throughout the range of the spectrum [172]. Walden et al. [172] were the first to perform extensive measurements of noise power spectra of an ICP under a variety of conditions. They emphasized in particular the measurement of general trends in the noise components with changes in the ICP operating parameters. In a subsequent study [164], Belchamber and Horlick reported the following results:

1. In the 0 to 5-Hz region noise power spectra were highly characteristic and diagnostically useful with respect to nebulizer performance and design (Fig. 4.18). Although a strict $1/f$ character was not observed, the spectra were clearly dominated by low frequency components, characteristic of analyte flicker noise.

2. In the 0 to 500-Hz region a peak in the noise spectrum occurred which seemed to result from rotation of the plasma discharge at a frequency ranging from 200 to 400 Hz. This feature of the plasma was considered interesting in

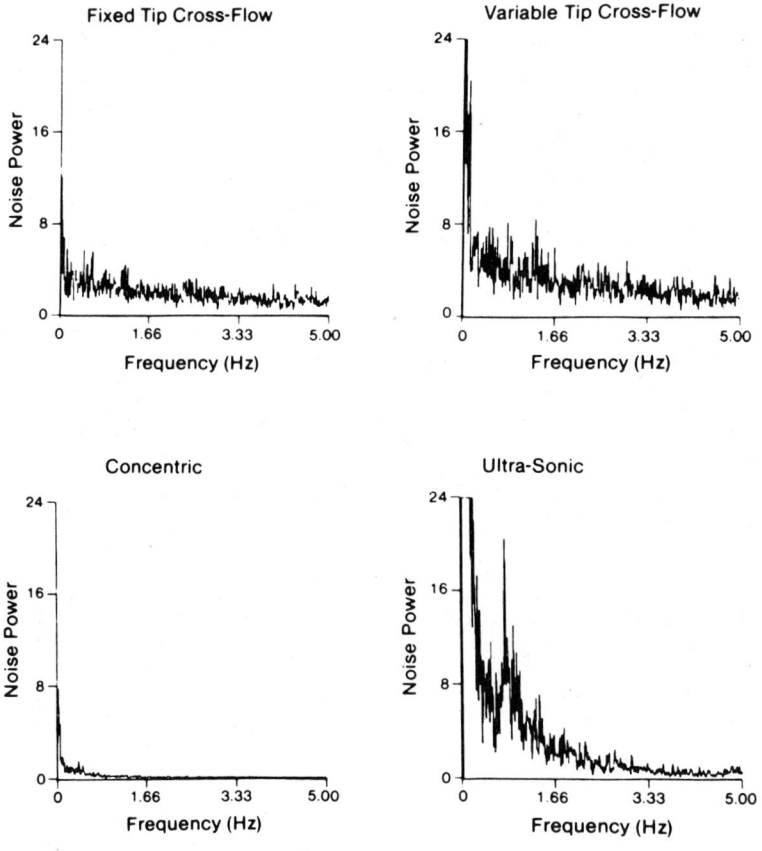

Figure 4.18. Low frequency (0–5 Hz) noise power spectra of the emission signal of Ca II 393.3 nm at 1 µg/mL Ca concentration for different nebulizers [164]. (Reprinted with permission from R. M. Belchamber and G. Horlick, "Noise-Power Spectra of Optical and Acoustical Emission Signals from an Inductively Coupled Plasma," *Spectrochim. Acta* **37B**, 20 (1982). Copyright (1982), Pergamon Journals, Oxford.)

its own right and as being important from a practical point of view when measurements are carried out on ICP systems with bandwidth responses in this region and when source modulation frequencies must be chosen for atomic absorption or fluorescence measurements using the plasma as atomizer.

In a separate study Belchamber and Horlick [177] investigated the effect of signal integration period on measurement precision in ICP–AES. It is well known [42–44, 64, 67, 178] that the improvement which can be achieved by increasing the integration time depends markedly on the exact nature of the limiting noise in the system. Improvements in the SNR in proportion to the

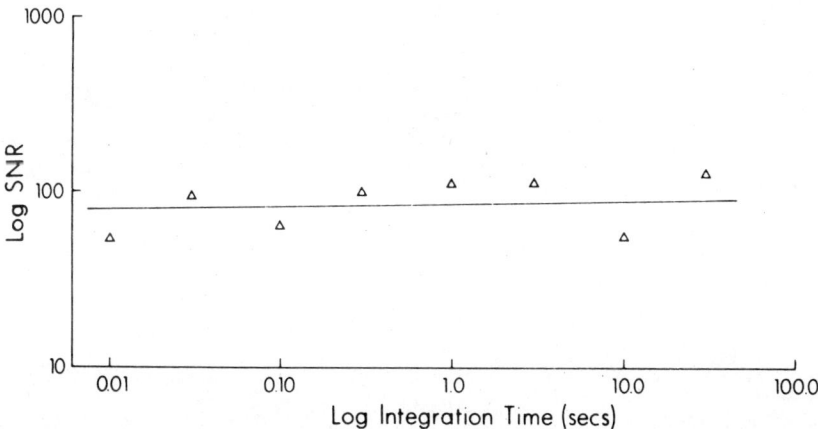

Figure 4.19. Dependence of SNR for Ca 393.3 nm (1 μg/mL Ca) on the integration time using a pneumatic concentric glass nebulizer. The SNR was calculated by dividing the mean of 32 measurements of the net analyte signal (x_A) by the standard deviation of x_A; thus the SNR is identical to the reciprocal of the RSD of the analyte signal [177]. (Reprinted with permission from R. M. Belchamber and G. Horlick, "Effect of Signal Integration Period on Measurement Precision in Inductively Coupled Plasma Emission Spectrometry," *Spectrochim. Acta* **37B**, 72 (1982). Copyright (1982), Pergamon Journals, Oxford.)

square root of the signal integration period will be achieved if the limiting noise has a white noise spectrum. However, if source flicker noise is the limiting noise, as is true for many spectrochemical measurements, then the noise tends to be characterized by a $1/f$ spectrum and, consequently, measurement precision cannot be improved by increasing the signal integration period [42–44, 64, 67, 178].

For illustration Figs. 4.19 and 4.20 show plots of log SNR for Ca II 393.3 nm (1 μg/mL Ca) versus integration time for a pneumatic concentric glass nebulizer and an ultrasonic nebulizer as reported by Belchamber and Horlick [177]. Results similar to those shown in Fig. 4.19 were found for cross-flow nebulizers. Belchamber and Horlick concluded that for pneumatic nebulizers little, if any, real improvement in SNR is realized by increasing the signal integration period from 10 ms to 30 s, which they considered to be a clear indication that the measurement at the intensity level used in their experiments is limited by source flicker noise and exhibited a $1/f^n$ type spectrum with $n > 1$. As has been shown by Boumans et al. [67] (cf. Section 4.1.4), this does not hold at a low radiant flux such as is encountered for background signals at relatively low wavelengths (<250 nm) and, thus, for net signals corresponding to concentrations of ≤100 times the detection limits (cf. Section 4.2.2). Then shot noise tends to dominate the noise characteristics of the system. This may lead to a different dependence of the measurement precision on the integration time than shown in Figs. 4.19 and 4.20.

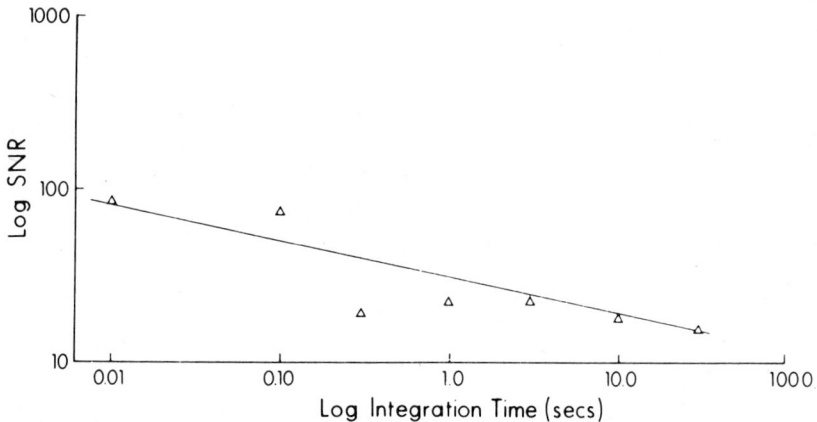

Figure 4.20. Dependence of SNR for Ca 393.3 nm (1 μg/mL Ca) on the integration time using an USN. See caption of Fig. 4.19 for the definition of the SNR [177]. (Reprinted with permission from R. M. Belchamber and G. Horlick, "Effect of Signal Integration Period on Measurement Precision in Inductively Coupled Plasma Emission Spectrometry," *Spectrochim. Acta* **37B**, 73 (1982). Copyright (1982), Pergamon Journals, Oxford.)

The change in the type of noise that dominates the measurements when the analyte concentration is varied is clearly revealed by the noise power spectra shown in Fig. 4.21: the shape of the curves tends to vary from a $1/f^n$ form to that of a frequency independent form when the concentration is decreased from 10 μg/mL (ppm) to 0.

The various studies including recent work of Benetti et al. [179] clearly demonstrate the usefulness of noise spectrum analysis for characterizing the stability of an ICP system in general and that of the nebulizer in particular. Such measurements, along with the determination of droplet size distributions (Part 2, Chapter 8), will eventually provide full insight into the factors which dictate nebulizer performance. This insight, in turn, is crucial for the further improvement of the precision of ICP–AES, generally believed to be limited by the instability of the nebulizer, including electrostatic effects [180] and spray chamber effects such as pressure fluctuations due to oscillations of the liquid level in the drain tube when droplets impact on it [181].

4.2.4. Precision of ICP–AES—Internal Standardization

In Section 1.6.2 we pointed to the use of calibration functions based on the ratio of the analyte signal (x_A) and a reference signal (x_R) emitted by a reference element or internal standard. This approach is generally required in AES to achieve satisfactory performance. Actually it was the introduction of internal

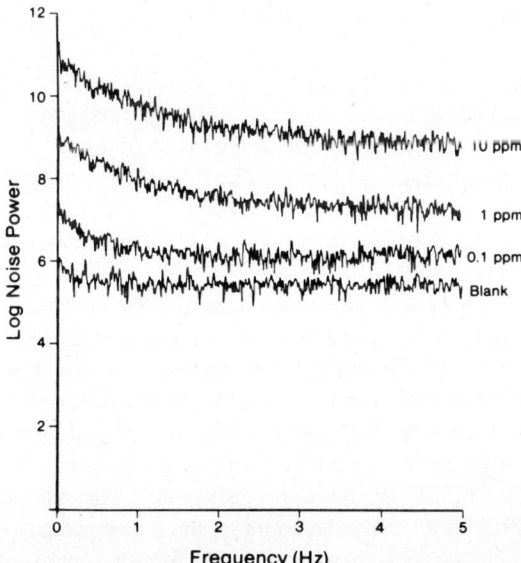

Figure 4.21. Noise power spectra in the 0–5 Hz region for the gross Ca II 393.3 nm signal illustrating the effect of the Ca concentration (ppm = µg/mL) on the shape of the noise spectrum [164]. (Reprinted with permission from R. M. Belchamber and G. Horlick, "Noise-Power Spectra of Optical Acoustic Emission Signals from An Inductively Coupled Plasma," *Spectrochim. Acta* **37B,** 21 (1982). Copyright (1982), Pergamon Journals, Oxford.)

standardization by Gerlach [182] which promoted AES to an instrumental method for quantitative analysis. Before that time AES was hardly more than a qualitative method.

Although internal standardization has been proposed in flame AES, it has never gained great popularity because the reasonable temporal stability of the emission signals and the use of short-term calibration ("bracketing") guaranteed satisfactory results, while it was not convincingly demonstrated that internal referencing would repay the higher investments. This view has proliferated in ICP–AES. However, recent work has shown that substantial improvements in analytical precision can be achieved if the internal standard principle is properly applied.

For an adequate assessment one should realize that a short-term precision of 0.5–2% in terms of RSD of net analyte signal can be attained at concentration levels equaling a few hundred times the detection limits. Which value in the range between 0.5 and 2% actually applies depends on the nebulizer, the ICP system, the spectrometer, the analyte, the sample type, the definition of "short term," and the "enthusiasm" of the analyst. Detailed data can be found in

Chapter 6 (dealing with sample introduction) and in the chapters on applications (Part 2, Chapters 1–6).

Clearly, reducing the RSD from, say, 2 to 0.2% in ICP–AES makes essentially higher demands upon the efficiency of the internal standard than reducing it from 50 to 5%, as may happen in dc arc spectrography. On the other hand, the level of multiplicative interferences in ICP–AES makes the choice of the internal standard less critical. This has been clearly recognized in the basic studies of internal standardization in ICP–AES by Belchamber and Horlick [183] and Myers and Tracy [184].

The starting point of both groups of authors is the common opinion that one major cause of ICP flicker noise is the nebulization step, in other words, that the emission noise arises primarily from fluctuations in aerosol density. Both groups also agree that internal referencing can be effective only for the reduction of flicker noise. Therefore, the intensity level of the analyte and reference signals should be high enough to make shot noise negligible. In addition, these signals should extend so far above the background that the latter does not essentially contribute to the noise behavior of the measured gross signal.

Belchamber and Horlick then argue that the two signals, x_A and x_R, must show sufficient correlation to provide an RSD in the ratio x_A/x_R lower than the RSD of x_A. They demonstrate theoretically and experimentally that this improvement is maximally a factor of 2.

Myers and Tracy, on the contrary, produce theoretical arguments that a high degree of correlation between the two signals (correlation coefficient ≈ 1) is a necessary but insufficient condition; also, the time-dependent relative amplitudes of the fluctuations in both signals should be the same. This theoretical argument is followed by a substantial amount of convincing experimental evidence. Myers and Tracy thus demonstrate that a judicious application of the internal standard principle in ICP–AES can improve the RSD by a factor of 10.

From the numerous results compiled by Myers and Tracy we reproduce here those shown in Figs. 4.22–4.25 as examples.

Figures 4.22 and 4.23 are for line pairs with similar (Cr II 205.5/Mn II 257.6 nm) and widely differing (Sr I 460.7/Cu II 224.7 nm) excitation characteristics, respectively. The different emission behavior of the lines is reflected in the spatial intensity profiles shown in the left-hand diagrams in the frames. The central part of each frame shows the variation of the signal amplitude over a 60-s period; the right-hand side shows the simultaneous recording of the ratio x_A/x_R and the reference signal, x_R, over a 10-min interval at a lower chart speed. Frames (a) to (c) are for three different carrier gas flows (F_c), as indicated in the upper left-hand corners.

For the lines with similar excitation characteristics (Fig. 4.22) internal standardization reduces the noise at all three carrier gas flows, the effect being largest (factor of 20) for $F_c = 0.7$ L/min. For lines with dissimilar excitation char-

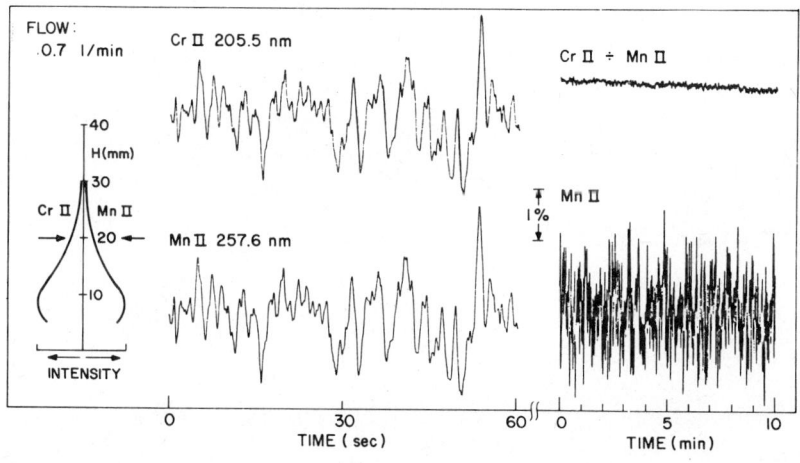
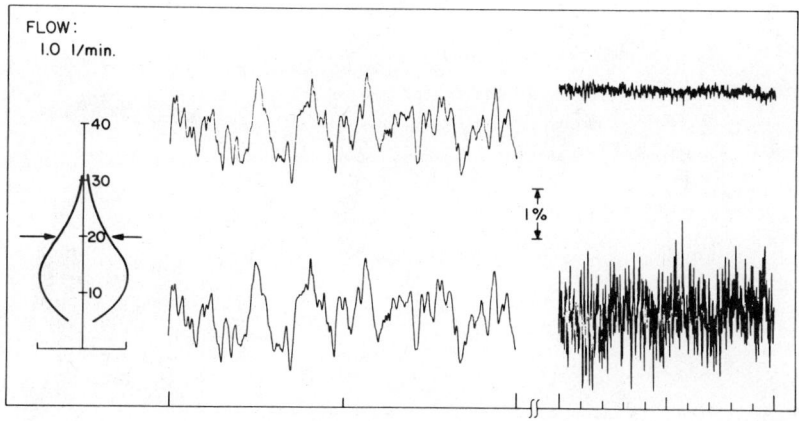
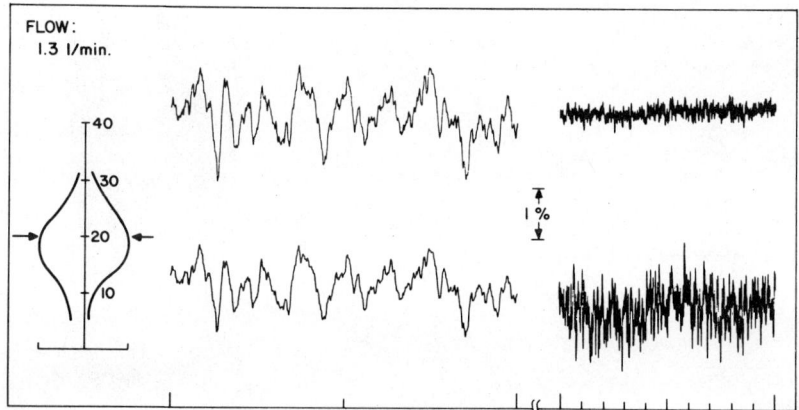

Figure 4.22. Correlation between the signals of two ionic lines. Successive frames show the change in correlation when the carrier gas flow is increased from 0.7 to 1.3 L/min. See text for further explanations [184]. (Reprinted with permission from S. A. Meyers and D. H. Tracy, "Improved Performance Using Internal Standardization in Inductively-Coupled Plasma Emission Spectroscopy," *Spectrochim. Acta* **38B**, 1237 (1983). Copyright (1983), Pergamon Journals, Oxford.)

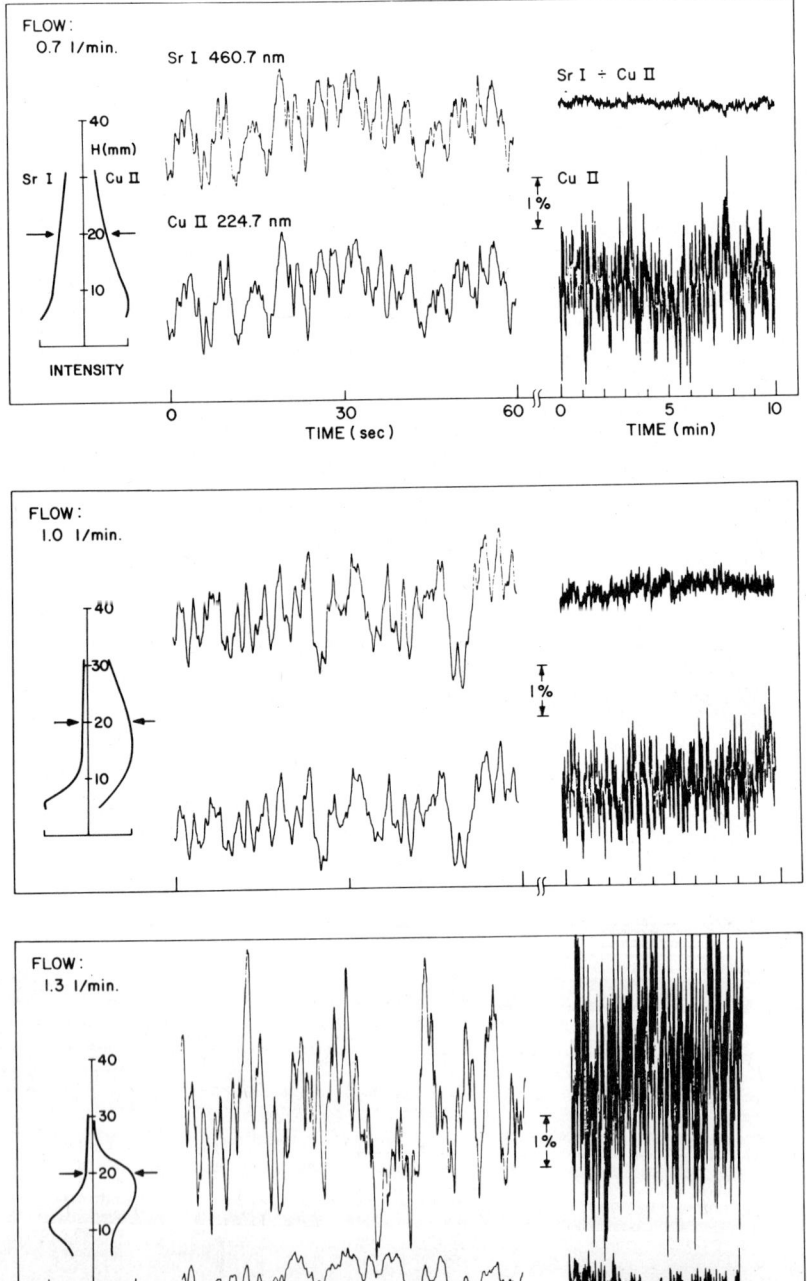

Figure 4.23. Correlation between the signals of an atomic line and an ionic line. Successive frames show the change in correlation when the carrier gas flow is increased from 0.7 to 1.3 L/min. See text for further explanations [184]. (Reprinted with permission from "Improved Performance Using Internal Standardization in Inductively-Coupled Plasma Emission Spectroscopy," *Spectrochim. Acta* **38B,** 1238 (1983). Copyright (1983), Pergamon Journals, Oxford.)

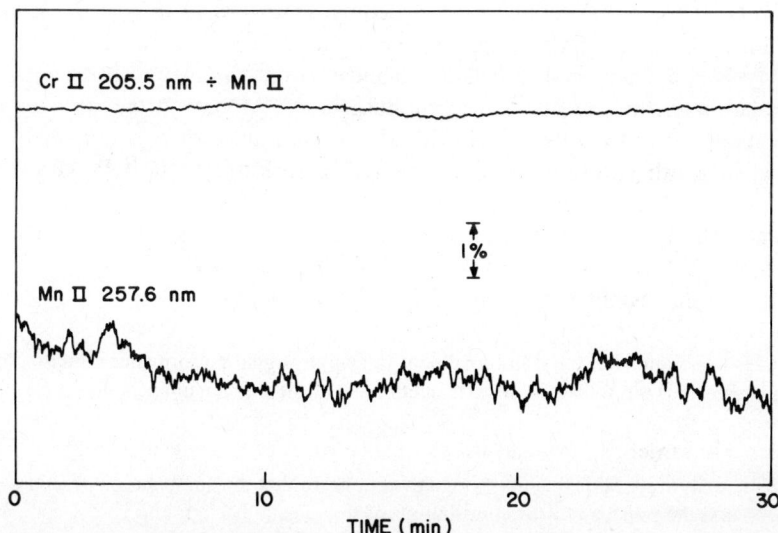

Figure 4.24. Reduction in drift using internal standardization. Upper curve: ratio of analyte and reference signal. Lower curve: reference signal. The time constant is 10 s [184]. (Reprinted with permission from S. A. Myers and D. H. Tracy, "Improved Performance Using Internal Standardization in Inductively-Coupled Plasma Emission Spectroscopy," *Spectrochim. Acta* **38B**, 1251 (1983). Copyright (1983), Pergamon Journals, Oxford.)

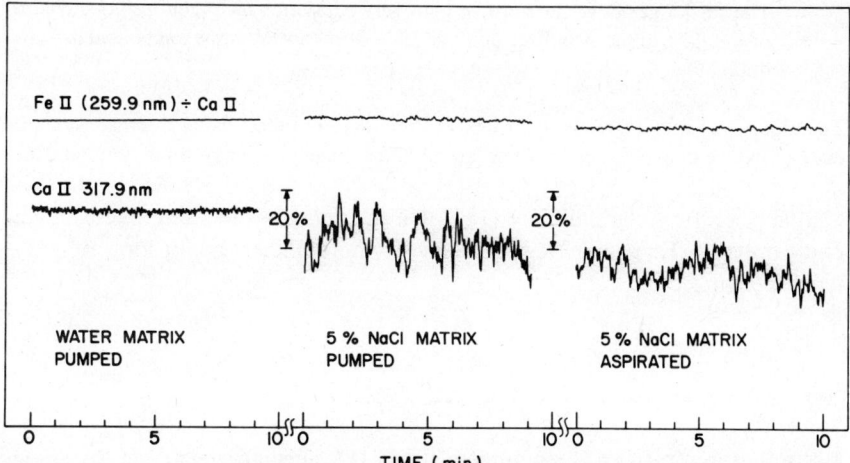

Figure 4.25. Improved signal stability using internal standardization (Fe II/Ca II) in a 5% (m/v) NaCl solution. Fe and Ca concentration: 200 and 100 μg/mL respectively [184]. (Reprinted with permission from S. A. Myers and D. L. Tracy, "Improved Performance Using Internal Standardization in Inductively-Coupled Plasma Emission Spectroscopy," *Spectrochim. Acta* **38B**, 1252 (1983). Copyright (1983), Pergamon Journals, Oxford.)

acteristics (Fig. 4.23) the noise is reduced by a factor of 11 if $F_c = 0.7$ L/min, but for $F_c = 1.3$ L/min no improvement is found.

Myers and Tracy made a detailed analysis of the various results in terms of correlation coefficients and noise amplitude ratios. Moreover, they provide explanations for analyte signal behavior and discuss the various potentials of internal standardization for improving analytical performance in ICP–AES. This discussion also covers the reduction of short-term drift in general and increased signal stability in the case that a concentrated solution is nebulized. This is illustrated in a self-explanatory way in Figs. 4.24 and 4.25.

In the conclusions Meyers and Tracy stress that

in order to obtain substantial improvement in analyte signal performance using internal standardization, the following requirements must all be met simultaneously.

1. The carrier gas flow rate and viewing height must be correctly chosen.
2. Analyte signal levels must be high enough so that the combination of shot noise and the noise and drift contribution of the plasma background is low compared to the remaining proportional noise.
3. The dual channel detection system must be very stable with identical response time in both channels.
4. Division must be performed on signals sampled at nearly the same instant in time.
5. Very good rf power regulation is required.
6. The carrier gas flow rate must be held very constant which requires good control of the flow to the nebulizer and adequate drip control in the spray chamber and drain.

The highly variable degree of improvement reported by others using internal standardization may be due to the inadvertent failure to meet one or more of these requirements.

It appears from the literature [185] that the potentials of using internal standards in ICP–AES and the requirements for their efficient application are gradually recognized.

4.3. ACCURACY

According to the IUPAC recommendations [1], accuracy relates to the agreement between the measured concentration and the "true value." The principal limitations on accuracy are then stated to be (1) randoms errors, (2) systematic errors due to bias in a given analytical procedure, (3) in multicomponent systems of elements, the treatment of interelement effects may involve some degree

4.3. ACCURACY

of approximation that leads to reproducible but incorrect estimates of concentrations (cf. [1] and Section 1.6.5).

Random errors are numerically expressed in terms of variance, standard deviation, RSD, and confidence limits about the mean. The numerical values of these statistics are based on repeated measurements under "identical" conditions. Therefore, they cannot themselves reveal bias. Bias represents the positive or negative deviation of the mean analytical result from the known or assumed true value. From the above statistics, the confidence limits are the most important since they indicate the range within which the true value is likely to be found with a given statistical risk in the case where bias is definitely absent. The confidence limits then give an indication of the maximum uncertainty in the result within the scope of the assumption that the procedure is free from bias.

The statement of analytical results in terms of means with confidence limits permits a straightforward comparison of means mutually or with certified values and therefore enables one to assess the presence or absence of statistically significant differences ("bias") between the results of an analysis and certified values or between results obtained in analyses with different analysis methods.

The calculation of confidence limits and associated quantities and the statistical analysis of results of measurements are treated in textbooks on statistics (e.g., [8, 153]) and in publications such as those noted in [157, 158, 163].

A narrow confidence interval indicates a high precision, but the result may yet deviate substantially from the true value if the analytical procedure is liable to systematic errors. The extent of the departure is a measure of the (in)accuracy of the procedure. Accuracy is tested by comparing the results with those obtained on the same sample types using other, independent methods or by comparing the results obtained on standard reference materials with the certified values. The literature on applications provides numerous examples of both types of comparisons. Illustrative and instructive examples are included in recent publications on

1. Routine water quality testing [186, 187].
2. Quality assurance in operating a multielement ICP spectrometer [188].
3. Quality assurance in the elemental analysis of foods [189].
4. Geostandards and geochemical analysis [190].
5. Reference samples as standards in the analysis of biological materials [191].
6. Preservation of accuracy and precision in the analytical practice of low power ICP–AES [192].
7. A method for correcting for acid and salt matrix interferences using the intensity of the H_β line [193].

A prominent cause of bias in physical methods of analysis is the inappropriateness of the calibration function. Inappropriateness means that the calibration function obtained with reference samples is not entirely correct for deriving concentrations from it using (net) analyte signals obtained on analysis samples. We briefly discussed this point in Section 1.6.4 for the simple case of a linear calibration function. A distinction was made between a deviation in the slope of the curve, attributed to multiplicative interference, that is, a change in sensitivity, and a deviation in the intercept attributed to additive interference, that is, spectral interference.

The cause of calibration errors is twofold: (1) drift in the measuring system during the measurements, and (2) the absence of perfect identity between analysis and reference samples ("standards"). Differences between samples and standards may be due to differences in their chemical composition and/or their physical properties, such as crystallographic structure in the case of solids, and density, viscosity, and surface tension in the case of liquid samples. Such differences may manifest in different behavior in sampling, sample preparation, transport to the nebulizer, nebulization, transport of aerosol to the ICP, atomization and transport in the ICP, ionization, excitation and emission in the ICP, and finally transfer of radiation to the detector. Discussions of effects that may occur in the various stages of the analytical procedure are woven in the chapters on applications in Part 2. In addition, Part 2, Chapter 8 specifically covers nebulizer effects, while Chapters 7 and 8 in this volume deal with spectral interferences.

4.4. DYNAMIC RANGE

4.4.1. Calibration Function Curvature

The term "dynamic range" denotes the concentration interval over which a calibration curve is linear. A large dynamic range is convenient because it permits simultaneous multielement analyses with one spectral line per element and with a single sample solution, thus without different dilutions for different analytes.

Since the early days of ICP-AES it has become common knowledge that ICP-AES is characterized by a large dynamic range, typically five orders of magnitude. Although this statement, in its generality, need not be questioned, it does require a more differentiated explanation.

Two questions beforehand: What is a calibration curve? and How linear is linear? To prove the linearity of a plot of analyte signal vs. concentration based on six concentrations that span five orders of magnitude is not difficult: the lower five points will clog together, so one actually proves the linearity of a straight line through two points! Aside from the statistical aspects, it is clear

4.4. DYNAMIC RANGE

that this approach is not viable. Therefore, if a large concentration range must be spanned, it has become customary to bring the Scheibe-Lomakin equation (Section 1.6).

$$x_A = s \times c^n \qquad (4.59)$$

into the logarithmic form:

$$\log x_A = n \log c + \log S \qquad (4.60)$$

and to prove that the slope n is unity.

A rigorous proof requires a statistical test involving the calculation of the confidence interval of the slope, which should contain 1 and be small enough to make the proof convincing. Using such tests, one has proved that in ICP-AES log-log calibration curves often have unit slope over five orders of magnitude.

If in practical applications slight departures from unit slope occur, this does not give problems. Computer procedures can easily accommodate for that.

Possible reasons for such departures are self-absorption, erroneous background correction, and nonlinear response of detector and/or readout system.

Self-Absorption. Radiation emitted by an atom or ion can be reabsorbed in the source by atoms or ions of the same kind and in a lower energy state than the excited emitters. This phenomenon is called "self-absorption." Self-absorption weakens the emission signal and destroys the proportionality between net analyte signal and concentration. The physical description of self-absorption is complex, and a comprehensive treatment of it is beyond the scope of this book. However, in Section 4.4.3 some general notions will be discussed along with experimental studies of self-absorption in ICPs.

Empirically, self-absorption is accounted for by an exponent $n < 1$ in the Scheibe-Lomakin equation (4.59), which then expresses that the degree of self-absorption, thus the relative weakening of the emission signal, increases with concentration. This empirical approximation links up with the results of physical descriptions, albeit that appropriate differentiations should be allowed for (Section 4.4.3). Accounting for self-absorption by logarithmic transformation of calibration Eq. (4.59) is effective only if self-absorption is relatively weak. Otherwise the logarithmic plot, too, will show curvature (see Fig. 4.31).

Erroneous Background Correction. If the analyte signal contains a spurious contribution, Δx, either positive or negative, due to an incorrect background reading, Eq. (4.60) assumes the form

$$\log (x_A + \Delta x) = n \log c + \log S \qquad (4.61)$$

This calibration curve bends towards the log c axis or log x axis depending on

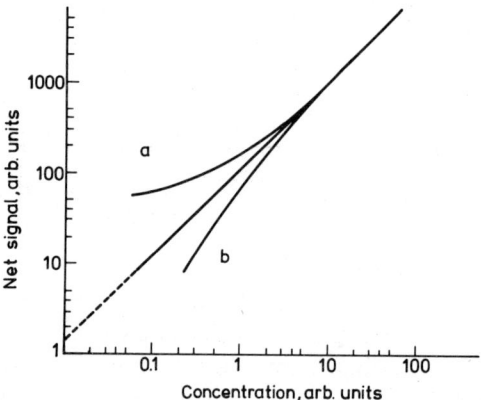

Figure 4.26. Double logarithmic plots of analyte signal versus concentration illustrating the effects of too small (a) or two large (b) a background correction (cf. Eq. (4-61)).

whether the background correction is too low ($\Delta x > 0$) or too high ($\Delta x < 0$) (see Fig. 4.26).

Nonlinear Response of Detector and/or Readout System. Exploitation of the intrinsically large dynamic range of ICP-AES makes high demands upon the measuring equipment. Great efforts have been made to accommodate the dynamic range of the measuring systems of ICP spectrometers to the possibilities offered by the ICP source. This is achieved by switching the PMT voltage and applying appropriate counting techniques in the PMT output (cf. Chapter 9). There is, however, an upper limit of the radiant flux that can be measured with a PMT. When this limit is approached, saturation effects occur. Then calibration functions may show curvature at high intensities in the same way as if this were produced by self-absorption. The two causes can be easily distinguished by making measurements at different intensity levels using filters.

Also connected with the nonlinear response of the detector are the departures from unit slope often found in spectrography. These departures are frequently attributable to the use of an inappropriate emulsion characteristic.

To conclude, there are various reasons why calibration curves in ICP-AES are not always linear. The curvature can be accounted for by the use of more complex calibration functions, inter alia, logarithm transformation of the Scheibe–Lomakin equation. Slope n of the logarithmic function thus becomes the "receptacle" of "mysteries," a panacea for the practicing analyst, who will hardly ever bother about the numerical value of n, provided that it is reproducible.

4.4.2. Dynamic Range—Analytical Range

Although ICP-AES is not entirely free from calibration function curvature due to self-absorption (Section 4.4.3), it remains true that one can often cover a wide concentration range with a single analysis line. This range starts at the detection limit and ends at a concentration some five orders of magnitude above the detection limit. Notable exceptions are the resonance ionic lines of Mg, Ca, and Sr. To determine high concentrations of these elements without sample dilution, one uses relatively weak lines, whence polychromators are often programmed with two lines for each of these elements. This is seen in the example of a polychromator program in Table 4.12, which lists the analysis lines along with the analytical ranges over which they are used.

Butler [195] proposed defining the "analytical range" of a method for a particular analyte in terms of the concentration interval within which a predefined value of the RSD in the concentration is not exceeded. Although the proposition has not been worked out rigorously and contains some minor errors, the principle deserves attention.

Figure 4.27 shows an error curve for AAS in terms of the RSD of the concentration as a function of concentration. At low concentration the RSD increases as the result of the decreasing SNR. At high concentration the calibration curve flattens off resulting in a poorer RSD. Symbolizing the minimum RSD—confusingly—as σ_B Butler then defines the analytical range as the concentration interval between the point on the error curve where the RSD is three or five times σ_B, depending on the precision requirements in the relevant analysis.

For the ICP this proposition is visualized in Fig. 4.28. Unfortunately Butler has overlooked that an AES calibration curve of intensity versus concentration does not bend to the concentration axis if the background is not corrected for. The curve will remain a straight line but does not pass through the origin (Section 1.6.4). It is the log–log curve which bends! However, the overall shape of the error curve in Fig. 4.28 is acceptable; the RSD rises when the concentration decreases and this is so because the net signal is found as the difference of two measurements whose values ever more approach each other (Section 4.2.2 and Fig. 4.16).

For the ICP, the analytical range is defined in the same way as for AAS. In agreement with general experience, Fig. 4.28 thus shows a distinctly larger range for ICP-AES than for AAS.

Butler's proposition is closely related to identifying the lower limit of the analytical range with the limit of determination instead of the limit of detection (Section 4.2.2); the essence of the proposal therefore is a more rigorous definition of the upper end of the range. It follows from Butler's experimental results that this approach is far more useful for AAS than for ICP-AES: ICP calibration curves do not show sufficient curvature to consider curvature as a

Table 4.12. Data Illustrating the Practical Analytical Range of a Polychromator for Multielement Analysis of a Variety of Samples Using Multicomponent Calibration and Interelement Interference Corrections (cf. Section 7.6) [194]. (Reprinted with permission from R. I. Botto, *Anal. Chem.* 54, 1657. Copyright (1982), American Chemical Society.)

Spectral Line (nm)		Analytical Limits[a]		Spectral Line (nm)		Analytical Limits[a]	
		Lower (ng/mL)	Upper (μg/mL)			Lower (ng/mL)	Upper (μg/mL)
Ag I	338.289	6.8	50	Mo II	202.030	12	100
Al I	308.215	13	100	Na I	588.995	5.1	75
Al I	237.312	28	500	Na I	330.298	220	2000
As I	193.696	25	100	Ni I	341.476	12	200
B I	249.773	5.6	100	P I	214.914	45	200
Ba II	455.403	0.9	50	Pb II	220.353	17	100
Be I	234.861	0.7	10	Pb I	283.306	32	100
Ca II	393.366	1.6	20	Pt I	265.945	77	100
Ca II	315.887	19	1000	Sb I	231.147	23	100
Cd II	214.438	5.6	100	Se I	196.026	55	100
Co II	228.616	6.3	100	Si I	288.158	14	100
Cr I	357.869	1.1	100	Si I	298.765	500	1000
Cu I	324.754	0.6	50	Sn II	189.980	16	100
Fe II	259.940	1.1	100	Sr II	407.771	0.1	50
Fe II	238.076	42	400	Ti II	334.941	0.5	100
K I	766.490	15	100	Tl I	377.572	26	50
Li I	670.784	0.5	100	U I	367.007	39	100
Mg II	279.553	0.9	50	V I	292.402	17	200
Mg I	383.231	18	1000	W II	207.911	60	200
Mn II	257.610	0.6	50	Zn II	206.200	4.9	100

[a]The lower limits were calculated from twice the standard deviation of the reagent blank background, the latter being determined from ten consecutive integrations.

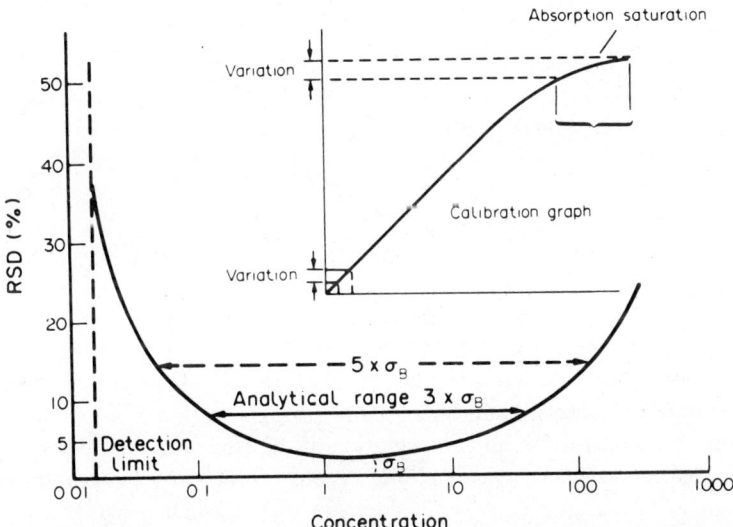

Figure 4.27. Typical graph for atomic absorption showing a calibration function and the change of the RSD of the analyte signal with concentration [195]. (Reprinted with permission from L. R. P. Butler, "Analytical Range," *Spectrochim. Acta* **38B,** 914 (1983). Copyright (1983), Pergamon Journals, Oxford.)

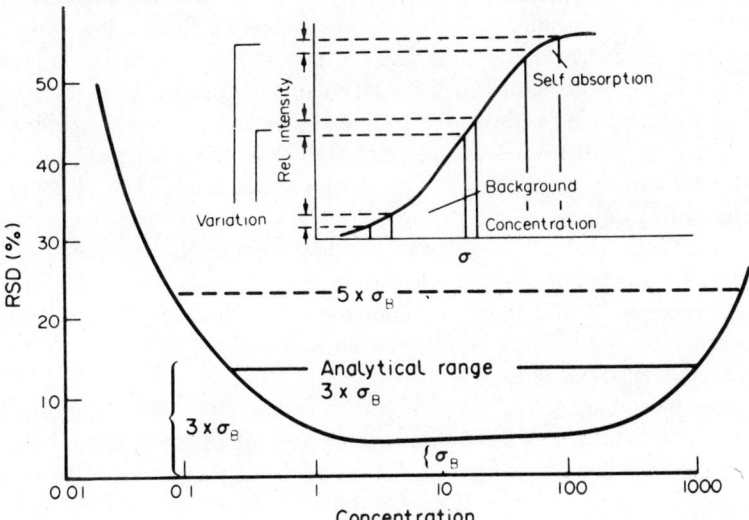

Figure 4.28. Typical graphs for emission spectroscopy in the case that self-absorption causes calibration function curvature at the upper end of the range. Bending of the calibration curve at the lower end of the range does not occur in a linear plot of intensity vs. concentration (cf. Section 1.6 and Fig. 4.26). However, at the lower end of the concentration range the RSD increases approximately in the way shown (cf. Fig. 4.16) [195]. (Reprinted with permission from L. R. P. Butler, "Analytical Range," *Spectrochim. Acta* **38B,** 915 (1983). Copyright (1983), Pergamon Journals, Oxford.)

real limitation of analytical precision. However, at high solute concentrations the RSD may increase for reasons other than calibration function curvature. Then Butler's definition of analytical range might also be useful in ICP–AES.

4.4.3. Self-Absorption

Self-absorption is treated in detail in standard textbooks, for example, in [196–198]. Only a few features will be considered here.

For a uniform source of given temperature and a spectral line of given shape and wavelength, the degree of self-absorption increases with the intensity of the emission line. It does not matter, for instance, whether a line appears relatively strong because it is an intrinsically strong line (large transition probability) with a relatively small concentration of the element, or whether the line is intrinsically weak but appears strong because of the larger concentration of the element in the source. Also, it does not matter whether a line originates from a transition between high levels or a transition between low levels. In a uniform source, two lines having the same intensity, wavelength, and line profile are similarly affected by self-absorption. However, a completely uniform source is an ideal not found in practice, the temperature usually declines radially. The radiation generated in the interior of the source thus passes through zones of lower temperatures. In these zones the population of low levels, especially the ground level, is large and the population of high levels is small compared to the populations in the interior of the source. As a result, lines having the ground state as the lower level tend to be more strongly self-absorbed in the outer zones than lines ending on a high level; also, the radiation added back for these lines by emission in the cooler vapor fringe is much less compensated for by emission.

Absorption is strongest in the peak of a line; hence self-absorption tends to flatten and consequently broaden the profile. Fig. 4.29 shows the modification of a Lorentz profile (Section 7.7.2) for different degrees of self-absorption. Extreme self-absorption causes a dip in the center of the line. This case is referred to as "self-reversal."

In a uniform source, self-reversal cannot occur [196, 199]; when, in contrast, the radiation passes through a lower temperature vapor fringe, self-absorption can lead to self-reversed lines. The final line profile can be pictured as being made up of radiation from the hot inner region modified by the radiation emitted by the cool vapor (Fig. 4.30). The absorption and emission lines in the cool envelope show less Doppler broadening because of the lower temperature; the wings of the original line profile consequently remain virtually unaltered, whereas the center changes greatly.

A relatively low-temperature vapor layer that envelops an excitation source is a typical example of self-reversal. Hence self-reversal is customarily referred

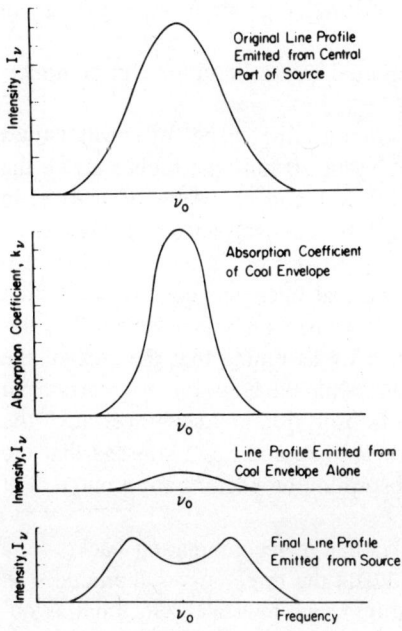

Figure 4.29. Modification of a line profile emitted from the (uniform) central portion of a source during passage trough the cool fringe. As a result of the lower temperature, the absorption line is narrower and the original line is selectively absorbed in the center of the profile. The contribution from emission in the cool envelope to the final line contour is small. (From Boumans [196].)

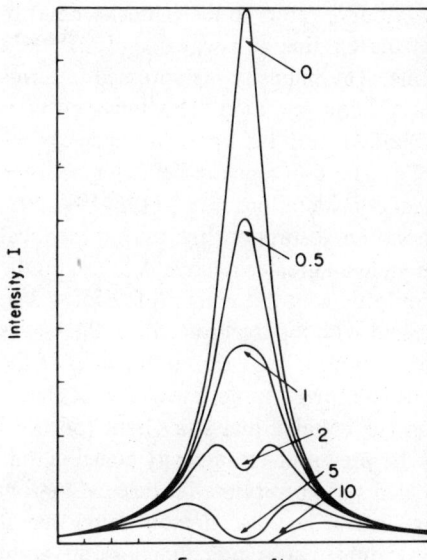

Figure 4.30. Modification of collisional-line profile by different degrees of self-absorption. The contour marked "0" is the original profile. The contour "1" marks the point beyond which self-reversal occurs. (From Boumans [196].)

to as self-absorption by an atomic vapor of low temperature rather than as an extreme case of self-absorption.

The observations of self-absorption reported in the literature can be understood in light of the above considerations.

Human and Scott [200] were the first to study the shapes of spectral lines emitted from an ICP. At observation heights of 20 mm and higher above the coil, they observed self-reversal of lines for high concentrations of analyte, in the same way as occurs in a flame [197, 198]. This self-absorption is caused by absorbing atoms around the emission region. Lower in the plasma (at 10-mm observation height), self-reversal of spectral lines was not found for high concentrations of analyte, but the lines were considerably broadened. The unusual shape of the lines could be explained by assuming that the core of the plasma is a predominantly absorbing region, while the emission originates from an annulus around the core. Although self-absorption and self-reversal of the Ca and Sr resonance lines were observed, Human and Scott stressed that the profiles of these lines did not show self-absorption broadening for a range of at least three orders of magnitude.

In their study of line broadening and radiative recombination background interferences, Larson and Fassel showed [201] the three curves reproduced in Fig. 4.31. The upper curve is the calibration curve for Ca II 396.8 nm as observed with a double monochromator, which precludes stray light contributions. This curve exhibits unit slope up to about 100 μg/mL Ca, above which the curvature toward the concentration axis indicates the presence of self-absorption. The 100-μg/mL level is more than five orders of magnitude above the detection limit of this Ca line. The lower curve refers to the intensity contributed by the wing (cf. Section 7.3.2) of the Ca line at a wavelength of 396.2 nm, that is the wavelength of an Al I line. The intensity is expressed in terms of the equivalent concentration of Al vs. Ca concentration. This lower curve is essentially a calibration curve for Ca obtained from the line wing and exhibits unit slope up to at least 10,000 μg/ml Ca. The difference in behavior between line peak and wing agrees with theoretical considerations (e.g., [196–199, 202, 203]) that self-absorption by atoms having an absorption line profile identical to or narrower than the emission profile yields a greater relative decrease in the central portion of the emission line profile than in the wings (cf. Fig. 4.29). The central curve in Fig. 4.31 was obtained with a polychromator at the wavelength of the Al I 396.2 nm line. It refers, in principle, also to the wing of the Ca II 396.8 nm line, in the same way as the lower curve. However, it shows curvature, from which Larson and Fassel concluded that stray light (Sections 7.3.1 and 8.5) from the Ca line peak is the predominant intensity contribution.

Thus Larson and Fassel not only showed self-absorption to cause calibration function curvature (upper curve in Fig. 4.31), but also demonstrated that if background enhancement is due to stray light, calibration function curvature

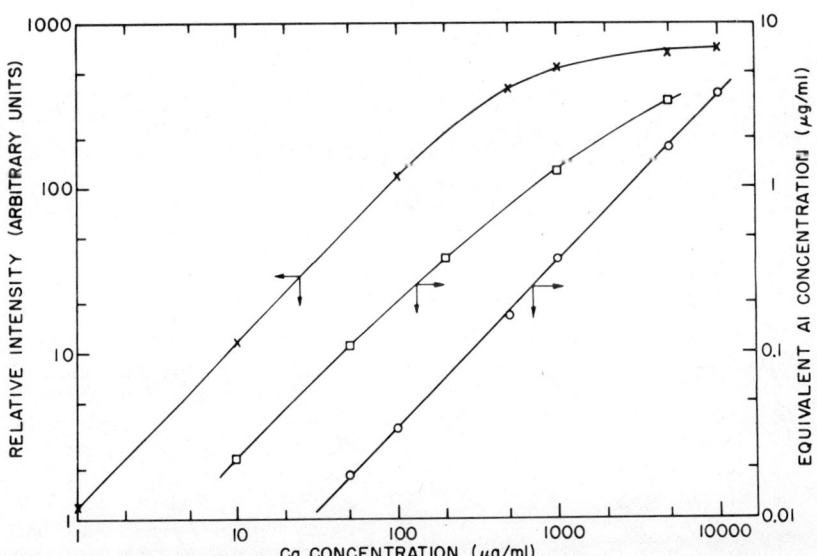

Figure 4.31. Intensity responses of Ca line and wing to concentration changes: x – double monochromator, line center at 396.8 nm; ○ – double monochromator, line wing at ~396.2 nm; □ – polychromator, 396.2 nm [201]. (Reprinted with permission from G. F. Larson and V. A. Fassel, *Appl. Spectrosc.* **33**, 596 (1979). Copyright (1979), Society for Applied Spectroscopy, Frederick, MD.)

manifests in that the background enhancement depends in a nonlinear way on the concentration of the relevant element. From the latter result, Larson and Fassel concluded that for elements which emit relatively few strong lines (alkalis, alkaline earths), a study of the dependence of background enhancement on the element concentration provides spectrochemists with a valuable method to determine whether stray light is a major contribution to the background enhancement.

Kawaguchi et al. [204] measured the widths of a variety of lines and found that in particular Ca II 393.4, Ba II 455.4, Sr II 407.8, Mg I 285.2, Mg II 279.5, Cd I 228.8, and Zn I 213.8 nm exhibited substantial broadening through self-absorption at 1000 μg/mL.

McLaren and Mermet [205] confirmed this for Cd I 228.8 and Zn 213.8 nm. They also showed (Fig. 4.32) that if a spectrometer with high resolving power is used, the linear portion of the calibration curve becomes smaller, the narrower the slit, or the higher the practical resolving power. This result confirms for the ICP what Laqua et al. [22] found in their early studies of high-resolution spectroscopy using dc arc and spark as excitation sources.

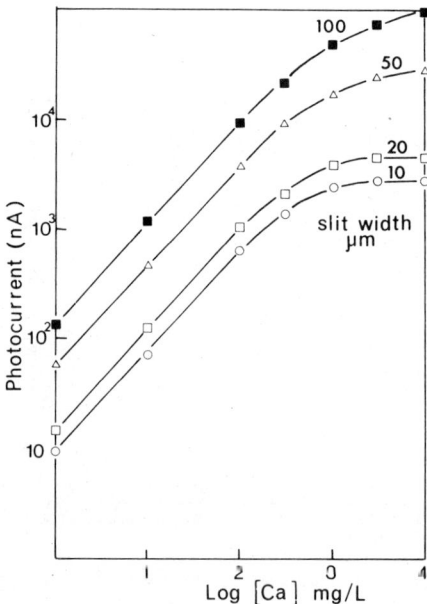

Figure 4.32. Influence of the resultant slit width on the shape of calibration curves for Ca II 393.367 nm observed with a high-resolution monochromator (1-m focal length, 3600 grooves/mm grating used in the first order) [205]. (Reprinted with permission from J. W. McLaren and J. M. Mermet, "Influence of the Dispersive System in Inductively Coupled Plasma Atomic Emission," *Spectrochim. Acta* **39B**, 1317 (1984). Copyright (1984), Pergamon Journals, Oxford.)

4.5. MULTIELEMENT CAPABILITY

The present section links up with the introductory discussions on atomic spectra in Section 1.2, the multielement capability of AES in Section 1.3, and spectroscopic instrumentation in Section 1.4.

It is easy and true to add that ICP–AES has an excellent multielement capability. Accurately specifying what this concretely means is not simple and even impossible because multielement capability cannot be generally specified in terms of a few numbers and, if one wishes to quantify it, this quantification is bound to sample type and instrumental facilities. Therefore, it is far easier to reverse the question, that is to (1) specify the analytical problems, (2) choose equipment with which the problems can be solved at an optimum performance to cost ratio, and (3) assess the multielement capability of the compromise reached between the initial requirements and the eventual performance.

Obviously such an approach forms part of the negotiations between potential users of ICP–AES and instrument manufacturers, but cannot be followed here. All that can be done is to provide a few guidelines in addition to those already given in Sections 1.2 to 1.4.

Let us take the Periodic Table in Fig. 4.13 as a starting point. The shaded boxes in this table mark the elements with good analysis lines in the 189–671-nm wavelength region. We first confine further discussion to this group of elements. The Periodic Table shows the distributions of the best detection limits and the numbers of lines per element having a detection limit within a factor of 3 from that of the best line. If we recall the dynamic range of four to five orders of magnitude (Section 4.4), then it is evident which elements and concentrations are accessible with a 27-MHz ICP operated with a pneumatic nebulizer and combined with an air-path spectrometer of medium resolving power. The detection limits may still be improved by a factor of up to 20 by the use of a higher frequency for the ICP and a high-resolution spectrometer or by replacing pneumatic by ultrasonic nebulization (Section 4.1.7.7).

Of particular interest is that the picture given in Fig. 4.13 applies to a single set of ICP compromise operating conditions (Section 4.7). However, the picture is idealized in that it refers to pure aqueous, single-element solutions. For real samples two points must be considered. First, the compromise operating conditions may require adaptations to minimize specific multiplicative interferences, and this may change the detection limits. Fortunately the picture is not often essentially affected by such adaptations. The second point, however, is far more important: the picture may be substantially distorted by spectral interferences from the major elements (Section 4.6 and Chapter 7). Then the detection limits will worsen (Section 4.1.7.8), and this holds even more strongly for the limits of determination (Section 7.7.7). This is a common penalty of AES methods: the intrinsically high multielement capability precludes high selectivity, and the relative lack of selectivity, in turn, limits the actual multielement capability for real samples.

The rather large variety of commercial instruments makes it possible to choose such facilities that the multielement capability required for solving a specified series of analytical problems is closely approached. Achieving a proper balance between cost and performance will always lead to some compromises. If many elements must be determined simultaneously, the spectroscopic apparatus will lack flexibility in line choice, which can be partly compensated for by complex software for spectral interference corrections (Section 7.6). Sequential systems have large flexibility in this respect and therefore have, in principle, a much larger multielement capability than simultaneous systems, but exploiting this capability demands a sacrifice from analysis time and sample consumption.

Looking now at Fig. 4.13 and the elements for which data are lacking or incomplete, we can add the following:

The determination of low concentrations of alkalis is difficult under compromise operating conditions for multielement analysis (cf. Section 1.2). Halogens are not normally covered (see end of Section 1.2 and Section 4.1.7.4, however). Boron, carbon, phosphorus, nitrogen, and sulfur are readily accessible with a vacuum or purged spectrometer (Sections 1.2 and 1.4.7.3). Hydride generation gives better access to hydride forming elements than is suggested in Fig. 4.13

(Section 4.1.7.6). Radioactive elements require practical precautions, but do not generally provide fundamental problems. Oxygen, hydrogen, and argon are not covered for obvious reasons, while few analysts, if any, will have much interest in He, Ne, Kr, or Xe as analytes.

On the whole, it can be stated that numerous successful applications of ICP–AES to multielement analyses of a large variety of sample types testify to the outstanding multielement capability of the method. There are boundaries, however.

4.6. LINE SELECTION AND SPECTRAL INTERFERENCES: AN OUTLINE

An analysis line is observed by centering a "spectral window" on the wavelength (λ_a) of that line in the dispersed spectrum. This is accomplished with an exit slit in the focal plane of the spectroscopic apparatus. The spectral window isolates a band of wavelengths ($\lambda_a \pm \Delta\lambda$) from the dispersed spectrum. The intensity distribution within this band results from the convolution of the apparatus profile and the intensity distribution as emitted from the source, aside from reflection and absorption effects in the light path.

The radiation in the selected wavelength band consists of the net analyte signal and a background signal. The omnipresent background, even in its simplest form of a smooth continuum, interferes with the determination of the true net analyte signal. Its fluctuations govern the detection limit of the net signal (Section 4.1). The presence of the background signal as such need not be considered as a spectral interference, however. As long as the background signal can be characterized by a constant mean value x_B with standard deviation σ_B, even its value is immaterial, since also a plot of the gross line signal versus concentration will yield a reproducible straight line, albeit with nonzero intercept. In spark spectrometers for routine metal analysis such gross signals are still often used. A problem arises if the calibration curve is plotted on a double logarithmic scale: the curve will bend and asymptotically approach a line parallel to the abscissa (Fig. 4.26). However, in view of the strong variation of the RSD of the net signal with the concentration when the detection limit is approached (Section 4.2.2), the use of log–log plots in this concentration range is irrational (Section 4.2.2).

The background spectrum of an ICP may comprise the following components:

1. Continuous radiation.
2. Stray light.
3. Quasi continua due to line wings.

4.6. LINE SELECTION AND SPECTRAL INTERFERENCES: AN OUTLINE

4. Molecular bands from molecules of the discharge gas or reaction products of this gas and the sample constituents, inter alia, N_2, O_2, OH, NH, NO, C_2, CN, LaO, ZrO, SiO, YO, and WO.
5. Spectral lines contributed by free atoms or ions of the discharge gas (Ar, H, O, N).
6. Spectral lines emitted by free atoms or ions of the concomitants of the sample, including contributions from reagent blanks.

Consequently, the background signal in the spectral window of an analysis line will appreciably depend on the sample composition and the instrumental operating conditions. It is primarily this dependence which gives rise to spectral interference and necessitates the reduction of gross signals to net signals by the use of background correction.

The concept of spectral interference in this sense thus refers to errors in concentration determinations due to the use of incorrect background signals, in particular, as a result of differences between the background signals for analysis and reference samples. This problem is a consequence of the fundamental inability of whatever instrument to measure the "true background signal under the spectral line." Any approach for background correction, therefore, remains an approximation, which will yield the more accurate results the greater the knowledge available on the sample composition and the closer the similarity between analysis and reference samples. As a consequence, there does not exist a unique method for background correction. The choice of the approach depends on both the sample type and the available instrumental facilities, while its viability can be assessed only from the quality of the eventual analytical results (see Section 7.5).

A second aspect of the concept of spectral interference is the enhancement of the background signal in the spectral window of an analysis line to such an extent that it becomes difficult or impossible to use this line for concentration determinations in the range required. Background enhancements with respect to, say, the level of the continuum in an ICP fed with pure water, are attributable to one or more of the spectral components (1)–(6) stated previously. These background enhancements lower the SBR and positively or negatively affect the background RSD, but the result always is that the limits of detection and the limits of determination (cf. Sections 4.2.2 and 7.7.7) are worsened while the accuracy is endangered.

Spectral interferences in this sense may severely limit the conditions under which an analysis line can be used. Consequently the problem of line selection in emission spectroscopy in general and in ICP–AES in particular is closely linked with the problem of spectral interferences. This comprehensive and important subject is treated in Chapter 7.

In summary, two aspects of "spectral interference" must be distinguished:

1. The variability of the background signal as a consequence of its dependence on the sample composition and the operating parameters; this variability primarily limits the accuracy with which net line signals can be determined.
2. Enhancements of the background signal essentially exceeding the limits set by random fluctuations; these enhancements raise the detection limits and the concentrations that can be reliably measured; in addition, it may become exceedingly more difficult to determine the background accurately, especially if the background enhancement is due to line overlap.

For illustration, we conclude this outline with an elementary numerical calculation of the magnitude of the error in the net line signal resulting from an erroneous background reading. Starting from the definition of the net line signal:

$$x_A = x_{A+B} - x_B \qquad (4.62)$$

and expressing the absolute systematic error dx_A in the corresponding errors dx_{A+B} and dx_B in gross line and background signal, respectively, by

$$dx_A = dx_{A+B} - dx_B \qquad (4.63)$$

one derives the expression

$$dx_A/x_A = (1 + 1/\text{SBR})\, dx_{A+B}/x_{A+B} - (1/\text{SBR})\, dx_B/x_B \qquad (4.64)$$

which is the equivalent of Eq. (4.49), the only difference being the linear instead of quadratic error combination required for random errors.

Rewriting Eq. (4.3) for the detection limit as

$$\text{SBR} = 0.01\, z \cdot (\text{RSD})_B \cdot c/c_L \qquad (4.65)$$

we have a set of equations, (4.64) and (4.65), describing the dependence of dx_A/x_A on c/c_L.

Figure 4.33 shows plots of this relationship for $z = 2\sqrt{2}$, $(\text{RSD})_B = 1\%$, $dx_{A+B}/x_{A+B} = -2\%$ and dx_B/x_B ranging from 1 to 20%. Obviously an error in the background correction as small as 2% propagates as a 30% error in the net signal at a concentration of 10 times the detection limit. An enhancement of the background by, say a factor of 10, does not affect the diagram in Fig. 4.33 as such. Three consequences must be borne in mind, however.

First, the concentration scale associated with the c/c_L scale shifts over one order of magnitude. Second, it is possible that a curve with a higher value of dx_B/x_B applies. The latter will happen in particular if the background enhancement is due to line overlap. Third, the concentration scale associated with the c/c_L scale in the figure for the random error (Fig. 4.16) also shifts upward by an order of magnitude as a result of the tenfold increase of the SBR. The shift may be even greater if also the background RSD is increased.

4.7. OPTIMIZATION OF THE ICP OPERATING CONDITIONS

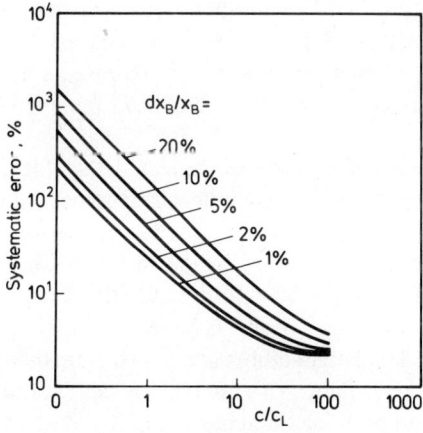

Figure 4.33. Systematic error in the net line signal resulting from a systematic error (dx_B/x_B) in the background measurement. The diagram shows the error in dependence on the ratio of the concentration to the detection limit for various values of dx_B/x_B.

To conclude, spectral lines of concomitants (including Ar and H) may contribute intensities exceeding the continuous background by orders of magnitude. If such lines appear in the spectral windows of analysis lines, the latter may become entirely unusable. The above elementary analysis of the problem shows the exact reasons for this.

The spectral interference problems encountered in ICP–AES do not essentially differ from those in emission spectroscopic methods using conventional sources such as arcs or sparks. However, spectral interferences tend to manifest themselves more severely in consequence of the intrinsically high detection power of ICP–AES, which, unfortunately applies to both the analysis lines and interfering lines alike. The ever growing knowledge on ICP spectra, the continuing improvement of experimental facilities, and the further development of software for instrument control and data acquisition and processing foster the expectation that ''living with spectral interferences'' will gradually become more comfortable. However, the physical characteristics of the spectra ultimately impose fundamental limits that cannot be bypassed.

4.7. OPTIMIZATION OF THE ICP OPERATING CONDITIONS

4.7.1. Definition of the Problem

Discussion here will be confined to the optimization of the analytical performance of existing ICP systems and not cover the optimum design of an ICP system. The latter aspect includes both performance and costs and belongs in the realm of instrument manufacturers.

Analytical performance encompasses a number of characteristics (Section 3.3) of which in particular detection power, accuracy, or precision may be optimized, either separately or in conjunction with each other. This is achieved by adjusting the freely selectable parameters of the ICP system. One may do so for each analyte separately ("single-element optimization") or find an acceptable "average" for a group of analytes ("compromise conditions for simultaneous multielement analysis").

Various investigators [58, 60, 61, 101, 206] have documented that for a given ICP generator, coupling configuration, and plasma tube arrangement the critical parameters for analysis are the power input to the discharge, the flow rates of the gases and the sample, and the observation region in the plasma. Dickinson and Fassel [58] and Boumans and de Boer [60] explicitly demonstrated that single-element optimization of detection limits is feasible by varying one or more of these critical parameters. The latter authors concluded from their experiments that at fixed power input two values of observation height and carrier gas flow would provide a good compromise for achieving high detection power for most elements. Fassel's group [101, 102] proceeded one step further and recommended a single set of compromise values of power, carrier gas flow, and observation height for all elements normally accessible by the ICP method. This recommendation was supported by the excellent detection limits they reached under such conditions and the experience that improvements larger than a factor of five could be hardly achieved by optimizing any of the above parameters. Boumans and de Boer [60, 103] independently arrived at the same conclusion for their experimental facilities, but their results were published slightly later.

It had also been established in the early works [58, 59, 207, 208] that a remarkably low degree of interelement interference could be observed under the operating conditions that yielded low detection limits. This point, too, was confirmed and further elaborated by Larson et al. [209], Boumans et al. [61, 63, 103, 210], Abdallah et al. [211], and others.

It was also shown by Fassel et al. [212] that organic samples could be analyzed under compromise conditions though with a parameter setting differing from that for aqueous solutions.

In retrospect we may state that for argon ICPs the use of a few sets (inorganics, organics) of compromise operating conditions for simultaneous multielement analysis has been corroborated by the following arguments:

1. Excellent detection limits, well equilibrated among the various elements, can be reached in both pure water and real samples of either inorganic or organic nature.
2. Multiplicative interference effects are at an acceptably low level under the same conditions where the excellent detection power is attained.
3. Compromise conditions ensuring that (1) and (2) can be found for different experimental facilities; only the recommended numerical values of the parameters may differ to some extent.

4.7. OPTIMIZATION OF THE ICP OPERATING CONDITIONS

4. The relative sensitivities, SBRs and the magnitudes of multiplicative interferences found under compromise conditions for different experimental facilities do not differ widely.
5. Also, for different experimental facilities, the dependencies of sensitivities, SBRs, and magnitudes of interferences on the ICP parameters generally show a high degree of congruency.

Statements (1)–(5) do not imply that refined optimizations for particular sample types would be entirely redundant. They do mean, however, that for argon ICPs (1) departing from what might be called "generally accepted" or "conventional" compromise conditions will hardly ever lead to a vast improvement in a detection limit, the alkalis being notable exceptions; and (2) the conventional compromise conditions yield a remarkably low level of multiplicative interferences compared with other sources, such as dc arcs, DCPs, CMPs, and MIPs; this rather surprising feature might have helped in creating the myth that the interferences in an ICP are by definition on an acceptably low level under all conditions and for all sample types. Evidently what seemed acceptable in the initial years need not be acceptable any longer in the present stage of development of the method; therefore, minor adaptations of the ICP operating conditions may be necessary and profitable to better meet the requirements upon a particular analysis. Finally, compromise conditions emphasize detection power, but take interferences, in particular those from alkalis, into account; however, an optimization may put more emphasis on the reduction of interferences of whatever type when achieving the best possible detection power is not the main problem.

Whatever approach is made in the optimization of a conventional argon ICP, one will always arrive in the vicinity of the conventional compromise conditions. This is so because these conditions have a firm basis in empirical relationships between analytical characteristics and the ICP parameters, while they find additional support in the results of fundamental characterizations of the ICP source. Therefore, one should not expect miracles from elaborate optimizations however useful they may be for defining the best conditions for a particular sample type. Before going further into this argument in Section 4.7.9, we shall first consider some notions and relationships of classical ICP optimization.

Unfortunately the data are scattered in the literature, while, in addition, the precise numerical values appearing in empirical relationships may be no longer valid in consequence of further developments and improvements of the components of ICP systems. Then, only those workers who have been directly involved in the early investigations will see through apparent inconsistencies and recognize the essential trends, whereas novices are left with confusion chiefly because numerical values have altered.

For these reasons we have attempted to build up in the subsequent sections a uniform picture of parameter optimization for argon ICPs such that it can reveal essential trends and may help with understanding the arguments upon which the conventional compromise conditions are founded. At the same time the discussion provides some guidelines that may be followed in those situations where the adequacy of a compromise setting must be checked or minor adaptations must be made to find a new optimum after some modification in the system.

4.7.2. Norm Temperature: A Useful Concept for Classifying Spectral Lines and Understanding Their Behavior

To understand empirical relationships between spectral line intensities and the ICP parameters the following points are of crucial importance.

1. Species from which spectral lines originate are distinguished as neutral atoms (denoted by the symbol "I"), singly charged ions ("II"), doubly charged ions ("III"), and so on. Lines then classify as atomic lines (e.g., Mg I 285.212 nm) and various types of ionic lines (e.g., Mg II 279.806 nm, La III 237.94 nm). Lines of neutral atoms and singly charged ions, with an emphasis on the latter, are of prime importance in an argon ICP.

2. The behavior of a spectral line is dictated by the excitation potential of the line and the ionization potential of the relevant species. The definitions of these potentials have been elucidated in Fig. 1.2.

3. If the intensity of a spectral line is considered as a function of temperature, the intensity will be seen to pass through a maximum at a certain temperature, called *"norm temperature."* This applies to the situation where a chemical element is present with a fixed atomic concentration in a homogeneous plasma in local thermal equilibrium (LTE), while the pressure or the volume of the plasma and its elemental composition are all fixed.

4. Using the concept of norm temperature one can make a distinction between "soft" and "hard" lines depending on their position on the norm temperature scale. This classification of spectral lines into "soft" and "hard" lines [213], based on norm temperatures, links up to a certain extent with the behavior of spectral lines in argon ICPs and, therefore, greatly facilitates the discussion. Therefore, we shall consider this point in some detail.

The term "norm temperature" is used after the German "Normtemperatur" (= "standard" or "type" temperature) and was introduced by Larenz [214]. The concept has been elaborated by Krempl et al. [215–219] and Boumans [196, 220] for high-temperature plasmas in general and spark [215–219] and arc [221] in particular. Calculations of norm temperatures for ICPs have been made by Greenfield [222], Boumans and de Boer [62], and Blades and Horlick [223].

4.7. OPTIMIZATION OF THE ICP OPERATING CONDITIONS

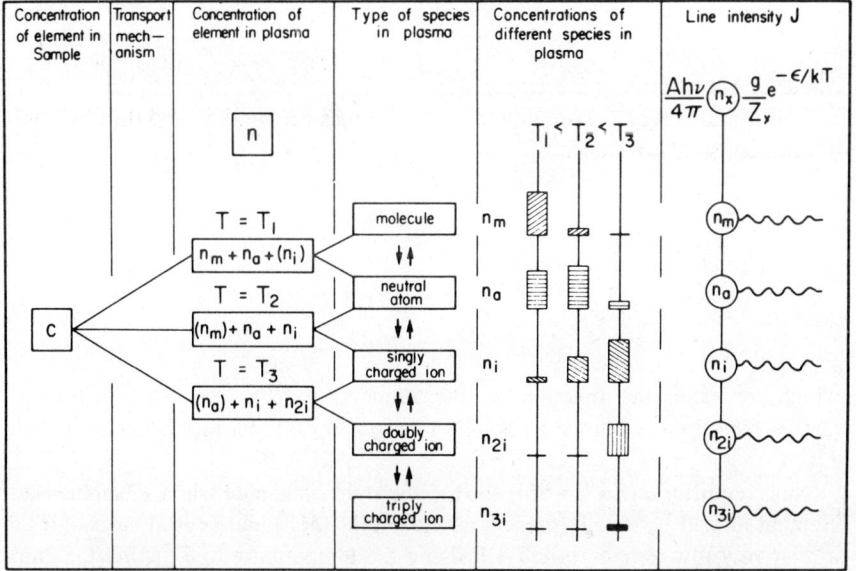

Figure 4.34. Interrelationships between the analyte concentration (c) in the sample, the corresponding concentration or number density (n) in the plasma, and the spectral-line intensities emitted at different temperatures by the various analyte species: molecules (n_m), neutral atoms (n_a), singly charged ions (n_i), doubly charged ions (n_{2i}), and triply charged ions (n_{3i}). (From Boumans [196].)

We elucidate the concept of norm temperature using Fig. 4.34. A chemical element M present with a concentration c (e.g., expressed in ng/mL) in a sample is transferred to the plasma state where it is assumed to be present with a concentration or number density n (cm^{-3}). This density represents the sum of the number densities of various types of free particles of M, namely the densities of the molecules (n_m), atoms (n_a), and ions of various charge (n_i, n_{2i}, n_{3i}, . . .). The transfer of M from the original solid or liquid state to the plasma state is dictated by transport and volatilization mechanisms, which, in the present context, are irrelevant.

The number densities of the molecules (MY), atoms (M), and ions (M^+, M^{2+}, M^{3+}, . . .) are linked by dissociation and ionization equilibria of the type

$$MY \leftrightharpoons M + Y \tag{4.66a}$$

$$M \leftrightharpoons M^+ + e^- \tag{4.66b}$$

$$M^+ \leftrightharpoons M^{2+} + e^- \tag{4.66c}$$

$$\vdots \qquad \vdots$$

$$M^{r+} \leftrightharpoons M^{(r+1)+} + e^- \tag{4.66d}$$

where e^- represents an electron.

Equilibria of type (1) are governed by the mass action law and thus by equilibrium constants of the form

$$K_{MY} = \frac{n_M n_Y}{n_{MY}} \tag{4.67a}$$

$$K_M = \frac{n_{M^+} n_{e^-}}{n_M} \tag{4.67b}$$

which are increasing functions of the temperature (T). The function for ionization equilibria is known as Saha's relationship (cf. Part 2, Chapters 10 and 11).

Since equilibria (Eq. (4.66)) shift to the right-hand side when T is increased, element M will be chiefly present as molecules (MY) and neutral atoms (M) at a relatively low temperature T_1. Raising the temperature to T_2 shifts the equilibria so that the neutral atoms and singly charged ions (M^+) will become the dominating species of M. At a still higher temperature T_3, singly and doubly (M^{2+}) charged ions will be predominant. In Fig. 4.34 these changes are symbolized by the variation in the areas representing the number densities of the species (column V). Figure 4.35 illustrates the situation for some real elements, as computed for a particular plasma model [220].

The intensity of a spectral line depends on the number density (n_X) of the relevant species and on the relative population of the upper level of the transition, as described by the Boltzmann equation, shown in the head of column VI of Fig. 4.34. Here n_X represents *one* of the possible number densities n_m, n_a, n_i, n_{2i} . . . . The relationship between spectral line intensity and temperature thus results from the superposition of curves of the type shown in Fig. 4.35 and the Boltzmann factor. This modifies the shapes of the curves and shifts the maxima to higher temperatures (Fig. 4.36). The temperatures at which these maxima occur are the norm temperatures (T_N) of the lines. Generally, the higher the ionization potential ($V_{i,j}$) of the species (j) and the higher the excitation potential ($V_{q,j}$) of the line, the higher the norm temperature.[7]

---

[7]From estimates of norm temperatures in the ICP, Boumans and Lux-Steiner [224] derived the relationship $T_N = 750 \, (V_{i,j} + V_{q,j}) - 940$, where the temperature is expressed in K and the potentials are expressed in V. This relationship applies to an electron number density of 10^{16} cm^{-3}, which value was thought to be a "suprathermal" value applicable to the ICP. However, values of 10^{14}–10^{15} cm^{-3} coupled to (LTE) plasma temperatures of 6500–7500 K in the "analytical zone" appear to be more appropriate [225]. This chiefly affects the norm temperatures of the atomic lines stated in [62]; they should be some 1000 K lower.

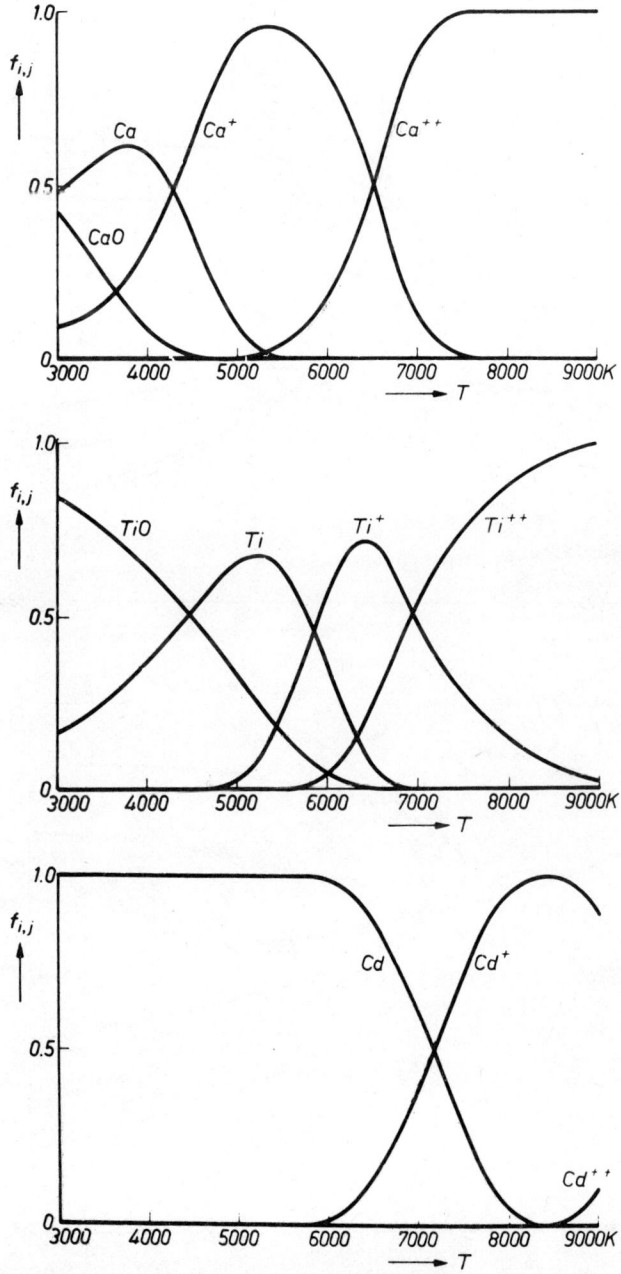

Figure 4.35. The relative number density or fraction $f_{i,j}$ of diatomic molecules, free atoms, singly charged ions, and doubly charged ions of Ca, Ti, and Cd as a function of the absolute temperature (T). It has been assumed that the elements are present as impurities in air at atmospheric pressure. (Reproduced with permission from *Philips Technical Review* [220].)

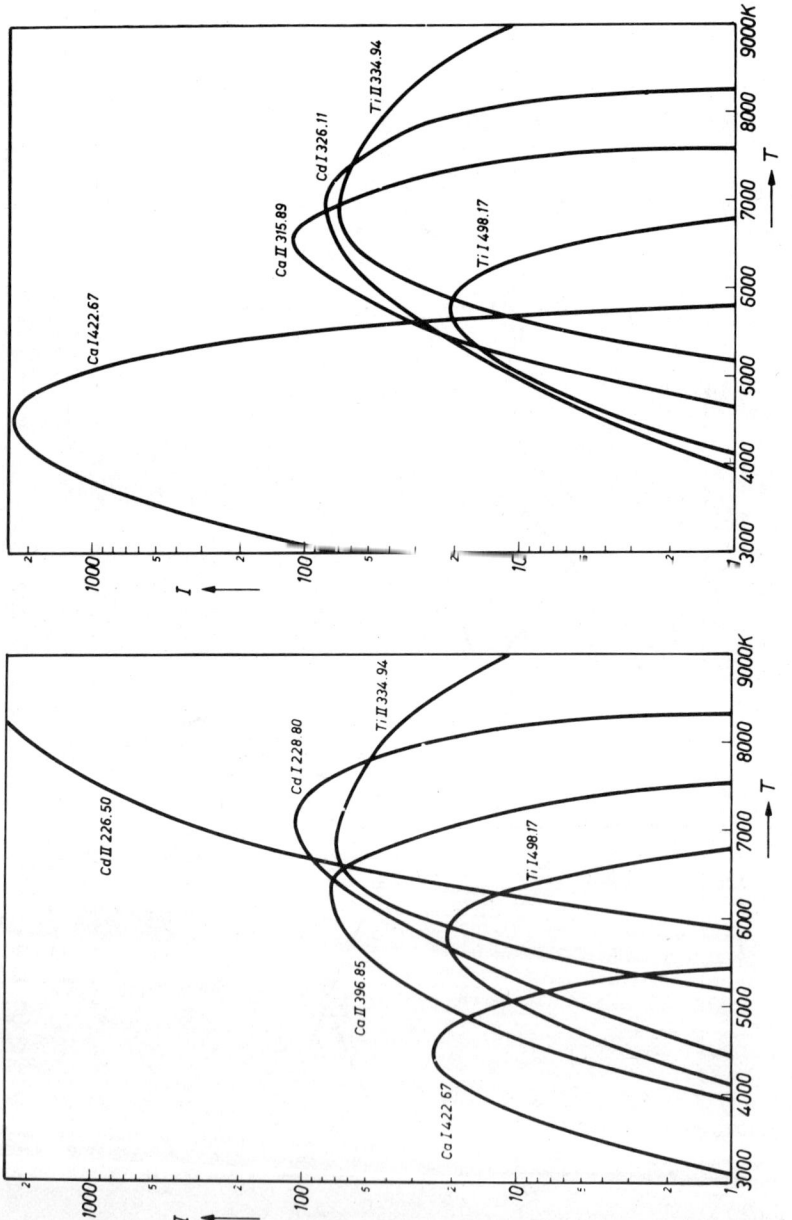

Figure 4.36. The intensity I (in arbitrary units) of some Ca, Ti, and Cd lines as a function of the temperature of an excitation source. This diagram is obtained by the superposition of the Boltzmann distributions of the upper levels of the lines on the curves for the fractions of the respective species shown in Fig. 4.35. It has been assumed that the three elements are present in equal amounts. (Reproduced with permission from *Philips Technical Review* [220].)

4.7. OPTIMIZATION OF THE ICP OPERATING CONDITIONS

New insights into the excitation mechanism of the ICP may have pushed the consideration of norm temperatures to the background; however, the concept has formed the basis of the distinction between "soft" and "hard" lines, which classification is still valid and useful. The distinction was first made by Boumans [213] on the basis of estimates of norm temperatures of lines and their intensity behavior in an argon ICP. The incentive was to facilitate the discussion of this behavior. Actually the distinction between "soft" and "hard" lines was not thought of as a sharp transition between two groups of lines but as a continuous ranking on the norm temperature scale. The term was derived from classical emission spectroscopy, which distinguishes between "soft" and "hard" sparks depending on whether the spectrum shows predominantly atomic or ionic lines [196].

Originally a norm temperature of about 9000 K was considered as the approximate boundary between hard and soft lines. Thus atomic lines of elements with a low to medium ionization potential (≤ 8 V) and ionic lines of elements with a low second ionization potential (e.g., Ba II and La II lines) were denoted as "soft" and all other atomic and ionic lines as "hard."

A spatial study of the ICP by Blades and Horlick [223] has confirmed that the distinction between soft and hard lines is most useful, also in the description of the spatial behavior of spectral lines. Blades and Horlick [223] as well as Kawaguchi et al. [226] concluded from their spatial studies that there exist two vertically distinct zones in the channel of an argon ICP. The first, denoted "thermal zone," is located at a relatively short distance from the coil; in this zone, soft lines have their emission maxima and show a behavior that can be described in terms of norm temperatures using classical LTE considerations based on excitation temperatures. The second region, denoted "nonthermal zone," is located higher up in the plasma; here all hard lines peak at virtually the same vertical position, while the classical concept of norm temperature loses much of its significance, that is, the hard lines behave more similarly than the differences in their norm temperatures would suggest. Blades and Horlick [223] also found that lines such as Ba II and La II lines show the spatial behavior of hard lines.[8]

In the subsequent sections we shall discuss ICP optimization in light of the distinction between soft and hard lines. In principle, one may consider as soft

---

[8]New insights into excitation mechanisms in the ICP (Part 2, Chapter 11) will require an update of the above considerations. The basis of the new vision on excitation mechanisms is the use of the electron temperature as the governing temperature and the coupling of this electron temperature with the electron concentration using the assumption of LTE with an allowance for relatively small departures ("partial LTE," p-LTE) [225]. This leads to a physically consistent description of ionization but takes the description of the populations of individual levels and, therefore, the interpretation of excitation temperatures to the realm of "departures" and "exceptions." Further work is needed to arrive at a complete and consistent picture capable of explaining *both* new and old experimental findings.

lines all atomic lines of elements with a first ionization potential below about 8 V (cf. Table 1.1); hard lines are atomic lines of elements with a first ionization potential above 8 V and all ionic lines.

4.7.3. Parameters To Be Optimized

For a defined ICP system six parameters have to be optimized:

1. Power input, P (kW)
2. Carrier gas flow, F_c (L/min)
3. Observation height, h (mm)
4. Outer gas flow, F_o (L/min)
5. Intermediate gas flow, F_i (L/min)
6. Sample solution feed rate, F_s (mL/min)

To conventional argon ICPs the following applies. Given the ICP configuration (generator, coil, torch), the outer gas flow has a lower limit. Too low a flow may result in overheating of the outer tube and consequent destruction of the torch, or in extinction of the plasma. For economic reasons one will generally work not too far above the lower limit for stable plasma operation, the more so since the effect of the outer gas flow on analytical performance is small. Only in the case of organic solvents does optimization of this flow play some part, and this is related to a higher power than needed.

The intermediate gas may be omitted with aqueous solutions containing inorganic matter only, but is indispensable to prevent carbon deposits in the torch when organic materials are analyzed [212].

The carrier gas flow is not only a very critical parameter of the ICP but also a nebulizer parameter that governs the amount of aerosol carried to the plasma. This holds a fortiori for pneumatic nebulizers, as is shown in Fig. 4.37, but also, to a smaller or larger extent, for ultrasonic nebulizers, as is shown in [61, Fig. 3]. Thus changes in spectral line intensities with the carrier gas flow rate will reflect the changes in both the aerosol flow and the plasma characteristics up to the point where the nebulizer reaches its saturation level. Frequently this point is at least closely approached when the carrier gas flow is optimum for the ICP so that the eventual optimum is dominated by the plasma rather than the nebulizer.

Observation height is usually defined as the distance from the center of the observation zone to the top of the coil. In the following sections we shall follow this definition, although a proposal for defining the observation height with respect to a definite emission zone is more adequate and rational [227, 228].

4.7. OPTIMIZATION OF THE ICP OPERATING CONDITIONS

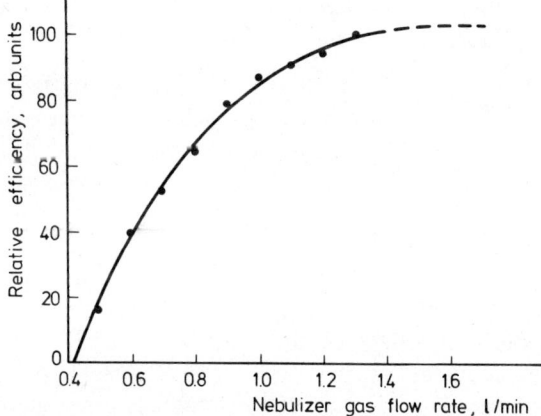

Figure 4.37. Dependence of the relative efficiency of a cross-flow pneumatic nebulizer including the spray chamber on the nebulizer gas flow rate as measured with a constant solution introduction rate controlled by a peristaltic pump.

However, since in the present context matters are put into a historical perspective, much confusion can be avoided by using the conventional definition. The observation region itself commonly is a 4–6-mm high zone containing the ICP axis as defined by projecting the ICP onto the spectrometer slit.

The importance of controlling the sample feed rate depends on the nebulizer and given the latter, on the physical properties (viscosity, surface tension, density) and chemical nature (inorganic, organic) of the sample. Control may be required to reduce nebulizer interferences in the case of liquids of varying physical properties [229] or to prevent extinguishing of the plasma in the case of several types of organic solvents (see Section 4.7.8).

Finally, when assessing optimization studies in the literature one should be aware that interelement interferences ("matrix effects") depend on the rate at which the aerosol is introduced into the plasma and on the properties of the aerosol, in particular the particle size distribution. Therefore, given the concentration of the major elements in the solution, the interference effects reported for ultrasonic nebulizers tend to be considerably larger and may be otherwise different from those found with pneumatic nebulizers. Thus one will see in the literature that interference effects of salts (e.g., alkali chlorides) could be made negligibly small up to 1-mg/mL salt concentration for ultrasonic nebulizers, whereas concentrations of up to five or ten times that magnitude still yield negligible interferences with pneumatic nebulizers. This is explained by the far lower efficiency of pneumatic nebulizers [100, 210].

4.7.4. Trends Observed in Argon ICPs and Compromise Conditions for Inorganic Samples in Aqueous Solutions

Figure 4.38 shows the dependence of net line signals on power at constant observation height (15 mm) and carrier gas flow (1.3 L/min), the latter being close to the eventual optimum values for the pertinent equipment. Notice the correspondence between the slopes of the curves and the "hardness" of the lines, as expressed by the norm temperatures stated in the figure caption. The intensities of the hard atomic and ionic lines (Cd I, Zn I, Mn II, Cd II, La III) initially increase sharply with power and do not reach their maxima in the present range. The soft atomic lines (Li I, Ba I, La I, Mn I) have a maximum at the lower end of the range, whereas the ionic lines of Ba II and La II approach the maximum in the middle of the range.

Figure 4.38 suggests that the characteristic trends in an argon ICP may be well followed by observing only a few spectral lines, for example, one or two representatives of each of the soft and the hard lines. This suggestion is supported by the observed dependence of net line signals on carrier gas flow shown in Fig. 4.39 for constant power (1.1 kW) and observation height (15 mm). As in Fig. 4.38, the data are plotted on a log scale so that the slopes of the curves indicate the relative changes in the net line signals. The curves have been ar ranged in order of increasing hardness of the lines from top to bottom. The intensities of outspoken soft lines (Li I, Ba I, Mn I) increase with carrier gas flow over the region shown and those of outspoken hard lines (Cd I, Zn I, V II, Mn II) decrease, while the group in between shows an intermediate behavior.

If the carrier gas flow is varied over a larger trajectory, all spectral lines are seen to pass through a maximum; soft lines of alkalis or alkaline earths even show two maxima. A few examples dating from the author's initial ICP work are shown in Fig. 4.40. High carrier gas flow rates not only decrease the temperature in the observation zone, but also markedly modify the structure and appearance of the ICP and lead to excitation conditions that were recognized in the early publications [61, 101–103] as unsuited for analytical purposes. It took some time before this notion spread generally, which for some part was due to the use of a pneumatic nebulizer that did not operate at flow rates below 1.5 L/min [230]. Although curves such as shown in Fig. 4.40 cover a small trajectory and seem to tell only part of the story, they do contain the essential information for analytical optimization, as will be illustrated below.

The dependence of *net* line signals on the ICP parameters is adequate for illustrating the relationship between the behavior of the various spectral lines and their spectroscopic properties, but net line signals alone do not constitute the only and prime characteristic for ICP optimization. For achieving high detection power, SBRs should be maximized and the RSD of the background signal should be minimized (cf. Section 4.1). Therefore, Figs. 4.41 and 4.42

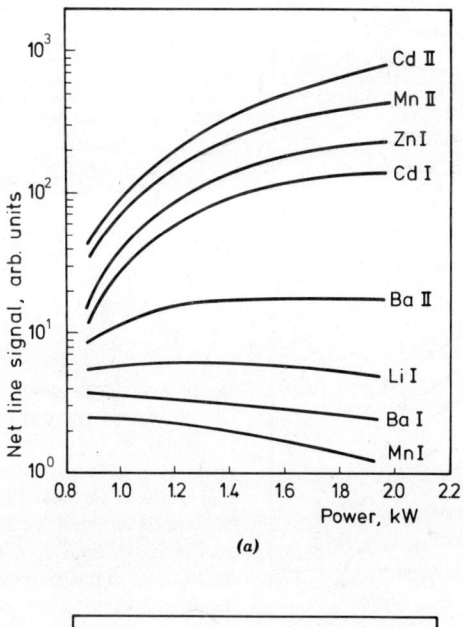

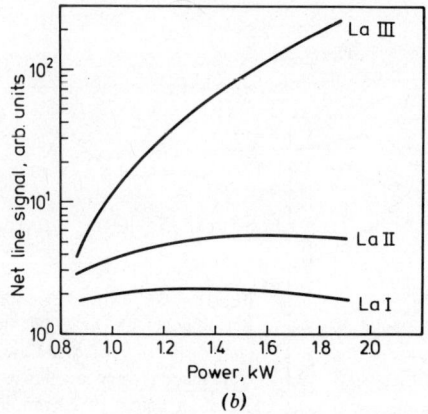

Figure 4.38. Dependence of the net line signals of various spectral lines on the power input to the ICP at constant carrier gas flow (1.0 mL/min) and observation height (15 mm). Spectral lines and estimated norm temperatures (in parentheses): Cd II 226.5 nm (15,000 K), Mn II 257.6 nm (14,000 K), Zn I 213.9 nm (11,000 K), Cd I 228.8 nm (11,000 K), Ba II 455.4 nm (8,000 K), Li I 670.8 nm (4,000 K), Ba I 553.5 nm (3,500 K), Mn I 403.1 nm (5,000 K), La III 237.9 nm (17,000 K), La II 408.7 nm (8,500 K), and La I 531.2 nm (4,000 K). (Reproduced with permission from *Spectrochimica Acta, Part B* [62].)

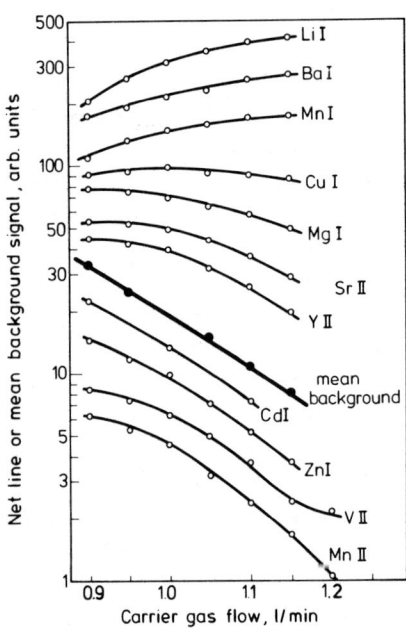

Figure 4.39. Dependence of the net line signals of various spectral lines and the mean background on the carrier gas flow at constant power (1.1 kW) and observation height (15 mm). The spectral lines of Ba, Li, and Mn were the same as in Fig. 4.38. The other lines were Mg I 285.2, Cu I 327.4, V II 309.3, Y II 371.0, and Sr II 407.8 nm. (Reproduced with permission from *Spectrochimica Acta, Part B* [224].)

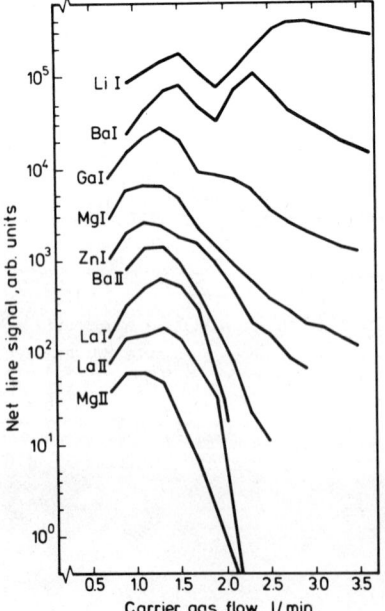

Figure 4.40. Typical examples of the trend of net line signals with carrier gas flow if observed over a large trajectory of this flow. These observations were made at very low power (~0.4 kW) at an observation height of 20 mm. The results have not been published but are covered by the data provided in [61]. The curves have been arranged so that the change in the shapes with the spectroscopic properties of the spectral lines is clearly revealed. These shapes are correlated with the norm temperatures. Note the peculiar position of the curve for the La I line. This peculiarity is probably related with the dissociation equilibrium LaO ⇌ La + O, the dissociation potential of LaO being 8 V. Spectral lines: Li I 670.8, Ba I 553.5, Ga I 417.2, Zn I 328.2, Ba II 455.4, La I 521.2, La II 408.7, and Mg II 280.3 nm.

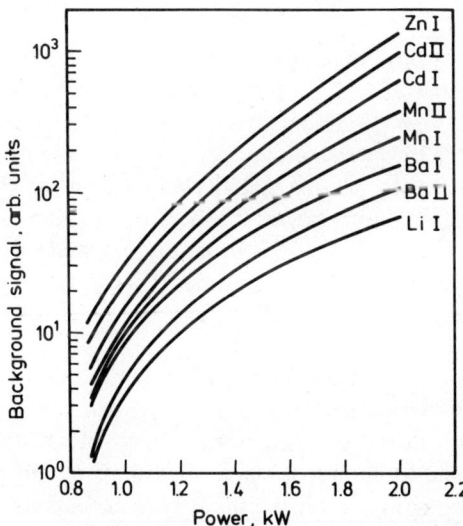

Figure 4.41. Background signal as a function of the power input to the ICP. The background was measured at the wavelength of the lines stated in the caption of Fig. 4.38. The carrier gas flow and observation height were the same as in Fig. 4.38. (Reproduced with permission from *Spectrochimica Acta, Part B* [62].)

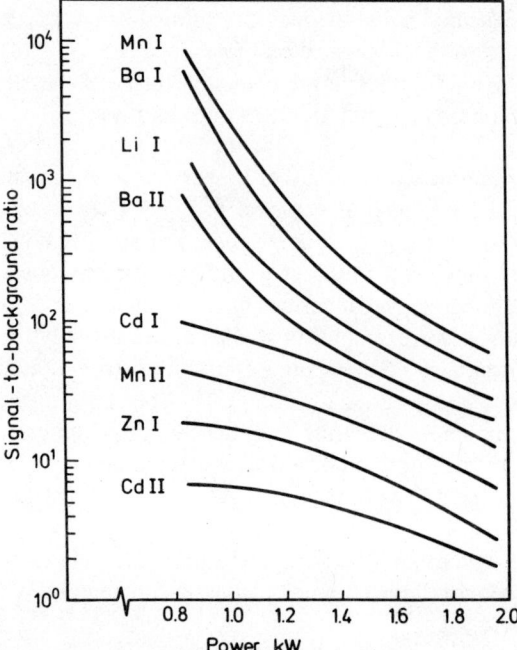

Figure 4.42. SBR as a function of the power input to the ICP as derived from Figs. 4.38 and 4.41. The SBRs refer to the following concentrations (μg/mL) in the solutions: Li I: 3.5; Ba I: 800; Ba II: 4.0; Mn I 150; Mn II 0.09; Zn I: 0.2; Cd I: 1.0; Cd II 0.1. The solutions were introduced with a pneumatic nebulizer. (Reproduced with permission from *Spectrochimca Acta, Part B* [62].)

show the dependence of the background signal and the SBR on the power for the conditions referred to in Fig. 4.38. The background signal is seen to increase more rapidly with power than the net line signals; therefore, the highest SBRs are found at the lower end of the power range. There is, however, a lower practical limit, dictated by plasma stability and shot noise considerations (cf. Section 4.1). A safe minimum power for the system under discussion is about 1.1 kW. A value of this magnitude is generally recommended for argon ICPs when aqueous solutions are dealt with. Evidently the stability requirement demands a small sacrifice from the SBRs (Fig. 4.42), and this sacrifice is the largest for soft lines.

The dependence of the SBR on carrier gas flow for the spectral lines considered in Fig. 4.39 is shown in Fig. 4.43. If SBR were the only criterion, one would opt for a high carrier gas flow to favor the SBRs of the soft lines. This would, however, lead to intolerably large matrix effects, in particular for soft lines, as is illustrated in Fig. 4.44. For the system under consideration, a carrier gas flow of 1.05 L/min offers a good compromise as to both detection limits and interferences. The trade-off with respect to the overall detection power actually is less serious than would appear from the figures, since at high carrier gas flow rates essentially better SBRs can be reached only for the softest lines, but these are generally not the lines that yield the highest SBRs for a chemical element. It has been shown that the majority of the elements accessible with an ICP have hard lines as their most prominent lines [65, 85, 101, 211, 232].[9] Since these lines primarily dictate the detection limits of the elements, an optimization should rationally refer to these lines and not to those of secondary importance.

Trends of net line signals with either power or carrier gas flow, while observation height is fixed, such as depicted in Figs. 4.38–4.40, appear to lead straightforwardly to a satisfactory compromise and actually do so. This should not be generalized, however, in that it would apply to any observation height. If measurements are performed at various observation heights, complex pictures may result, as shown, for example, in studies aimed at analytical optimization [61, 233, 234] or studies of the spatial emission distributions undertaken for diagnostic purposes [223, 226–228, 235–237]. The straightforwardness illustrated above chiefly sprang from the fact that the measurements were made in the close vicinity of the optimum observation height, about which the literature shows a curious consensus as to the "magic value" of about 15 mm above the coil.

Recent spatial studies of Blades and Horlick [223] and Kawaguchi et al. [226] have clarified part of the complex picture encountered in ICP optimization

---

[9]The only serious exceptions are the alkalis, but these elements require a separate treatment anyway if their detection limits should be optimized (cf. Section 4.1.7.5), and it would be irrational to let them carry too much weight in choosing compromise conditions for all other elements.

4.7. OPTIMIZATION OF THE ICP OPERATING CONDITIONS

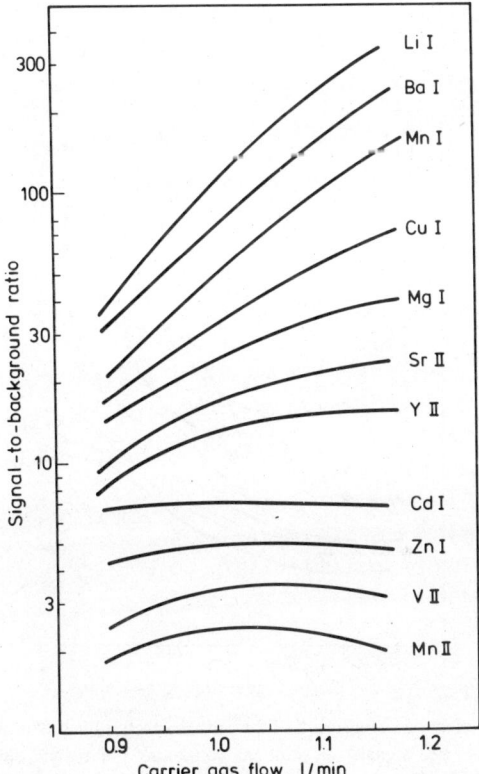

Figure 4.43. SBR as a function of carrier gas flow for the lines stated in the caption of Fig. 4.39. The curves have been arranged to fit the order of the curves of the net line signals in Fig. 4.39 and do not refer to equal concentrations of the analytes. (Reproduced with permission from *Spectrochimica Acta, Part B* [224].)

and further rationalized and supported the use of compromise conditions such as those selected by earlier workers. Only a few essentials of these spatial studies will be mentioned here.

First, as discussed earlier in this section, there exist two distinct zones in the channel of an argon ICP: a "thermal" zone relatively close to the coil and a "nonthermal" zone higher up in the plasma. The (excitation) temperature in the thermal zone increases with increasing distance to the coil, presumably because of diffusion of hot argon from the annulus region toward and into the channel region. In agreement herewith atomic lines peak in this thermal region (Fig. 4.45) and show their maxima at a greater distance from the coil the higher their norm temperatures (Fig. 4.46). A nonthermal mechanism, which has not yet been completely cleared up, governs excitation-ionization in the nonthermal

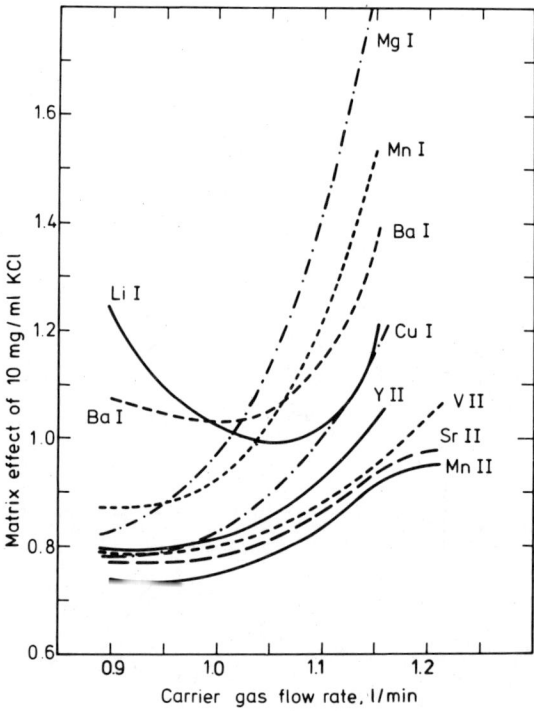

Figure 4.44. Matrix effect of 10 mg/mL KCl for various spectral lines. The matrix effect is defined as the ratio of the net lines signals for solutions containing the analyte plus the matrix and the analyte alone respectively. The wavelength of the lines are the same as those stated in the caption of Fig. 4.38. (Reproduced with permission from *Spectrochimica Acta, Part B* [231].)

region and primarily leads to excitation of ionic lines so that these lines peak in this region within fairly small spatial boundaries. The mechanism in the nonthermal region [223] appears to favor the emission of particular atomic lines (inter alia those with high excitation potentials [238]); consequently the intensity maxima of all lines that are classified as "hard" according to the norm temperature criterion are found in the nonthermal region.

If we now consider that the background intensity is high in the region close to the coil, as a result of continuous emission from the annulus, and decreases with increasing observation height, we understand that maximum SBRs for a great many *hard* lines will be found in the nonthermal region, which is located at about 15 mm above the coil when a power of about 1 kW and a carrier gas flow of about 1.0 L/min are adopted. The soft lines then have their peaks lower in the plasma in the thermal region. Here, however, the background is too high to make the SBRs of soft lines competitive with those of the hard lines in the nonthermal region.

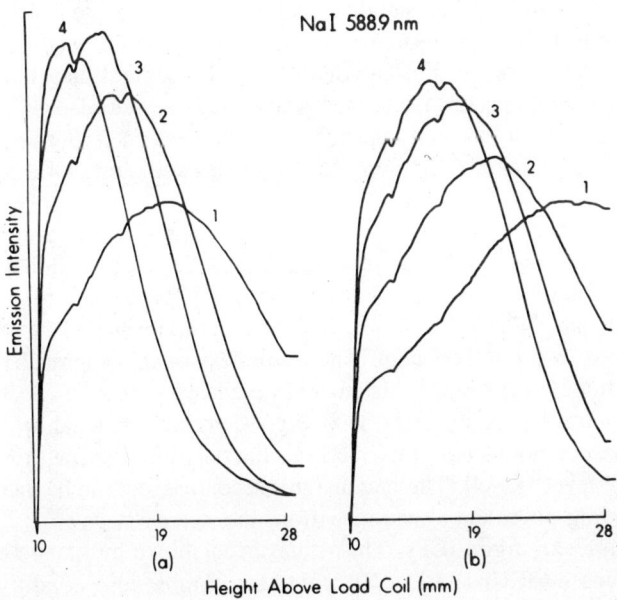

Figure 4.45. Vertical spatial profiles for Na I 588.9 nm. Aerosol flow rates: (*a*) 0.8 L/min, (*b*) 0.9 L/min. Power: *1*, 1.25 kW, *2*, 1.50 kW, *3*, 1.75 kW, and *4*, 2.0 kW [223]. (Reprinted with permission from M. W. Blades and G. Horlick, "The Vertical Spatial Characteristics of Analyte Emission in the Inductively Coupled Plasma," *Spectrochim. Acta* **36B**, 867 (1981). Copyright (1981), Pergamon Journals, Oxford.)

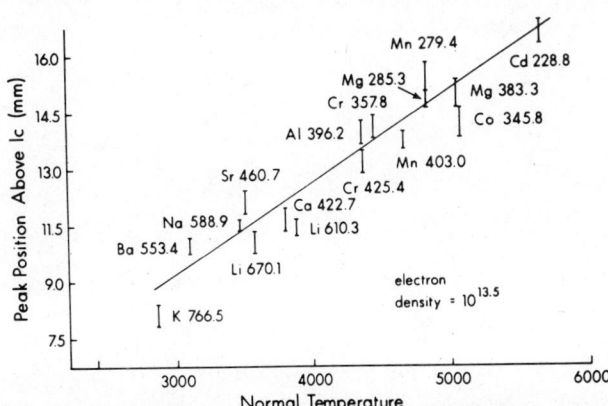

Figure 4.46. Position of vertical profile peak as a function of the norm temperature (K) [223]. (Reprinted with permission from M. W. Blades and G. Horlick, "The Vertical Spatial Characteristics of Analyte Emission in the Inductively Coupled Plasma," *Spectrochim. Acta* **36B**, 876 (1981). Copyright (1981), Pergamon Journals, Oxford.)

Accordingly we can understand (1) that optimization of the SBRs in an argon ICP always leads to a well-defined optimum observation height, and (2) the majority of the prominent lines are hard lines. These hard lines not only show their intensity maxima in a confined spatial region, but also behave closely similarly with respect to their dependence on power and carrier gas flow, as is illustrated by the spatial maps reported by Blades and Horlick [223], of which Figs. 4.47 and 4.48 are examples.

This diagnosis, which is now finding a firm basis in the results of spatial studies, has been more or less implicit in the guidelines followed in optimization studies. In this light, Boumans and Lux-Steiner [224] argued for the use of a single hard line, Mn II 257.6 nm, if a new optimum has to be sought after modifications in the ICP system. The usefulness of this approach was shown by Moore et al. [140] to apply also to a nitrogen–argon ICP (cf. Section 4.7.9).

As was noted in conjunction with Fig. 4.44, it is rational when defining compromise conditions not to use SBRs as the only criterion, but to include the interference level as well. The fact that this second aspect can be included without demanding an essential sacrifice from the detection power is one of the unique features of argon ICPs. Thus it has been shown by Larson et al. [209] and Boumans et al. [61, 62, 210, 231] that compromise conditions can be found so that (a) the detection limits among the elements are well equilibrated, and (2) at the same time a good balance between the overall detection power and interferences is achieved. In this connection interferences from alkalis are

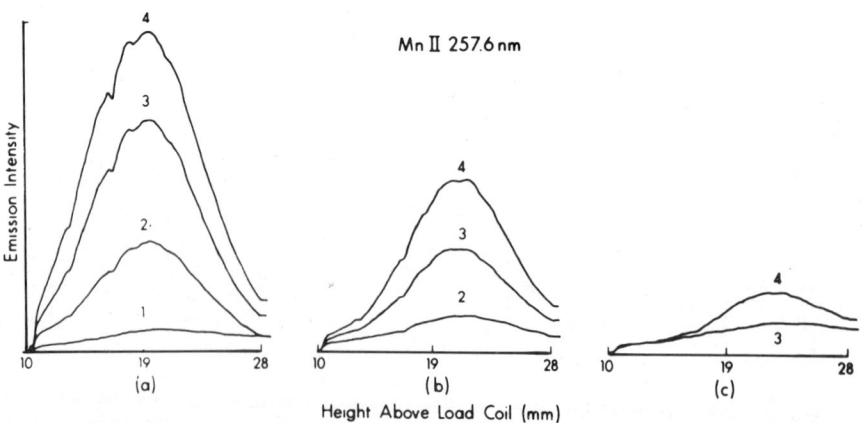

Figure 4.47. Vertical spatial profiles for Cd I 228.8 nm. Aerosol flow rates; (a) 0.9 L/min, (b) 1.0 L/min, (c) 1.1 L/min. Power as in Fig. 4.45 [223]. (Reprinted with permission from M. W. Blades and G. Horlick, "The Vertical Spatial Characteristics of Analyte Emission in the Inductively Coupled Plasma," *Spectrochim. Acta* **36B**, 866 (1981). Copyright (1981), Pergamon Journals, Oxford.)

4.7. OPTIMIZATION OF THE ICP OPERATING CONDITIONS

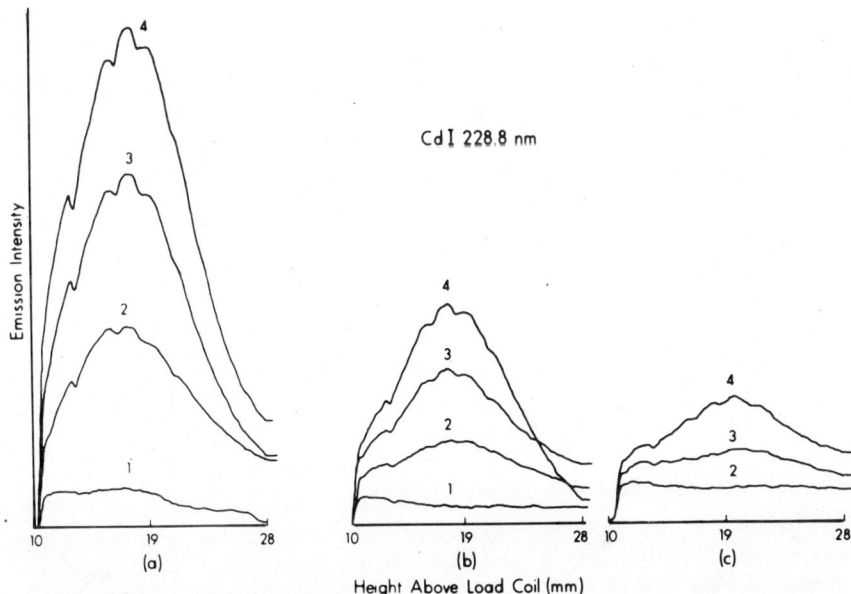

Figure 4.48. Vertical spatial profiles for Mn II 257.6 nm. Aerosol flow rates and power as in Fig. 4.47 [223]. (Reprinted with permission from M. W. Blades and G. Horlick, "The Vertical Spatial Characteristics of Analyte Emission in the Inductively Coupled Plasma," *Spectrochim. Acta* **36B**, 866 (1981). Copyright (1981), Pergamon Journals, Oxford.)

stressed because these interferences typically occur in the classical dc carbon arc [221], which has an excitation temperature of the same magnitude (6000 K) as those found in the analytical zone of argon ICPs. Solute vaporization interferences, such as the Ca–PO$_4$ and Ca–Al interferences typically found in flames [197], were recognized in early work [58, 59, 207, 208] to be exceptionally low in ICPs, which was ascribed to the high temperature, long residence time, and inert gas surroundings. The Ca–PO$_4$ and Ca–Al interferences have been extensively studied by Larson et al. [209], and it was found that these interferences are minimized under essentially the same conditions that yield minimum interference from alkalis. Therefore, solute vaporization interferences will generally not require special attention in ICP optimization. If, however, conditions far from the analytically optimum working point are explored, for example, for fundamental interference studies [239], this type of interference will manifest.

Figure 4.49 shows the interference of NaCl on net line signals such as observed by Larson et al. [209]. At the optimum carrier gas flow (1.0 L/min) for this system the effect of NaCl on the hard lines is a suppression, which varies little with observation height, whereas the effect on soft lines (Ca I, Cr I) changes from a suppression to an enhancement if the observation zone moves upwards.

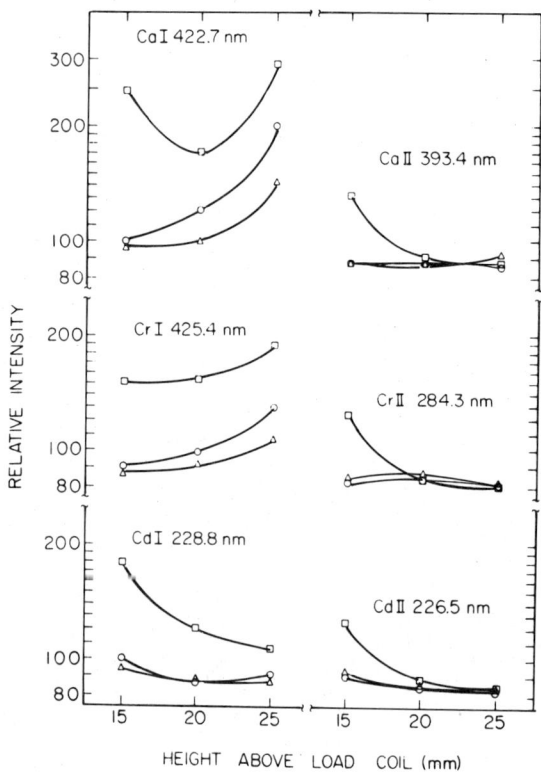

Figure 4.49. Matrix effect of Na (6.9 mg/mL) on the net line signals of various spectral lines. The net line signal for a given species in the absence of Na has been normalized to 100 arbitrary units for each height, power (P), and carrier gas flow rate (F_c).

○ $P = 1025$ W, $F_c = 1.0$ L/min;
△ $P = 1250$ W, $F_c = 1.0$ L/min;
□ $P = 1025$ W, $F_c = 1.0$ L/min [209].

(Reprinted with permission from G. F. Larson, V. A. Fassel, R. H. Scott, and R. N. Kniseley, Inductively Coupled Plasma-Optical Emission Analytical Spectrometry. A Study of Some Interelement Effects, *Anal. Chem.* **47**, 242 (1975), American Chemical Society, Washington, D.C.)

A minimum interference effect was found at 20 mm in this ICP and it was reported that this 20 mm height yielded at the same time excellent detection limits for those facilities. Figure 4.49 also shows that too high a carrier gas flow (1.3 L/min for those facilities) led to a substantially larger interference effect, in particular for soft lines (cf. Fig. 4.44, in which 1.15 L/min corresponds to 1.0 L/min in Fig. 4.49). The effect of power (at the optimum carrier gas flow) is seen to be negligible for hard lines, whereas the higher power yields an improvement for the soft lines. Therefore, by choosing the correct observation height a satisfactory interference level can be achieved at low power.

4.7. OPTIMIZATION OF THE ICP OPERATING CONDITIONS

The observations of Larson et al. [209] were essentially confirmed by Boumans and de Boer [62, 103] with their experimental facilities, as is exemplified in Fig. 4.50. Accordingly the power of 1.1 kW considered optimum for attaining low detection limits (Fig. 4.42) is compatible with a low interference level if an observation height of 15 mm is chosen for this ICP. For hard lines the dependence of the interference effect of alkalis on observation height was found to be far less critical than for soft lines (see also [61, Fig. 6]). For the latter group of lines the effect could be easily shifted from a severe suppression when measured relatively low in the plasma to a severe enhancement when observed at a relatively large distance from the coil, and a point could be found where it was essentially absent (cf. Fig. 4.50). The height at which this "zero point" was found differed from line to line, but it was possible to find such a height that the interference effect for all lines remained within acceptable margins, for example, between ±15% for 10 mg/mL KCl.

A superficial comparison of interferences of easily ionizable elements such as reported by Blades and Horlick [223] and Kawaguchi et al. [226] with results

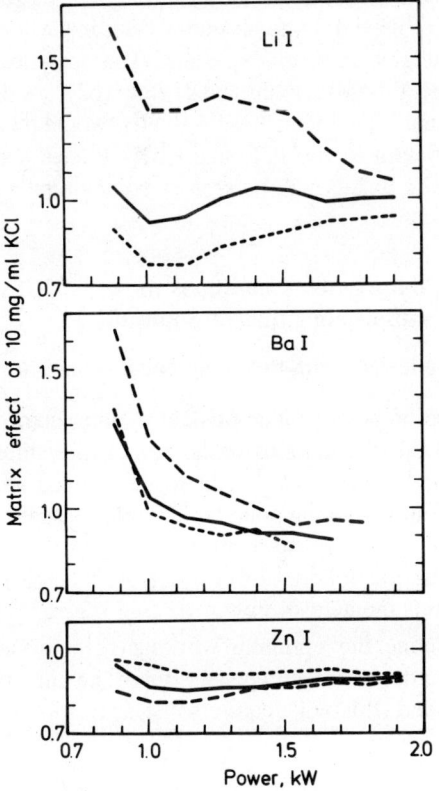

Figure 4.50. Matrix effect of 10 mg/mL KCl for Li I 670.8, Ba I 553.5, and Zn I 213.9 nm as a function of the power for three observation heights (h) and a fixed carrier gas flow rate of 1.3 L/min.
- - - - $h = 17$ mm
———— $h = 15$ mm
------ $h = 13$ mm
(Reproduced with permission from *Spectrochimica Acta, Part B* [62].)

found by Boumans and de Boer [62] might seduce one to see major discrepancies. Blades and Horlick, for instance, reported the matrix effect of Na on lines of various hardness (Ca I, Ca II, Mg I, Mg II, Cd I, Cd II) to be an enhancement low in the plasma, to pass through a "zero point" high in the plasma (>15 mm), and to become negative at still greater distances from the coil. The shapes of the curves of matrix effect vs. height depended on the analyte and the type of line (I, II), and the "zero point" shifted upwards in the plasma with the ionization potential of the analyte. Those results were for 1.5-kW power and 0.85-L/min carrier gas flow.

These observations seem to contrast with findings of Boumans and de Boer [62] for the matrix effect of KCl.

The various apparent discrepancies may be understood, however, in light of Edmonds and Horlick's remark [235] about the relative nature of spatial data: "Small changes in torch geometry and coil design certainly alter the absolute position and intensity axes of these plots. However, even over six different torches and one change in coil geometry (a three-turn coil rather than a two-turn coil) the basic nature of these plots has been very consistent and reproducible." Thus a difference between the experimental conditions (torch configuration, coil, frequency, etc.) used by different authors may have given rise to different spatial characteristics when projected on an absolute scale, but which are essentially similar in nature if judged on a relative scale. The analytical working point at 15 mm observation height of Boumans and de Boer [62], yielding slight to medium suppressions of hard line intensities by alkalis, would then correspond to a working point at 20 mm in the ICP studied by Blades and Horlick [223]. If this is considered, the results will be seen to coincide rather than to contrast.

4.7.5. Rules for the Selection of Compromise Conditions in Argon ICPs for Analysis of Inorganic Samples in Aqueous Solutions

The discussion in the previous section leads to the following conclusions:

1. If it is required that high detection power and a satisfactory interference level are concomitantly achieved, the margins in the numerical values of power, carrier gas flow, and observation height to be adopted are small. Departing from the optimum working point rapidly worsens either the detection limits (higher power) or the interference level (higher carrier gas flow or observation height), or both (lower observation height). Too low a carrier gas flow limits the aerosol flow.
2. The numerical values that define the optimum working point differ somewhat from system to system, and if a component (torch, nebulizer) in an existing system is modified, this will require some adaptation of the parameter setting.

3. An optimum working point can be rapidly found by observing (a) the SBR variations for a single hard line (Mn II 257.6 nm) to optimize the detection power, and (b) the magnitude of the interference effect of KCl for a few soft and hard lines (e.g., Li I 670.7, Ba I 553.5, Zn I 213.9, and Mn II 257.6 nm) to minimize interferences. The following rules may then be used [224].

   a. *Power:* choose the lowest possible power compatible with stable operation of the generator. This rule will usually lead to a power level of about 1 kW.
   b. *Carrier gas flow:* having fixed the power, select an observation zone of approximately 5 mm in height, the center of which is located at 15 mm above the work coil. Vary the carrier gas flow rate while observing the SBR of Mn II 257.6 nm. Fix the carrier gas flow at the point of maximum SBR.
   c. *Observation height:* observe the matrix effect of 10 mg/ml KCl on the net line signals of some soft and hard lines, such as stated above, for observation heights within a range of ± 2 or 3 mm from the starting point of 15 mm, and adapt the observation height, should the matrix effect depart appreciably from the $\pm 15\%$ level.
   d. *Final adjustment:* check whether the matrix effect of KCl on the lines used in (c) may be further reduced by slight adaptations of the carrier gas flow and choose the eventual flow rate accordingly.

If further optimization for the suppression of interferences is required, then the use of a simplex technique can be recommended (cf. Section 4.7.9).

4.7.6. Selection of Compromise Conditions for Aqueous Solutions Containing both Inorganic and Organic Matter (Argon ICPs)

Organic matter in aqueous solutions, such as biological fluids, does not essentially affect the relationships discussed in the preceding section and therefore a compromise setting virtually equal to that for aqueous solutions with inorganic matter only can be used. This is illustrated in Table 4.13 for a typical ICP system [224, 240, 241]. The only difference is the higher flow rate of the intermediate gas. This is required to prevent the formation of carbon-like deposits on the rims of the inner and intermediate tubes [212] as well as the crystallization of substances such as albumine near the orifice of the inner tube. For this flow to be effective under all working conditions, including ICP operation with organic solvents, the precise configuration of the torch is critical [212, 224].

In view of this, further discussion of the choice of ICP conditions for aqueous solutions with organic constituents is redundant. Other factors, such as sample pretreatment, nebulizer performance, stray light elimination, and background correction deserve chief consideration for samples of this type (see, e.g., [242–244], Chapters 6–8 and Part 2, Chapters 3–6).

4.7.7. Some Trends Observed in Argon ICPs Fed with MIBK as Organic Solvent

Fassel et al. [212] successfully employed an argon ICP for analysis of oils (1:10) diluted in methyl–isobutyl–ketone (MIBK) using the following parameter setting for their ICP: $P = 2.1$ kW, $F_0 = 17.0$ L/min, $F_i = 1.0$ L/min, $F_c = 0.9$ L/min, or $F_c = 1.0$ L/min, $F_s = 1.0$ mL/min, and an observation height of 18–20 mm above the coil [245]. Wavelength scans of the background spectrum (cf. Section 7.3.5) illustrating the occurrence of continua, molecular bands, and lines of C, H, and Ar were reported [245] for various observation heights, referred to as the tail, the analytical region, the zone between the analytical region and the coil, and the coil region ("toroid"). These results were used to argue the choice of the observation zone to be a region where the spectral background was much lower than in the hotter regions below the analytical region. The carbon lines at 193.09, 199.36, 247.86, 258.29, 296.72 nm, the C_2 Mulliken system (232.5 nm), the C_2 Swan system (436.5–667.7 nm), the violet CN bands (main system and tail bands [246] and the omnipresents OH bands (281.1 and 306.4 nm)) were identified (cf. Section 7.3.5). The precise choice of the operating conditions was supported by the excellent analytical results, but was not further argued.

Boumans and Lux-Steiner [224, 240, 241] reported trends of background, net line signals and SBRs with the ICP parameters also operated with MIBK, and they discussed in this light the choice of the compromise conditions specified in Table 4.13. They optimized the system for detection power only and did not include measurements of interferences.

The trends found show both similarities and discrepancies with those discussed in Section 4.7.4 for aqueous solutions. Some main features will be considered here; for details as well as arguments in favor of limiting the working range to that shown here the original paper [224] should be consulted.

Figure 4.51 shows that the background signal measured at various wavelengths as a function of power passes through a minimum. With the higher outer gas flow the minimum is reached at a higher power, but the background level does not change essentially.

To explain the minimum the authors suggest that the contributions from true background continua and molecular bands (including possible stray light contributions from these bands) change with power owing to modifications in the spatial structure of the ICP and a consequent change in the "temperature" that governs the emission in the observation zone.

Net line signals steadily rise with power (Fig. 4.52a), the relative increase being generally steeper the harder the line.

Table 4.13. Compromise Operating Conditions Used in a Typical Argon ICP System for Three Groups of Sample Types [224]

Parameter	Aqueous Solutions with Inorganic Matter[a]	Aqueous Solutions with Inorganic and Organic Matter[b]	Organic Solvents (MIBK)[b]
Power (kW)	1.1	1.1	1.7
Outer gas flow (L/min)	14	14	18
Intermediate gas flow (L/min)	0.2	0.7	0.9
Carrier gas flow (L/min)	1.0	0.9	0.8
Observation height (mm)	15	15	15
Sample feed rate (mL/min)[c]	1.4	1.4	0.8–1.4

[a] Detection limits and interference level optimized.
[b] Detection limits optimized.
[c] Controlled by peristaltic pump.

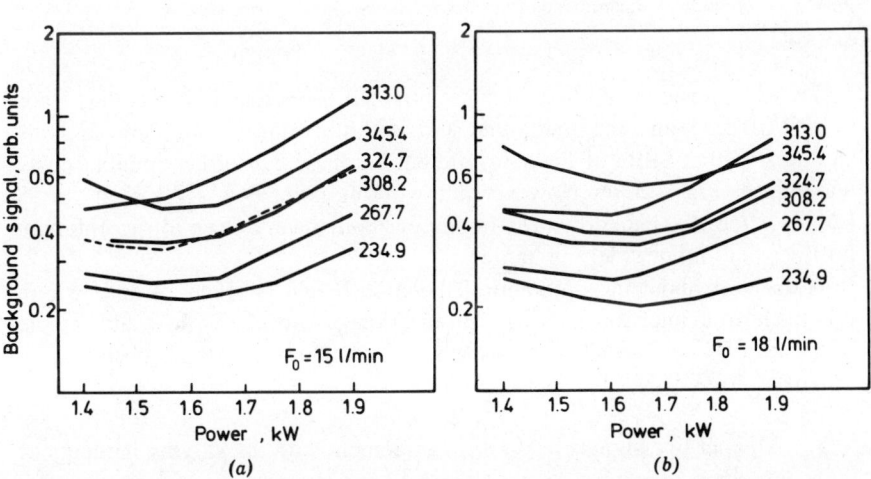

Figure 4.51. Dependence of the background signal at various wavelengths (nm) on the power for an outer gas flow rate (F_o) of (a) 15 L/min, and (b) 18 L/min. Intermediate gas flow: 0.9 L/min. Carrier gas flow: 0.86 L/min. Sample feed rate: 0.8 mL/min. Observation height: 14 mm. Sample: MIBK. (Reproduced with permission from *Spectrochimica Acta, Part B* [224].)

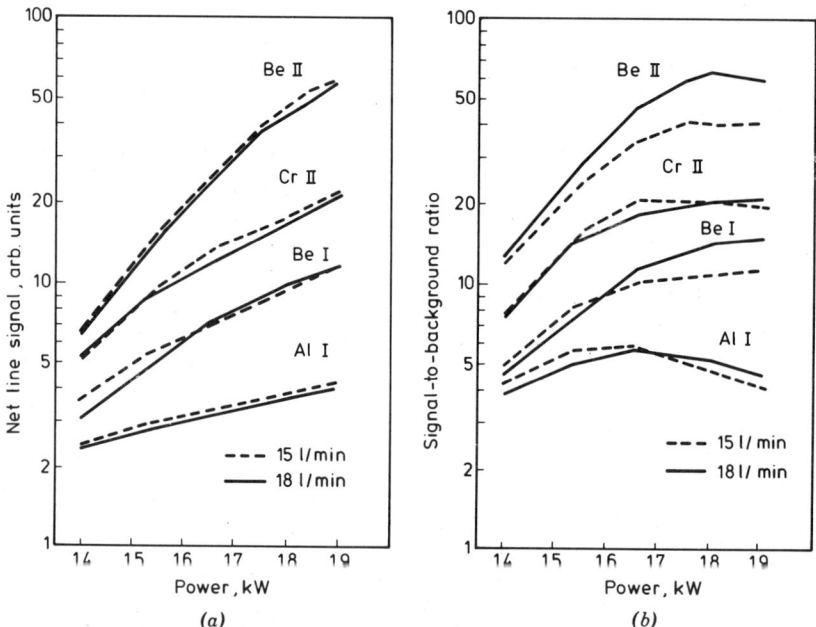

Figure 4.52. (*a*) Net line signal as a function of power for spectral lines of different "hardness" as measured in an ICP fed with methyl-isobutyl ketone (MIBK). Outer gas flow 15 and 18 L/min. Setting of the other parameters as in Fig. 4.51. Wavelengths: Be II 313.0, Cr II 267.7, Be I 234.9 and Al 308.2 nm. (*b*) SBR as a function of power for the same spectral lines and operating conditions. (Reproduced with permission from *Spectrochimica Acta, Part B* [224].)

The superposition of curves for net line and background signals thus leads to SBR curves with a maximum that shifts to higher power with increasing line hardness so that SBRs of lines of different hardness cannot be simultaneously maximized (Fig. 4.52*b*). However, the working point $P = 1.7$ kW, $F_o = 18$ L/min, where the background is minimum, comes very close to the optimum for lines that differ widely in hardness.

A decision about the optimum observation height between 14 and 18 mm was made in connection with the optimization of carrier gas flow and sample feed rate. Fig. 4.53 shows a representative example of the trends observed. In summary, for MIBK

1. The net line signals behaved in agreement with the varying hardness of the lines.
2. The background signal had a minimum in the vicinity of $F_c = 0.8$ L/min under all conditions.
3. If the carrier gas flow associated with the minimum background was

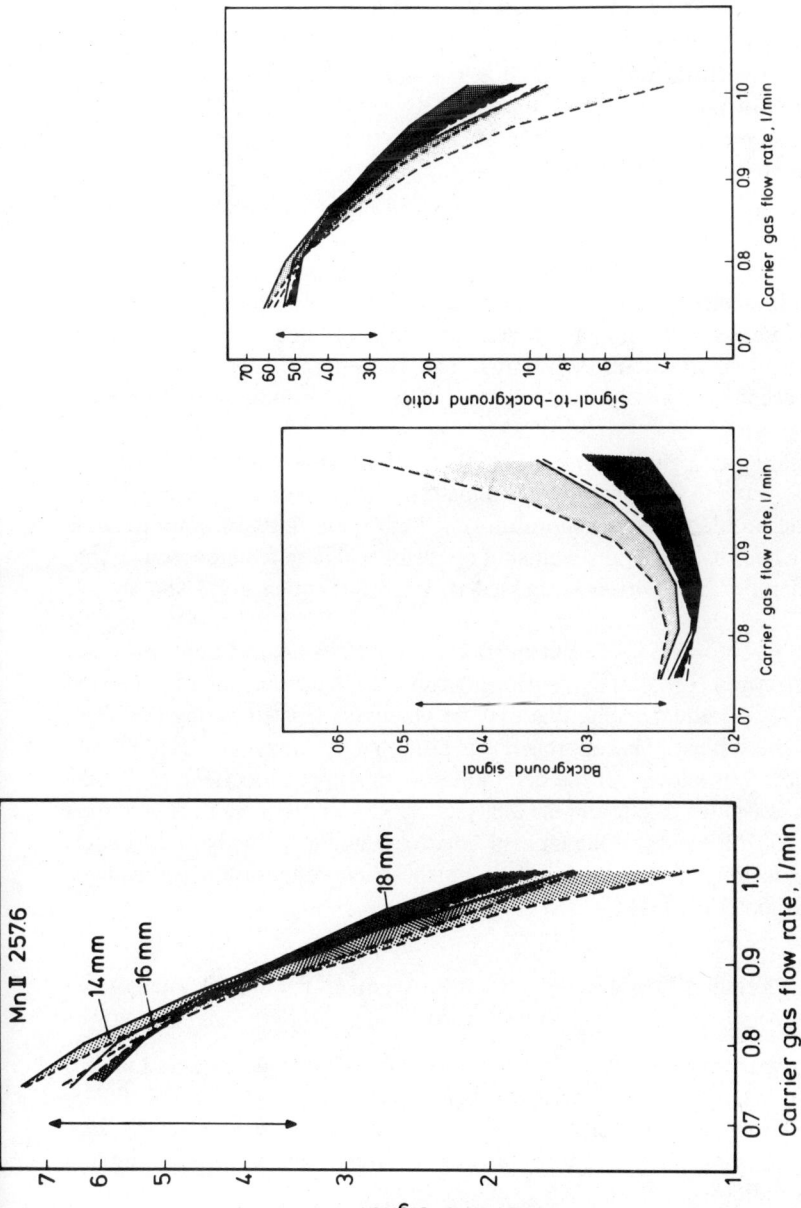

Figure 4.53. Net line signal, background signal, and SBR as a function of the carrier gas flow for Mn II 257.6 nm in an ICP fed with MIBK. Power: 1.7 kW. Outer gas flow: 18 L/min. Intermediate gas flow: 0.9 L/min. Observation height: 14, 16, and 18 mm. Sample feed rate: ——— 0.8 mL/min, ------ 1.4 mL/min. The length of the arrows corresponds to a factor of 2 on the ordinate scales. (Reproduced with permission from *Spectrochimica Acta, Part B* [224].)

chosen as "optimum," the SBRs did not essentially differ from the values maximally attainable.
4. The best SBRs were achieved at the lowest observation height considered (14 mm), but the differences between the 14 and 16 mm heights were marginal; the authors chose 15 mm to have the same value for all sample types (Table 4.13).
5. The sample feed rate had an insignificant effect with MIBK as solvent.

As will be discussed in Section 4.7.8, MIBK and xylenes (generally, solvents with low vapor pressure) belong to the category of solvents that show a behavior closely similar to that of water, the prime difference being that higher power is required to sustain the plasma. This is so because molecular constituents enhance the thermal conductivity and, therefore, the enthalpy of the plasma (cf. Section 4.8).

The handling of organic solvents of high vapor pressure and ICP optimization for these solvents is a far more delicate matter. Such solvents may cause initial introduction problems (e.g., chloroform) or both initial introduction problems and plasma overloading (e.g., methanol or ethanol). These features require special consideration and will be discussed in detail in Section 4.7.8 and Part 2, Chapter 6.

Finally, when organic solvents are fed into the ICP, the addition of a small amount of oxygen (1% v/v) to the carrier gas results in combustion of molecular species (C_2, CN) and graphite that may be otherwise formed during the introduction of the solvent. This approach was proposed by Magyar et al. [247] and its usefulness was shown for the determination of sulphur in xylene [248], the analysis of waste oil [249], and the analysis of petroleum products diluted with kerosene [250]. Recently, Magyar et al. made a quantitative study of the effects of oxygen feeding and aerosol drying on the background and SBR using a nitrogen–argon ICP [251].

4.7.8. Feeding Organic Solvents into Argon ICPs: Significance of Solvent Load[10]

Optimum experimental conditions for the analysis of nonaqueous solutions are still elusive, except for a few solvents such as xylenes [252] and MIBK [224]. This has two causes: (1) nothing is known about the amount of solvent that actually reaches the plasma, and (2) the introduction of several types of organic solvents is difficult.

This section summarizes the results of recent work [253, 254] aimed at the

---

[10]This section was contributed by Dr. F. J. M. J. Maessen, University of Amsterdam, whose cooperation is gratefully acknowledged.

4.7. OPTIMIZATION OF THE ICP OPERATING CONDITIONS

experimental control of the solvent plasma load and consequently the plasma stability and the excitation conditions. This control was achieved by varying the liquid uptake rate and, more importantly, external cooling of the aerosol.

Figure 4.54 shows the solvent plasma load determined by the continuous weighing method [253], as a function of the liquid uptake rate for chloroform, water, and methanol, and two types of nebulizers. Above an uptake rate of about 1 mL/min the solvent load either reaches a plateau or increases only very slowly. Evidently uptake rate is an insensitive parameter for characterizing the

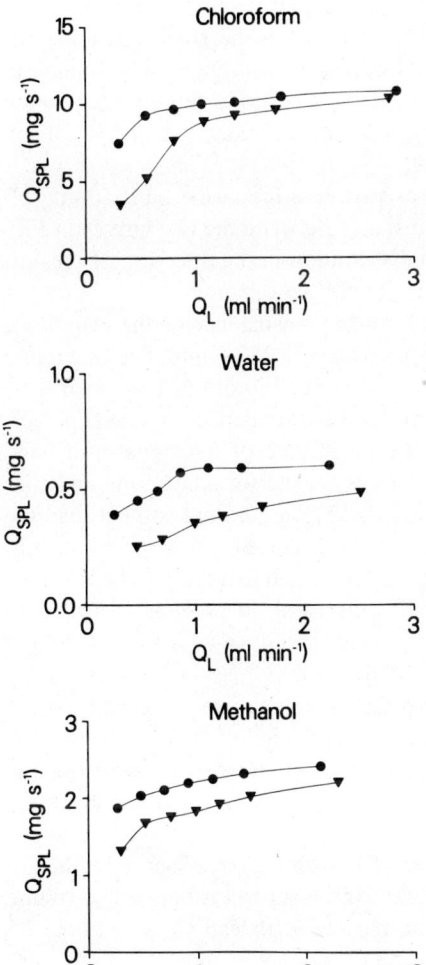

Figure 4.54. The solvent plasma load (Q_{SPL}) as a function of the liquid uptake rate (Q_L). ●: concentric nebulizer; ▼: V-groove nebulizer. Carrier gas flow: 0.7 L/min [254]. (Reprinted with permission from F. J. M. J. Maessen, G. Kreuning, and J. Balke, "Experimental Control of the Solvent Load of Inductively Coupled Argon Plasmas and Effects of the Chloroform Plasma Load on Their Analytical Performance," *Spectrochim. Acta* **41B,** 7 (1986). Copyright (1986), Pergamon Journals, Oxford.)

actual conditions in the plasma, in particular with respect to plasma stability. Thus, for example, for methanol a "maximum tolerable uptake rate" of more than 6.5 mL/min has been reported [255], but equally well an "upper limiting aspiration rate" of no more than 0.1 mL/min [256]. This discrepancy is understandable when considering the effect of the liquid uptake rate on the solvent load. Maessen et al. [253, 254] varied this load also for a fixed liquid uptake rate by means of a condenser between spray chamber and ICP.

Figure 4.55 contrasts experimental results (continuous curve) with values calculated on the basis of saturated vapor. Clearly, above $-20°$ C (chloroform) or above $-10°C$ (methanol) the larger part of the solvent load originates from vapor. The increasing distance between the continuous and broken curves toward lower temperature means that the fraction of the solvent leaving the condenser in the form of aerosol droplets decreases with increasing condenser temperature. The continuous curves indicate that below a certain temperature aerosol cooling does not contribute any more to the reduction of the solvent load. This can be understood as follows.

The saturated organic vapor condenses partly on the walls, partly on desolvated aerosol particles. The increasing distance between the two curves in Fig. 4.55 reflects a shift in the condensation distribution in the direction of the aerosol particles.

Figure 4.56 shows the effect of the chloroform plasma load on the magnitude and the axial position of the net line signal of Mn II 257.6 nm. The maximum shifts to a lower position in the plasma when the chloroform load increases. Interestingly, the axial positions of the maxima hardly depend on the rf power, while, in agreement with expectation, the magnitude of the signal at a fixed chloroform load increases with power. This is typical for a hard line and contrasts with the behavior of a soft line (Fig. 4.57): at constant solvent load the maximum shifts downwards when the power is increased. From the results obtained for representative spectral lines it can be concluded that the behavior of lines as to their "hardness" is essentially the same in plasmas loaded with chloroform or water: compare Figs. 4.45, 4.47, and 4.48 with Figs. 4.56 and 4.57.

Table 4.14 lists background RSDs for plasmas with various chloroform loads. The RSDs indicate that the stability of plasmas loaded with chloroform need not be worse than that of plasmas loaded with water. However, stable organic plasmas can be maintained only if carbon deposition in the torch is completely prevented, as is further illustrated in [254].

Figure 4.58 shows the variation of the SBR with power, observation height (h), and chloroform load for Cr II 267.7 nm. If the aerosol is not cooled (frame a), the SBR rapidly increases with h and reaches a plateau value at about 25 mm. Reduction of the plasma load to 5 mg/s by aerosol cooling (frame b) leads to a notable improvement in SBR for $h \leq 20$ mm. A further reduction of the plasma load to 3 mg/s (frame c) does not produce significant changes in SBR

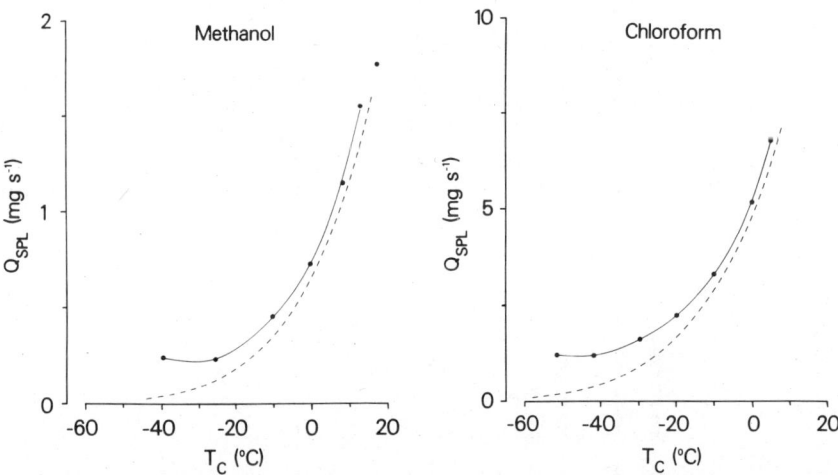

Figure 4.55. The solvent plasma load (Q_{SPL}) as a function of the condenser temperature (T_C). —— experimental; ------ calculated for saturated vapor. Carrier gas flow: 0.7 L/min. Liquid uptake rate: 0.8 mL/min [254]. (Reprinted with permission from F. J. M. J. Maessen, G. Kreuning, and J. Balke, "Experimental Control of the Solvent Load of Inductively Coupled Argon Plasmas and Effects of the Chloroform Plasma Load on Their Analytical Performance," *Spectrochim. Acta* **41B**, 9 (1986). Copyright (1986), Pergamon Journals, Oxford.)

and creates a situation where the SBR hardly depends on power and observation height. This situation remains so for still lower plasma loads.

The basic conclusion from Fig. 4.58 is that, independent of the solvent plasma load, SBRs of comparable magnitudes can be obtained provided that proper observation heights are applied.

Generally it should be realized that one uses for organic solvents the same assemblies of nebulizer, spray chamber, and torch as have been originally developed for aqueous solutions. However, these systems are not necessarily tailored to fit whatever type of organic solvent.

When such assemblies are employed under common conditions (i.e., liquid uptake rates and aerosol carrier gas flow rates of the order of 1 mL/min and 1 L/min, respectively, and rf powers in the range of 1 to 2 kW), then, depending on the nature of the solvent, stability problems can arise as a result of plasma overloading or mismatching of the plasma impedance during the initial solvent introduction stage. Taking stable plasma operation as the criterion, different categories of solvents can be distinguished: first those yielding "aqueous plasmas." Although the performance of these plasmas is extremely sensitive to water loading [257], this hardly plays a role in analytical practice. The reason is that under common operating conditions the water plasma load varies only within small limits and amounts to about 0.5 mg/s, irrespective of the type of pneumatic nebulizer employed [258]. A group of solvents, popularly called the

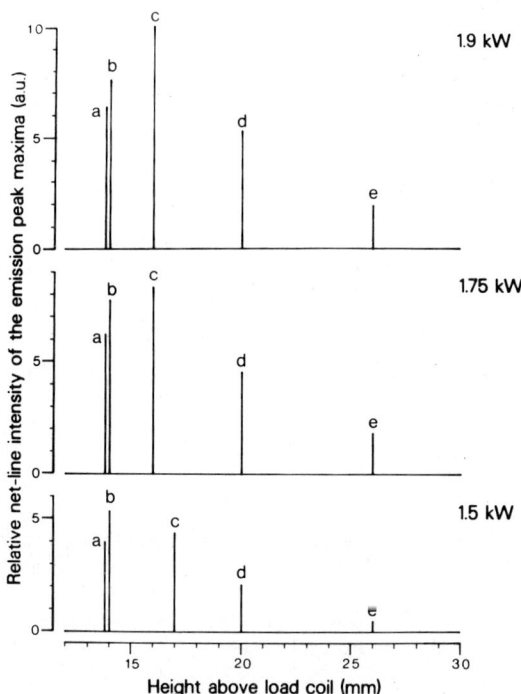

Figure 4.56. Axial position and relative magnitude of net line signal of Mn II 257.610 nm for three power levels and five values of the chloroform plasma load (mg/s): (a) 7.5; (b) 5; (c) 3; (d) 1.8; (e) 1.2. Carrier gas flow and liquid uptake rate as in Fig. 4.55 [254]. (Reprinted with permission from F. J. M. J. Maessen, G. Kreuning, and J. Balke, "Experimental Control of the Solvent Load of Inductively Coupled Argon Plasmas and Effects of the Chloroform Plasma Load on Their Analytical Performance," *Spectrochim. Acta* **41B**, 9 (1986). Copyright (1986), Pergamon Journals, Oxford.)

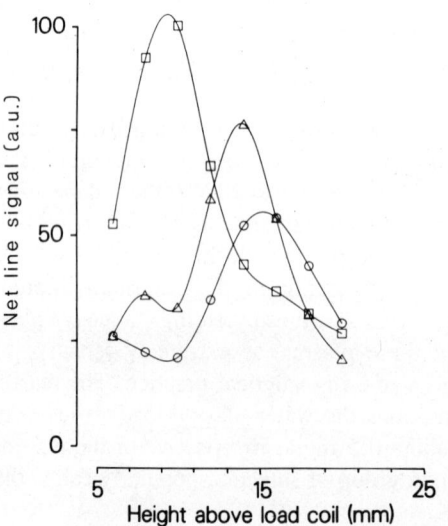

Figure 4.57. Net line signal of Li I 670.784 nm as a function of observation height for various rf powers. ○: 1.5 kW; △: 1.75 kW; □: 1.9 kW. Chloroform plasma load 7.5 mg/s. Carrier gas flow and uptake rate as in Fig. 4.55 [254]. (Reprinted with permission from F. J. M. J. Maessen, G. Kreuning, J. Balke, "Experimental Control of the Solvent Load of Inductively Coupled Argon Plasmas and Effects of the Chloroform Plasma Load on Their Analytical Performance," *Spectrochim. Acta* **41B**, 14 (1986). Copyright (1986), Pergamon Journals, Oxford.)

Table 4.14. **Background RSD (%) for Various Chloroform Plasma Loads.**
Power: 1.75 kW, Observation Height: 20 mm [254]. (Reprinted with permission
from F. J. M. J. Maessen, G. Kreuning, and J. Balke, *Spectrochim. Acta* 41B, 18.
Copyright (1986), Pergamon Journals, Oxford.)

Wavelength (nm)		Chloroform Plasma Load (mg s^{-1})				
		7.5	5	3	1.8	1.2
Pt II	203.6	0.6	0.2	0.4	0.6	0.4
Mn II	257.6	1.0	0.3	0.6	0.6	0.5
Cr II	267.7	1.1	0.3	0.4	0.3	0.3
Cu I	324.8	1.4	0.2	0.3	0.3	0.3
Ba II	455.4	2.0	0.2	0.3	0.4	0.2
Li I	670.8	2.2	1.1	0.6	0.9	1.0
K I	766.5	2.0	0.4	0.4	0.3	0.4
Condenser temperature (°C)		20	0	−12	−26	−52

"easy solvents," show a behavior very similar to that of water. The only noticeable difference is that plasmas fed with solvents of this type require about 0.5 kW more rf power to achieve an analytical performance comparable to that of aqueous plasmas [259]; examples are xylenes and MIBK. Obviously, in analogy to water, the highest solvent plasma loads achieved under common operating conditions do not exceed the plasma tolerances to solvents of this category.

The second group comprises solvents of which chloroform is a typical representative. These solvents behave similarly to the "easy" ones but cause problems when introduced into an ICP maintained with a commercial crystal-controlled generator with 2 kW maximum operating power [253, 260]. As a result of carbon deposition at the top of the torch, which occurs during the initial introduction stage, no stable plasmas can be generated with this type of solvents, unless special procedures are applied [253].

Finally, there is a third category comprising solvents that require provisions to prevent plasma overloading, for example, aerosol cooling, as well as special procedures to overcome problems during the initial introduction stage. Examples of solvents of this type are methanol and ethanol [254, 260].

If one considers the evaporation factors [256] and the vapor pressures of the solvents under discussion (Table 4.15), along with their specific behavior in ICPs as outlined in [254], it follows that solvent volatility in terms of mass density (E factors) offers no basis for the classification of organic solvents. However, if volatility is expressed in terms of vapor pressure, which is a measure of the particle density, the solvents can be categorized into two distinct

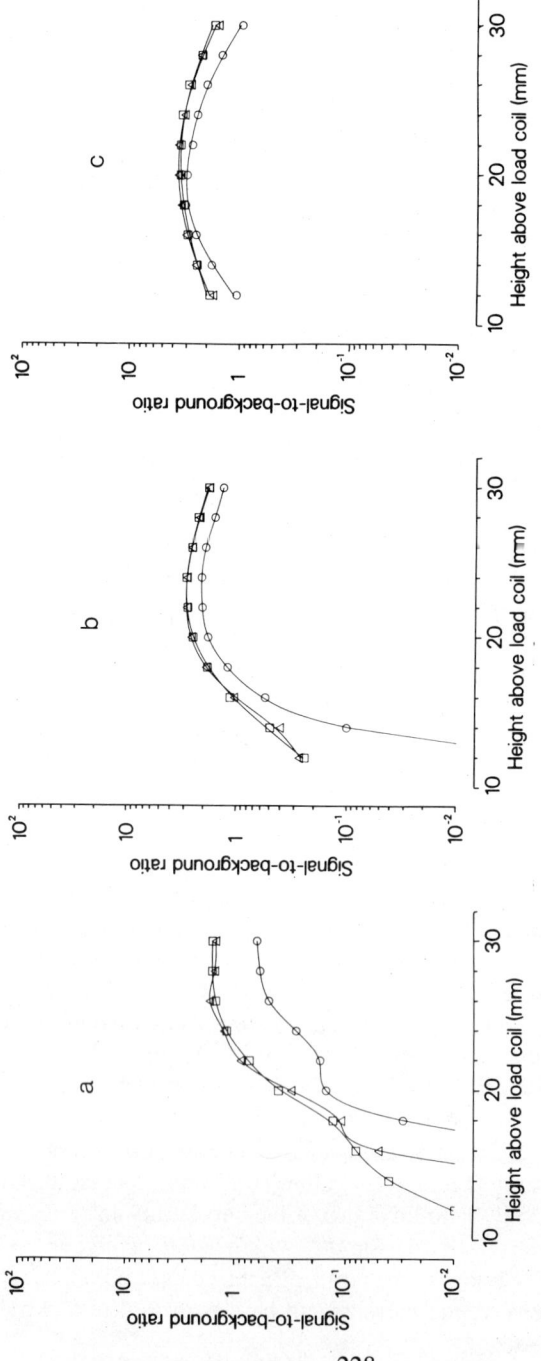

Figure 4.58. SBR for Cr II 267.716 nm as a function of observation height for various rf powers and chloroform plasma loads (mg/s). ○: 1.5 kw; △: 1.75 kw; □: 1.9 kw. (*a*) 7.5; (*b*) 5; (*c*) 3. Carrier gas flow and uptake rate as in Fig. 4.55 [254]. (Reprinted with permission from F. J. M. J. Maessen, G. Kreuning, and J. Balke, "Experimental Control of the Solvent Load of Inductively Coupled Argon Plasmas and Effects of the Chloroform Plasma Load on Their Analytical Performance," *Spectrochim. Acta* **41B**, 16 (1986). Copyright (1986), Pergamon Journals, Oxford.)

Table 4.15. Evaporation Factor (E-Factor) and Saturated Vapor Pressure for Selected Solvents [254] (Reprinted with permission from F. J. M. J. Maessen, G. Kreuning, J. Balke, *Spectrochim. Acta* 41B, 18. Copyright (1986), Pergamon Journals, Oxford.)

Solvent	E-Factor $(\mu m^3\ s^{-1})^a$	E-Factor $(\mu g\ s^{-1})$	Vapor Pressure at 20 °C $(mm \cdot Hg)^b$
Water	13.1	13.1	18
Xylene	18.5	16.0	4
MIBK	77.3	61.7	5
Chloroform	321	478	105
Methanol	47.2	37.4	105
Ethanol	45.6	36.0	120

^{a} From Boorn and Browner [256]. ^{b} From Miyazaki et al. [260].

groups. It is conspicuous that the representatives of the group with low vapor pressure, that is, water, xylene, and MIBK, do not raise any problem at all, whereas the representatives of the group with high vapor pressure cause either initial introduction problems (chloroform) or both initial introduction problems and plasma overloading (methanol, ethanol). The different behavior in ICPs of solvents that produce similar particle densities may be explained in terms of the thermal conductivity of the plasma. This is so because the thermal conductivity strongly depends on the nature of the dissociation products formed in the plasma. This view is supported by the fact that the thermal conductivity associated with the transport of dissociation energy constitutes a major contribution to the total thermal conductivity of hot gases [221] (cf. Section 4.8). However, differences in the distribution of the solvents over the entrance paths to the plasma may play a role as well.

In conclusion, the effect of the volatility of organic solvents in ICP–AES is twofold. The evaporation rate affects, through a shift in the droplet size distribution [255], the rate at which the analyte is delivered to the plasma. The saturation vapor pressure affects, through the solvent load of the plasma, the excitation conditions in the plasma.

4.7.9. Other Views on and Approaches to Optimization of the ICP Parameters: Simplex Optimization

In the light of the arguments produced in Section 4.7.1 in favor of compromise operating conditions, the "intuitive approach" to ICP optimization (Sections 4.7.4–4.7.7) may be considered adequate for many purposes. The approach then finds its main justification in the satisfactory analytical performance and the convenience of the working conditions to which its leads. However, the result should be seen as an overall optimization in which one does not bother about whether the detection limit of a particular element is a factor of two lower or higher, nor whether a particular interference effect measured with respect to a simple reference solution, such as a pure aqueous solution, is, say 10 or 15%. Therefore, the compromise conditions thus found can be considered only as a starting point from which, by further slight variations of the ICP parameters, conditions can be derived which better fit the analytical requirements in a specified situation than the generalized conditions do. We shall illustrate this with two examples.

Maessen et al. [154] determined for a particular set of analysis lines and specified matrices in an argon ICP the compromise conditions that yielded the best reproducibility of net line signals. The resulting conditions, viz. $P = 1.05$ kW, $F_o = 19$ L/min, $F_c = 0.75$ L/min, $h = 16$ mm above the top of the coil, are recognized as being close to the "world-wide standard values" commonly recommended for the type of ICP used. They were, however, somewhat different from the conditions yielding optimum SBRs.

Leary et al. [261] used a sequential simplex technique [262–265] along with a composite response function (or objective function) to find such compromise conditions for a group of analytes that the anticipated analyte concentrations were taken into account. More specifically, the objective function, formulated in terms of SBRs gave more significance to the analytes present at concentrations close to their detection limits than to elements present at relatively high concentrations. Using fixed gas flows Leary et al. optimized observation height and power. The result is shown in Table 4.16. The technique is seen to lead rapidly to the rejection of extreme observation heights and powers, after which the further search is limited to small ranges of the two parameters, 13.1–15.4 mm, and 1.12–1.27 kW. The variation of the SBRs within this range is not shocking, but the approach will at least give the comfort of a mathematically backed final result rather than one decided upon by intuition. The authors indicated the intention to expand the approach to a multiple-variable optimization for variables such as observation height, forward power, solution pumping rate, nebulizer and outer gas flow, and concentration of an ionization suppressing element. It will be interesting to follow these efforts since the approach may be helpful in improving the precision and accuracy for a given sample type. Although the resulting compromise conditions are not likely to be essentially different from those found by the rather rough "intuitive approach," they will differ sufficiently to yield more reliable analysis results. In this light, too, one should consider the arguments of de Galan et al. [266] in favor of single-element optimizations using a specified interference effect as the objective function. The example given in that context may raise some questions, however, as to the optimum design of the ICP used.

Table 4.16. Data Illustrating the Coverage of the Objective Function Using a Sequential Simplex Optimization for Determining the Optimum Observation Height in a Multielement Analysis Problem [261]. (Reprinted with permission from J. J. Leary, A. E. Brokes, A. F. Dorrzapf, Jr., and D. W. Golightly, *Appl. Spectrose.* 36, 39. Copyright (1982), Society for Applied Spectroscopy, Frederick, MD.)

Simplex Vertex no.	Forward Power (kW)	Observation Height (mm)	(S/B) Ratios for Elements in Standards Solution					Objective Function (F)
			Al	Na	Ti	P	Mn	
1	0.80	8.0	2.24	2.44	0.04	0.17	0.07	0.104
2	1.25	25.0	7.24	2.64	0.75	*	0.77	*
3	1.70	8.0	2.17	1.50	0.03	0.32	0.16	0.106
4	1.25	9.0	Simplex boundary violation					...
5	1.25	16.5	12.51	2.66	0.53	0.18	2.00	0.598
6	2.15	16.5	Simplex boundary violation					...
7	1.14	10.1	7.32	2.76	0.09	0.34	0.38	0.299
8	0.69	18.6	Simplex boundary violation					...
9	1.45	10.7	4.96	2.06	0.14	0.32	0.51	0.381
10	1.25	16.5	12.74	2.47	0.56	0.24	1.76	0.708
11	1.56	17.0	6.83	1.58	0.41	*	1.63	*
12	1.24	11.9	8.01	2.53	0.20	0.95	0.71	0.624
13	1.05	17.7	17.28	2.86	0.71	0.14	1.44	0.514
14	1.15	15.9	14.19	2.83	0.50	0.29	1.50	0.766
15	1.15	20.6	15.05	2.85	0.76	*	1.50	*
16	1.22	14.0	11.48	2.62	0.41	0.50	1.32	0.877
17	1.12	13.5	12.15	2.74	0.36	0.45	0.97	0.767
18	1.19	11.6	8.85	2.83	0.17	0.48	0.60	0.503
19	1.16	14.8	13.04	2.83	0.46	0.36	1.34	0.821
20	1.22	14.0	12.04	2.88	0.40	0.52	1.25	0.884
21	1.26	15.4	11.83	2.60	0.48	0.30	1.60	0.763
22	1.15	14.0	12.38	2.79	0.38	0.40	1.14	0.782
23	1.22	13.1	10.49	2.72	0.32	0.45	1.04	0.744
24	1.20	13.6	11.60	2.87	0.36	0.44	1.13	0.784
25	1.22	14.0	10.77	2.35	0.40	0.44	1.29	0.852
26	1.27	13.6	10.13	2.63	0.34	1.44	1.18	0.758
27	1.18	13.9	12.03	2.99	0.37	0.43	1.16	0.788
28	1.20	14.3	12.08	2.70	0.43	0.43	1.39	0.857
29	1.20	14.7	12.41	2.84	0.43	0.37	1.39	0.811
30	1.24	14.5	11.74	2.71	0.42	0.38	1.35	0.802

*Asterisk indicates that random noise caused background to exceed signal.

Table 4.17. Optimum Conditions for Oil Analysis Using Different Solvents as Diluents (Argon ICP) [267]. (Reprinted with permission from J. J. Brocas, *Analusis* 10, 388. Copyright (1982), Société française de Chimie, Paris.)

Parameter	Solvent			
	Toluene	Kerosene	MIBK	Xylene
Flow rate (L/min)				
Outer gas	17	17.5	19.5	19
Intermediate gas	1.1	1.1	1.1	1.1
Carrier gas	0.4	0.8	0.7	0.5
Power (kW)	1.6	1.5	1.5	1.45
Observation height (mm)[a]	22	22	22	23
Nebulizer[b]	CF + P	CF + P	CONC	CF + P

[a] Center of a 4-mm high axial zone.
[b] CF + P = cross-flow pneumatic nebulizer with peristaltic pump; CONC = glass pneumatic nebulizer with free uptake.

Tables 4.17 and 4.18 show results of a simple optimization used to decide upon the "best" diluent for oil samples to be analyzed with an argon ICP and a polychromator [267]. The optimization was carried out with the three gas flows, the power and the observation height as parameters, and the SBR of Mn II 257.6 as criterion. The upper and lower limits of the parameters as imposed by the characteristics of the equipment and the correct operation of the ICP were borne in mind in the optimization. Oil samples (Conostan standards) were diluted 1:10 with each of the organic solvents stated in the tables, of which Table 4.17 shows the eventual optimum conditions. Table 4.18 lists the detection limits attained under these conditions. Accordingly this approach led to definitely better results for kerosene compared with the other solvents. Kerosene has, in contrast to the second best solvent, MIBK, the additional advantage that it can be easily fed to a nebulizer using a peristaltic pump without rapidly corroding the tube.

The starting point and outcome of the optimizations as previously discussed keep closely in line with those of the intuitive approach. In contrast, Greenfield and Thorburn Burns [268] and Ebdon et al. [269] followed approaches in which any preknowledge about the (un)suitability of particular ranges of the ICP parameters, as demonstrated by previous experience, is disregarded so that the search for an optimum is extended over a very wide range of the ICP parameters in order to include the true optimum and to take interactions between the variables into account.

Greenfield and Thorburn Burns [268] used an alternating variable search method [270] and optimized the SBR for eight spectral lines separately (Table

Table 4.18. Detection Limits (μg/g) in Oil Using Different Diluents [267][a]. (Reprinted with permission from J. J. Brocas, Analusis 10, 389. Copyright (1982), Société française de Chimie, Paris.)

Analyte	Wavelength (nm)	Solvent			
		Toluene	Kerosene	MIBK	Xylene
Zn I	213.8	0.14	0.03	0.07	0.12
Pb II	220.3	2.6	0.6	1.5	2.3
Pb I	283.3		0.5		1.5
Cd II	226.5	0.18	0.03	0.08	0.13
Si I	251.6	0.6	0.09	0.1	0.2
Mn II	257.6	0.04	0.005	0.009	0.016
Fe II	259.9	0.22	0.025	0.04	0.07
Mg II	279.5		0.01	0.01	0.008
Cu I	324.7	0.28	0.015	0.03	0.035
Ag I	338.2	0.9	0.06		0.21
Ni I	341.4	3.3	0.12	0.3	0.35
Ca II	393.3	0.02	0.002	0.004	0.005
Al I	396.1	3.7	0.25	0.7	0.75
Na I	588.9		0.3		1.3

[a] The results are for the optimum conditions stated in Table 4.17. The detection limits are on a "3σ basis" (with a background RSD of 1%) and refer to the original, undiluted oil.

4.19). Results were collected for a Fassel torch (Fig. 3.7) and a Greenfield torch (Fig. 3.6), each operated with three argon flows, and for a Greenfield torch operated with nitrogen as outer gas and argon as intermediate and carrier gas. The SBR of each line was optimized using the three gas flows, the power and the observation height as variables. The optimum values of the variables differed considerably from line to line for each of the three torch systems, as is illustrated for the Fassel torch in Table 4.19. These results show that if a rigorous optimization with SBR as the only criterion is attempted, it may become difficult to define compromise conditions. In practice, however, irrespective of whether a compromise for many elements is being sought or single-element optimization is attempted, one will include additional boundary conditions, such as an upper limit to the outer argon flow rate with a view to economizing gas consumption and—more importantly—a lower and an upper limit to the carrier gas flow rate with a view to nebulizer operation and interferences. The generator will often dictate a lower limit to the power as will the background radiant flux that must reach the detector to prevent excessive shot noise. The work of Greenfield and Thorburn Burns does not consider these aspects, since the prime purpose was the comparison of plasmas on the basis of the SBRs achieved under truly optimized conditions. These results are discussed in Section 4.8.

Table 4.19. Operating Conditions Optimized for Different Spectral Lines Using SBR as Criterion. The Results Are for an Argon ICP Produced with a Fassel Torch [268]. (Reprinted with permission from S. Greenfield and D. Thorburn Burns, *Anal. Chim. Acta* 113, 215. Copyright (1980), Elsevier Science Publishers, Amsterdam.)

Element	Na	Cs	Al	Zn	Zn	Ba	V	Al
Wavelength (nm)	589.1	455.5	396.1	307.6	307.2	455.4	309.3	281.6
Concn (µg/mL)	1	1,000	1	1,000	1,000	1	1,000	1,000
Flow rate (L/min)								
Outer gas	22.0	20.0	26.0	19.0	19.0	20.0	19.0	22.0
Intermediate gas	0.2	0.0	0.0	0.0	0.0	0.0	0.25	1.0
Carrier gas	2.2	1.7	1.75	1.0	0.5	0.95	0.75	1.0
Power (kW)	1.0	0.66	1.12	0.89	0.53	0.66	0.65	1.16
Observation height (mm)[a]	56	37	48	33	20	33	23	35
SBR	20.5	129	17	15.6	1.09	232	1340	0.7

[a] Defined with respect to the base of the plasma.

4.7. OPTIMIZATION OF THE ICP OPERATING CONDITIONS

Ebdon et al. [269] applied the simplex technique [262-265] to ICP optimization using a modified torch configuration [271]. They demonstrated the viability of this technique for this purpose and this was confirmed by Terblanche et al. [272]. Comparing the results obtained with a pure argon ICP and a nitrogen-argon ICP using their new torch configuration, Ebdon et al. [269] concluded that "in practical terms it appears that compromise multielement operating conditions will be less detrimental to individual element sensitivities with the argon-cooled plasma (pure argon ICP), or that the greatest improvements in analytical sensitivity for a given element may be achieved by judicious line and power selection with the nitrogen-cooled plasma." The gas flows identified by Ebdon et al. [271] as being optimum for their new torch as well as for the Greenfield torch differ rather drastically from those normally used in ICP-AES and so are the reported optimum SBRs for the Mn II 257.6-nm line, which are over two orders of magnitude poorer than those reported by authors using conventional argon ICPs under compromise conditions (cf. Section 4.1.7).

More recently, Moore et al. [140] applied simplex optimization to a nitrogen-argon ICP using either a composite SBR function to optimize detection limits or composite "MII" function to "minimize ionization interferences." Attempts to apply a combined SBR-MII response function were found to be impracticable because a relation between the change in SBR and the change in ionization interferences could not be established, and a balanced shift in emphasis from SBR to MII could not be achieved during optimization.

Moore et al. found the following approach to be the most useful to optimization of multielement analysis: first the use of a composite SBR function and then a composite MII function. In practice, losses of up to 25% in SBR were suffered to achieve a reduction in ionization interferences down to 1-8%. In principle, this approach does not differ so much from the "intuitive optimization" discussed in the preceding sections, where the SBRs were maximized to the extent that interferences from alkalis could be kept at a tolerable level. However, the simplex approach such as proposed by Moore et al. has the advantage of being an "ad hoc" optimization for a particular sample type, so that, in a rigorous, convenient and straightforward way, proper attention can be paid to the analytes and concomitants that are of main interest in a particular situation.

Moore et al. also established for the nitrogen-argon ICP that optimizing the SBR of a representative line such as Mn II 257.610 nm is a rapid way of establishing compromise conditions for multielement analysis. Only for a few soft lines did individual optimization lead to essentially different operating conditions and vital improvements in detection limits.

On the whole, it appears even more that the strength of simplex optimization lies in the possibility of establishing conditions with minimum interferences rather than maximum detection power [140, 266]. This is understandable for the following reasons. First, in optimizing SBRs it generally does not matter very much whether the SBRs will be, say, 50% lower or higher, whereas reducing an interference effect from, say, 25 to 5% is analytically significant. Second, the dependence of SBRs on the ICP operating conditions is straightforward and predictable, while the range of conditions yielding maximum SBRs for a great many lines is relatively small. In contrast, interference effects behave whimsically, even within this relatively small range of operating conditions. Therefore, the use of approaches such as simplex techniques may repay to minimize interferences without essentially jeopardizing the detection power.

Finally, for sample types whose major elements emit line-rich spectra SBRs may depend far less critically on the ICP parameters than for pure water, at least in spectral regions with strong lines of the major elements. The wings of these lines substantially contribute to the background, whence the background behaves closely similarly to a line spectrum. Consequently analysis lines and background will respond rather similarly to changes in the ICP operating conditions (see Section 7.3.3). These circumstances may provide more freedom in the choice of the ICP parameters in order to emphasize the minimization of interferences without sacrificing the detection power.

4.8. THE ETERNAL CONTROVERSY: HIGH POWER VERSUS LOW POWER, NITROGEN-ARGON VERSUS ARGON ICPS

The literature bears witness of an eternal controversy between ICP spectroscopists as to the rewards and penalties of high or low power ICPs, and nitrogen-argon or pure argon ICPs, where usually high power is linked with nitrogen-argon and low power with pure argon ICPs (cf. Section 3.2). This association of power level and type of gas is qualitatively understandable in terms of the difference in heat content (enthalpy) between molecular and atomic gases, as set out by Reed [273, 274]. The dissociation involves absorption of energy, which gives rise to a sharp increase in the curve of heat content as a function of temperature (Fig. 4.59), whence a molecular gas will require more energy than an atomic gas to reach a temperature where it is sufficiently ionized to sustain a discharge. A quantitative description involves energy balance considerations from which a spatial temperature distribution is derived (see, e.g., [275–278] and Part 2, Chapter 9). The effect of the dissociation of molecular gases enters these considerations via the thermal conductivity enhancement due to the transport of dissociation energy [221]: molecules diffuse from colder to hotter regions, absorb energy and dissociate; the atoms formed diffuse to colder regions, associate, and release energy. This process involves a net transport of energy from the hotter to the colder regions. Considerations of this type are implicit in explanations of phenomena observed with ICPs generated in molec-

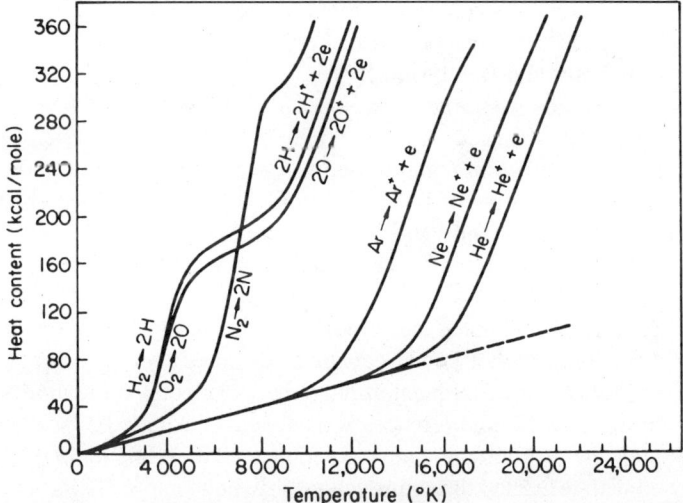

Figure 4.59. Heat content of various gases under LTE conditions as a function of temperature at atmospheric pressure. The straight line forming the lower envelope of the curves represents the dependence according to the ideal gas law, the slope being $C_p = 5/2(R)$, where C_p is the specific heat and R the gas constant. The curves for the diatomic gases show two "jumps": one for the dissociation step and one for the (first) ionization step. A thermal plasma can exist only above the temperature where ionization sets in. The figure illustrates that an appreciably larger energy input is required for molecular gases than for atomic gases [274]. (Reprinted with permission from T. B. Reed, in L. Eyring, ed., "Plasmas for High Temperature Chemistry," *Advances in High Temperature Chemistry*, Vol. 1, p. 263. Copyright (1967), Academic Press, Orlando, FL.)

ular gases or mixtures of such gases with argon [273, 274, 279–282]. It would be an oversimplification, however, to conclude that the use of a molecular gas would necessarily require a higher power input than argon to generate an ICP useful for analytical purposes. Many more parameters, in particular the spatial confinement, gas flow conditions, plasma stability, and electrical conductivity, play an essential role. In addition, once the physical conditions for obtaining a stable plasma are fulfilled, it is the outcome of measurements of analytical performance characteristics, such as detection limits and interference levels, which is decisive for the suitability of a particular ICP configuration for spectrochemical analysis. The effects of the composition of the gases, the gas flows, and the power on the spatial structure of the ICP and the excitation conditions are so complex that it is not possible to discuss the dependence of analytical characteristics on the ICP parameters from a uniform, fundamental point of view. The present status of our knowledge primarily permits a phenomenological and historical treatment of the subject.

Greenfield and his associates [89, 90, 206, 222, 279, 283, 284] were the first to use mixtures of argon and diatomic gases in the ICP and to study their effects on its spectrochemical behavior. Later studies were made by Capitelli et al. [286] and Truitt and Robinson [287].

Greenfield and his co-workers developed the nitrogen–argon ICP to a useful analytical tool and advanced various arguments in favor of the use of nitrogen as outer gas in combination with high-power, viz, high temperature, high electron density, the prevailing of LTE conditions, high ''sensitivity,'' better freedom from interferences, and better tolerance for samples of high salt content, concentrated acids and organics, in comparison with low-power argon ICPs [89, 206, 268, 287, 288].

In 1977, Fassel [289] noted that ''most of the advantages cited by Greenfield et al. [89] for using higher power cannot be disputed, but documentation that greater 'sensitivity', more frequent improvement in precision and more effective elimination of chemical interferences is achieved appears to be lacking.''

Reviewing the present and future position of ICP–AES, in 1979 Boumans [159] stated with respect to the controversy ''low-power argon ICPs versus high-power nitrogen–argon ICPs'': ''The advantages and disadvantages of either group of ICPs are not clear and are a matter of dispute, but excellent analytical results obtained with real samples have proved the viability of both approaches. The future will have to show (1) whether one of the two types of ICPs can do something that the other cannot, and (2) which of the two will ultimately provide the best performance-to-cost ratio. It is not likely that in this respect the gas consumption will be a decisive factor, at least not in industrial laboratories: nitrogen–argon ICPs consume 20–70 L/min of nitrogen and 10–35 L/min of argon; the gas consumption of argon ICPs is 10–20 L/min.''

During an ICP conference in 1978 Greenfield [68] challenged (low-power argon) ICP spectroscopists and asked them to determine detection limits not in distilled water, but in samples with difficult matrices: 10% m/v solutions of sodium chloride, calcium chloride, a 1:1 mixture of nickel and cobalt nitrates, and olive oil. This challenge resulted in the installation of an ICP Detection Limits Committee and a report [51] in which the rules for unambiguous measurements of detection limits were stated (cf. section 4.1). In this scope Boumans et al. [67] published experimental results as determined with a low-power argon ICP for the stated inorganic matrices and concluded that ''for the high-power nitrogen–argon ICP it will be useful that experimental values of detection limits and related quantities, such as produced here for pure water, sodium chloride, calcium chloride and nickel–cobalt nitrates can be collected under compromise conditions for simultaneous multielement analysis. Such data will provide information not only about the detection limits, but also about the convenience with which these detection limits are achieved, and about the magnitude of the matrix effects that must be tolerated under these conditions.''

In an attempt to arrive at an unambiguous assessment Greenfield and Thor-

4.8. NITROGEN-ARGON VERSUS ARGON ICPS

burn Burns [69] introduced the concept of "intrinsic merit" of ICPs, which in fact implies the rather obvious proposal of comparing ICPs on the basis of SBRs measured on one and the same spectrometer so that the properties of the spectrometer and the detector(s) are not involved. It was at the same time recognized that detection limits, not SBRs, are the figures of merit of interest to the analyst and, therefore, Greenfield and Thorburn Burns performed an extensive analysis of the noise properties of their multichannel spectrometer in connection with the ICP. However useful this analysis has been in the scope of the investigation, a recent analysis of noise by Boumans et al. [67] (cf. Section 4.1) led the latter authors to the conclusion "that many complications inherent in the treatment of Greenfield and Thorburn Burns can be circumvented if measuring equipment is used that eliminates the detector noise contribution." Since under the latter conditions the detection limit can be expressed by Eq. (4.30):

$$c_L = \left(\alpha_B^2 + \frac{g\beta}{x_B}\right)^{1/2} \cdot \frac{c_0}{\text{SBR}} \qquad (4.30)$$

Boumans et al. proposed SBR/α_B as an intrinsic characteristic for the comparison of plasmas under conditions at which such a background radiant flux reaches the detector where the shot noise term ($g\beta/x_B$) in Eq. (4.30) is negligible. Evidently, if it is proven experimentally that the flicker noise (α_B) in the two plasmas is the same, a comparison on the basis of SBRs alone is sufficient.

Boumans et al. further argued that a situation with entirely negligible shot noise is not likely to be found, particularly at low wavelengths, so that the shot noise term in Eq. (4.30) should be involved in the intrinsic figure of merit. Then the background radiant flux reaching the detector ("sensitivity") also contributes. The associated "sensitivity advantage" of the more powerful plasma is implicit in the treatment of the problem by Greenfield and Thorburn Burns. This advantage will, however, play an increasingly smaller part the higher the optical conductance of the spectrometer and the higher the sensitivity of the PMT; in other words, the more powerful plasma will lose more of its "sensitivity advantage" the better the spectrometer.

We should further note here that a comparison of the detection limits reached with the *same* lines in different ICPs provides interesting information, but is not necessarily decisive for judging the capabilities of the ICPs for detecting *elements*. Then the detection limits attainable with the most prominent lines of the one ICP must be compared with those of the most prominent lines in the other ICP.

Greenfield and Thorburn Burns [268] compared experimentally two argon ICP systems and a nitrogen-argon ICP system (cf. Section 4.7.9). The comparison was based on optimizations of the SBRs for each of eight spectral lines separately. The results confirmed the general experience of other workers (see Section 4.7.4) that pure argon ICPs operated with a Fassel torch (Fig. 3.7) require low power to achieve optimum SBRs, that is, 0.53–1.16 kW in [268]. Higher values, viz. 0.8–2.6 kW, were found optimum for the nitrogen–argon ICP generated with the Greenfield torch (Fig. 3.6). Substantially higher power levels of up to 3.92 kW were needed to sustain an argon plasma with the latter torch, but owing to generator limitations those results were not considered decisive. Because the latter system requires high argon outer flow rates (16–40 L/min) and yields SBRs similar to or poorer than those obtained with the Fassel torch, this system is hardly of any practical interest and does not deserve further consideration here.

From the optimized SBRs for the eight lines studied (Table 4.19) it was concluded that five of them gave the better SBRs for the Greenfield nitrogen–argon system. The poorer results for the other three lines were attributed to molecular band interference (V, Zn) or the impossibility to reach a sufficiently low-power level with the particular generator (Na). It was eventually concluded that "of the thousands of useful spectral lines, most would be expected to give better results with the Greenfield torch with a nitrogen coolant." This expectation has not yet been verified by experimental evidence. In contrast, Montaser, Fassel, and Zalewski [115] produced extensive experimental results for nitrogen–argon and pure argon ICPs operated with a Fassel torch rather than the Greenfield torch commonly used for nitrogen–argon ICPs. These results are most interesting because they are comprehensive and unambiguous, provide useful information in the scope of present trends to reduce instrument and operating cost of ICPs, and link up closely with considerations on compromise experimental conditions usually followed by analytical spectroscopists. We shall deal with this and related work in Section 4.9.

Concluding the present section we must point out that the worldwide application of conventional argon ICPs to the solution of an ever increasing number and variety of analysis problems extensively proves their value. The usefulness of high-power nitrogen–argon ICPs has also been documented in particular by Greenfield and his group [59, 89, 206, 290] and by Watson [140, 291–293], Ohls [294–296], Broekaert [141, 297–300], and their associates (cf. Chapter 6 and Part 2, Chapter 1). However, most manufacturers appear to see no reason for replacing or supplementing their present conventional argon ICP instruments by equipment suitable for generating high-power nitrogen–argon ICPs, except that it would raise the initial investment for the users.

4.9 NEW SYSTEMATIC STUDIES OF ICPS IN WHICH ARGON IS PARTLY OR WHOLLY REPLACED BY MOLECULAR GASES

Montaser and Montazavi [281] considered the realization and evaluation of an ICP running on an economic diatomic gas, such as N_2, primarily with a view to the acceptance of ICPs for spectrochemical analysis in those countries where argon is expensive or not produced locally. Montaser and Montazavi used a 2.5 kW, 27.12 MHz crystal-controlled generator in combination with a Fassel torch, normally used for operating argon ICPs. They introduced N_2 only as the outer gas, the carrier gas still being argon. For initiating the ICP they used the general experience mentioned in the early (physical) literature on induction plasmas (cf. Section 3.2) and stated the following procedure. The plasma is first generated in the conventional way in an all-argon atmosphere. The forward power is increased to about 1 kW, the carrier gas flow is initiated, and an intermediate argon flow of 1.5L/min is introduced and maintained. Nitrogen is then introduced in the outer gas flow by gradually decreasing the outer argon flow to zero, while increasing the nitrogen flow so that the total outer flow remains constant at 15 L/min. The change-over from 100% argon to 100% N_2 extends over a time interval of about 2 min because the automatic servo-driven matching network of the generator should follow the impedance alteration of the plasma.

A similar procedure has been described by Barnes and Meyer [301] for generating an all N_2 ICP with a Fassel torch, operating on 25 L/min outer, 3.5 L/min intermediate, and 1.5 L/min carrier gas flows and about 1.5 kW forward power.

Figures 4.60 and 4.61 show the effects of N_2 and the power on the net line

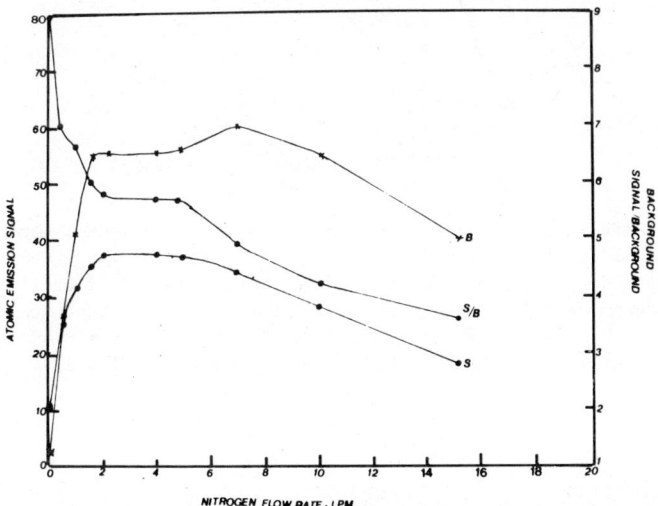

Figure 4.60. Effect of N_2 flow rate in the outer gas flow on the net line signal (S), background signal (B), and SBR (S/B) of Cd I 228.8 nm. Forward power = 1200 W; total outer gas flow rate = 15 L/min [281]. (Reprinted with permission from A. Montaser and J. Mortazavi, "Optical Emission Spectrometry with an Inductively Coupled Plasma Operated in Argon–Nitrogen Atmosphere," *Anal. Chem.* **52,** 258 (1980). Copyright (1980), American Chemical Society, Washington, DC.)

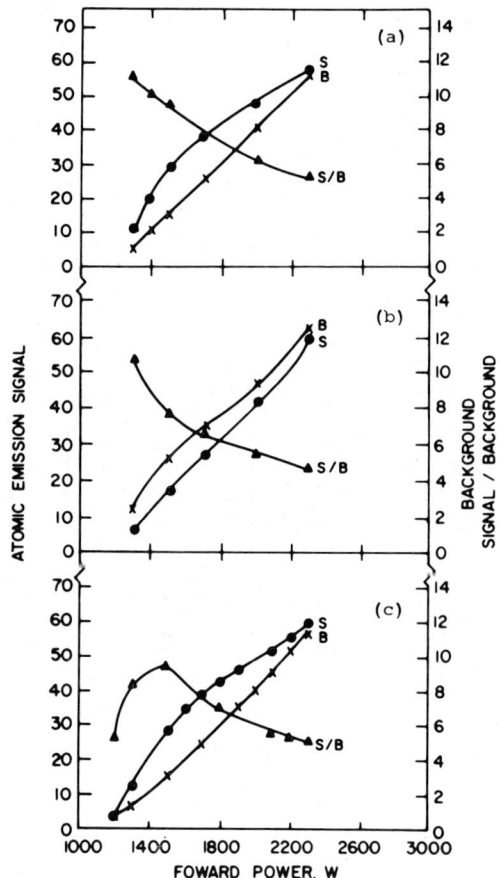

Figure 4.61. Influence of forward power on the net line signal (*S*), background signal (*B*), and SBR (*S/B*) of Cd I 228.8 nm. The outer gas was 100% N_2 of which the flow rate was varied as follows: (*a*) 5 L/min, (*b*) 10 L/min, and (*c*) 15 L/min. The observation height was 17 mm [281]. (Reprinted with permission from A. Montaser and J. Mortazavi, "Optical Emission Spectrometry with an Inductively Coupled Plasma Operated in Argon–Nitrogen Atmosphere," *Anal. Chem.* **52** 258 (1980). Copyright (1980) American Chemical Society, Washington, DC.)

signal of Cd I 228.8 nm, the background signal and the SBR, as observed by Montaser and Montazavi [281]. Compared to the argon ICP, the N_2–Ar plasmas were physically smaller in size, their spectra exhibited greater intensity, and the optimum observation heights were located lower in the plasma. An initial comparison of the detection limits of 16 elements in the Ar and N_2–Ar ICPs operated at identical power (1200 W) and flow rates indicated that the pure argon plasma yielded generally better detection limits.

In a subsequent investigation of Montaser et al. [115], substantially broader ranges of the ICP parameters were covered. In addition, they used an ultrasonic nebulizer with desolvation facilities in order to have a greater freedom in the choice of the carrier gas flow rate and sample feed rate. The first part of this work involved optimization studies on outer and carrier gas flow compositions at relatively low-power levels (1.0–1.6 kW).

Figures 4.62 and 4.63 show typical results of the dependence of background and net line signals on outer gas flow composition and power at an approximately optimum observation height. The generalized conclusions were as follows:

1. Outer gas flows containing 5 to 15% N_2 provided the highest net line signals at all forward power investigated.
2. Net line signals increased with power, although to a lesser extent for (soft) atomic lines than for ionic lines.
3. SBRs of hard lines showed maxima for outer gas flows containing 5 to 15% N_2; for soft lines a rather sharp maximum occurred at 5% N_2 in the outer gas.

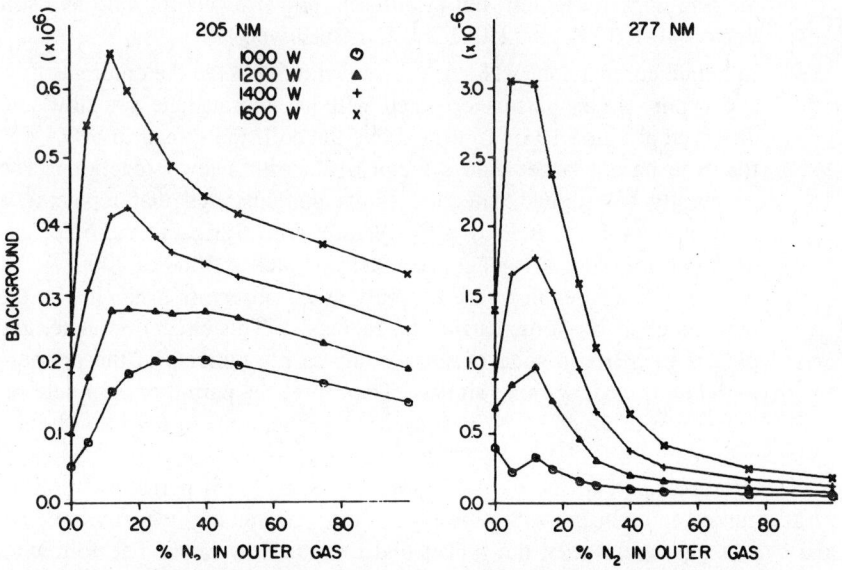

Figure 4.62. Influence of outer gas composition on the background emission at an observation height of 10 mm. The outer, intermediate, and carrier gas flow rates were 15 (Ar + N_2), 2.5 (Ar), and 1.0 (Ar) L/min respectively. A USN with desolvation facilities was used. Sample uptake rate = 4 mL/min [115]. (Reprinted with permission from A. Montaser, V. A. Fassel, and J. Zalewski, *Appl. Spectrosc.* **35**, 297. Copyright (1981) Society for Applied Spectroscopy, Frederick, MD.)

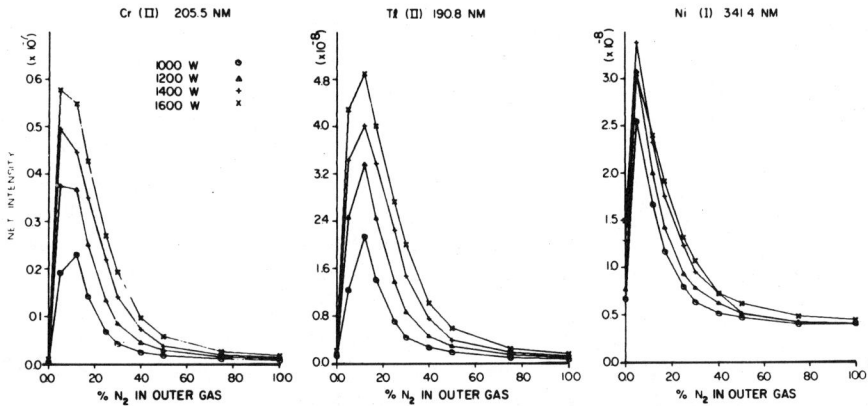

Figure 4.63. Effect of forward power and percent N_2 in the outer gas flow on the net line signals of three spectral lines. The conditions were similar to those stated in the caption of Fig. 4.62 [115]. (Reprinted with permission from A. Montaser, V. A. Fassel, and J. Zalewski, *Appl. Spectrosc.* **35**, 296. Copyright (1981), Society for Applied Spectroscopy, Frederick, MD.)

4. Although S/N ratios showed some randomness, maxima were found in the 5 to 25% N_2 region, but at different power levels for hard and soft lines (1.4–1.6 kW and 1.0–1.2 kW, respectively).

5. In a final comparison, a N_2–Ar plasma with 5% N_2 in the outer gas flow and a pure argon plasma operated without intermediate gas flow and observed at either 10 or 15 mm above the coil, the spread in SBRs was found to be not larger than a factor of 2, with a few exceptions. The eventually determined detection limits demonstrated that replacing a conventional argon ICP by a N_2–Ar ICP with 5% N_2 in the outer gas does not offer any advantage, whether operated at 1000 or 2000 W.

6. Net line intensities decreased rapidly at all observation heights as the percentage of N_2 in the carrier gas increased. This effect dominated the picture of SBRs in spite of some compensatory effects of the pressure of N_2 on the background signals. Therefore, the partial or complete replacement of argon by N_2 in the carrier gas led to poor detection limits.

In the second part of their work, Montaser et al. [115] performed optimization studies on a plasma with pure N_2 as outer gas and argon as intermediate and carrier gas, and varied the power and the gas flow rates. The following results should be mentioned (cf. Table 4.9 in Section 4.1.8).

1. In agreement with the experience reported for soft lines in Section 4.7.4 for an argon ICP, an increase in carrier gas flow from 1 to 2 L/min resulted in better SBRs for the neutral atom lines of Cr, Co, Ni, Mo, and Tl, all of which

can be considered as soft lines; evidently low-power conditions (1000 W) were the best for these lines. This is reflected in the detection limits in column VI of Table 4.9, which for the soft lines stated are up to a factor of seven better than for the conventional argon ICP operated with a 1 L/min carrier gas flow. A comparison with the latter ICP operated at 2 L/min carrier gas flow is lacking, however, as are data on the interferences that must be tolerated at this high carrier gas flow.

2. The highest SBRs and the lowest detection limits for the soft lines considered in either the pure argon or N_2 outer flow ICPs were obtained at 30–40-mm observation heights (1000–1200-W power). This seems understandable in light of the observation that the background in the 340 to 390-nm region decreased by one to two orders of magnitude as the observation height was increased from 10 to 40 mm; for the soft lines considered all have wavelengths in the 340 to 390-nm region. For all other lines observed the background intensities in the N_2–Ar ICP decreased by factors of 2 to 3 when the observation height was increased from 10 to 20 mm, but showed an increasing trend above 20 mm observation height.

It appears to the present author that much of the behavior of the background intensity as a function of power, observation height, and carrier gas flow will be understandable if it is analyzed in terms of the contributions from continua and band spectra and their dependence on the ICP parameters. This will require measurement of SBRs from wavelength scans and observation of the behavior of the background structure in the vicinity of the analysis lines. The usual approach, in which SBRs are determined by observing the signals in the spectral window of an analysis line for a solution containing the analyte and a blank solution respectively, conceals any differentiations between the various contributions.

3. In contrast to the conventional argon ICP, an ICP with N_2 as outer gas required power levels in a range between 2 and 3 kW to reach optimum SBRs for hard lines, but even then the detection limits lagged substantially behind those achieved in the conventional argon ICP, that is, if the N_2–Ar is operated with an ultrasonic nebulizer plus desolvation apparatus, the detection limits are of the same magnitude as those in the conventional argon ICP operated with a pneumatic nebulizer (cf. Table 4.9).

Therefore, if a generator capable of delivering 2 to 3 kW is available and ultrasonic nebulization and desolvation [210] are used, an acceptable detection power (Table 4.9) can be reached with a reasonably low argon consumption of 2 L/min carrier gas and 2.5 L/min intermediate gas, while a 40 L/min N_2 outer flow is required. It is not known, however, what interference effects will be encountered at the 2 L/min carrier gas flow to be used under these conditions.

On the whole, Montaser et al. [115] (1) agreed with Ebdon et al. [269] that for N_2–Ar ICPs it is more difficult to find good compromise conditions than for an argon ICP, (2) found a "sensitivity advantage" (higher net line intensities) in N_2-Ar ICPs operated with a Fassel torch, at all forward powers and observations heights, provided that the percentage of N_2 in the outer gas did not exceed 5 to 15%; this sensitivity advantage did not translate into improved detection limits because of the higher background.

Meyer and Barnes [302] investigated analytical inductively coupled nitrogen and air plasmas using a specially designed torch having approximately the same diameter as a Fassel torch (Fig. 3.7). Unfortunately these very extensive explorations, including temperature profile measurements [303] and computer simulations [304], did not yet result in the design of an ICP source competitive with conventional argon ICPs or ICPs running on molecular outer gases with a Greenfield torch (Fig. 3.6), as has been illustrated in Section 4.1.8 by the detection limits quoted from Meyer and Thompson [302]. However, these efforts should be seen as one of the possible approaches to reduce the running costs for ICPs (cf. Chapter 5).

Meyer and Barnes [302] also investigated the direct injection of fine powders (20–40 μm) into air and argon ICPs. The results were consistent with a previous prediction [305] that the molecular gas ICP should be more effective than the argon ICP for sample decomposition. This is attributable to the high thermal conductivity of a molecular gas plasma (cf. Section 4.8).

In a series of papers, Choot and Horlick [309] recently published the results of an extensive study of mixed gas ICPs. This study covered the spatial emission and spectral characteristics and an evaluation of the analytical performance.

REFERENCES

1. I.U.P.A.C., "Nomenclature, Symbols, Units and Their Usage in Spectrochemical Analysis-II. Data Interpretation," *Pure Appl. Chem.* **45,** 99 (1976), and *Spectrochim. Acta* **33B,** 241 (1978).
2. H. Kaiser, *Z. Anal. Chem.* **209,** 1 (1965).
3. H. Kaiser, *Spectrochim. Acta* **3,** 40 (1947).
4. H. Kaiser, *Z. Anal. Chem.* **216,** 80 (1966).
5. H. Kaiser, *Spectrochim. Acta* **33B,** 551 (1978).
6. H. Kaiser, *Two Papers on the Limit of Detection of a Complete Analytical Procedure.* English Transl. of [66] and [68]. Hilger, London (1968).
7. H. Kaiser and H. Specker, *Z. Anal. Chem.* **149,** 46 (1956).
8. V. V. Nalimov, *The Application of Mathematical Statistics to Chemical Analysis.* Pergamon, Oxford (1963).
9. G. Ehrlich, *Reinstoffe in Wissenschaft und Technik, 3. Int. Symp., Plenar- und Hauptvortraege.* Akademie-Verlag, Berlin (1972), p. 861.

10. G. Ehrlich, *Z. Anal. Chem.* **232,** 1 (1967).
11. G. Ehrlich and H. Mai, *Z. Anal. Chem.* **218,** 1 (1966).
12. L. A. Curry, *Anal. Chem.* **40,** 586 (1968).
13. V. Svoboda and R. Gerbatsch, *Z. Anal. Chem.* **242,** 1 (1968).
14. O. G. Koch and G. A. Koch-Dedic, *Handbuch der Spurenanalyse,* Vol. I, Springer, Berlin (1974), p. 12.
15. J. D. Winefordner, Appendix A in J. D. Winefordner, ed., *Trace Analysis, Spectroscopic Methods for Elements.* Wiley, New York (1976), p. 435.
16. Kh. I. Zilbershtein, ed., *Spectrochemical Analysis of Pure Substances.* Hilger, Bristol (1977).
17. K. Zimmer, *Proc. 20th Coll. Spectr. Int. and 7th Int. Conf. Atomic Spectr., Prague 1977, Invited Lectures,* Vol. 2. Statni Pedagogicke Nakladatelstvi, Prague (1977), p. 17.
18. G. L. Long and J. D. Winefordner, *Anal. Chem.* **55,** 712A (1983).
19. P. W. J. M. Boumans, *Spectrochim. Acta* **33B,** 625 (1978).
20. H. Kaiser, *Optik* **21,** 309 (1964).
21. K. Laqua, *Z. Anal. Chem.* **221,** 44 (1966).
22. K. Laqua, W.-D. Hagenah, and H. Waechter, *Z. Anal. Chem.* **225,** 142 (1967).
23. U. Haisch, *Spectrochim. Acta* **25B,** 597 (1970).
24. U. Haisch, K. Laqua, and W.-D. Hagenah, *Spectrochim. Acta* **26B,** 651 (1971).
25. U. Haisch, *Z. Anal. Chem.* **259,** 1 (1972).
26. H. Bubert, W.-D. Hagenah, and K. Laqua, *Spectrochim. Acta* **34B,** 19 (1979).
27. J. A. C. Broekaert, F. Leis, and K. Laqua, *Spectrochim. Acta* **34B,** 73 (1979).
28. K. Laqua, "Emissionsspektroskopie," in *Ullmanns Encyklopaedie der technischen Chemie,* 4th ed., Vol. 5. Verlag Chemie GmbH, Weinheim (1980), p. 441.
29. J. D. Winefordner and T. J. Vickers, *Anal. Chem.* **36,** 1939 (1964).
30. J. D. Winefordner, *Appl. Spectrosc.* **17,** 109 (1963).
31. T. L. Chester and J. D. Winefordner, *Anal. Chem.* **49,** 119 (1977).
32. J. D. Winefordner and C. Veillon, *Anal. Chem.* **37,** 416 (1965).
33. J. D. Winefordner, W. J. McCarthy, and P. A. St. John, *J. Chem. Education* **44,** 80 (1967).
34. M. L. Parsons, W. J. McCarthy, and J. D. Winefordner, *J. Chem. Education* **44,** 214 (1967).
35. M. L. Parsons and J. D. Winefordner, *Appl. Spectrosc.* **21,** 368 (1967).
36. L. de Galan and J. D. Winefordner, *Spectrochim. Acta* **23B,** 277 (1968).
37. E. D. Prudnikov, *Spectrochim. Acta* **34B,** 293 (1979).
38. E. D. Prudnikov, *Spectrochim. Acta* **36B,** 385 (1981).
39. P. W. J. M. Boumans, *Proc. 21st Coll. Spectr. Int. and 8th Int. Conf. Atomic Spectr., Cambridge 1979, Keynote Lectures.* Heyden, London (1979), p. 49; *Spectrochim. Acta* **35B,** 57 (1980).

40. J. D. Winefordner, J. J. Fitzgerald, and H. Omenetto, *Appl. Spectrosc.* **29**, 369 (1975).
41. J. D. Winefordner, ed., *Trace Analysis, Spectroscopic Methods for Elements.* Wiley, New York (1976).
42. C. Th. J. Alkemade, W. Snelleman, G. D. Boutilier, B. D. Pollard, J. D. Winefordner, T. L. Chester, and N. Omenetto, *Spectrochim. Acta* **33B**, 383 (1978).
43. G. D. Boutilier, B. D. Pollard, J. D. Winefordner, T. L. Chester, and N. Omenetto, *Spectrochim. Acta* **33B**, 401 (1978).
44. C. Th. J. Alkemade, W. Snelleman, G. D. Boutilier, and J. D. Winefordner, *Spectrochim. Acta* **35B**, 261 (1980).
45. P. W. J. M. Boumans and F. J. M. J. Maessen, *Z. Anal. Chem.* **220**, 241 (1966); **225**, 98 (1967).
46. S. El Alfy, Ph.D. thesis, University of Dortmund (1978).
47. H. Bubert and W.-D. Hagenah, *Spectrochim. Acta* **36B**, 489 (1981).
48. M. S. Epstein and J. D. Winefordner, *Progr. Anal. Atom. Spectrosc.* **7**, 67 (1984).
49. S. W. McGeorge and E. D. Salin, *Spectrochim. Acta.* **40B**, 447 (1985).
50. S. W. McGeorge and E. D. Salin, *Spectrochim. Acta.* **40B**, 435 (1985).
51. "ICP Detection Limits Committee, Publication of an ICP Detection Limits Program," *ICP Information Newslett.* **6**, 295 (1979).
52. G. Toelg, *Talanta* **19**, 1489 (1972).
53. G. Toelg, *Pure Appl. Chem.* **44**, 645 (1975).
54. G. Toelg, *Naturwissenschaften* **63**, 99 (1976).
55. G. Toelg, *Mikrochim. Acta.*, Suppl. 7, 1 (1977).
56. P. D. LaFleur, ed., *Accuracy in Trace Analysis: Sampling, Sample Handling, Analysis*, Vols. 1, 2. N.B.S. Spec. Tech. Publ. 422. U.S. Government Printing Office, Washington, DC (1976).
57. R. H. Wendt and V. A. Fassel, *Anal. Chem.* **37**, 920 (1965).
58. G. W. Dickinson and V. A. Fassel, *Anal. Chem.* **41**, 1021 (1969).
59. S. Greenfield, I. Ll. Jones, and C. T. Berry, *Analyst* **89**, 713 (1964).
60. P. W. J. M. Boumans and F. J. de Boer, *Spectrochim. Acta* **27B**, 391 (1972).
61. P. W. J. M. Boumans and F. J. de Boer, *Spectrochim. Acta* **30B**, 309 (1975).
62. P. W. J. M. Boumans and F. J. de Boer, *Spectrochim. Acta* **32B**, 365 (1977).
63. P. W. J. M. Boumans, F. J. de Boer, F. J. Dahmen, H. Hoelzel, and A. Meyer, *Spectrochim. Acta* **30B**, 449 (1975).
64. P. W. J. M. Boumans, *ICP Information Newslett.* **4**, 232 (1978).
65. R. K. Winge, V. J. Peterson, and V. A. Fassel, *Appl. Spectrosc.* **33**, 206 (1979).
66. C. D. Allemand, R. M. Barnes, and C. C. Wohlers, *Anal. Chem.* **51**, 2392 (1979).
67. P. W. J. M. Boumans, R. J. McKenna, and M. Bosveld, *Spectrochim. Acta* **36B**, 1031 (1981).

REFERENCES

68. S. Greenfield, *ICP Information Newslett.* **4,** 199 (1978).
69. S. Greenfield and D. Thorburn Burns, *Spectrochim. Acta* **34B,** 423 (1979).
70. J. D. Ingle, Jr., and S. R. Crouch, *Anal. Chem.* **43,** 1331 (1971).
71. J. D. Ingle, Jr., and S. R. Crouch, *Anal. Chem.* **44,** 1709 (1972).
72. J. D. Ingle, Jr., and S. R. Crouch, *Anal. Chem.* **44,** 785 (1972).
73. R. E. Santini, *Anal. Chem.* **44,** 1708 (1972).
74. T. C. O'Haver, "Analytical Considerations," in J. D. Winefordner, ed., *Trace Analysis, Spectroscopic Methods for Elements,* Ch. 2. Wiley, New York (1976).
75. P. W. J. M. Boumans and J. J. A. M. Vrakking, *Spectrochim. Acta.* **39B,** 1261 (1984).
76. H. Siedentopf, *Physik. Z.* **38,** 454 (1937).
77. H. Kaiser, *Spectrochim. Acta.* **3,** 159 (1948).
78. P. W. J. M. Boumans, *Preprints 16th Coll. Spectr. Int., Heidelberg 1971,* Vol. 2, Adam Hilger, London (1971), p. 247.
79. U. Haisch and K. Jungmann, *Spectrochim. Acta* **39B,** 1221 (1984).
80. D. H. Tracy and S. A. Myers, *Spectrochim. Acta* **37B,** 1055 (1982).
81. I. B. Brenner, A. E. Watson, T. W. Steele, E. A. Jones, and M. Goncalves, *Spectrochim. Acta* **36B,** 785 (1981).
82. B. T. N. Newland and R. A. Mostyn, *ICP Information Newslett.* **1,** 183 (1976).
83. B. T. N. Newland and R. A. Mostyn, *ICP Information Newslett.* **2,** 135 (1976).
84. A. E. Watson, G. M. Russell, and G. Balaes, *ICP Information Newslett.* **2,** 205 (1976).
85. P. W. J. M. Boumans and M. Bosveld, *Spectrochim. Acta* **34B,** 59 (1979).
86. P. W. J. M. Boumans, F. J. de Boer, A. W. Witmer, and M. Bosveld, *Spectrochim. Acta.* **33B,** 535 (1978).
87. P. W. J. M. Boumans, *Spectrochim. Acta* **38B,** 747 (1983).
88. R. M. Barnes, *CRC Crit. Rev. Anal. Chem.* **7,** 203 (1978).
89. S. Greenfield, H. McD. McGeachin, and P. B. Smith, *Talanta* **23,** 1 (1976).
90. S. Greenfield, I. L. W. Jones, C. T. Berry, and L. G. Bunch, *Proc. Soc. Anal. Chem.* **2,** 111 (1965).
91. S. Greenfield, C. T. Berry, and L. G. Bunch, *Spectroscopy with a High Frequency Plasma Torch.* Radyne International, Wokingham, England (1965).
92. G. Pforr and V. Kapicka, *Collection Czech Chem. Comm.* **31,** 4710 (1966).
93. H. C. Hoare and R. A. Mostyn, *Anal. Chem.* **39,** 1153 (1967).
94. M. E. Britske, V. M. Borisov, and Yu. S. Sukah, *Ind. Lab.* **33,** 301 (1967).
95. C. Veillon and M. Margoshes, *Spectrochim. Acta* **23B,** 503 (1968).
96. S. Greenfield, I. L. W. Jones, and C. T. Berry, U.S. Patent 3,467,471, September 16, 1969.
97. J. C. Souilliart and J. Robin, *Analusis* **1,** 427 (1972).
98. J. M. Mermet, *C.R. Acad. Sci. Ser. B* **281,** 273 (1975).

99. R. N. Kniseley, H. Amenson, C. C. Butler, and V. A. Fassel, *Appl. Spectrosc.* **28**, 285 (1974).
100. K. W. Olson, W. J. Haas, Jr., and V. A. Fassel, *Anal. Chem.* **49**, 632 (1977).
101. R. H. Scott, V. A. Fassel, R. N. Kniseley, and R. N. Nixon, *Anal. Chem.* **46**, 75 (1974).
102. V. A. Fassel and R. N. Kniseley, *Anal. Chem.* **46**, 1110A, 1155A (1974).
103. P. W. J. M. Boumans and F. J. de Boer, *Proc. Anal. Div. Chem. Soc.* **12**, 140 (1975).
104. P. W. J. M. Boumans, *Line Coincidence Tables for Inductively Coupled Plasma Atomic Emission Spectrometry*. Pergamon Press, Oxford (1980).
105. P. W. J. M. Boumans, *Spectrochim. Acta* **36B**, 169 (1981).
106. W. F. Meggers, C. H. Corliss, and B. F. Scribner, "Tables of Spectral-Line Intensities," *N.B.S. Monograph 145*. U. S. Government Printing Office, Washington, DC (1975).
107. P. W. J. M. Boumans, *Line Coincidence Tables for Inductively Coupled Plasma Atomic Emission Spectrometry*, 2nd revised ed. Pergamon Press, Oxford (1984).
108. P. W. J. M. Boumans and J. J. A. M. Vrakking, *Spectrochim. Acta* **40B**, 1437 (1985).
109. P. W. J. M. Boumans and J. J. A. M. Vrakking, *Spectrochim. Acta* **39B**, 1239 (1984).
110. C. C. Wohlers, "ICP-AES Wavelength Table," *ICP Information Newslett.* **10**, 601 (1985).
111. L. A. Fernando, *Spectrochim. Acta* **37B**, 859 (1982).
112. R. K. Winge, V. A. Fassel, R. N. Kniseley, E. DeKalb, and W. J. Haas, Jr., *Spectrochim. Acta* **32B**, 327 (1977).
113. C. E. Taylor and T. L. Floyd, *Appl. Spectrosc.* **35**, 408 (1981).
114. P. W. J. M. Boumans, *ICP Information Newslett.* **1**, 222 (1976).
115. A. Montaser, V. A. Fassel, and J. Zalewski, *Appl. Spectrosc.* **35**, 292 (1981).
116. M. A. Floyd, V. A. Fassel, and A. P. D'Silva, *Anal. Chem.* **52**, 2168 (1980).
117. M. A. Floyd, V. A. Fassel, R. K. Winge, J. M. Katzenberger, and A. P. D'Silva, *Anal. Chem.* **52**, 431 (1980).
118. G. F. Kirkbright, A. F. Ward, and T. S. West, *Anal. Chim. Acta* **62**, 241 (1972).
119. G. F. Kirkbright, A. F. Ward, and T. S. West, *Anal. Chim. Acta* **64**, 353 (1973).
120. G. F. Wallace, *Atom. Spectrosc.* **1**, 38 (1980).
121. R. D. Ediger, *Atom. Spectrosc.* **1**, 59 (1980).
122. G. F. Wallace, D. W. Hoult, and R. D. Ediger, *Atom. Spectrosc.* **1**, 120 (1980).
123. G. F. Wallace, *Atom. Spectrosc.* **2**, 61 (1981).
124. G. F. Wallace and R. D. Ediger, *Atom. Spectrosc.* **2**, 169 (1981).
125. T. Nakahara, *Spectrochim. Acta* **40B**, 293 (1985).
126. T. Hayakawa, F. Kikui, and S. Ikeda, *Spectrochim. Acta* **37B**, 1069 (1983).
127. J. Lee and M. W. Pritchard, *Spectrochim. Acta* **36B**, 591 (1981).

128. C. Butler-Sobel and M. Cass, Paper, *9th Int. Conf. At. Spectr./22nd Coll. Spectr. Int.*, Tokyo, *1981*, Abstract No. 7A05.
129. D. R. Heine, J. S. Babis, and M. B. Denton, *Appl. Spectrosc.* **34,** 595 (1980).
130. J. A. C. Broekaert and P. B. Zeeman, *Spectrochim. Acta.* **39B,** 851 (1984).
131. S. R. Ellebracht and C. M. Fairless, in R. M. Barnes, ed., *Developments in Atomic Plasma Spectrochemical Analysis.* Heyden, London/Philadelphia (1981), p. 392.
132. R. L. Kelly, "A Table of Emission Lines in the Vacuum Ultraviolet for All Elements." University of California, Lawrence Radiation Laboratory, Livermore, CA, UCRL 5612 (1959).
133. R. L. Kelly and L. J. Palumbo, "Atomic and Ionic Emission Lines below 2000 Angstroms—Hydrogen through Krypton," NRL Report 7599. NRL, Washington, DC (1973).
134. S. K. Hughes and R. C. Fry, *Appl. Spectrosc.* **35,** 493 (1981).
135. P. W. J. M. Boumans, *Paper No. 50, Int. Winter Conf. 1980, Developments in Atomic Plasma Spectrochemical Analyses,* San Juan, Puerto Rico, 1980.
136. P. W. J. M. Boumans, R. F. Rumphorst, L. Willemsen, and F. J. de Boer, *Spectrochim. Acta.* **28B,** 227 (1973).
137. H. Bubert, W.-D. Hagenah, and K. Laqua, *Spectrochim. Acta* **33B,** 701 (1978).
138. H. Bubert, *Spectrochim. Acta* **37B,** 533 (1982).
139. T. Nakahara, *Prog. Anal. Atom. Spectrosc.* **6,** 163 (1983).
140. G. L. Moore, P. J. Humphries-Cuff and A. E. Watson, *Spectrochim. Acta* **39B,** 915 (1984).
141. A. Aziz, J. A. C. Broekaert, and F. Leis, *Spectrochim. Acta* **37B,** 369 (1982).
142. G. A. Meyer and M. D. Thompson, *Spectrochim. Acta* **40B,** 195 (1985).
143. K. Ohls and D. Sommer, *ICP Information Newslett.* **9,** 555 (1984).
144. N. Omenetto, H. G. C. Human, P. Cavalli, and G. Rossi, *Spectrochim. Acta* **39B,** 115 (1984).
145. H. G. C. Human, N. Omenetto, P. Cavalli, and G. Rossi, *Spectrochim. Acta* **39B,** 1345 (1984).
146. D. R. Demers, *Spectrochim. Acta* **40B,** 93 (1985).
147. R. L. Lancione and D. M. Drew, *Spectrochim. Acta* **40B,** 107 (1985).
148. E. B. M. Jansen and R. D. Demers, *Analyst* **110,** 541 (1985).
149. A. L. Gray, *Proc. 1985 European Winter Conf. Plasma Spectrochemistry,* Leysin, Switzerland, *Spectrochim. Acta* **41B,** 151 (1986).
150. D. J. Douglas and R. S. Houk, *Progr. Anal. Atom. Spectrosc.* **8,** 1 (1985).
151. A. L. Gray, *Proc. Symp. Instr. Multielement Anal., Juelich 1984,* B. Sansoni, ed., Verlag Chemie, Weinheim, F.R.G. (1985), p. 227.
152. U. Haisch, K. Laqua, W.-D. Hagenah, and H. Waechter, *Fresenius Z. Anal. Chem.* **316,** 157 (1983).
153. J. Mandel, *The Statistical Analysis of Experimental Data.* Wiley, New York (1964).

154. F. J. M. J. Maessen and H. Balke, *Spectrochim. Acta* **37B,** 37 (1982).
155. J. Agterdenbos, *Anal. Chim. Acta.* **108,** 315 (1979).
156. F. J. M. J. Maessen, J. W. Elgersma, and P. W. J. M. Boumans, *Spectrochim. Acta.* **31B,** 179 (1976).
157. H. Bubert and R. Klockenkaemper, *Spectrochim. Acta* **38B,** 1087 (1983).
158. H. Bubert, R. Klockenkaemper, and H. Waechter, *Spectrochim. Acta.* **39B,** 1465 (1984).
159. P. W. J. M. Boumans, *Fresenius Z. Anal. Chem.* **299,** 337 (1979).
160. P. W. J. M. Boumans, Spectrochim. Acta **31B,** 147 (1976).
161. S. Greenfield, in R. M. Barnes, ed., *Developments in Atomic Plasma Spectrochemical Analysis.* Heyden, London/Philadelphia (1981), p. 1.
162. R. M. Ajhar, P. D. Dalager, and A. L. Davison, *Amer. Lab.* **8** (3), 71 (1976).
163. R. Klockenkaemper and H. Bubert, *Spectrochim. Acta* **37B,** 127 (1982).
164. R. M. Belchamber and G. Horlick, *Spectrochim. Acta* **37B,** 17 (1982).
165. Yu. J. Belyaev, L. M. Iranstov, A. V. Karyakin, P. H. Phi, and V. V. Shimet, *J. Anal. Chem. USSR* **23,** 855 (1968).
166. Y. Talmi, R. Crosmun, and N. M. Larson, *Anal. Chem.* **48,** 326 (1976).
167. J. D. Ingle, Jr., *Anal. Chem.* **49,** 339 (1977).
168. C. Th. J. Alkemade, H. P. Hooymayers, P. L. Lijnse, and T. J. M. J. Uienberger, *Spectrochim. Acta* **27B,** 249 (1972).
169. G. M. Hieftje and R. I. Bystroff, *Spectrochim. Acta* **30B,** 187 (1975).
170. C. Th. J. Alkemade, T. Hollander, and K. E. J. Zijlstra, *Spectrochim. Acta* **34B,** 85 (1979).
171. N. W. Bower and J. D. Ingle, Jr., *Spectrochim. Acta* **34B,** 275 (1979).
172. G. L. Walden, J. N. Bower, S. Nikdel, D. L. Bolten, and J. D. Winefordner, *Spectrochim. Acta* **35B,** 535 (1980).
173. R. D. Deutsch and G. M. Hieftje, *Appl. Spectrosc.* **39,** 19 (1985).
174. S. Cova and A. Longoni, in N. Omenetto, ed., "Analytical Laser Spectroscopy," Ch. 7, *Chemical Analysis*, Vol. 30. Wiley, New York (1979), p. 411.
175. C. Th. J. Alkemade, Tj. Hollander, J. Jansen, H. Snippe, and R. J. J. Zijlstra, *Spectrochim. Acta* **38B,** 669 (1983).
176. R. King, *Electrical Noise.* Chapman and Hall, London (1966).
177. R. M. Belchamber and G. Horlick, *Spectrochim. Acta* **37B,** 71 (1982).
178. J. D. Winefordner, R. Avni, T. L. Chester, J. J. Fitzgerald, L. P. Hart, D. J. Johnson, and F. W. Plankey, *Spectrochim. Acta* **31B,** 1 (1976).
179. P. Benetti, A. Bonelli, M. Cambiaghi, and P. Frigieri, *Spectrochim. Acta* **37B,** 1047 (1982).
180. D. H. Tracy, S. A. Myers, and B. G. Balistee, *Spectrochim. Acta* **37B,** 739 (1982).
181. R. M. Belchamber and G. Horlick, *Spectrochim. Acta* **37B,** 1075 (1982).
182. W. Gerlach and E. Schweitzer, *Die chemische Emissionsspektralanalyse*, Vol. I. Leopold Voss, Leipzig (1930).

183. R. M. Belchamber and G. Horlick, *Spectrochim. Acta* **37B,** 1037 (1982).
184. S. A. Myers and D. H. Tracy, *Spectrochim. Acta* **38B,** 1227 (1983).
185. A. Lorber, Z. Goldbart, and M. Eldan, *Anal. Chem.* **56,** 43 (1984).
186. J. R. Garbardino and H. E. Taylor, *Spectrochim. Acta* **38B,** 323 (1983).
187. J. R. Garbardino and H. E. Taylor, *Appl. Spectrosc.* **33,** 220 (1979).
188. R. I. Botto, *Spectrochim. Acta* **39B,** 95 (1984).
189. K. A. Wolnik, F. L. Fricke, and C. M. Gaston, *Spectrochim. Acta* **39B,** 649 (1984).
190. K. Govindaraju and G. Mevelle, *Spectrochim. Acta* **38B,** 1447 (1983).
191. P. Braetter, K. P. Berthold, and P. E. Gardiner, *Spectrochim. Acta* **38B,** 221 (1982).
192. F. J. M. J. Maessen, J. Balke, and J. M. de Boer, *Spectrochim. Acta* **37B,** 517 (1982).
193. R. I. Botto, *Spectrochim. Acta* **40B,** 397 (1985).
194. R. I. Botto, *Anal. Chem.* **54,** 1654 (1982).
195. L. R. P. Butler, *Spectrochim. Acta* **38B,** 913 (1983).
196. P. W. J. M. Boumans, "Excitation of Spectra," in E. L. Grove, ed., *Analytical Emission Spectroscopy*, Part 2, Ch. 6. Dekker, New York (1972), p. 155.
197. C. Th. J. Alkemade and R. Herrmann, *Fundamentals of Flame Spectroscopy.* Adam Hilger, Bristol (1979).
198. C. Th. J. Alkemade, Tj. Hollander, W. Snelleman, and P. J. Th. Zeegers, *Metal Vapours in Flames.* Pergamon Press, Oxford (1982).
199. R. D. Cowan and G. H. Dieke, *Rev. Mod. Phys.* **20,** 418 (1948).
200. H. G. C. Human and R. H. Scott, *Spectrochim. Acta* **31B,** 459 (1976).
201. G. F. Larson and V. A. Fassel, *Appl. Spectrosc.* **33,** 592 (1979).
202. J. D. Winefordner, W. W. McGee, J. M. Mansfield, M. L. Parsons, and K. E. Zacha, *Anal. Chim. Acta* **36,** 25 (1966).
203. C. S. Rann, *Spectrochim. Acta* **23B,** 245 (1968).
204. H. Kawaguchi, Y. Oshio, and A. Mizuike, *Spectrochim. Acta* **37B,** 809 (1982).
205. J. W. McLaren and J. M. Mermet, *Spectrochim. Acta* **39B,** 1307 (1984).
206. S. Greenfield, I. Ll. Jones, H. McD. McGeachin, and P. B. Smith, *Anal. Chim. Acta* **74,** 225 (1975).
207. V. A. Fassel, *Proc. 16th Coll. Spectr. Int., Heidelberg 1971, Plenary lectures and reports.* Adam Hilger, London (1972), p. 63.
208. R. H. Wendt and V. A. Fassel, *Anal. Chem.* **38,** 337 (1966).
209. G. F. Larson, V. A. Fassel, R. H. Scott, and R. N. Kniseley, *Anal. Chem.* **47,** 238 (1975).
210. P. W. J. M. Boumans and F. J. de Boer, *Spectrochim. Acta* **31B,** 355 (1976).
211. M. H. Abdallah, J. M. Mermet, and C. Trassy, *Anal. Chim. Acta* **87,** 329 (1976).
212. V. A. Fassel, C. A. Peterson, F. N. Abercrombie, and R. N. Kniseley, *Anal. Chem.* **48,** 516 (1976).

213. P. W. J. M. Boumans and F. J. de Boer, *20th Coll. Spectr. Int. and 7th Int. Conf. Atomic Spectr., Prague 1977*, Paper No. 19. Cf. *ICP Information Newslett.* **3,** 228 (1977).
214. R. W. Larenz, *Z. Physik* **200,** 239 (1951).
215. H. Krempl, *Z. Physik* **167,** 302 (1962).
216. H. Krempl, "Equilibria at Very High Temperatures," in W. Jost, ed., *Physical Chemistry*, Vol. I, Ch. 8. Academic Press, New York (1971), p. 545.
217. H. Krempl, *Z. Anal. Chem.* **198,** 21 (1963).
218. R. Diermeier and H. Krempl, *Z. Physik* **200,** 239 (1967).
219. W. Boegershausen, J. Hingshammer, and H. Krempl, *Ber. Bunsenges. Phys. Chem.* **71,** 64 (1967).
220. P. W. J. M. Boumans, *Philips Tech. Rev.* **34,** 305 (1974).
221. P. W. J. M. Boumans, *Theory of Spectrochemical Excitation*. Adam Hilger, London/Plenum, New York (1966).
222. S. Greenfield, *Proc. Anal. Div. Chem. Soc.* **13,** 279 (1976).
223. M. W. Blades and G. Horlick, *Spectrochim. Acta* **36B,** 861 (1981).
224. P. W. J. M. Boumans and M. Ch. Lux-Steiner, *Spectrochim. Acta* **36B,** 97 (1982).
225. I. J. M. M. Raaijmakers, P. W. J. M. Boumans, B. van der Sijde, and D. C. Schram, *Spectrochim. Acta* **38B,** 697 (1983).
226. H. Kawaguchi, T. Ito, and A. Mizuike, *Spectrochim. Acta* **36B,** 615 (1981).
227. S. R. Koirtyohann, J. S. Jones, and D. A. Yates, *Anal. Chem.* **53,** 1965 (1981).
228. S. R. Koirtyohann, J. S. Jones, C. P. Jester, and D. A. Yates, *Spectrochim. Acta* **36B,** 49 (1981).
229. S. Greenfield, H. McD. McGeachin, and P. B. Smith, *Anal. Chim. Acta* **84,** 67 (1976).
230. K. Ohls, K. H. Koch, and H. Grote, *Z. Anal. Chem.* **284,** 177 (1977).
231. P. W. J. M. Boumans, L. C. Bastings, F. J. de Boer, and L. W. J. van Kollenburg, *Fresenius Z. Anal. Chem.* **291,** 10 (1978).
232. J. Robin, *Analusis* **6,** 89 (1978); *ICP Information Newslett.* **4,** 495 (1979).
233. G. Dube and M. I. Boulos, *Can. J. Spectrosc.* **22,** 68 (1977).
234. S. S. Berman and J. W. McLaren, *Appl. Spectrosc.* **32,** 372 (1978).
235. T. Edmonds and G. Horlick, *Appl. Spectrosc.* **22,** 68 (1977).
236. M. W. Blades and G. Horlick, *Spectrochim. Acta* **36B,** 881 (1981).
237. M. Franklin, C. Baber, and S. R. Koirtyohann, *Spectrochim. Acta* **31B,** 589 (1976).
238. P. W. J. M. Boumans, *Spectrochim. Acta* **37B,** 75 (1982).
239. G. R. Kornblum and L. de Galan, *Spectrochim. Acta* **32B,** 455 (1977).
240. M. Ch. Lux-Steiner, Ph.D. thesis, Swiss Federal Institute of Technology, Zürich, Switzerland (1980).
241. P. W. J. M. Boumans, *Second ICP Conf. Noordwijk aan Zee 1978, ICP Information Newslett.* **4,** 124 (1978).

242. W. J. Haas, Jr., V. A. Fassel, F. Grabau IV, R. N. Kniseley, and W. L. Sutherland, "Simultaneous Determination of Trace Elements in Urine by Inductively Coupled Plasma—Atomic Emission Spectrometry," in *Ultratrace Metal Analysis in Science and Environment*, Ch. 8. American Chemical Society, Washington, DC (1979), p. 91.

243. W. J. Haas and V A. Fassel, "Inductively Coupled Plasma Atomic Emission Spectrometry," in *Elemental Analysis of Biological Materials*, Ch. 9. Int. Atomic Energy Agency, Tech. Rept 197, Vienna, Austria (1979), p. 167.

244. R. L. Dahlquist and J. W. Knoll, *Appl. Spectrosc.* **32**, 1 (1978).

245. C. A. Peterson, Ph.D. thesis, Iowa State University, Ames, IA (1977).

246. R. W. B. Pearse and A. G. Gaydon, *The Identification of Molecular Spectra*, 4th ed. Chapman and Hall, London (1976).

247. B. Magyar, P. Lienemann, and S. Wunderli, *GIT Fachz. Lab.* **26**, 541 (1982).

248. M. W. Blades and P. Hauser, *Anal. Chim. Acta* **157**, 163 (1984).

249. K. Ohls and D. Sommer, *Erdoel und Kohle, Petrochemie vereinigt mit Brennstoff-Chemie* **37**, 177 (1984).

250. R. L. Lancione and D. M. Drew, *ICP Information Newslett.* **9**, 527 (1984).

251. B. Magyar, P. Lienemann, and H. Vonmont, *Proc. 1985 European Winter Conf. Plasma Spectrochemistry, Leysin, Spectrochim. Acta* **41B**, 27 (1986).

252. A. F. Ward, H. R. Sobel, R. L. Crawford, *ICP Information Newslett.* **3**, 94 (1977).

253. F. J. M. J. Maessen, P. J. H. Seeverens, and G. Kreuning, *Spectrochim. Acta* **39B**, 1171 (1984).

254. F. J. M. J. Maessen, G. Kreuning, and J. Balke, *Proc. 1985 European Winter Conf. Plasma Spectrochemistry, Leysin, Spectrochim. Acta* **41B**, 3 (1986).

255. A. W. Boorn, M. S. Cresser, and R. F. Browner, *Spectrochim. Acta* **35B**, 823 (1980).

256. A. W. Boorn and R. F. Browner, *Anal. Chem.* **54**, 1402 (1982).

257. R. F. Browner and A. W. Boorn, *Anal. Chem.* **56**, 786A (1984).

258. F. J. M. J. Maessen, P. Coevert, and J. Balke, *Anal. Chem.* **56**, 899 (1984).

259. M. W. Blades and B. L. Caughlin, *Spectrochim. Acta* **40B**, 579 (1985).

260. A. Miyazaki, A. Kimura, K. Bansho, and Y. Umezaki, *Anal. Chim. Acta* **144**, 213 (1982).

261. J. J. Leary, A. E. Brokes, A. F. Dorrzapf, Jr., and D. W. Golightly, *Appl. Spectrosc.* **36**, 37 (1982).

262. W. Spendley, G. R. Hext, and F. R. Himsworth, *Technometrics* **4**, 441 (1962).

263. J. A. Nelder and R. Mead, *Comput. J.* **7**, 308 (1965).

264. S. N. Deming and L. R. Parker, *CRC. Crit. Rev. Anal. Chem.* **7**, 187 (1978).

265. L. A. Yarbro and S. M. Deming, *Anal. Chim. Acta* **73**, 391 (1974).

266. L. de Galan, G. R. Kornblum, and M. T. C. de Loos-Vollebregt, in K. Fuwa, ed., *Recent Advances in Analytical Spectrometry*. Pergamon Press, Oxford (1982), p. 33.

267. J. J. Brocas, *Analusis* **10**, 387 (1982).
268. S. Greenfield and D. Thorburn Burns, *Anal. Chim. Acta* **113**, 205 (1980).
269. L. Ebdon, M. R. Cave, and D. J. Mowthorpe, *Anal. Chim. Acta* **115**, 179 (1980).
270. M. J. Box, D. Davies, and W. H. Swann, *Non-Linear Optimization Techniques*. Oliver and Boyd, Edinburgh (1969).
271. L. Ebdon, D. J. Mowthorpe, and M. R. Cave, *Anal. Chim. Acta* **115**, 171 (1980).
272. S. P. Terblanche, K. Visser, and P. B. Zeeman, *Spectrochim. Acta* **36B**, 293 (1981).
273. T. B. Reed, *Int. Sci. Technol.* **6**, 42 (1962).
274. T. B. Reed, "Plasmas for High Temperature Chemistry," in L. Eyring, ed., *Advances in High Temperature Chemistry*, Vol. 1. Academic Press, New York (1967), p. 259.
275. H. U. Eckert, *High Temp. Sci.* **6**, 99 (1974).
276. R. M. Barnes and R. G. Schleicher, *Spectrochim. Acta* **30B**, 109 (1975).
277. H. U. Eckert, *J. Appl. Phys.* **48**, 1467 (1977).
278. S. V. Dresvin, ed., *Physics and Technology of Low-Temperature Plasmas*. Atomizdat, Moscow (1972); Engl. Transl.: The Iowa State University Press, Ames, IA (1977).
279. S. Greenfield and H. McD. McGeachin, *Anal. Chim. Acta* **100**, 101 (1978).
280. M. L. Thorpe, "NASA Contractors Report CR-1143" (1968).
281. A. Montaser and J. Mortazavi, *Anal. Chem.* **52**, 255 (1980).
282. R. C. Miller and R. I. Ayen, *Ind. Engng. Chem. Proc. Design. Dev.* **8**, 370 (1969).
283. S. Greenfield and P. B. Smith, *Anal. Chim. Acta* **57**, 209 (1971).
284. S. Greenfield and P. B. Smith, *Anal. Chim. Acta* **59**, 341 (1972).
285. M. Capitelli, F. Cramarossa, L. Triolo, and M. Molinari, *Combust. Flame* **15**, 23 (1970).
286. D. Truitt and J. W. Robinson, *Anal. Chim. Acta* **49**, 401 (1970); **51**, 61 (1970).
287. S. Greenfield, *The Spex Speaker,* Vol. 22, No. 3, (1977), p. 1.
288. S. Greenfield, *Analyst* **105**, 1032 (1980).
289. V. A. Fassel, *Pure Appl. Chem.* **49**, 1533 (1977).
290. S. Greenfield, *ICP Information Newslett.* **1**, 3 (1975).
291. A. E. Watson and G. M. Russell, Report No. 1907. National Institute of Metallurgy, Randburg, South Africa (1977); *ICP Information Newslett.* **3**, 273 (1977).
292. A. E. Watson and G. M. Russell, Report No. 1934. National Institute of Metallurgy, Randburg, South Africa (1977); *ICP Information Newslett.* **3**, 409 (1978).
293. A. E. Watson and T. W. Steele, Report No. 2029. National Institute of Metallurgy, Randburg, South Africa (1977); *ICP Information Newslett.* **5**, 553 (1980).
294. K. Ohls and D. Sommer, *ICP Information Newslett.* **4**, 532 (1979).
295. D. Sommer and K. Ohls, *Fresenius Z. Anal. Chem.* **295**, 337 (1979).
296. K. Ohls and D. Sommer, *Fresenius Z. Anal. Chem.* **296**, 241 (1979).

297. J. A. C. Broekaert, F. Leis, and K. Laqua, *Fresenius Z. Anal. Chem.* **301,** 105 (1980).
298. A. Aziz, J. A. C. Broekaert, and F. Leis, *Spectrochim. Acta* **36B,** 251 (1981).
299. J. A. C. Broekaert, F. Leis, and K. Laqua, in R. M. Barnes, ed., *Developments in Atomic Plasma Spectrochemical Analysis.* Heyden, London/Philadelphia (1981), p. 84.
300. J. A. C. Broekaert, F. Leis, and G. Dincler, *Analyst* **108,** 717 (1983).
301. R. M. Barnes and G. A. Meyer, *Anal. Chem.* **52,** 1523 (1980).
302. G. A. Meyer and R. M. Barnes, *Spectrochim. Acta* **40B,** 893 (1985).
303. N. Kovacic, G. A. Meyer, Lin Ke-ling, and R. M. Barnes, *Spectrochim. Acta* **40B,** 943 (1985).
304. R. M. Barnes, N. Kovacic, and G. A. Meyer, *Spectrochim. Acta* **40B,** 907 (1985).
305. R. M. Barnes and S. Nikdel, *Appl. Spectrosc.* **30,** 310 (1976).
306. P. W. J. M. Boumans, *Fresenius Z. Anal. Chem.* **324,** 397 (1986).
307. P. W. J. M. Boumans and J. J. A. M. Vrakking, *Spectrochim. Acta, Part B,* in press for Vol. **42B** (1987).
308. P. W. J. M. Boumans and J. J. A. M. Vrakking, *Spectrochim. Acta* **41B,** No. 12 (1986).
309. E. H. Choot and G. Horlick, *Spectrochim. Acta* **41B,** No. 9 (1986).

CHAPTER

5

TORCHES FOR INDUCTIVELY COUPLED PLASMAS

P. W. J. M. BOUMANS

Philips Research Laboratories
Eindhoven, The Netherlands

and

G. M. HIEFTJE

Department of Chemistry
Indiana University
Bloomington, Indiana

5.1. Torch Design Considerations
5.2. Demountable Versus Prealigned Torches
5.3. Special-Purpose Torches
5.4. New Trends in Torch Development
5.5. Influence of the Frequency on the Spectral Characteristics of Conventional ICPs
References

5.1. TORCH DESIGN CONSIDERATIONS

In the discussion of the principles of ICPs (Section 3.2) it has been indicated that the torch dictates the flow patterns of the gas streams in the ICP and must be considered as the most critical component of the ICP assembly. The construction and performance of torches for engineering purposes have been reviewed by various authors [1–5]. Although aim and operating conditions often differ widely from those in AES, a look at this literature may provide ideas useful to the design of torches for "analytical ICPs." The latter should generally be designed so that they provide an optimum answer to the following performance requirements [6, 7]:

1. Easy ignition of the plasma.
2. Continuous, stable plasma generation with a minimum influence of the injected sample, primarily the absence of risks of extinguishing the plasma and formation of deposits in the torch.

3. A sufficiently high sample flow through the plasma tunnel to the observation zone.
4. An optimum sample heating efficiency by a long residence time of the sample in the plasma.
5. A low gas consumption rate.
6. Minimal power requirements to reduce size and cost of the rf power supply.

Ultimately the quality of a torch is judged, not on the basis of what it is believed to do as a result of design considerations, but on its analytical behavior: detection limits, interferences, stability of background and analysis signals, maintenance during routine operation, and gas consumption.

Allemand and Barnes [6] studied torch design and torch shapes experimentally and by computer simulations involving power input, gas velocity and temperature distributions, and particle decomposition. Thirty different torches resulting from the combination of two different outer tubes, five different intermediate tubes, and three different inner tubes were evaluated. Ease of ignition, gas flow necessary to maintain stable operation, confinement of the sample in a narrow channel necessary for it to be optimally heated, atomized, and excited, and the prevention of clogging of the torch by organic samples were the main criteria. These studies indicated that (1) ignition was facilitated by a reverse flow produced by a Venturi effect resulting from a suction hole in the intermediate tube, (2) argon consumption could be drastically reduced by careful torch design, and (3) the torch with the lowest flow rate showed the best sample confinement. The ratio of intermediate-to-outer tube diameters was defined as the "configuration factor." It was derived from the computation that it is easier to maintain a plasma in a torch arrangement with a large configuration factor. Allemand's and Barnes' calculations showed also that torches with significantly smaller diameters can be operated if a sufficiently high magnetic flux density is used and the nozzle of the central tube is kept small enough to force the sample into the plasma, but also as large as possible to produce a sufficiently slow sample flow through the plasma tunnel.

The study of Allemand and Barnes [6] led to the streamlined torch configuration shown in Fig. 5.1, which required precisely machined dimensions obtainable with materials such as boron nitride, but not with quartz. This torch showed increased heating efficiency, greater ease of ignition and operation at lower outer gas flows than required by the Fassel torch, that is, 9 instead of 18 L/min.

Genna, Barnes, and Allemand [7] improved further on this design by incorporating into their torch a constricted nozzle on the outer gas inlet port. The resulting increase in the swirl velocity of the outer gas produced a more stable plasma, offered simpler and more reliable ignition, and permitted operation at

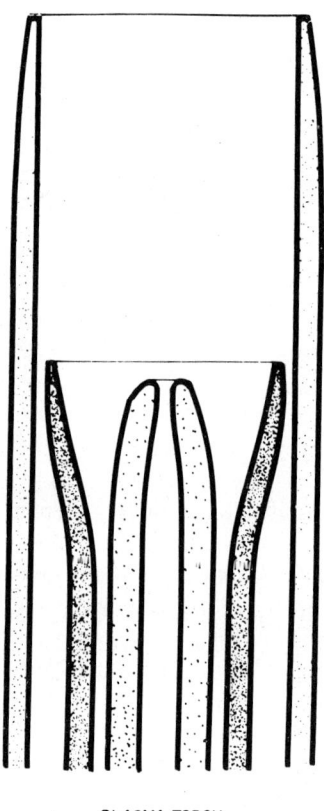

Figure 5.1. Configuration of a streamlined boron nitride torch [6]. (Reprinted with permission from C. D. Allemand and R. M. Barnes, "A Study of Inductively Coupled Plasma Torch Configurations," *Appl. Spectrosc.* **31**, 441 (1977). Copyright (1977), Society for Applied Spectroscopy, Frederick, MD.)

PLASMA TORCH

30–40% lower gas flow. It was found that a stable plasma could be maintained on as little as 5 L/min outer argon flow, although a higher flow (12 L/min) was favored for routine analytical use.

An interesting torch arrangement that incorporated the foregoing design considerations was described by Windsor et al. [8]. Not only was the configuration factor adjustable in their demountable torch, but swirl velocity was increased by adding a threaded spacer between the outer and intermediate tubes. Outer gas flowing upward between the tubes was forced to spiral through these threads, thereby increasing both the gas and swirl velocities. Importantly, the threaded spacer also helped maintain firm alignment and concentricity of the torch tubes.

A refinement in this torch design was published by Boumans and Lux–Steiner [9], who redesigned the original torch used by the Boumans group [10]. This redesign exploited the experiences of Greenfield [11–13], Scott et al. [14], Allemand [15], Allemand and Barnes [6], and Genna et al. [7], and considered

5.1. TORCH DESIGN CONSIDERATIONS

also the effect of variations in torch tube shapes and dimensions on gas flow patterns, the plasma configuration, and the extent of carbon deposition when the ICP was operated with an organic solvent.

Boumans and Lux–Steiner varied the extension of the outer tube above the load coil, the dimensions of the cup of the intermediate tube, the slope of its edges and the distance from the top of the tube to the coil, the outer diameter, shape and orifice diameter of the inner tube, as well as its axial position with respect to the intermediate tube. This empirical approach led to the torch geometry shown in Fig. 3.8. The analytical performance has been documented for both inorganic aqueous solutions [16] and an organic liquid, methylisobutylketone [9].

Lux–Steiner [17] also described the effects that the separate plasma parameters (power, gas flow rates, liquid uptake rate) exert on the visible spatial emission distribution. These effects were studied by varying each parameter separately while organic matter was being introduced. Part of these observations are discussed in [9] in connection with the behavior of the background intensity in dependence on the ICP parameters.

Recently, Boumans [18] described a demountable torch with a plastic base using threaded silica spacers to maintain alignment and to control the flow patterns of the gases. This torch, shown schematically in Fig. 5.2, is used in connection with a new type of free-running generator operating at 100 MHz. An outer gas flow of 7.5 L/min is used at a power input of about 1 kW (see Tables 5.4 and 5.5, and Section 5.5).

Ebdon et al. [19] described a versatile torch with a flared ("tulip-shaped") intermediate tube, a capillary injector tube, and interchangeable jets at the gas inlets. The torch permits operation over a wide range of gas flows, which is useful in simplex optimization studies [20] (cf. Section 4.7.9). The operating conditions were optimized for the SBR of the line Mn II 257.6 nm with nitrogen or argon as outer gas. The optimum gas flows are rather unusual, and the optimized SBR of the order of 1 compares unfavorably with the values of > 100 achieved with conventional torches under "compromise conditions" [16, 21]. This large difference cannot be explained from the difference in the spectral resolution of the equipment employed.

Although empiricism has dominated in the initial approaches to the design of torches for the analytical ICP—and continues to do so—a number of workers have attempted to lay a better fundamental foundation for future improvements. Notable among these efforts have been the studies of Mermet, et al. [22, 23], Ripson and de Galan [24], and those who have considered the importance of "skin depth" in the plasma. The Mermet group [22, 23] recognized that the same cold argon that flows in the annular region between the outer and intermediate tubes must, after heating, occupy the plasma "fireball" volume confined by the outer tube. These considerations can be formulated in the following way:

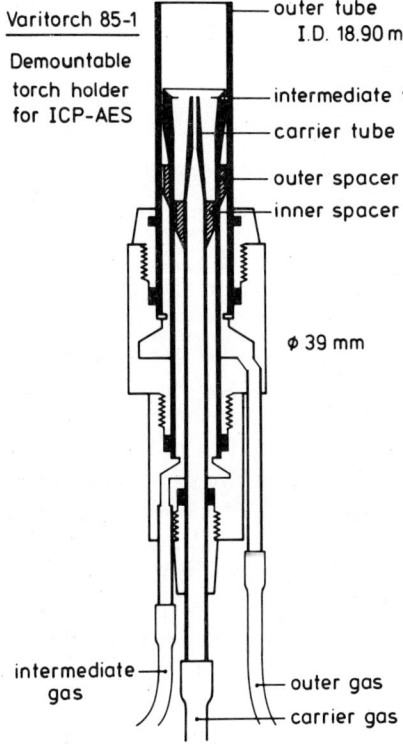

Figure 5.2. Configuration of a demountable high-efficiency torch with threaded silica spacers to maintain alignment and control the gas flows; the latter is achieved with the aid of channels in the spacers. This torch is used in conjunction with a 100-MHz ICP [18]. Further details are given in Tables 5.4–5.5 and in Section 5.5.

$$F_p > S_p v_c \tag{5.1}$$

$$F_p > S_T v_c (T_{\text{gas}}/T_{\text{plasma}}) \tag{5.2}$$

where F_p is the flow of room-temperature (T_{gas}) argon needed to sustain a stable plasma, S_p is the cross-sectional area of the annulus between the outer and intermediate tubes, S_T is the cross-sectional area of the outer tube itself, T_{plasma} is the argon temperature in the discharge, and v_c is the critical velocity of room-temperature argon in the annulus required to stabilize the plasma. Moreover, because the flow F_p must be the same in both torch regions, Eqs. (5.1) and (5.2) can be related as

$$S_p v_c = S_T v_c (T_{\text{gas}}/T_{\text{plasma}})$$

or

$$S_p = S_T (T_{\text{gas}}/T_{\text{plasma}}) \tag{5.3}$$

Equation (5.3) can be expressed in terms of the annular spacing e and the outer-

5.1. TORCH DESIGN CONSIDERATIONS

tube diameter D_i as

$$e = D_i \cdot 0.5[1 - (1 - T_{gas}/T_{plasma})^{1/2}] \quad (5.4)$$

Thus for a stable plasma the diameter of the torch cannot be reduced without reducing also the annular spacing. This finding is important when miniaturized torches are being studied [25–28]. Because v_c has been found to be 3.3 m/s [22] the limiting flows (F_p) and the required annular spacing (e) can be calculated from Eqs. (5.2) and (5.4) for any desired torch size (D_i). The relationship among these variables is shown in graphical form in Fig. 5.3, redrawn from [22].

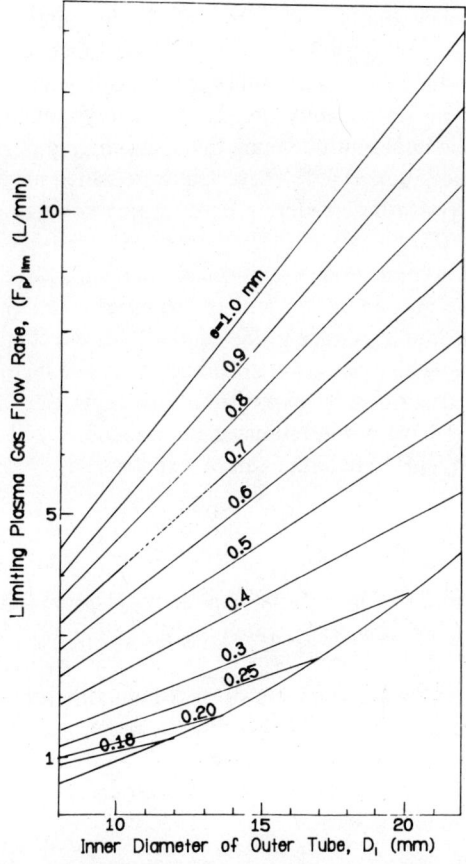

Figure 5.3. Relationship among the inner diameter (D_i) of a torch, the annular spacing between the coolant and intermediate tubes (e), and the total Ar flow (F_p) necessary to sustain a stable plasma [22]. (Reprinted with permission from C. Angleys and J. M. Mermet, *Appl. Spectrosc.* **38**, 649 (1984). Copyright (1984), Society for Applied Spectroscopy, Frederick, MD.)

Ripson and de Galan [24] have contributed to the understanding of torch design by calculating how applied power is dissipated in a plasma. From an empirical power balance computation, they found that no more than 75% of the applied rf power is actually consumed by a conventional plasma; of that power, most is used to heat the argon. For example, to heat a 22 L/min argon flow to a gas temperature of 3500 K requires a continuous power input of 1100 W. Additional applied power beyond this minimum then goes toward ionization of the argon. Of course, the outer argon is probably not heated to the mean gas temperature in the plasma, a fact that contributes to the success of new lower power plasmas, which will be described later.

An important factor that bears upon torch design is the so-called "skin depth" of the plasma. Conceptually, the skin depth is the outer zone of the plasma where coupling of rf energy takes place. Because this coupling occurs mostly in the outer part of the discharge, sample aerosol can be directed into the center of the plasma without significantly altering the energy-coupling process. In contrast, if the aerosol should intrude upon this coupling region, substantial matrix interferences would be expected. These features, which have long been recognized [10, 14], are worth considering here in greater detail. For a lucid theoretical discussion refer to Dresvin [4].

If an rf current is induced in a cylindrical conductor placed in the work coil of an rf generator, an annular current is produced, the density (J) of which declines radially from the surface to the center. This effect is known as the skin effect of rf currents. The total power absorbed in the conductor and the coupling efficiency can be described by expressions involving Bessel functions of the ratio R/δ, where R (cm) is the radius of the conductor and δ (cm) a quantity called "skin depth" [4]. The latter can be expressed as

$$\delta = \frac{5030}{\sqrt{\sigma \mu f}} \tag{5.5}$$

where σ (mho · cm^{-1}) is the specific conductivity, μ the (dimensionless) relative permeability of the material, which for gases is unity, and f(Hz) the frequency of the rf field.

For illustrating characteristic features of rf induction heating one usually introduces approximations to avoid the complications inherent in dealing with Bessel functions. These approximations are valid only if $R \gg \delta$ and if calculations are limited to distances $r \gg \delta$. Under these conditions the current density (A · cm^{-2}) can be written as

$$J(r) = J_0 \exp\left(-\frac{R - r}{\delta}\right) \tag{5.6}$$

where J_0 is the current density at the edge of the cylinder.

Although approximate only, this equation can be used to illustrate that the

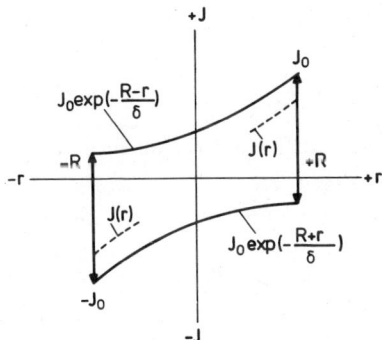

Figure 5.4. Exponential decline of the current density $J(r)$ along the diameter of a cylinder in which an rf current is induced. This picture applies only in the case that the radius (R) of the cylinder is large compared to the skin depth (δ) of the rf current.

condition $R \gg \delta$ is required also for efficient heating. A decline of $J(r)$ according to Eq. (5.6) implies that in any point of the cylinder two opposite current components will be acting, as schematically depicted in Fig. 5.4. Thus the resultant current density would have to be written as

$$J(r) = J_0 \left[\exp\left(-\frac{R-r}{\delta}\right) - \exp\left(-\frac{R+r}{\delta}\right) \right] \quad (5.7)$$

To make the second term in square brackets in Eq. (5.7) negligible, say, smaller than 0.01 of the first term, the condition $R > 2.3\delta$ must be fulfilled. However, Eq. (5.7) already assumes that $R \gg \delta$ and $r \gg \delta$.

By integrating Eq. (5.6) one computes the total current I_t (A) per unit of length of the cylinder to be $I_t = J_0\delta$, and the current in the skin to be

$$I_\delta = J_0\delta[1 - \exp(-1)]. \quad (5.8)$$

Therefore, 63% of the total current flows in the skin. For a shell of 2δ thickness this figure is 86.5%.

Calculations of the power dissipation $P(J)$ per unit length involve integrals of the form

$$P = \int \frac{[J(r)]^2}{\sigma} 2\pi \, r dr \quad (5.9)$$

and generally are complex even if the conductivity is considered constant over the cross section (see e.g., [4]).

Approximations for $R \gg \delta$, $r \gg \delta$ and $\sigma = $ constant lead to integrals of the simple form

$$P = \frac{2\pi R}{\sigma} (J_0) \int \exp\left(-2\left[\frac{R-r}{\delta}\right]\right) dr \quad (5.10)$$

from which follows that the power developed per unit of length in the skin is

$$P = \frac{\pi R \delta}{\sigma} J_0 [1 - \exp(-2)] \quad (5.11)$$

which corresponds to about 86.5% of the total power, while about 98% of the total power is dissipated in a shell of 2δ thickness.

Arguments of this type have been used to define a minimum plasma diameter of 4 to 4.5 times the skin depth [10, 14, 27]. Although such reasonings use quantitative calculations, they should not be taken too rigorously, since assumptions have to be made about the temperature of the gas and the electron mobility, and because the plasma is considered as a homogeneous medium. Nevertheless, the arguments have been and still are most useful in torch design.

For the conductivity σ one may use a dc limit if [29]

$$f \ll \nu_{c,e,i} \ll \omega_{\text{plasma}} \quad (5.12)$$

where $\nu_{c,e,i}$ is the collision frequency for momentum transfer between ions and electrons and ω_{plasma} the plasma frequency:

$$\omega_{\text{plasma}} = \left[\frac{n_e e^2}{m_e \epsilon_0}\right]^{1/2} \quad (5.13)$$

where n_e, e, m_e, and ϵ_0 have the usual meaning.

From the theory of wave propagation [29, 30] in plasmas one obtains

$$\sigma = \frac{n_e e^2}{m_e \nu_{c,e,i}} \quad (5.14)$$

According to Braginskii [31], for a Maxwellian velocity distribution the collision frequency $\nu_{c,e,i}$ can be approximated by

$$\nu_{c,e,i} = \frac{Z \ln \Lambda n_e}{3.5 \times 10^{11} \hat{T}_e^{3/2}} \quad (5.15)$$

where Z is the charge of the ions in the plasma, $\ln \Lambda$ the Coulomb logarithm (~ 10), $\hat{T}_e$ the electron temperature in eV and n_e the electron density in m^{-3}.

Heald and Wharton [29] gave an elaborate treatment of wave propagation in plasmas and obtained the following semi-empirical formula for σ in the dc limit

$$\sigma = 3.3 \times 10^2 \frac{\hat{T}_e^{3/2}}{Z \ln \Lambda} \quad (5.16)$$

Table 5.1 lists numerical values of σ calculated with Eqs. 5.14–5.16 and values reported in the literature in connection with torch design. The value of Dresvin et al. [4], based on calorimetric and spectroscopic measurements, agrees well with the value calculated by Boumans and de Boer [10] and that computed with Braginskii's formulas [31]. The conductivity values used by Weiss et al.

5.1. TORCH DESIGN CONSIDERATIONS

Table 5.1. Values of the Conductivity (σ) Calculated with Eqs. (5.14) to (5.16) or Reported in the Literature in Connection with Torch Design

Origin	References	σ (Ω^{-1} cm^{-1})
Formulas (5.14)–(5.15)	[31]	4.7[a]
Formula (5.16)	[29]	15[a]
Boumans and de Boer	[10]	6.5
Greenfield et al.	[13]	30
Dresvin et al.	[4]	6
Scott et al.	[14]	10
Weiss et al.	[27]	22

[a] Calculated with $T_e = 7000$ K and $n_e = 5 \times 10^{15}$ cm^{-3} [83].

[27] and Greenfield et al. [13] are much higher, probably because they used higher values for the electron temperature (e.g., 10,000 K [32]). Although these values were originally thought to pertain to the core of the discharge, recent evidence [32] suggests that tail-flame electron temperatures are indeed in the range of 10,000–12,000 K.

Table 5.2 lists skin depths for various frequencies, all computed with $\sigma = 5$, 10, or 20 Ω^{-1} cm^{-1}. Thus for the most common frequency of 27 MHz, δ is estimated to be 0.2–0.4 cm.

Using a value of 2 mm for the skin depth ($f = 27.12$ MHz), Weiss et al. [27] argued that the rf energy addition at a distance of 4 mm from the discharge surface should be only $(0.37)^2 = 0.13$ of its surface value. A sample sent into the plasma at a distance of 4 mm from its surface should then have less than a 13% effect on the energy coupling. They take this degree of perturbation to be the maximum permissible, so the plasma must have a diameter at least 8 mm

Table 5.2. Skin Depth for Various Frequencies and Three Values of the Conductivity, σ (Ω^{-1} cm^{-1}) [28]

Frequency (MHz)	Skin Depth (cm)		
	$\sigma = 20$	$\sigma = 10$	$\sigma = 5$
7	0.43	0.60	0.85
13	0.31	0.44	0.62
27	0.22	0.31	0.43
40	0.18	0.25	0.36
50	0.16	0.22	0.32
100	0.11	0.16	0.22
2450	0.023	0.032	0.045

greater than that of the aerosol stream sent into it, which, under the assumption of a laminar aerosol flow is taken to have a 0.75 to 1.00 mm diameter. Under these conditions they calculated a minimum plasma diameter of 9 mm, the size of the torch developed in their work.

In view of these power calculations [Eqs. (5.9) and (5.11)] the argument of Weiss et al. [27] can be disputed. Actually a power dissipation of 13% will apply to a central region of radius $R - \delta$, not $R - 2\delta$. Weiss et al., in fact, overlooked that a power calculation should involve the integration of a squared current density rather than the squaring of an integrated current density. Therefore, the power dissipated in a central channel of radius r_c will be less than 2% if $R > \delta + r_c$, but this condition is not sufficient for efficient heating, for which $R > 2.3\delta + r_c$ should be satisfied. Therefore, the arguments, though different, eventually lead to the same conclusion.

Weiss et al. [27] did not find negligible interferences under the adopted conditions and concluded that a higher frequency would be required in order to reduce further the power dissipation in the central channel. Perhaps more importantly, increasing the frequency would also make it easier to "bore" a tunnel in the plasma using a lower carrier gas *speed* at the same flow. Reduction of the carrier gas speed is known to be beneficial for the suppression of interferences [33]. For illustration, Table 5.3 covers results obtained [28] with a torch similar to the one described by Weiss et al., except for the orifice diameter in the carrier gas tube which was 1.35 instead of 0.75 mm (cf. Section 5.4). This torch was operated at 80 MHz instead of the 27 MHz as used by Weiss et al. [27]. The interferences of phosphate on calcium and of KCl on chromium are seen to be at the low level expected for a conventional ICP.

The preceding argument emphasizes that the skin effect is a necessary but not sufficient condition for creating a toroidal ICP. The gas flow conditions, especially the flow pattern and velocity of the inner (carrier) gas stream, dictate the extent to which the inner stream forms a tunnel in the plasma. Whether this result can be achieved more easily at higher frequencies has been a matter of dispute [13]. Recent work [18, 44] has shown, however, that high frequencies (100 MHz) permit the generation of toroidal ICPs with rather large orifice diameters (2.5–3.0 mm) in the carrier gas tube. This means that at a high frequency the tunnel in the ICP can be produced with a relatively low carrier gas velocity (see also Section 5.5).

From the discussion thus far and the experience detailed in the literature, there are certain rules that should be observed in torch design, but no complete theoretical picture of the ideal torch exists. The requirement of "careful torch design," often echoed in publications, emphasizes the empirical character of the final stages of torch fabrication. The dimensions, precise shapes, and mutual positions (including manufacturing tolerances) all can have significant effects on the gas flow patterns; it is the latter that dictate the plasma configuration, gas consumption requirements, and the spectroanalytical properties of the plasma.

5.2. DEMOUNTABLE VERSUS PREALIGNED TORCHES

For routine work one generally prefers a torch with prealigned tubes, consisting of one or two pieces such as exemplified in Figs. 3.6 and 3.7. Mechanical manufacturing of such torches is desirable to ensure perfect concentricity of the tubes and rotational symmetry of the assembly, which are vital requirements for prolonged ICP operation and analytical reproducibility.

In contrast, demountable torches are convenient in research, particularly if torch development is the primary goal. A demountable torch consists of a torch base into which the separate tubes can be fitted and aligned, and which might or might not allow final adjustments to be made during operation. Thus different assemblies of tubes can be easily tested, and the separate tubes can be conveniently changed or replaced in the case of failure or damage.

Bases of demountable torches have been described by various authors (see [8, 9, 18]. Figs. 5.5a and 5.5b show a base used by Boumans and Lux–Steiner [9]. The base has some features in common with those proposed by Abdallah et al. [45], Mermet and Trassy [23], and Windsor et al. [8] (see also Fig. 5.2).

Ebdon et al. [19] described a support machined from brass (Fig. 5.6) in which the tubes were cemented with an epoxy resin and held in position with a jig while the glue hardened. The tubes could be released simply by heating the brass support in an oven. This approach has the advantage that the critical alignment can be maintained better than with O-ring seals. However, no adjustments during operation are possible.

The novel demountable torch of Windsor, Heine, and Denton [8] was mentioned earlier. The threaded spacer they used to separate the outer and plasma tubes served not only to provide firm alignment and concentricity of the tubes, at a point well above the torch base, but also imparted a higher swirl velocity to the outer gas (cf. Fig. 5.2).

5.3. SPECIAL-PURPOSE TORCHES

A number of torches of unusual design and constructed for special applications has been described in the literature. Included in this group are extended torches, torches with purged side arms to enable plasma radiation in the VUV region to be viewed, torches for atomic fluorescence, torches for operation at reduced pressure, and others. In the following paragraphs, some of the important features of these designs will be described.

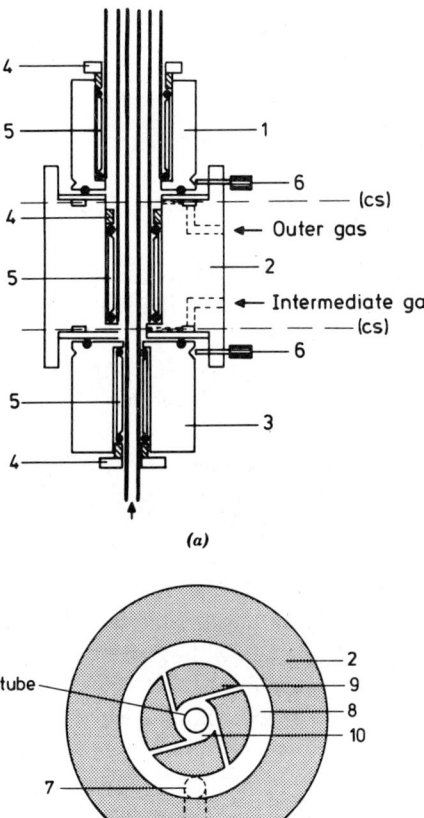

Figure 5.5 Schematic drawings of the base of a demountable torch: (*a*) axial cross section; (*b*) cross section perpendicular to the axis at the positions "cs" in Fig. 5.5*a*. *1*, holder of outer tube; *2*, holder of intermediate tube; *3*, holder of inner tube; *4*, concentric tubes for fixing the tubes into the holders; *5*, cylindrical washers and "O" rings (●); *6*, screws positioned radially at 120° (these screws fix the holders together and provide for radial adjustment of the tubes); *7*, gas inlet into the concentric channel; *8*, concentric channel; *9*, tangential gas inlets into the central channel; and *10*, central channel. (Reproduced with permission from *Spectrochimica Acta, Part B* [9].)

Inductively coupled plasma torches with extended outer tubes have been employed in a number of applications where exclusion of atmospheric gas is desired. For example, atomic fluorescence detection limits have been found to be reduced if the potentially important quenching effects of atmospheric nitrogen and oxygen are avoided [46]. Interestingly, in one such application it was found that the plasma shape and character changed dramatically when the torch was moved upward slightly within the load coils [47]. When the torch was raised so that its aerosol tube was located between the first and second turn of the load coil, and when the outer gas flow was lowered and the aerosol gas flow raised somewhat, a thin "pencil plasma" developed. The pencil plasma was predicted to be useful in atomic fluorescence applications (cf. Section 3.4).

In another application, where viewing of the ICP from the top was explored,

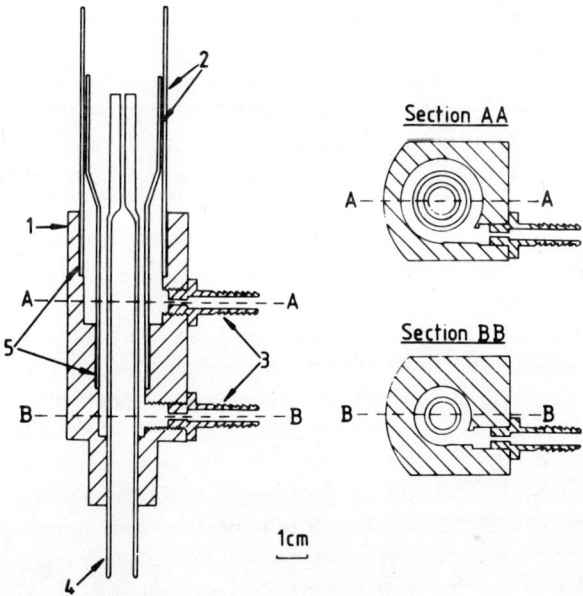

Figure 5.6. Scale diagram of the construction of a demountable torch. *1*, Brass support; *2*, silica tubes; *3*, threaded brass gas-inlet jets; *4*, borosilicate glass injector tube; and *5*, epoxy resin bonds [19]. (Reprinted with permission from L. Ebdon, D. J. Mowthorpe, and M. R. Cave, "A Versatile New Torch for Inductively Coupled Plasma Spectrometry," *Anal. Chim. Acta* **115**, 174 (1980). Copyright (1980), Elsevier Science Publishers, Amsterdam.

extended-tube torches were also desirable [48]. With a longer coolant tube, air entrainment was minimized and an air "cutoff stream" could be blown across the top of the extended tube to protect viewing optics and yet minimize plasma disturbance. An alternative approach was to position a brass tee above a conventional torch to vent exhaust gases to the side before they could damage optical systems [49].

Of course, when atmospheric gases themselves are to be determined, it is important that they be excluded from the viewing region. Sobel [50] utilized a commercial torch with an extended coolant tube for this purpose. However, Broekaert et al. [51] found that even with an extended coolant tube and purging of the optical path, nitrogen background contributions were undesirably high (cf. Section 4.1.7.3).

Because many elements, particularly nonmetals, emit their strongest lines in the VUV region, special torches have been fabricated with purged side arms to permit their direct connection to a VUV spectrometer. A typical design was described by Heine, Babis, and Denton [52]. This design was similar to the one they described earlier [8], but with an extended outer tube having a helium-purged side arm.

For those who wish to explore the use of extended-tube or side-arm torches, the method of Devine, Brown, and Fry might be explored [53]. In their approach, an extension tube having the same diameter as the torch coolant tube is cut and fitted with a quartz collar having an inside diameter equal to the outside diameter of the extension tube. This assembly, which can be melted together if desired, then fits snugly over the existing outer tube. A similar technique can be used to repair torches in which the coolant tube has been damaged or melted.

An interesting new torch has been designed especially for permitting an inductively coupled plasma to be sustained at reduced pressures [54]. This torch, which is shown in Fig. 5.7, is not of the conventional three-tube arrangement. Instead, only a single plasma chamber exists and is fed by a capillary injection tube. This discharge chamber is surrounded by a concentric cooling sleeve through which a flowing liquid or gas is sent. The authors prefer water for a cooling medium since it is transparent to rf fields and has a high cooling efficiency. However, gases such as air, argon, and nitrogen also serve at flow rates of 0.3–1.5 L/min as long as rf powers are less than 300 W. This assembly permits a stable ICP to be sustained at power levels between 5 and 500 W and plasma-gas flows from less than 1 mL/min to several hundred milliliters per minute. Plasmas can be sustained in a range of gases (Ar, He, Ne, H_2, and N_2 have been employed) and at pressures from 10 mtorr to several tens of torr. Obviously, the plasma must be viewed through the torch wall. However, the

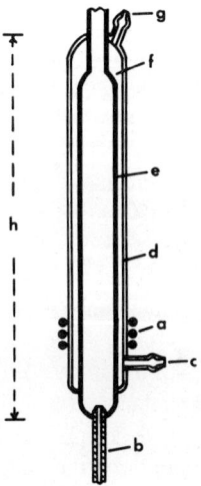

Figure 5.7. New torch for operation at reduced pressure and described by Seliskar and Warner [54]. a, 27.12-MHz load coil; b, capillary injection tube (1-mm i.d.); c, coolant gas inlet tube; d, cooling jacket (typical outer diameter, 28.6 mm); e, plasma containment tube (typical outer diameter 19 mm); f, constriction of plasma containment tube for exit of plasma gas; g, coolant exit tube; and h, typical torch length 30.5 cm. (Reprinted with permission from C. J. Seliskar and D. K. Warner, "A New Reduced Pressure ICP Torch," *Appl. Spectrosc.* **39**, 182 (1985). Copyright (1985), Society for Applied Spectroscopy, Frederick, MD.)

authors have observed very little etching over extended periods of use and have indicated that even strong silicon lines are barely visible in the plasma background spectrum.

This reduced-pressure torch has [54] been employed for hydrogen isotope analysis [55] in which an argon plasma was used at pressures between 0.3 and 0.8 torr and at rf powers between 50 and 400 W. Apparently, complete hydrogen fragmentation takes place, as evidenced by the absence of molecular emission. When used with helium as a support gas, the reduced-pressure torch has been used also for the measurement of bromine and chlorine emission [55]. At a power less than 300 W, an excitation temperature of 6600 (+2900, −1600) K was found for Cl I; for Cl II, the temperature was measured to be 11,100 (+1700, −1300) K.

Finally, a new kind of demountable torch has recently been described for fundamental studies in the ICP [57]. Most blown or demountable torches seem not to be sufficiently reproducible from laboratory to laboratory thus enabling the reliable spatial mapping of species believed to be important in plasma sustenance and in analyte excitation. In part, these variations are caused not by the plasma itself, but by changes in the sample introduction system and in the positioning of the torch with respect to the load coils. A completely machined assembly to support load coils, torch, and sample introduction system has been developed and has shown excellent reproducibility from torch to torch and from laboratory to laboratory [57].

5.4. NEW TRENDS IN TORCH DEVELOPMENT

The principal motivation in the development of new torches is cost reduction, primarily the operating cost (argon consumption) and secondarily the apparatus cost and size (rf generator). Table 3.1 summarized data relevant to conventional argon and argon–nitrogen ICPs.

Approaches used to achieve lower argon consumption and/or lower power input comprise the following means:

1. Increasing the swirling velocity of the outer gas.
2. Reducing the diameter of the torch.
3. Optimizing the torch dimensions.
4. Cooling the outer tube externally with water.
5. Cooling the outer tube externally with air.
6. Replacing the argon by nitrogen or air.

These approaches have been recently reviewed [58, 59]. Tables 5.3 to 5.5

Table 5.3. Effect of Interferents, High Salt Concentrations and Organic Solvents on Several High-Efficiency ICP Discharges[a]

Interferent	Savage [26, 34]	Allemand [25]	Weiss [27]	Rezaaiyaan [35]	Kornblum Water-Cooled [36]	Kawaguchi Water-Cooled [37]
Phosphate	-5%	-8%	-25% (Ca II) -5% (Ca I)	<5%	-21%	<5%
Al	-4% (Ca I) -14% (Ca II)			<5%	-8%	
Na	+6%	-19%	-5% (Ca II) +40% (Ca I)	<5% (Ca II) +8% (Ca I)	-11%	-20%
NaCl (>1%)	OK			OK		
Organic				OK		

Interferent	Ripson [38]	Ripson Air-Cooled [39][b] 16 mm	Ripson Air-Cooled [39][b] 13.5 mm	Ripson Water-Cooled[b] [39]	van der Plas Radiatively Cooled (BN) [40]	Kawaguchi Water-Cooled [61]
Phosphate	-12%	None	None	None	None	None
Al	+15%					None
Na	-11%	5%	50%	-15%	None	None
NaCl (>1%)		OK	OK	OK	OK	
Organic solvents		NG[c]	NG[c]	NG[c]		

Interferent	Raaijmakers [28]	Rezaaiyaan [41, 42]	Rezaaiyaan [60]	Montaser [62]	van der Plas Improved Air-Cooled [73]
Phosphate	−6% (Ca I, II)	−6%	<5%	<1%	
Al		None	<5%		
Na	−4%[d]	None	None	−12%	
NaCl (>1%)		OK		−20%	OK
Organic solvents		OK		OK[e]	OK

[a] Unless otherwise indicated, interference is on Ca II emission and is calculated as the percentage increase (+) or decrease (−) in emission signal caused by 1 g/L of interferent.
[b] Interference measured on the Mn 403.3-nm line, calculated as above.
[c] Not easily used with organic solvent under listed operating conditions (Table 5.4).
[d] Effect of 10-g/L KCl on Cr II or Cr I.
[e] See [43].

Table 5.4. Dimensions, Operating Conditions and Characteristics of Several High-Efficiency ICP Torches

Parameter	Savage [26]	Allemand [25]	Weiss [27]	Genra [7]	Rezaaiyaan [35]	Kornblum Water-Cooled [36]	Kawaguchi Water-Cooled [37, 69]
Outer tube i.d. (mm)	13	13	9	18	18	18	17
Injection tube i.d. (mm)	0.75	1.0	0.75	—	1.0	0.5	1.5
Outer flow (L/min)	7.9	12	6.4	12	5^a	0.9	2.5^b
Intermediate flow (L/min)	0	0	0	0.5	0.5	0.8	4.0
Carrier gas flow (L/min)	1.0	0.76	0.75	0.5	0.5	0.1	0.8
Sample uptake (mL/min)	0.4	1.0	1.2	1.4	1.0	—	1.5
Observation height (mm above coil)	15	13	15	20	10^a	4	12
Applied RF power (kW)	1.0	0.85	0.5	1.2	0.35^a	0.7	1.1
Frequency (MHz)	27	27	27	27	27	50	27
Excitation temperature (K)	4300	—	4000	—	4000	4500	7000

Parameter	Ripson [38]	Ripson Air-Cooled [39]	Ripson Water-Cooled [39]	van der Plas Radiatively Cooled (BN) [40]	Kawaguchi Water-Cooled [61]
Outer tube i.d. (mm)	13.5	16	13.5	15	17.4
Injection tube i.d. (mm)	0.5	0.5	0.6	0.5	1.5
Outer flow (L/min)	50^c	62^c	1.3^b	0.8	4^b
Intermediate flow (L/min)	0.75	0.8	1.0	0	4.0
Carrier gas flow (L/min)	0.15	0.12	0.17	0.2	1.0
Sample uptake (mL/min)	1.5	2.0	2.0	2.0	—
Observation height (mm above coil)	10	11	10	2	12
Applied RF power (kW)	0.5^d	0.5^d	1.1	0.6	1.2
Frequency (MHz)	50	50	50	40	27
Excitation temperature (K)	5500	—	—	—	6500

Parameter	Raaijmakers [28]	Rezaaiyaan [41, 42]	Rezaaiyaan [60]	Montaser [62][e]	van der Plas Improved Air-Cooled [73]	Boumans 100 MHz [18]
Outer tube i.d. (mm)	9	18	13	18	15	19
Injection tube i.d. (mm)	1.35	1.0	0.75	1.0	0.5	2.4
Outer flow (L/min)	9.0	6.0	7.5	4.0	50[b]	7.5
Intermediate flow (L/min)	0.3	0.7	0.3	0.5	1.0	0.15
Carrier gas flow (L/min)	0.95	0.64	0.48	0.5	0.15	0.95
Sample uptake (mL/min)	1.4	1.0	1.0	0.5	2.0	1.4
Observation height (mm above coil)	12	12	13	11	2–5	10
Applied RF power (kW)	0.7	0.45	0.45	0.72	0.6–0.9	1.0
Frequency (MHz)	30	40	40	27	40 or 50	100
Excitation temperature (K)		4000			4800–5500	

[a] R. Rezaaiyaan, Indiana University, unpublished results (1983).
[b] Flow rate of cooling water; argon flow listed as "intermediate flow."
[c] Flow rate of cooling air; argon flow listed as "intermediate flow."
[d] Adjusted from author's values cited for power dissipated in the plasma.
[e] Commercial "MAK" torch employed.

Table 5.5. Detection Limits (ng/mL) Obtained with Miniaturized, Optimized, or Externally Cooled Torches[a,b]

Element	Savage [26]	Allemand [25]	Weiss [27]	Genna [7]	Rezaaiyaan [35]	Kornblum Water-Cooled [36]	Kawaguchi Water-Cooled [37, 69]
Al (I)	4	3	6	18	25^c	2300	45
B (I)				2.5			25
Ba (II)		0.2	0.9	0.9	0.45	30	2
Ca (II)	0.1	6	0.12	4.5	0.11	9	0.45
Cd (II)	60 (I)	14	14	11	3^c		
Co (II)					9^c		10
Cr (II)		4.5 (II)		15	10^c		8
		3 (I)					
Cu (I)	11	2	1.4	6	3^c	250	
Fe (II)	14 (I)	6	70	7	5^c	1000	
La (II)						300	9
Mg (II)	8 (I)	0.3	3		0.3	15	0.2
Mn (II)		1.2		20	0.3	80	1.3
Na (I)	1.0		0.7				
Ni (II)	5.5 (I)		1.2 (I)				20
Pb (II)	45 (I)	30	35 (I)	110			
Ti (II)		1.5		4.5		140 (I)	2.5
Zn (I)	100	1.2	1.4				5.5

278

Element	Ripson [38]	Ripson, Air-Cooled 16 mm [39]	13.5 mm [39]	Ripson Water-Cooled [39]	van der Plas Radiatively Cooled (BN) [40]	Kawaguchi Water-Cooled [61]
Al (I)	40	40	40	70	50	9
B (I)	90	10		35	d	
Ba (II)						
Ca (II)	0.5	0.13			0.1	
Cd (II)		16	16	45		2.5
Co (II)	70 (II) / 18 (I)	45		35	10	9
Cr (II)	10	10	11	17	8	2.5
Cu (I)	60	5	2	55	5	2
Fe (II)	5	10	11	50	15	
La (II)						
Mg (II)	0.6	0.3	0.3	1	0.4	
Mn (II)		2	2		2	0.5
Na (I)	2	7	7	25	5	
Ni (II)		25			6	
Pb (II)		250	250		100	25
Ti (II)					5	0.9
Zn (I)		12	8		9	5

Table 5.5. (Continued)

Element	Raaijmakers [28]	Rezaaiyaan [41, 42]	Rezaaiyaan [60]	Montaser [62]^g	van der Plas Improved Air-Cooled [73]	Boumans 100 MHz [18]	Winge^h Conventional [63]
Al (I)		25	11		14	1.2^e	30
B (I)				11	3	0.3^f	5
Ba (II)		4.5		0.7	0.4	0.02^e	1.3
Ca (II)		0.35	0.3	20	0.06		0.2
Cd (II)		25 (I)		3.5	2	1.3^f	3.5
Co (II)				5.0	6	1.1^f	6
Cr (II)	6.5 (II)			5.5	4	0.7^e	7
	7 (I)						
Cu (II)		3	7		2	0.2^e	5
Fe (II)	3.5 (II)	30 (I)	40 (I)	2.5	3	1.1^f	6
	55 (I)						
La (II)					9	0.3^e	10
Mg (II)		0.65			0.1	0.02^e	0.15
Mn (II)	0.9	2.5	4 (I)	2.5	1.0	0.12^e	1.5
Na (I)		3.5	2		1.0		30
Ni (II)		17 (I)	35 (I)	5.5	6	3^f	10
Pb (II)	20		120 (I)	30	40	20^f	40
Ti (II)				1.5	1.3	0.12^e	4
Zn (I)	2.5	20	35		2.5	0.8^f	2

^{a}All detection limits have been adjusted to 3σ.
^{b}Roman numerals in parentheses refer to atom (I) or ion (II) line.
^{c}K. A. Wolnik, Food and Drug Administration, Cincinnati, OH, personal communication (1983).
^{d}Boron nitride outer tube segment prevented B determination.
^{e}Spectral bandwidth: 4–7 pm.
^{f}Spectral bandwidth: 9–12 pm.
^{g}Detection limits derived from reported SBRs using a value of 1% for the background RSD.
^{h}Spectral bandwidth: 17 pm (cf. [81]).

5.4. NEW TRENDS IN TORCH DEVELOPMENT

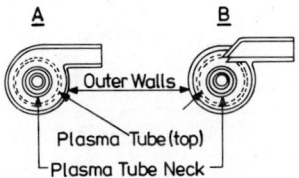

Figure 5.8. Gas inlet geometries. (*a*) Conventional torch. (*b*) Modified torch. [7]. (Reprinted with permission from J. L. Genna, R. Barnes, and C. D. Allemand, "Modified Inductively Coupled Plasma Arrangement for Easy Ignition and Low Gas Consumption," *Anal. Chem.* **49,** 1451 (1977). Copyright (1977), American Chemical Society, Washington, DC.)

summarize the basic features and performance of the torches designed using means (1)–(5); the approach using (6) has been considered in Section 4.9.

Genna et al. [7] modified the position and shape of the gas inlet of the outer gas of a conventional torch to increase the swirl in the annular space between the outer and intermediate tubes (Fig. 5.8). This change enabled them to reduce the outer gas flow from 18 to 12 L/min, with only a slight (factor of 1.6 on average) deterioration in detection limits. The authors also gave a useful picture about the action of the swirl velocity on gas consumption and ease of discharge ignition.

Allemand, Barnes, and Wohlers [25] compared the performance of 18-, 13-, and 9-mm (inner diameter of outer tube) versions of their new boron nitride torch (Fig. 5.1). All versions incorporated the constricted gas inlet nozzle (Fig. 5.8) to increase the swirl velocity of the outer gas. Different coil configurations were required to increase the magnetic flux density, particularly for the 9-mm torch. For the 18- and 13-mm torches, the optimum outer gas flows and rf powers were found to be 15 and 12 L/min and 1100 and 850 W, respectively. Detection limits and the interference effect from 1 mg/mL Na did not differ between these torches. More difficulties were experienced with the 9-mm torch, which could not be operated below 850 W without having to divert part of the aerosol flow to waste. For an outer gas flow of 8 L/min and power levels between 850 and 1000 W the detection limits were poorer than for the 18- and 13-mm torches.

Savage and Hieftje [26, 34, 64] constructed a 13-mm "mini torch" (Fig. 5.9), which featured constricted gas inlet tubes with their central axes precisely oriented on tangential lines to the outer tube, an injection tube constricted to 0.75 mm at its upper end to create a laminar flow [6, 15] that readily penetrates the plasma, and an overall construction aided by hydrodynamic observations using streaming water [65]. Operated at an outer gas flow of 8 L/min and a power of 1 kW, the mini ICP showed similar analytical capabilities as a conventional ICP operated at 16 L/min outer gas flow and 1.5-kW power.

Subsequently, Weiss et al. [27] further reduced the 13-mm mini torch to a

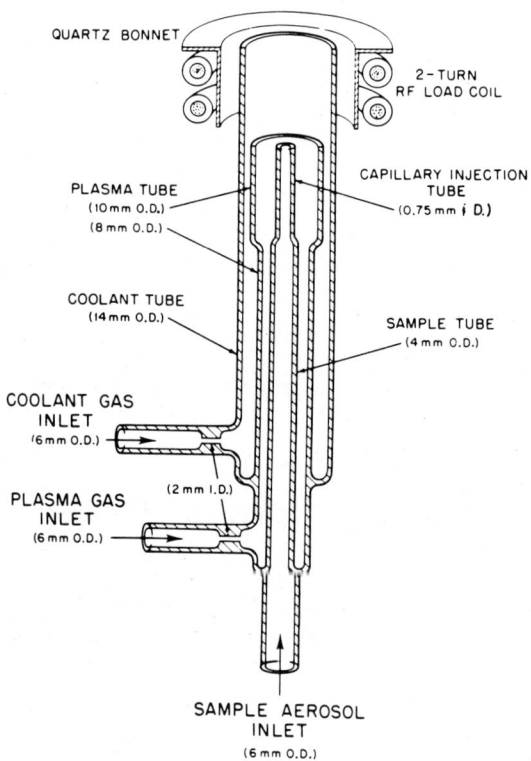

Figure 5.9. Configuration of 13-mm i.d. "mini torch." (From Savage and Hieftje [26].) (Reprinted with permission from R. N. Savage and G. M. Hieftje "Development and Characterization of a Miniature Inductively Coupled Plasma Source for Atomic Emission Spectrometry," *Anal. Chem.* **51,** 409 (1979). Copyright (1979), American Chemical Society, Washington, DC.)

9-mm micro torch, which has basically the same configuration as the mini torch. However, some critical features such as the diameter of the constricted gas inlets and the outer diameter of the flared-out portion of the intermediate tube had to be modified to ensure proper operation. The micro torch could be operated at an outer gas flow of 6.4 L/min and a power of 500 W, and yielded detection limits similar to its larger brothers (Table 5.5). The 9-mm torch readily accepted sample solutions with concentrations up to 10 mg/mL, but salt buildup in the tip of the capillary tube occasionally occurred. Interferences of Al, Cs, Na, and phosphate on Ca I and Ca II lines were somewhat worse in the micro plasma, and even at more elevated power levels (750 W) were fairly persistent. The authors believed this to arise in part from the effect of the sample on the efficiency of energy coupling (cf. Section 5.1) and, therefore, suggested

5.4. NEW TRENDS IN TORCH DEVELOPMENT

that a higher frequency should improve the performance since this change would reduce the skin depth (see Section 5.1).

Operating the torch of Weiss et al. [27] at 80 MHz and using a 1.35-mm-diameter orifice in the carrier gas tube, Raaijmakers and Boumans [28] obtained detection limits (Table 5.5) which compare favorably with the Winge values [63] as listed by Boumans [66, 67] and are not more than a factor of three worse than those for a conventional 50-MHz ICP [9, 10] when measured with the same equipment.

Kornblum and de Galan [36] added a water-cooled jacket to a conventional three-tube torch, which enabled them to run the ICP on a total argon flow of 2 L/min, provided that a carrier gas flow not exceeding 0.1 L/min was used. This necessitated rejecting the major portion of the carrier gas and aerosol when the system was operated with a conventional pneumatic nebulizer operating on 1 L/min of argon. Under these conditions the detection limits were about two orders of magnitude poorer than those of Winge et al. [63]. Interferences measured up to a 1-mg/mL level of the interfering ions were higher than in a conventional ICP. Coupling the water-cooled torch with an improved sample introduction device using a type of Babington nebulizer operating on 0.1-L/min gas flow [68] did not lead to acceptable detection limits. It was shown that this defect was not attributable to the nebulizer. As the cause of the deficiency Ripson and de Galan [38] suggested inefficient coupling of rf power into the plasma when the work coil and the torch are separated by a water jacket.

With a somewhat smaller separation between the torch and coil and a different design of the water-cooling jacket (Fig. 5.10) Kawaguchi et al. [37] obtained better detection limits and a lower level of interferences using an outer gas flow of 4 L/min and a carrier gas flow of 0.8 L/min. The main difference from the design of Kornblum and de Galan [36] is that the cooling water inlets and outlets are located above instead of below the coil so that gas and steam bubbles could leave with the flowing water and not be trapped in the cooling jacket. Interferences still were noted, however; a 20% suppression of a Ca II signal of 10 μg/mL Ca by 1 mg/mL Na and a 15% suppression by 10 mg/mL of phosphate were reported [37].

In later communications, Kawaguchi et al. [61, 69] described a torch similar to their earlier one [37] but with a third, flared intermediate tube. With this modification, and with the plasma operated at higher power (1200 W), interferences of Na, Al, and phosphate on Ca II signals virtually disappeared when a viewing height of 12 mm above the load coil was selected.

Britske et al. [70] constructed a water-cooled torch for a high-power ICP (4 kW, 40 MHz) with a 40-mm inner diameter of the outer tube. The argon consumption was 4 L/min, but for obtaining good analytical results "end-on" observation was necessary [71].

Because of the somewhat disappointing performance of their water-cooled

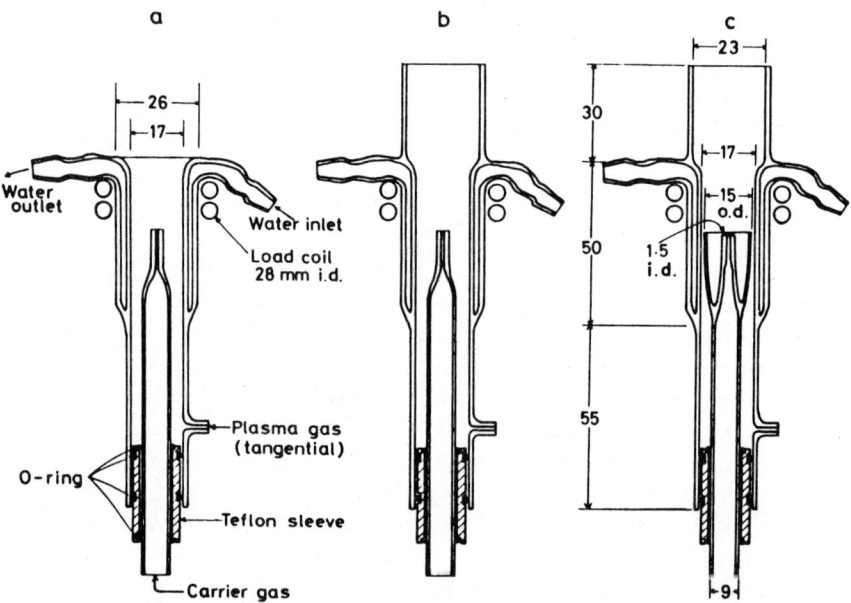

Figure 5.10. Alternative configurations of torches with a water-cooling jacket. [37]. (Reprinted with permission from A. Kawaguchi, I. Ito, S. Ruhi, and A. Mizuike, "Modified Inductively Coupled Plasma Using 1 L/min of Argon," *Anal. Chem.* **52**, 2441 (1980). Copyright (1980), American Chemical Society, Washington, DC.)

torch [36], Ripson, de Galan, and de Ruiter turned to a design that was externally cooled by pressurized air [38]. Figure 5.11 shows a schematic drawing of their work coil, which consists of two copper plates, each with a cylindrical bore and a saw cut. Air at a flow rate of 50 L/min is blown against the torch through five inlets between the copper plates (only two are shown in the figure), in such a way that the air is forced upwards to pass the gap between the coil and the torch. Three- and two-tube torch configurations were studied and optimum results were obtained with a two-tube torch having a fairly large separation between the outer tube and the rf coil. To achieve easy ignition and a stable tail flame it was found necessary to extend the coil, which in turn required for the plasma to be viewed through the silica tube (at 5 to 15 mm above the coil). A carrier gas flow of 0.15 L/min, an outer argon flow of 0.75 L/min, and a power level less than 1 kW (at 50 MHz) were used in the eventual analytical evaluation.

The detection limits (Table 5.5) compared reasonably with the results reported for the conventional ICP of Winge et al. [63], although there is an indication that the plasma is cooler than a conventional ICP, as evidenced by the

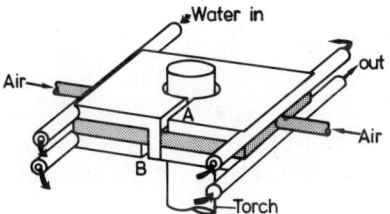

Figure 5.11. Schematic drawing of a work coil incorporating water-cooling of the coil and air-cooling of the torch. Points A and B are connected to the rf generator. The actual system has five symmetrically positioned air inlets [38]. (Reprinted with permission from P. A. M. Ripson, L. de Galan, and J. W. de Ruiter," An Inductively Coupled Plasma Using 1 L/min of Argon," *Spectrochim. Acta* **37B** 734 (1982). Copyright (1982), Pergamon Journals, Oxford.)

rather low detection limit of sodium and the worsening of the detection limits with increasing excitation energy of the analysis lines (Table 5.5). The interference effect of Al, Na, and phosphate at the 1 mg/mL level on the signal of 1 µg/mL Ca was found to be +15, +5, and −12% respectively (Table 5.3).

Ripson, Jansen, and de Galan pursued further the use of the same flat-plate, two-turn load coil but with both air and water external cooling [72]. In accordance with their prediction of conductive losses of heat through the torch, the air-cooled unit operated at 300 W whereas the water-cooled system required 600 W. The air-cooled system was clearly superior in terms of detection limits, matrix interferences, and other figures of merit (see Tables 5.3 and 5.5). In an expanded description of this same work [39], Ripson et al. compared air- and water-cooled torches of two different diameters. Air-cooled systems of 13.5 and 16 mm inside diameter were compared to a 13.5-mm i.d. water-cooled torch. Of the group, the 16-mm air-cooled torch performed best and produced figures of merit approximately equal to those expected from a conventional torch (see Tables 5.3–5.5). Unfortunately, because of the low argon flow (1 L/min), a tail flame did not develop and it was necessary to view the torch through its side wall. Presumably, end-on viewing would also be possible but was not explored by the authors. Results of an analytical evaluation of an improved version of the air-cooled 1-L/min argon ICP were recently reported by van der Plas, de Waaij, and de Galan [73] (cf. Tables 5.3–5.5).

Lowe [74] designed a torch with which the following argon gas flows were used: 1.7 L/min (outer flow), 0.3 L/min (intermediate flow), and 5 L/min (carrier flow). The input power was estimated to be about 2 kW. These flows and the power level are rather unusual, as is the torch configuration without flared intermediate tube and a carrier gas tube extending 4 mm above the intermediate tube. The few preliminary analytical results do not yet permit an assessment of the analytical performance.

The success of workers who used externally cooled torches convinced Hieftje

[58] to speculate that it might be possible for a torch to be sufficiently cooled simply by radiative heat loss. Using the data of Kawaguchi et al. [37] for the amount of heat carried away by their cooling water, he calculated that a sufficiently robust cooling tube that could reach a temperature of 1400 K would dissipate enough energy radiatively to prevent a further temperature rise. It was suggested that some refractory ceramics might serve this purpose well.

These predictions were verified by van der Plas and de Galan [40], who showed that a radiatively cooled torch could be operated at powers of 600 W and a total gas flow of 1 L/min. These authors employed the same power-balance calculation as Ripson and de Galan [24] to show that such an approach was not only feasible but that specific ceramics were possible torch materials. Among the ceramics chosen for this study were alumina, translucent Al_2O_3, ZrO_2, SiC, and BN. They constructed torches having external tubes of these ceramics and found the most successful design to employ a segmented tube that contained two pieces of silica with a central piece of 93% BN. The BN segment was located in the load-coil region and was able to dissipate energy radiatively. With this segmented tube, analyte emission could be viewed through the top silica piece, again necessary since the low argon flow used by this torch yielded only a very small tail flame. Other ceramics were judged to be unsuitable; SiC was too conductive and produced sparks when present within the load coil, while alumina and ZrO_2 developed tiny cracks over time.

This important study proves the feasibility of a radiatively cooled torch which requires only a low flow of argon and no external cooling medium. However, the present design still has limitations. Because the boron nitride reached a temperature of 1600 K, it volatilized slowly as evidenced by boron lines in the plasma background spectrum. Nonetheless, detection limits and vaporization interferences were as low as expected from conventional plasmas.

One of the more recent developments in torch design involves optimization of the torch dimensions to permit operation at unusually low rf powers and argon coolant flow rates [58]. In many ways, these studies are an outgrowth of the pioneering work of Barnes et al. [6] in that the most important parameter has been found to be the configuration factor. Theoretical limits to such an approach have been defined by Angleys and Mermet [22], as discussed earlier in this chapter. The first such published study [35] described clearly the general approach that has been taken.

In this approach, the influence of a particular characteristic of a torch is ascertained by the development of a "plasma stability curve," examples of which are shown in Figs. 5.12 and 5.13. To generate a plasma stability curve, a stable plasma in the designated torch is first established. Afterward, either the outer flow or sustaining rf power is incrementally reduced until the discharge spontaneously quenches. The combination of flow and power at which quenching occurs then identifies a point on the stability curve. In some regions of

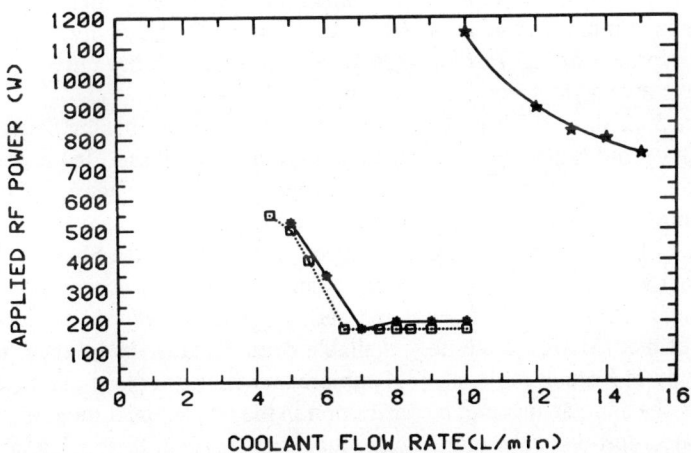

Figure 5.12. Plasma stability curves from [35] such as used in developing "optimized" torches. Here the effect of annular spacing between the outer and intermediate tubes is dramatically illustrated ★, 1-mm annular spacing; *, 0.7-mm annular spacing; □, 0.5-mm annular spacing. (Reprinted with permission from R. Rezaaiyaan, G. M. Hieftje, H. Anderson, H. Kaiser, and B. Meddings, "Design and Construction of a Low-Flow, Low-Power Torch for Inductively Coupled Plasma Spectrometry," *Appl. Spectrosc.* **36,** 629 (1982). Copyright (1982), Society for Applied Spectroscopy, Frederick, MD.)

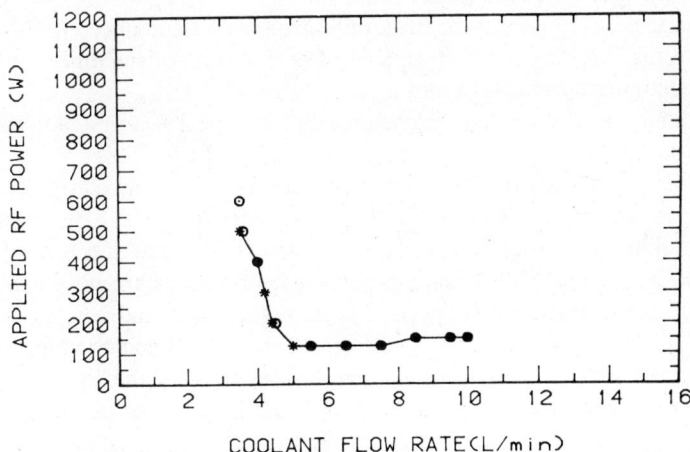

Figure 5.13. Comparison of plasma stability curves obtained on a fully optimized torch [35] for a discharge into which is aspirated distilled water (*) and a 1% sodium chloride solution (⊙). (Reprinted with permission from R. Rezaaiyaan, G. M. Hieftje, H. Anderson, H. Kaiser, and B. Meddings, "Design and Construction of a Low-Flow, Low-Power Torch for Inductively Coupled Plasma Spectrometry," *Appl. Spectrosc.* **36,** 631 (1982). Copyright (1982), Society for Applied Spectroscopy, Frederick, MD.)

operation, it is more convenient to decrement the coolant flow; in others, where torch damage might result, it is best to reduce rf power gradually. The family of points formed in this fashion then yield a plasma stability curve of which examples are shown in Figs. 5.12 and 5.13.

Figure 5.12 shows clearly the importance of the annular spacing between the intermediate and outer tubes, defined by Allemand and Barnes [6] in terms of the configuration factor. A greater configuration factor (smaller annular spacing) permits operation both at lower flows and lower powers.

In the original work on optimized torches, carried out at the Sherritt-Gordon Research Laboratories in Fort Saskatchewan, Alberta [75], it was principally this annular spacing that was explored and which ultimately led to the development of the "MAK" torch now available commercially. In contrast, the detailed study of Rezaaiyaan et al. [35] explored a number of plasma dimensions, including the annular spacing, a constriction in the coolant inlet tube, the length of the flared portion of the intermediate tube, the length of the cooled tube, the inner diameter of the aerosol tube, and the length of the capillary extension on the aerosol injection tube. Of these dimensions, the most critical as far as plasma stability was concerned were found to be the annular spacing and the inside diameter of the aerosol injection tube. Recommended values for these dimensions were 0.5 and 1 mm, respectively. All other parameters were found to be relatively unimportant for plasma stability, although a constriction on the outer argon inlet was found to simplify plasma ignition. The diameter of this constriction was recommended to be 1 mm.

Plasma stability curves of the final, optimized torch, reproduced in Fig. 5.13, show that the resulting torch is capable of performing satisfactorily with high salt concentrations introduced into it.

The dimensions, operating characteristics, analytical features, and interferences pertaining to this fully optimized torch are listed in Tables 5.3–5.5. For minimal interferences [41], it is recommended that the optimized torch be operated at an applied rf power of 450 W and an outer flow of 6 L/min. Interferences are minimized when the analyte emission is observed at a point 12 mm above the load coil. Under such conditions, interferences are no greater than found in a conventional torch operated at far higher flows and powers. Analytical characteristics under these conditions [41] are also comparable to those experienced with a conventional system. However, with the low-flow, low-power torch, background spectra are simpler and the background continuum intensity is lower. At the lower powers, analyte emission is reduced also, so that SBRs are comparable to what is encountered in conventional systems.

These characteristics are consistent with the fundamental features of the low-flow, low-power plasma [42]. Excitation temperatures, measured with iron as the thermometric species, are somewhat lower than those in a conventional plasma (5575 versus 4036 K at 15 mm above the load coil) and electron number

5.4. NEW TRENDS IN TORCH DEVELOPMENT

densities are somewhat lower (1.7×10^{14} vs. 1.6×10^{15} cm^{-3} at 15 mm above the load coil). This optimized torch has recently been coupled with an unusually inexpensive rf system constructed from radio transmitter components [76]. This system operates at 27.12 MHz and reportedly costs only 10–20% as much as commercial equipment.

Not surprisingly, characteristics of the original MAK torch differ only slightly from those of the fully optimized system. Montaser et al. [62] used a simplex algorithm, geared to a weighted SBR summed over 20 elements, to optimize gas flows, rf power, observation height, and sample uptake in the MAK system. Optimal values were initially found to be 920 W, 11-L/min outer flow, and an observation height of 14 mm above the load coil. Because these values were judged to be impractical, the authors fixed the outer flow at 4 L/min and the intermediate flow at 0.5 L/min and optimized the remaining variables. Under these conditions an optimal rf power was found to be 720 W and the best observation height 11 mm above the load coil. Under these conditions, the SBRs for the low-flow system were equivalent to or inferior to those of the conventional plasma by a factor of 1.5–2.0. However, phosphate- and sodium-induced interferences were as small as are ordinarily expected in ICP systems. In agreement with the values of Rezaaiyaan et al. [42], an electron density of 1.4×10^{14} cm^{-3} was determined under the lower power conditions. This system was found to be satisfactory for the measurement of real samples, as evidenced by results obtained from a USGS reference geological standard.

The MAK torch is good also for samples dissolved in organic solvents [43]; a finding which presumably pertains also to the fully optimized system. Xylene can be aspirated directly into the torch at powers as low as 300 W. However, solvents with higher vapor pressure (e.g., methanol, acetone) require powers of 2.5 kW. The authors show plasma stability curves for several solvents and demonstrate that SBRs are improved at lower usable powers. Precision and detection limits obtained with organic solvents are comparable to those found with aqueous solutions. Excellent agreement is found with cited values for a standard reference material of trace elements in fuel oil (NBS 1634a). Overall, powers recommended for use with organic solvents are somewhat higher than those ordinarily used for the MAK torch (750 W and 7.5-L/min coolant flow).

Recently, Rezaaiyaan and Hieftje [60] explored the possibility of both optimizing and reducing the size of an ICP torch. However, in accordance with the prediction of Angleys and Mermet [22], little further reduction in either gas flow or rf power requirements was found. The 13-mm i.d. torch that was employed required approximately the same outer flows and power levels required by a larger optimized 18-mm torch and an original, nonoptimized 13-mm torch [26]. Moreover, vaporization interferences were somewhat greater than found in the other systems although detection limits and linearity were comparable.

In a study quite unrelated to reducing gas flow or power requirements of the

ICP, Davies and Snook [77] attempted to reduce noise in ICP emission by abandoning the "vortex stabilization approach" employed by most workers and using instead an axial flow of argon supporting gas. Although a vortex has been found to promote plasma ignition and to stabilize the discharge, it also causes problems. For example, Belchamber and Horlick [78] found that the swirling argon produces a rotation in any plasma asymmetry that exists. As a result, both background and analyte features are often modulated at frequencies between 200 and 500 Hz.

The "laminar-flow" torch of Davies and Snook [77] appears to overcome these problems by sending the outer gas axially into the annular space between the outer and intermediate tubes through 21 holes arranged circumferentially. The Reynolds number for the resulting flow was calculated to be approximately 650[1]. As a result of their efforts, the authors achieved a tenfold lower background noise when the system was operated without nebulizer. Unfortunately, when a nebulizer was used, a signal-to-noise improvement was hardly perceptible. It is surprising, therefore, that the authors could achieve a 10- to 30-fold improvement in detection limits: if the background RSD is only marginally reduced, then the gain in detection limits should stem from a vast increase in SBRs. This point is not addressed in the preliminary communication. More results will be required before the merits of this approach can be further assessed.

5.5 INFLUENCE OF THE FREQUENCY ON THE SPECTRAL CHARACTERISTICS OF CONVENTIONAL ICPS

Capelle et al. [44] studied the effect of frequency on some spectral characteristics of argon ICPs operated under compromise conditions for multielement

---

[1]The term "laminar" contrasts with "turbulent." Both a vortex and an axial flow may be either "laminar" or "turbulent." It seems that Davies and Snook identify "vortex flow" with "turbulent flow" and that their "laminar flow torch" could be better called "axial flow torch."

Further, the Reynolds number gives an indication about whether a flow is laminar or turbulent and its use in the context of Davies' and Snook's work is irrelevant, primarily because the authors aim at creating an axial flow instead of a vortex flow, which has nothing to do with the Reynolds number. On the other hand, if the aim is to avoid turbulence in either an axial or a vortex flow, then the Reynolds number is relevant. However, one meets the difficulty of accurately computing this number for the gas flow above the rim of the intermediate tube, which flow, eventually, dictates the "spectrochemical behaviour" of the ICP. One may calculate the Reynolds number for the flow below this rim to assess whether this flow is turbulent or not, but the result is not decisive as to what happens higher up in the plasma.

Davies and Snook computed the Reynolds number for the flow below the rim of the intermediate tube in the "laminar flow torch" to be about 650, which value is well below the critical value of 2000 that would indicate turbulence. If, on the contrary, in the "laminar flow torch" the gas would be introduced tangentially, the Reynolds number would be also found to be smaller than 2000, thus indicating the laminarity of the vortex flow at the rim. Unfortunately, for neither of the two configurations ("axial" or "vortex") does this calculation permit it to draw a rigorous conclusion about the turbulence of the flow above the rim of the intermediate tube.

5.5. INFLUENCE OF THE RF FREQUENCY

analysis. Generators from various manufacturers were used and covered frequencies of 5, 27, 40, 50, and 56 MHz.

From the results reported by Capelle et al. we point out the following items.

1. Both the excitation temperature (measured with iron atomic lines) and the electron number density were found to decrease with increasing frequency.

2. As a result of (1) both net line intensities and the argon continuum decreased with increasing frequency. However, the lowering of the continuum intensity exceeded that of the line intensities so that there remained a net gain in SBR when the frequency was raised. The magnitude of this gain depended on the element and the wavelength of the spectral line and generally was larger the higher the wavelength.

3. Detection limits obtained with a 40-MHz ICP were up to a factor of 60 better than those found with a 5-MHz ICP.

4. In a comparison of the detection limits achieved at 27 or 56 MHz (Table 5.6) it was found that the results at 56 MHz were a factor of 1.3 to 8 better for deionized water and solutions containing 10 mg/mL of Na or Fe. No improvement in the detection limit was found with oil as a matrix, which the authors attributed to dissociation problems. For elements that form refractory oxides (W, Al, Ti) the detection limits improved in a way similar to that demonstrated in Table 5.6.

For deriving detection limits from SBRs, Capelle et al. used mean values of the RSD of the background, averaged over the wavelength region between 200 and 300 nm. The improvement in the detection limits with increased frequency (Table 5.6) must, therefore, be completely attributed to an increase in SBR, which agrees qualitatively with predictions by Barnes et al. [79, 80] and the experimental results reported by Boumans et al. [16], and, more recently, by

Table 5.6. Ratio of the Detection Limits Attained with a 27-MHz ICP and a 56-MHz ICP, Respectively [44] (Reprinted with permission from B. Capelle, J. M. Mermet, and J. Robin, *Appl. Spectrosc.* 36, 104. Copyright (1982) Society for Applied Spectroscopy, Frederick, MD.)

Spectral Line (nm)		Matrix			
		Deionized Water	1% m/v Na in Water	1% m/v Fe in Water	Oil in Xylene (1:1)
B I	208.9	1.7	1.5	1.3	
Co II	228.1	3.3	2.0	1.8	
Ni II	231.6	3.3	2.7	2.0	
V II	311.1	5.5	4.0	5.0	0.5
Cu I	327.4	7.0	8.3	6.0	

Boumans and Vrakking [81], for their 50 MHz ICP, which were compared with those of Winge et al. [63] for a 27-MHz ICP (cf. Section 4.1.7.2).

Some caution must be exercised as to several apparently contradictory or at least confusing statements by Capelle et al. about the behavior of the noise and their sometimes confusing the terms SNR and SBR (cf. Section 4.1.3 et seq.). It is striking also to see that they found a constant RSD of the background signal (0.3% for deionized water) to apply to the entire wavelength region between 200 and 300 nm at both 27 and 56 MHz. This result should imply that the shot noise contribution was entirely negligible down to the lowest wavelength either in both systems or at least in the 27-MHz system. In the former case, both generators should show the same stability, in contrast to what the authors believe. In the latter case, the expectedly greater shot noise contribution to the RSD at 56 MHz would have to be perfectly balanced by an increase in generator stability depending on wavelength.

5. The change in frequency from 27 to 56 MHz did not affect the level of ionization and atomization interferences, which remained low. A low level of atomization interferences at 56 MHz was considered compatible with the relatively low temperature of the ICP because a fairly long residence time of the analyte was achieved by the use of a wide orifice (3 mm) in the injector tube and a low carrier gas flow rate (0.5 L/min).

Recently, Michaud-Poussel and Mermet [82] examined the influence of the generator frequency and the intertube annular area on torch design. They employed rf frequencies of 27, 40, 56, 64, and 102 MHz and coupled energy to the plasma via load coils with between three and six turns. They found that the use of higher frequencies makes torch dimensions less critical and permits stable operation at lower plasma-gas flow rates. For example, operation at 102 MHz makes it possible to operate routinely with gas flows below 10 L/min, regardless of the annular spacing between the outer and intermediate tubes. Similarly, the power required to sustain a discharge is reduced at higher frequencies; 400 to 600 W is sufficient at 102 MHz even with a 26-mm i.d. torch. As far as precision or detection limits are concerned, the high-frequency, low-flow systems perform as well as conventional ones. However, the authors have not yet examined closely interelement interferences, although they note that a toroidal-shaped plasma can be formed only above gas flow rates of 3 to 3.5 L/min.

Combining the torch shown in Fig. 5.2 with a 100-MHz free-running generator, Boumans et al. [18] operated an ICP with an outer argon flow of 7.5 L/min and achieved virtually the same detection limits as with their 50-MHz ICP (see Table 5.5). Interestingly, the high frequency of 100 MHz permitted the use of carrier gas tubes having orifice diameters up to 3 mm. A value of 2.4 mm was chosen as a good compromise to maximize the detection power and concomitantly minimize interferences. The system was operated with a cross-flow nebulizer requiring a nebulizer gas flow of about 1 L/min for optimum performance. With a 2.4-mm orifice diameter the cold gas velocity of the carrier gas is 3.7 m/s for a 1 L/min flow. This figure compares with 6.5 m/s for the torch used in conjunction with the 50-MHz ICP (Fig. 3.8), also for a 1-L/min flow, and with 10.6 m/s for the torch described by Rezaaiyaan et al. [35] (''MAK torch'') used with 27- or 40-MHz generators and operated with a carrier gas flow of only 0.5 L/min.

REFERENCES

1. G. R. Jordan, *Rev. Phys. Technol.* **2,** 128 (1971).
2. A. Czernikowski and J. Jurewics, *ICP Information Newslett.* **1,** 1 (1976).
3. H. U. Eckert, *High Temp. Sci.* **6,** 99 (1974).
4. S. V. Dresvin, ed., *Physics and Technology of Low-Temperature Plasmas*. Atomizdat, Moscow (1972); Eng. Transl.: The Iowa State University Press, Ames, IA (1977).
5. M. L. Thorpe, NASA Contractors Report CR-1143 (1968).
6. C. D. Allemand and R. M. Barnes, *Appl. Spectrosc.* **31,** 434 (1977).
7. J. L. Genna, R. M. Barnes, and C. D. Allemand, *Anal. Chem.* **49,** 1450 (1977).
8. D. L. Windsor, D. R. Heine, and M. B. Denton, *Appl. Spectrosc.* **33,** 56 (1979).
9. P. W. J. M. Boumans and M. Ch. Lux-Steiner, *Spectrochim. Acta* **37B,** 97 (1982).
10. P. W. J. M. Boumans and F. J. de Boer, *Spectrochim. Acta* **27,** 391 (1972).
11. S. Greenfield, I. L. W. Jones, and C. T. Berry, U.S. Patent 3,467,471, September 16, 1969.
12. S. Greenfield, I. Ll. Jones, and C. T. Berry, *Analyst* **89,** 713 (1964).
13. S. Greenfield, I. Ll. Jones, H. McD. McGeachin, and P. B. Smith, *Anal. Chim. Acta* **74,** 225 (1975).
14. R. H. Scott, V. A. Fassel, R. N. Kniseley, and R. N. Nixon, *Anal. Chem.* **46,** 75 (1974).
15. C. D. Allemand, *ICP Information Newslett.* 2, 1 (1976).
16. P. W. J. M. Boumans, R. J. McKenna, and M. Bosveld, *Spectrochim. Acta* **36B,** 1013 (1981).
17. M. Ch. Lux-Steiner, Ph.D. thesis, Swiss Federal Institute of Technology, Zuerich, Switzerland (1980).
18. P. W. J. M. Boumans and J. J. A. M. Vrakking, *1986 Winter Conference on Plasma Spectrochemistry, Hawaii, 1986, Plenary lecture, Abstr. 62*. To be published in *Spectrochim. Acta*, Part B.
19. L. Ebdon, D. J. Mowthorpe, and M. R. Cave, *Anal. Chim. Acta* **115,** 171 (1980).
20. M. E. Britske, V. M. Borisov, and Yu. S. Sukah, *Ind. Lab.* **33,** 301 (1967).
21. M. A. Floyd, V. A. Fassel, and A. P. D'Silva, *Anal. Chem.* **52,** 2168 (1980).
22. G. Angleys and J.-M. Mermet, *Appl. Spectrosc.* **38,** 647 (1984).
23. J. M. Mermet and C. Trassy, *Appl. Spectrosc.* **31,** 237 (1977).

24. P. A. M. Ripson and L. de Galan, *Spectrochim. Acta* **38B**, 707 (1983).
25. C. D. Allemand, R. M. Barnes, and C. C. Wohlers, *Anal. Chem.* **51**, 2392 (1979).
26. R. N. Savage and G. M. Hieftje, *Anal. Chem.* **51**, 408 (1979).
27. A. D. Weiss, R. N. Savage, and G. M. Hieftje, *Anal. Chim. Acta.* **124**, 245 (1981).
28. I. J. M. Raaijmakers and P. W. J. M. Boumans, unpublished results (1983).
29. M. A. Heald and J. B. Wharton, *Plasma Diagnostics with Microwaves*. Wiley, New York (1965).
30. V. L. Ginzburg and A. V. Gurevich, *Sov. Phys. Uspekhi* **3**, 115 (1960).
31. S. I. Braginskii, "Transport Processes in a Plasma," in M. A. Leontovich, Reviews of Plasma Physics, Vol. 1. Consultants Bureau, New York (1965), p. 215.
32. M. Huang, K. A. Marshall, and G. M. Hieftje, *Anal. Chem.* **58**, 000 (1986).
33. G. R. Kornblum and L. de Galan, *Spectrochim. Acta* **32B**, 455 (1977).
34. R. N. Savage and G. M. Hieftje, *Anal. Chem.* **52**, 1267 (1980).
35. R. Rezaaiyaan, G. M. Hieftje, H. Anderson, H. Kaiser, and B. Meddings, *Appl. Spectrosc.* **36**, 627 (1982).
36. G. R. Kornblum, W. van de Waa, and L. de Galan, *Anal. Chem.* **51**, 2378 (1979).
37. H. Kawaguchi, T. Ito, S. Rubi, and A. Mizuike, *Anal. Chem.* **52**, 2440 (1980).
38. P. A. M. Ripson, L. de Galan, and J. W. de Ruiter, *Spectrochim. Acta* **37B**, 733 (1982).
39. P. A. M. Ripson, E. B. M. Jansen, and L. de Galan, *Anal. Chem.* **56**, 2329 (1984).
40. P. S. C. van der Plas and L. de Galan, *Spectrochim. Acta* **39B**, 1161 (1984).
41. R. Rezaaiyaan, J. W. Olesik, and G. M. Hieftje, *Spectrochim. Acta.* **40B**, 73 (1985).
42. R. Rezaaiyaan and G. M. Hieftje, *Anal. Chem.* **57**, 412 (1985).
43. R. C. Ng, H. Kaiser, and B. Meddings, *Spectrochim. Acta* **40B**, 63 (1985).
44. B. Capelle, J. M. Mermet, and J. Robin, *Appl. Spectrosc.* **36**, 102 (1982).
45. M. H. Abdallah, R. Diemiaszonek, J. Jarosz, J. M. Mermet, J. Robin, and C. Trassy, *Anal. Chim. Acta* **84**, 271 (1976).
46. D. R. Demers and C. D. Allemand, *Anal. Chem.* **53**, 1915 (1980).
47. G. L. Long and J. D. Winefordner, *Appl. Spectrosc.* **38**, 563 (1984).
48. J. Davies, J. R. Dean, and R. D. Snook, *Analyst* **110**, 535 (1985).
49. L. M. Faires, T. M. Bieniewski, C. T. Apel, and T. M. Niemczyk, *Appl. Spectrosc.* **39**, 5 (1985).
50. C. Butler-Sobel, *Appl. Spectrosc.* **367**, 691 (1982).
51. J. A. C. Broekaert and P. B. Zeeman, *Spectrochim. Acta* **39B**, 851 (1984).
52. D. R. Heine, J. S. Babis, and M. B. Denton, *Appl. Spectrosc.* **34**, 595 (1980).
53. D. J. Devine, R. M. Brown, and R. C. Fry, *Appl. Spectrosc.* **35**, 332 (1981).
54. C. J. Seliskar and D. K. Warner, *Appl. Spectrosc.* **39**, 181 (1985).
55. D. C. Miller, C. J. Seliskar, and T. M. Davidson, *Appl. Spectrosc.* **39**, 13 (1985).

56. K. A. Wolnick, D. C. Miller, C. J. Seliskar, and F. L. Fricke, *Appl. Spectrosc.* **39**, 930 (1985).
57. G. D. Rayson, K. A. Marshall, R. Rezaaiyaan, and G. M. Hieftje, *1985 ACS Meeting, Chicago, IL, Abstr. 85*; *ICP Information Newslett.* **11**, 344 (1985).
58. G. M. Hieftje, *Spectrochim. Acta* **38B**, 1465 (1983).
59. M. Kubota, *Bunseki*, **(9)**, 669 (1984).
60. R. Rezaaiyaan and G. M. Hieftje, *Anal. Chim. Acta* **173**, 63 (1985).
61. H. Kawaguchi, T. Tanaka, S. Miura and A. Mizuike, *Spectrochim. Acta* **38B**, 1319 (1983).
62. A. Montaser, G. R. Huse, R. A. Wax, S. K. Chan, D. W. Golightly, J. S. Kane, and A. F. Dorrzapf, Jr., *Anal. Chem.* **56**, 283 (1984).
63. R. K. Winge, V. J. Peterson, and V. A. Fassel, *Appl. Spectrosc.* **33**, 206 (1979).
64. R. N. Savage and G. M. Hieftje, *Anal. Chim. Acta* **123**, 319 (1981).
65. E. Sexton, R. N. Savage, and G. M. Hieftje, *Appl. Spectrosc.* **33**, 643 (1979).
66. P. W. J. M. Boumans, *Line Coincidence Tables for Inductively Coupled Plasma Atomic Emission Spectrometry.* Pergamon Press, Oxford (1980).
67. P. W. J. M. Boumans, *Spectrochim. Acta* **36B**, 169 (1981).
68. P. A. M. Ripson and L. de Galan, *Spectrochim. Acta* **36B**, 71 (1981).
69. H. Kawaguchi, T. Tanaka, S. Miura, and A. Mizuike, *Spectrochim. Acta.* **36B**, Supplement, 176 (1983).
70. M. E. Britske, J. S. Sukach, and L. N. Filimov, *Zh. Prikl. Spektrosk.* **25**, 5 (1976).
71. K. I. Zilbershtein, *ICP Information Newslett.* **5**, 508 (1980).
72. P. A. M. Ripson, E. B. M. Jansen, and L. de Galan, *Spectrochim. Acta* **38B**, Supplement, 283 (1983).
73. P. S. C. van der Plas, A. C. de Waaij, and L. de Galan, *Spectrochim. Acta* **40B**, 1457 (1985).
74. M. D. Lowe, *Appl. Spectrosc.* **35**, 126 (1981).
75. B. Meddings, personal communication (1981).
76. G. M. Allen and D. M. Coleman, *Anal. Chim. Acta* **158**, 267 (1984).
77. J. Davies and R. D. Snook, *Analyst* **110**, 887 (1985).
78. R. M. Belchamber and G. Horlick, *Spectrochim. Acta* **37B**, 17 (1982).
79. R. M. Barnes and R. G. Schleicher, *Spectrochim. Acta* **30B**, 109 (1975).
80. R. M. Barnes, *CRC Crit. Rev. Anal. Chem.* **7**, 203 (1978).
81. P. W. J. M. Boumans and J. J. A. M. Vrakking, *Spectrochim. Acta* **40B**, 1437 (1985).
82. E. Michaud-Poussel and J.-M. Mermet, *Spectrochim. Acta* **41B**, 125 (1986).
83. I. J. M. M. Raaijmakers, P. W. J. M. Boumans, B. van der Sijde, and D. C. Schram, *Spectrochim. Acta* **38B**, 697 (1983).

CHAPTER 6

SAMPLE INTRODUCTION TECHNIQUES IN ICP-AES

J. A. C. BROEKAERT

Institut für Spektrochemie und angewandte Spektroskopie
Dortmund, Federal Republic of Germany

P. W. J. M. BOUMANS

Philips Research Laboratories
Eindhoven, The Netherlands

6.1. Introduction
6.2. Pneumatic Nebulizers
    6.2.1. The Concentric Nebulizer
    6.2.2. The Cross-Flow Nebulizer
    6.2.3. The Babington Nebulizer
    6.2.4. The Fritted Disc Nebulizer
    6.2.5. Stability of Pneumatic Nebulizers
    6.2.6. Perspectives
    6.2.7. Discrete Sampling Using Pneumatic Nebulization
        6.2.7.1. Injection Methods Combined with a Low-Power Argon ICP
        6.2.7.2. Injection Methods Combined with a High-Power Argon–Nitrogen ICP
        6.2.7.3. Flow Injection Methods
        6.2.7.4. Element-Specific Detection in Liquid Chromatography
6.3. Ultrasonic Nebulization
    6.3.1. Principles of Operation
    6.3.2. Analytical Performance
        6.3.2.1. Power of Detection
        6.3.2.2. Operation Stability and Memory Effects
        6.3.2.3. The Sample Nature
    6.3.3. Perspectives
6.4. Electrothermal Nebulization Devices
    6.4.1. Metal Boats and Filaments
    6.4.2. Graphite Yarn, Graphite Rod, and Graphite Furnace Systems
    6.4.3. Direct Sample Insertion Devices
6.5. Hydride Techniques
6.6. The Introduction of Gases and Vapors

6.7. Direct Solids Sampling
  6.7.1. Powder Sampling
  6.7.2. Arc and Spark Devices
  6.7.3. Laser Evaporation of Solid Samples
6.8. Conclusion
  References

6.1. INTRODUCTION

A wide variety of sample introduction devices is used in ICP-AES in order to generate an argon-aerosol mixture from samples of varying nature, form, and size. These samples may be

- Liquids of varying physical properties such as surface tension, viscosity, density, and salt content
- Solids, compact metallic samples, nonconducting samples, or powder samples
- Permanent gases or gaseous products arising from evolution techniques

Most sample introduction devices used in ICP-AES originate from flame atomic emission spectroscopy (FAES) [1, 2] or arc emission spectroscopy [3] and are not essentially new. Their description and adaptation as well as their parametric optimization for ICP-AES and typical applications form the subject of this chapter.

A survey of the sample introduction devices proposed for ICP-AES is shown in Fig. 6.1. Pneumatic nebulization is of prime importance for the analysis of liquids. However, pneumatic nebulizers known from FAES and flame atomic absorption spectroscopy (FAAS) had to be tailored to the requirements of analytical ICP-AES. Ultrasonic nebulization equipment often comprises a desolvation unit. Electrothermal atomization devices (carbon rod, tantalum filament, and graphite furnace system) are useful for the analysis of microsamples and are similar to those used in AAS work. Evolution techniques, such as hydride techniques, allow it to increase the aerosol generation efficiency and the power of detection for a number of elements as known from AAS (see e.g., [5, 6]). ICP-AES may be used for element-specific detection in chromatography. Despite the fact that the prime merits of ICP-AES lie in the multielement trace analysis of solutions aiming at high analytical precision and accuracy, the direct analysis of solids with ICP-AES is possible. For special problems, such as samples which are difficult to handle or dissolve, direct solids nebulization with the aid of a laser, arc, or spark may be useful. The direct introduction of powders into the ICP, although of interest for routine analysis, copes with a number of fundamental and practical problems.

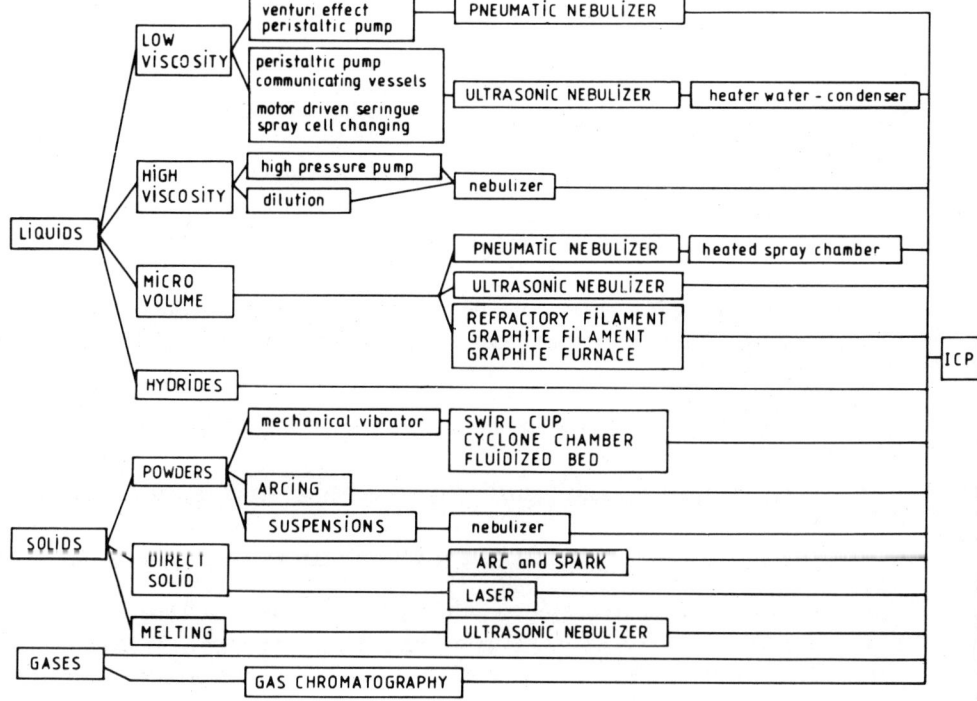

Figure 6.1. Schematic diagram for sample introduction in ICP–AES. Similar to given in [4].

The breakdown of an aerosol introduced into an ICP proceeds in the following stages:

- Desolvation, that is, evaporation of the solvent
- Evaporation of the dry aerosol particles
- Thermal dissociation of the molecules
- Partial ionization of the free atoms

A sufficiently high gas temperature, small particle size, and a high residence time (and, thus, low carrier gas velocity) are required for adequate thermal evaporation and dissociation of aerosol particles.

While carrier gas velocity and flow are interrelated via the cross section of the orifice of the carrier gas tube, the gas flow has always been considered as the prime parameter chiefly because it can be easily measured. Moreover, already in early ICP work it was found that the temperature depends considerably on the gas flow [7–10]. Given the orifice cross section, the carrier gas flow is one of the primary parameters of an ICP. It has been recognized in the early ICP work [11] that keeping this flow at about 1 L/min is of paramount importance to achieve analytically useful conditions in argon ICPs (cf. Section 4.7). Apart from a low argon carrier gas flow, the aerosol particles must be sufficiently small. Barnes and Schleicher [12], for example, show that in the case of Al_2O_3 only particles with a diameter below 1 μm introduced with an argon flow of 1 L/min into an ICP of 6000 K completely evaporated during their transit to the analytical zone. The need for an efficient production of an aerosol having a small mean particle size and a well-defined particle size distribution, as well as the condition of a low carrier gas flow, have governed the development of all sample introduction devices listed in Section 6.2.

6.2. PNEUMATIC NEBULIZERS

Pneumatic nebulization is well known from FAES [1, 13]. Several types of nebulizers are used in ICP–AES, namely,

- The concentric nebulizer [14]
- The cross-flow nebulizer [15–23]
- The Babington nebulizer [24–29]
- The fritted disc nebulizer [30–32]

Their operation principle is illustrated in Fig. 6.2 and will be detailed in Sections 6.2.1 to 6.2.4. In all devices, the sample solution is fed to the nebulizer nozzle by forced feed (e.g., with a pump) or aspiration resulting from the Venturi-effect, that is, the effect causing a reduced pressure at the nozzle. The solution is split into droplets under the influence of a high-speed gas flow. This gas serves, at the same time, as aerosol transport gas. Since the flow rate of the latter should be about 1 L/min, a high pressure is generally required at the gas nozzle to generate an aerosol. Only a small portion of the introduced liquid is nebulized, the larger fraction flows off to the waste. The aerosol, which consists of droplets of various size, is blown into a spray chamber where the larger droplets are sorted out. A multiplicity of spray chamber configurations has been described in the literature (e.g., [32, 33]). As an example, Fig. 6.3 shows the spray chambers proposed by Schutyser and Janssens [33], who described their optimization in some detail. Other types of spray chambers are shown in some of the figures discussed later or in Part 2, Chapter 8, which deals with the fundamental aspects of aerosol generation and transport, including spray chamber effects.

The stability of the signals emitted from the ICP depends on the nebulizer and spray chamber design and on the stability of the gas and liquid supplies.

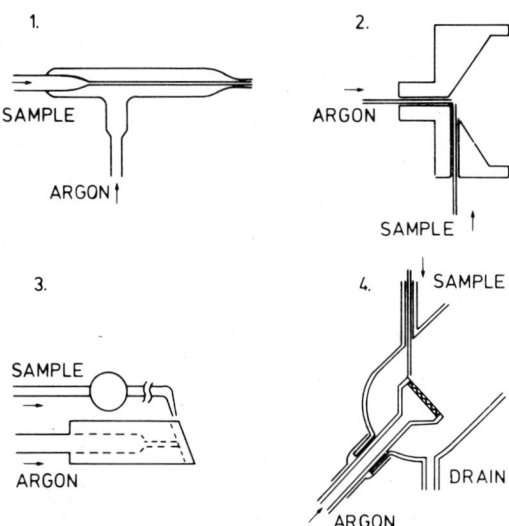

Figure 6.2. Types of pneumatic nebulizers: 1, concentric glass nebulizer; 2, cross-flow nebulizer; 3, Babington nebulizer; and 4, fritted disc nebulizer.

To achieve a stable gas flow it is advisable to use a separate supply for the aerosol gas and the plasma gas. High-precision manostats are used and the gas flow through the nebulizer is monitored with the aid of a rotameter. However, as the amount of gas flowing through the nebulizer is of importance, the use of mass-flow meters and mass-flow control is adequate. The latter devices are optionally available from manufacturers of ICP instruments and are indispensable for high-precision work.

Pumps may be used to ensure a constant supply of liquid. Peristaltic pumps with a gear having at least 10 rolls can be used in order to obtain a pulse-free sample feeding. The use of types with a planet gear avoids tube distortions and makes the frequent replacement of tubes and the use of connection spares superfluent. Also, mechanically driven syringes have been proposed and may be of interest in the analysis of organic liquids [18] or corrosive samples deteriorating flexible tubes.

For cross-flow nebulizers, the Babington nebulizers, and the fritted-disc nebulizers, the pump is indispensable for sample transport as there is no Venturi effect. A pump is also advantageous with the concentric glass nebulizer, which is self-aspirating, because it ensures a stable sample uptake as well as reduces nebulization effects that arise from viscosity differences between the samples. These effects constitute a substantial part of the matrix effects in ICP–AES. Differences in viscosity may be related to the presence of tensids as happens in

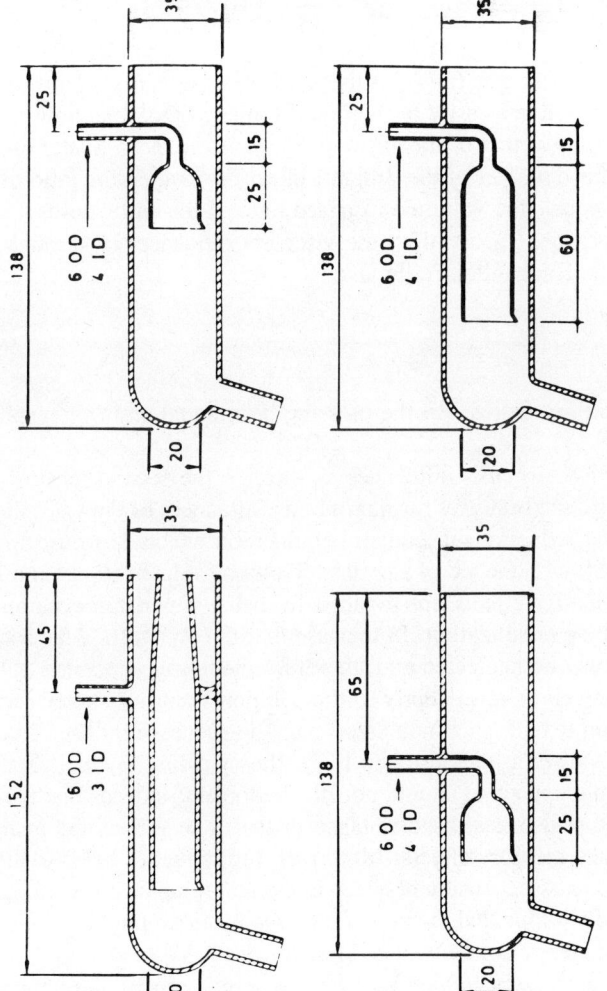

Figure 6.3. Various spray chambers proposed for ICP-AES [33]. [Reprinted with permission from P. Schutyser and E. Janssens, "Evaluation of Spray Chambers for Use in Inductively Coupled Plasma-Atomic Emission Spectrometry," *Spectrochim. Acta* **34B**, 443 (1979). Copyright (1979), Pergamon Journals, Oxford.]

waste waters, varying total salt or acid contents, or the presence of different acids [34]. The influence of the viscosity may be understood from the Nukijama–Tanasawa equation [2, 35, 36], which relates the aerosol droplet size to the physical properties of the sample solution and the aerosol gas flow:

$$d_0 = \frac{585}{V_G} \left(\frac{\sigma}{\rho}\right)^{0.5} + 597 \frac{\eta}{(\sigma\rho)^{0.5}} \left(1000 \frac{Q_L}{Q_G}\right) \tag{6.1}$$

V_G is the velocity of the gas (m/s), σ the surface tension (dyne/cm), ρ the density (g/cm^3) and η the viscosity of the liquid (poises). Q_L is the flow rate of the liquid (cm^3/s) and Q_G that of the gas (cm^3/s), d_0 is the mean Sauter droplet diameter, that is, the diameter of the droplet (in μm) for which the ratio of the volume to surface equals the volume to surface ratio of the entire aerosol.

At low gas flows ($Q_L/Q_G > 10^{-3}$) the viscosity influences d_0 through Q_L, being defined by the Hagen–Poiseuille law:

$$Q_L = \frac{\pi R^4}{8 \cdot \eta \cdot l} \cdot \Delta p \tag{6.2}$$

where R is the capillary radius, Δp the pressure differential and l the capillary length.

However, the viscosity also influences d_0 through the second term of Eq. (6.1). Consequently, the best way for minimizing influences of the viscosity on the droplet size and reducing subsequent nebulization effects is to use forced feed (Q_L = constant) with the aid of a peristaltic pump [36]. Broekaert and Leis [37] showed that matrix effects are reduced by using a peristaltic pump for forced feed in discrete nebulization. In the case that $Q_L/Q_G < 10^{-3}$ the second term in Eq. (6.1) may be neglected and, therefore, nebulizers operated at high gas flow work relatively independently of the solution viscosity at free sample uptake. To a certain extent, such nebulizers may be successfully used in conjunction with high-power argon–nitrogen ICPs, thus yielding low nebulization effects at free sample uptake. It should not be overlooked, indeed, that the use of a peristaltic pump also has disadvantages in that it lengthens the analysis cycle and may evoke memory effects. Moreover, the influence of large differences in viscosity or surface tension between individual samples on nebulization cannot be completely eliminated, since not only the mean droplet size, but also the nebulization efficiency is involved as known from AAS work [39].

The pneumatic nebulizers, which will be discussed, can be used for both aqueous and organic solutions. As with AAS nebulizers [5, 40], their nebulization efficiency for organic solvents is higher than for water. This effect may be used to increase the analytical sensitivity and the power of detection. One accordingly can analyze oils with ICP–AES after a 10-fold dilution with solvents such as xylene and MIBK. In the case of samples with a complex matrix, one can isolate the elements to be determined as complexes in an organic phase (see e.g., [41–43] and Part 2, Chapters 3 and 5).

6.2.1. The Concentric Nebulizer

Concentric pneumatic nebulizers were already described by Gouy in 1879 [44]. The analyte solution is fed through a capillary surrounded by a second capillary, while the nebulizer gas flows through the slit between them and produces the aerosol. Several types are used in flame AAS and AES work; however, all need relatively high gas flows. For ICP work requiring efficient aerosol generation at low gas flows (1–3 L/min) only high-pressure types (operated at 2–7 bar) are of interest. Concentric glass nebulizers are widely used in ICP–AES. They are robust and allow high operation stability due to their monolithic construction. A widely spread commercial type is the concentric glass nebulizer manufactured by Meinhard [14, 45]. The device is offered as calibrated type (sample uptake: 1–2 or 3 mL/min) (Fig. 6.4) with orifice dimensions of 15–35 μm. Concentric nebulizers with a plastic base and platinum capillaries are also offered and designed for work with solutions containing fluoric acid. However, here the nebulization chamber must be also machined in plastics and the aerosol inlet tube must be made from an HF resistant material, e.g., boron nitride. The surface conditions and the material used for the various parts of the nebulizer assembly dictate the extent of memory effects.

Concentric nebulizers are self-aspirating and can be operated without a peristaltic pump. However, differences in the physical properties of the sample solutions can only be neglected when a high aerosol gas flow can be applied as happens with a high-power argon–nitrogen ICP. Generally, forced sample feed at a pumping rate just below the free sample uptake rate of the nebulizer (1–3 mL/min) leads to the best analytical precision. The optimum rate depends on the nebulizer type, the pressure applied, and the physical properties of the sample solution. Although forced feed is highly desirable with organic liquids such as MIBK, xylene, or CCl_4, free aspiration is often used because flexible tubes required in peristaltic pumps tend to be easily affected by organic solvents. Controlled introduction of organic aerosols into the ICP can be achieved by thermostatting the spray chamber at low temperature (cf. Section 4.7.8).

The maximum permitted salt content of a sample solution (20–40 g/L) depends on the aerosol gas flow and the sample uptake cycle. However, it seems to be limited by risks of deposits at the tip of the aerosol inlet tube in the torch rather than by nebulizer blockage. It also strongly depends on the type of salt. Easily hydrolyzable compounds such as aluminium salts and phosphates more easily cause nebulizer blockage. By long-term experiments it has been shown that concentric glass nebulizers are also sensitive to high alkali and alkaline earth concentrations. The changes in nebulization efficiency and sample uptake

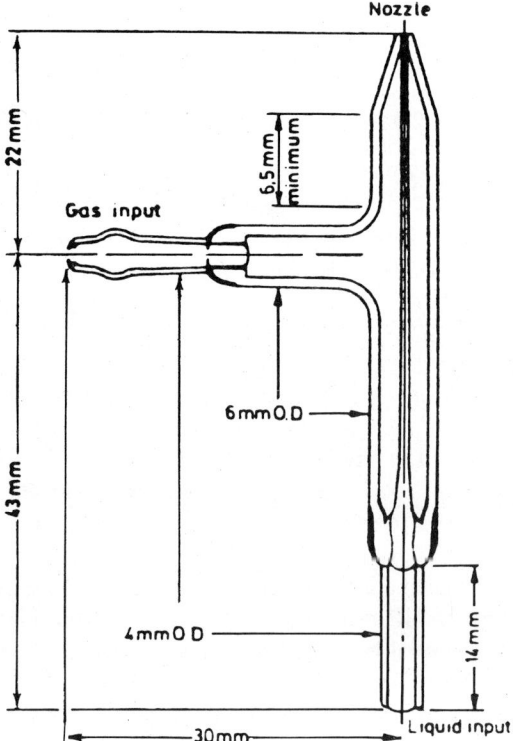

Figure 6.4. Concentric glass nebulizer for ICP–AES. (courtesy of Meinhard Assoc.)

rates are connected with ion-exchange phenomena at the glass wall, as mentioned in [46]. They are also due to the deposition of salt crystals at the nebulizer nozzle diminishing the free orifice and the gas flow. The latter not only influences the net line intensities but also the background intensity level as shown in Fig. 6.5.

Several measures may be taken to improve the operation stability of the concentric glass nebulizer. Meinhard [45] reported that types with rounded capillaries and types with a redrawn capillary are less sensitive to high salt concentrations especially at high gas pressure (up to 7 bar). He also introduced a type with a dual nozzle for continuous rinsing of the sample capillary (Fig. 6.6). Wetting of the aerosol gas is also effective. The use of a computer-controlled tip desalter for preventing nebulizer instabilities has been described [48].

The clean-up or washout time is defined as the time after sample change required to let the analytical signal decline to about 1% of its original value. Usually, this is less than 1 min. However, the concentric glass nebulizer was found to have a larger washout time than a cross-flow nebulizer with Pt-capil-

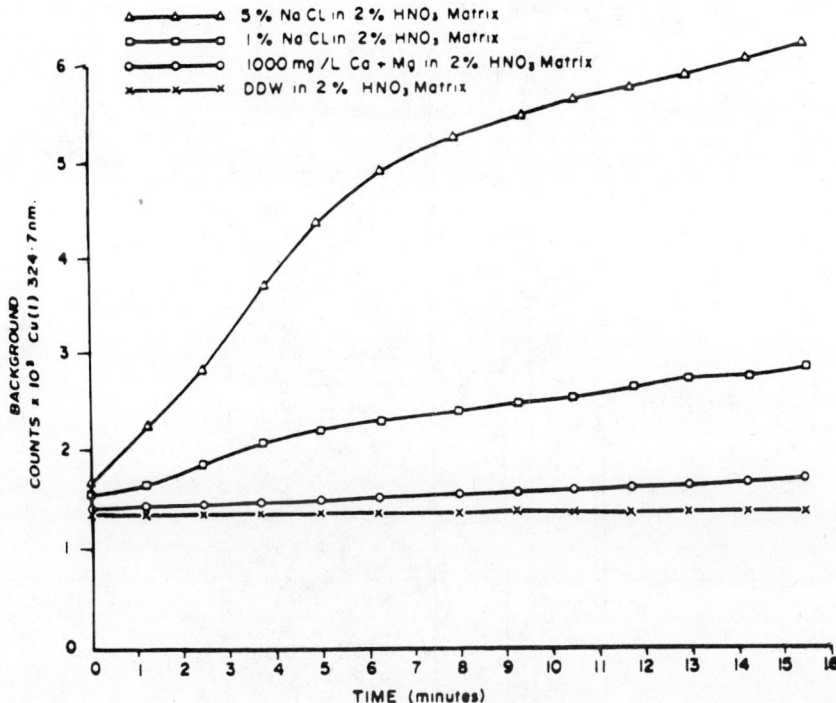

Figure 6.5. Influence of the sample matrix on the stability of the background intensities when using a concentric glass nebulizer [28]. [P. W. McKinnon, K. C. Giess, and T. V. Knight, in B. M. Barnes, ed., *Developments in Atomic Plasma Spectrochemical Analysis*, p. 287. Copyright (1981), Heyden & Son Ltd., London. Reprinted by permission of John Wiley & Sons, Ltd.]

laries [49]. The volume of the nebulizer, that is, the volume inside the sampling capillary and the connection tube to the aspiration sleeve, considerably influences the washout time. Ramsey et al. [50] showed that the cleanup curve consists of a linear part with high slope and a hyperbolic portion declining asymptotically to the baseline. When the nebulizer volume is small, which in the case of a concentric glass nebulizer was attained by fixing the aspiration sleeve to a metal capillary cemented in the sample capillary, the hyperbolic part was practically eliminated. Thus he obtained total analysis times of 45 s, that is, 9 s to reach steady state, 15 s to measure, and 21 s as cleanup time.

6.2.2. The Cross-Flow Nebulizer

In the cross-flow nebulizer, the sample solution is fed through a vertically mounted capillary and nebulized by the gas flow entering through a horizontal capillary which ends close to the tip of the former one.

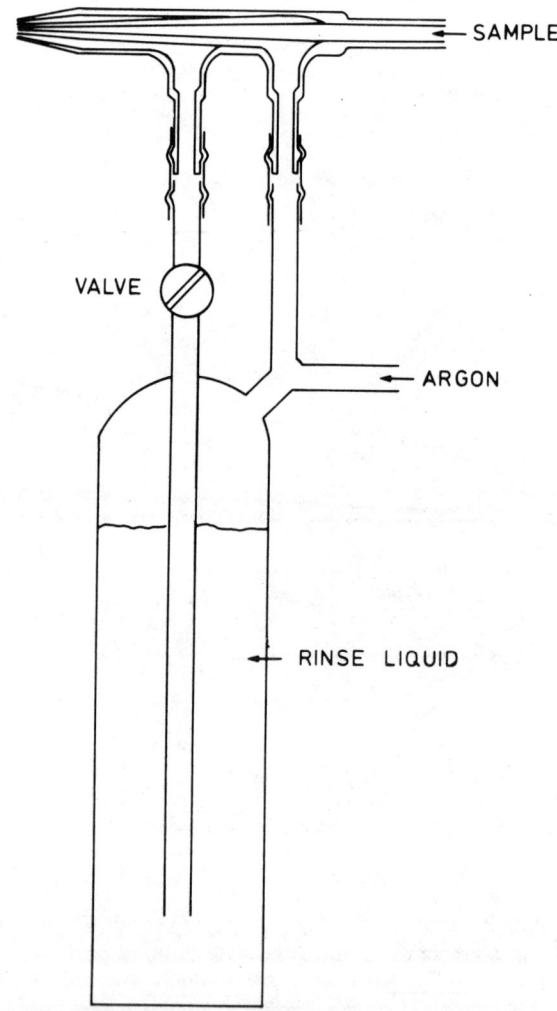

Figure 6.6. Feed arrangement for dual concentric glass nebulizer [45]. [J. E. Meinhard, in R. H. Barnes, ed., *Applications of Plasma Emission Spectroscopy*, p. 1. Copyright (1979), Heyden & Sons, Ltd., London. Reprinted by permission of John Wiley & Sons, Ltd.]

The design of cross-flow nebulizers has been discussed by Kranz [3]. A cross-flow nebulizer for dc arc solution analysis and AAS work was described by Valente and Schrenk [17]. However, the gas consumption was too high for ICP work. Types for ICP-AES should have capillary diameters of < 0.2 mm and a distance of 0.05–0.5 mm between the tips. Then aerosol gas flows will range from 1–2 L/min at 2–5 bar. This was confirmed by recent optimization of the design, as described by Fujishiro et al. [23]. They found that by varying the inside and outside diameter of the sample capillary from 0.15 to 0.9 mm and from 0.5 to 1.5 mm, respectively, the pressure drop across the sample tube decreases from 150 to 50 mbar. By increasing the gas flow from 0.75 to 1.75 L/min, the pressure drop across the sample tube was found to increase from 150 to 350 mbar. As shown in Fig. 6.7, the pressure drop variations considerably influence the droplet size. Most types of cross-flow nebulizers require forced sample feeding with the aid of a peristaltic pump.

Cross-flow nebulizers for ICP work are made of glass or plastics and have metal (e.g., in [19]) or glass capillaries (see e.g., [16]). The first work was done with types having adjustable capillaries. For routine work, manufacturers now offer types with prefixed capillaries and calibration.

ICP detection limits obtained with concentric glass nebulizers and cross-flow nebulizers do not significantly differ; also, the operation stability and the analysis cycle are similar (see e.g., [49]). Maximum salt concentrations in the case

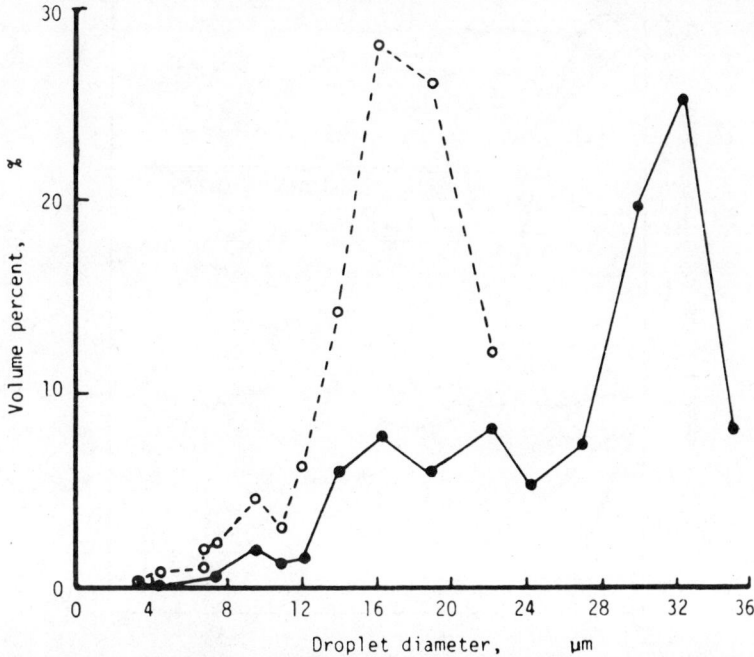

Figure 6.7. Influence of pressure drop across the sample tube of a cross-flow nebulizer on droplet size distribution. Pressure drop -●- 95 mbar; -○- 250 mbar. Gas needle; 0.3 mm id. [23]. [Reprinted with permission from Y. Fujishiro, M. Kubota, and R. Ishida, "A Study of a Cross Flow Nebulizer for ICP Atomic Emission Spectrometry," *Spectrochim. Acta* **39B**, 620 (1984). Copyright (1984), Pergamon Journals, Oxford.]

of a cross-flow nebulizer may be up to 20–50 mg/mL, and nebulizer blockage and long-term drifts due to high alkali concentrations are lower than in the case of other nebulizer types (Fig. 6.8).

The long-term precision obtainable with a cross-flow nebulizer could be improved by manufacturing a monolithic assembly of the capillaries and the support, so that the distances at the nozzles are fixed. Such types were described by Novak et al. [21] and Anderson et al. [22]. The latter type was introduced as a MAK nebulizer and is shown with its original spray chamber in Fig. 6.9. This type has capillaries with an internal diameter below 0.1 mm and requires aerosol gas pressures of 15 bar in order to realize a flow of 500 mL/min. The capillaries have thick walls so as to minimize vibrations. It has been shown that

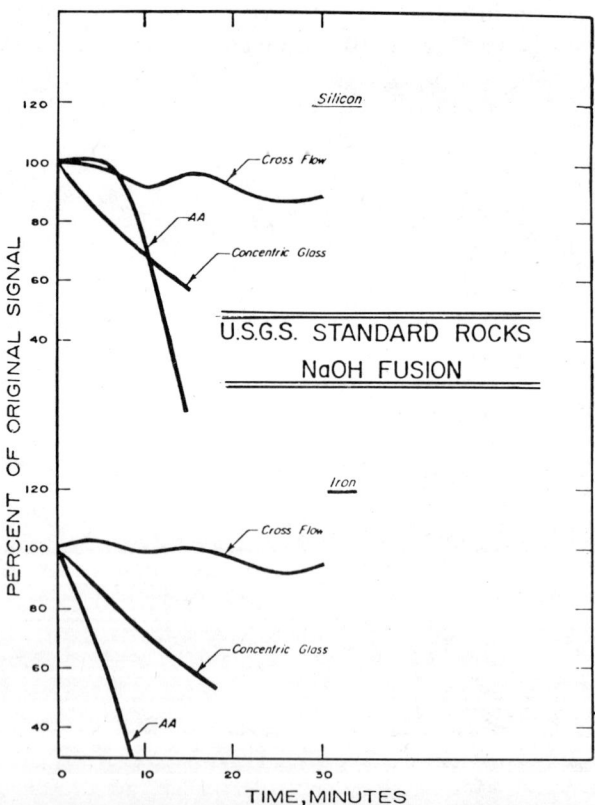

Figure 6.8. Operation stability of different types of pneumatic nebulizers for solutions having high salt contents. *AA*, concentric nebulizer as used in AAS-work. Solution: ~16 g dissolved salts (mainly NaCl) per liter [46]. [Reprinted from *ICP Information Newsletter* **3**, 37 (1977), and with the permission of C. C. Wohlers.]

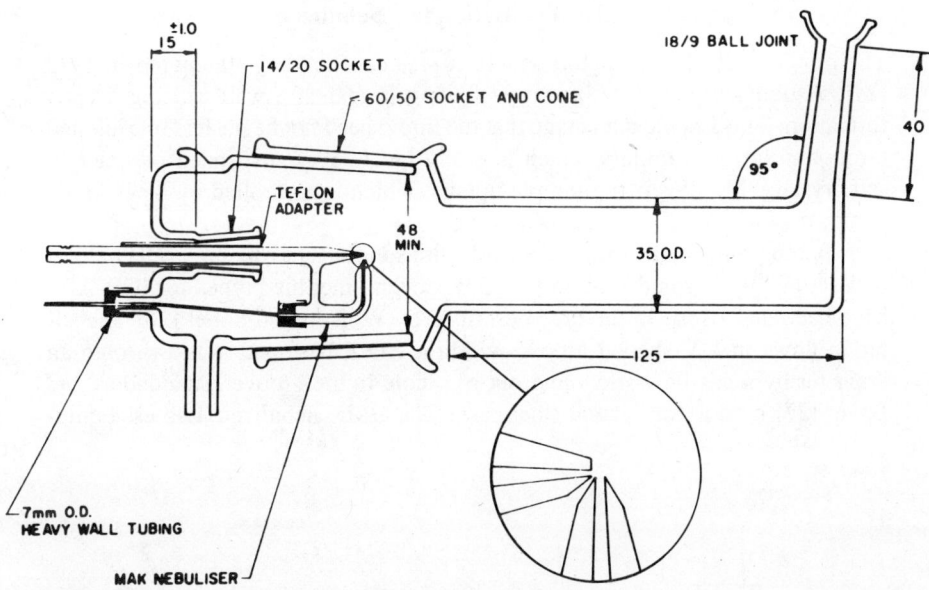

Figure 6.9. MAK nebulizer and expansion chamber [22]. [H. Anderson, H. Kaiser, and B. Meddings, in R. M. Barnes, ed., *Developments Atomic Plasma Spectrochemical Analysis*, p. 251. Copyright (1981), Heyden & Son, Ltd., London. Reprinted by permission of John Wiley & Sons, Ltd.]

RSDs of measured intensities for a series of elements at the microgram per milliliter level were 0.3 to 1.5%, they could be brought to 0.1% by using a reference line. The maximum salt concentrations are similar to those of a conventional cross-flow nebulizer. However, the high pressure required by the MAK nebulizer is unfavorable from a safety point of view. Further, this nebulizer requires a high stabilization time, that is, the time required to reach steady-state conditions. It has been shown by Barnes et al. [47] that in the case of aluminium samples (analyte solutions: 50 g/L), the stabilization time of the MAK nebulizer was 46 s instead of 20 s for the concentric glass nebulizer and the conventional cross-flow nebulizer.

Cross-flow nebulizers may be entirely made of plastics. A wholly PTFE (*polytri*fluor*e*thylene) adjustable cross-flow nebulizer has been designed by Boumans and Lux-Steiner [18]. As PTFE tubing is used for the capillaries, the nebulizer can be used for the analysis of solutions containing HF and for the analysis of organic solutions.

A Ryton made cross-flow nebulizer with replaceable capillary tips is commercially available and incorporated in a sample introduction system resistant to HF for ICP–AES [51].

6.2.3. The Babington Nebulizer

The principle of the Babington nebulizer was described by Babington in 1973 [24]. A liquid film is nebulized by blowing it against a wall causing droplet formation. This has the advantage that the liquid need not be pushed or aspirated through a narrow capillary which is easily blocked by suspended particles or salt crystals. Moreover, the sample uptake, which is controlled by a peristaltic pump, may be varied within a wide range.

A Babington nebulizer for AAS work with solutions having high salt contents is described by Fry and Denton [52, 53]. Several modified types for ICP-AES have been described. In the type described by Wolcott and Sobel [26], the solution flows in a V-shaped groove within a PTFE base and is blown onto an impacter by a gas flow streaming out of a hole in the groove. Suddendorf and Boyer [27] used a gold plated steel base. The GMK-nebulizer (Labtest Equip-

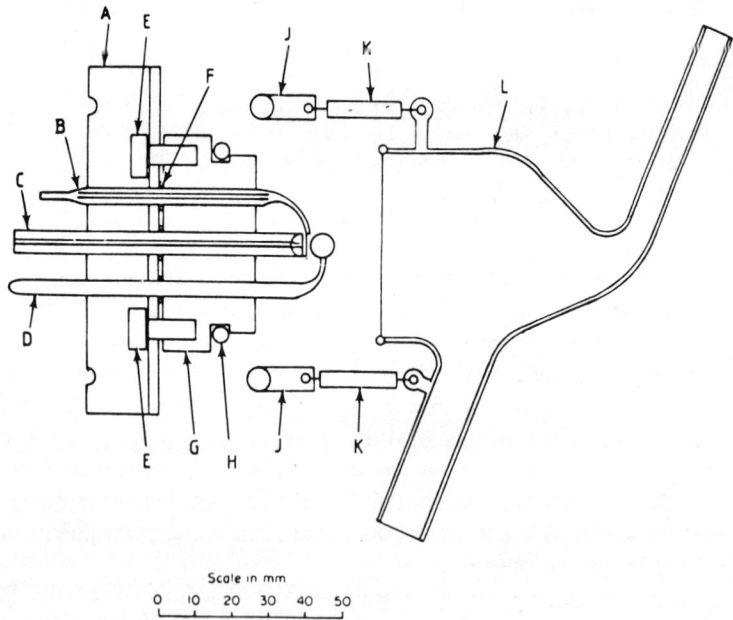

Figure 6.10. GMK nebulizer for ICP-AES (Labtest Equipment Co). A, Aluminium mounting; B, sample injector; C, gas injector; D, glass impactor bead; E, nylon screws; F, H, O-rings; G, PTFE block; J, retaining bars; K, springs; L, spray chamber [28]. [P. W. McKinnon, K. C. Giess, and T. V. Knight, in R. M. Barnes, ed., *Developments in Atomic Plasma Spectrochemical Analysis*, p. 287. Copyright (1981), Heyden & Son, Ltd., London. Reprinted by permission of John Wiley & Sons, Ltd.

ment) is a commercial version of the Babington type (Fig. 6.10). However, types using a flat base instead of a base bearing a V-groove are also available.

Gas flows used may vary from 0.05–2 L/min and are related to the capillary diameter. By using a steel nebulizer base with a 100-μm hole for gas supply, Ripson and de Galan [54] could operate the nebulizer with 0.05–0.2 L argon/min. When combined with a 10-mL spray chamber, the system could be used for sample introduction into a low consumption argon ICP (operating at a total gas consumption of 2 L argon/min). Samples could be forced-fed by gas pressure or introduced via a sample loop.

As the solution feeding tube may have a large diameter, the nebulization of solutions with suspended particles is possible. The GMK nebulizer, for example, could be used for the analysis of TiO_2 suspensions where the mean particle size is about 1 μm [55]. Wickman et al. [56] proposed a PTFE-based Babington nebulizer for the analysis of slurries. This has the advantage of being corrosion resistant. The low wettability was found to be of minor importance as the aerosol produced was similar to the one obtained with a metal base type. Independent of the nebulizer performance, however, the direct nebulization of suspensions is limited by the mean particle size which determines the degree of atomization. It was shown by Fuller et al. [57] that the atomization of rutile and ilmenite in ICP–AES similar to flame AAS became fairly incomplete above a particle size of a few micrometers; this leads to systematic errors.

The main advantage of the Babington nebulizer is realized in the analysis of solutions with high salt contents. Thelin [29] reported that solutions of steel samples with up to 100 g/L dissolved salts may be reproducibly nebulized with a V-shaped Babington nebulizer. The detection limits with respect to the solution were similar to those of a concentric pneumatic nebulizer. Because of its freedom from clogging, the nebulizer is very appropriate for water analysis [58] where suspended matter or high salt contents may hamper the use of other types of nebulizer.

6.2.4. The Fritted Disc Nebulizer

The use of a porous glass plate for the nebulization of limited amounts of sample liquid was reported by Apel et al. [30]. Nebulization efficiencies of up to 60% could be obtained. In early work, however, memory effects limited the use of this nebulizer for continuous nebulization purposes. This difficulty could be partly overcome by upside-down operation and applying continuous washing (Fig. 6.11). As the nebulizer can be operated with down to 30 μL/min sample solution—and at that rate any organic solvent may also be brought into the ICP—this nebulizer has become of interest for the coupling of liquid chromatography and ICP–AES.

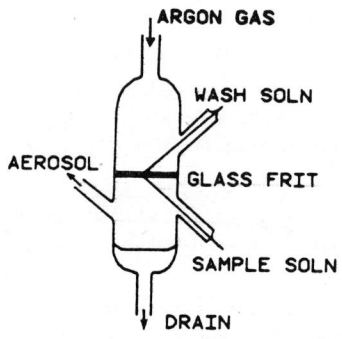

Figure 6.11. Glass-frit nebulizer assembly. Frit diameter: 20 mm. Wash and sample solution are introduced through 0.5 mm i.d. capillaries [31]. [Reprinted with permission from L. R. Layman and F. E. Lichte, "Glass Frit Nebulizer for Atomic Spectroscopy," *Anal. Chem.* **54**, 639 (1982). Copyright (1982), American Chemical Society, Washington, DC.]

6.2.5. Stability of Pneumatic Nebulizers

Since the larger part of the fluctuations in the ICP signal is concerned with the aerosol generation (see Part 2, Chapter 8), considerable efforts were made to improve nebulization systems (sample feeding device, nebulizer, spray chambers) in this respect. The nebulizers previously discussed were compared from various points of view:

- SNR
- Short-term stability
- Long-term stability

These comparisons are made after the nebulizer is attached to an appropriate spray chamber and after sample feeding and analysis cycle have been optimized.

In the case of a well optimized pneumatic nebulizer Belchamber and Horlick [208] showed that the SNR cannot be much improved by increasing the measurement time. Indeed, both with the concentric glass nebulizer and the cross-flow nebulizer, source flicker noise is the limiting noise. This type of noise is characterized by a $(1/f)^\eta$ spectrum ($\eta \approx 1$). In the abscence of detector noise and at analyte concentrations sufficiently far above the detection limit so that shot noise is negligible, the short-term precision at increasing measurement times was constant from 30 ms onwards. This was shown to contrast with ultrasonic nebulization (see Section 6.3) where the measurement precision decreased with the integration time and severe flicker noise limitations occurred ($\eta > 1$) (cf. Section 4.2.3). Olsen and Strasheim [59] measured the aerosol droplet size distributions of several pneumatic nebulizers (Fig. 6.12). They found that the MAK nebulizer had the narrowest size distribution and the highest SNR, whereas a conventional fixed cross-flow nebulizer (Jarrell–Ash) had the broadest size distribution and, therefore, the lowest SNR. It turned out that the SNR correlated

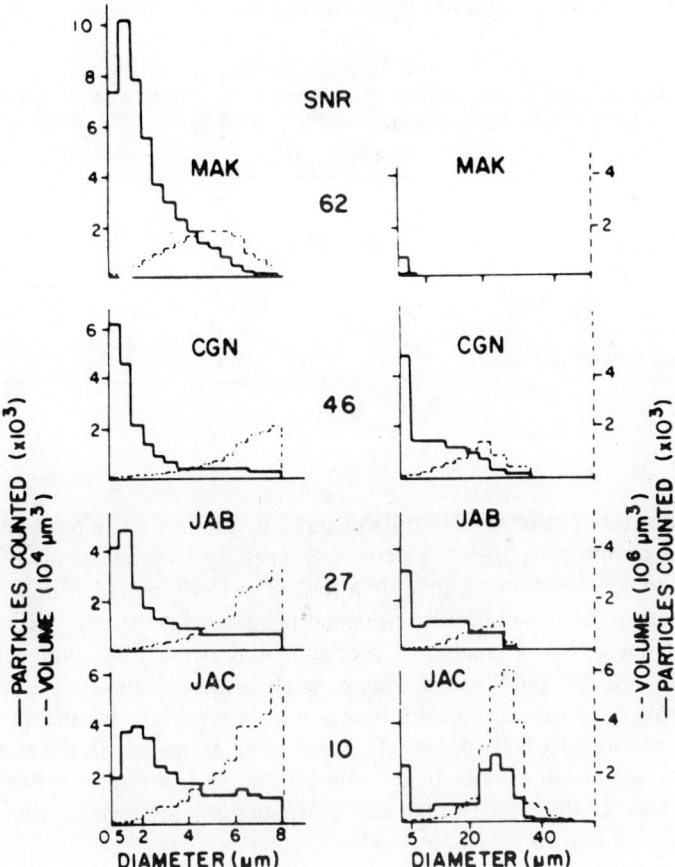

Figure 6.12. Particle size distributions of ICP-pneumatic nebulizers tested. MAK, High-pressure cross-flow nebulizer; CGN, Concentric glass nebulizer; JAB, Jarrell–Ash Babington nebulizer. JAC, Jarrell–Ash fixed cross-flow nebulizer [59]. [Reprinted with permission from S. D. Olsen and A. Strasheim, "Correlation of the Analytical Signal to the Characterized Nebulizer Spray," *Spectrochim. Acta* **38B,** 974 (1983). Copyright (1983) Pergamon Journals, Oxford.]

with the total volume of the droplets smaller than 8 μm produced by a nebulizer. However, as shown by Ebdon and Cave [60] for the case of a concentric glass and a plastics based Babington nebulizer, the SNR considerably depends on the spray chamber used. Short and long-term stability of different nebulizers were compared by many investigators. Fujishiro et al. [23] reported that after the previously described optimization of the cross-flow nebulizer RSDs were somewhat lower than those obtained with a concentric glass nebulizer. Indeed, the pressure drop across a concentric glass nebulizer was lower than for a cross-

Table 6.1. Determination of Copper in High-Purity Aluminium Solution (50 g/L)[a,b]

(Present 0.012 mg/g.)

[Reprinted with permission from R. M. Barnes, H. S. Mahanti, M. R. Cave, and L. Fernando, *ICP Information Newslett.* **8**, 562. Copyright (1983), R. M. Barnes, Amherst, MA.]

Nebulizer	Found (mg/g)	RSD (%)
Concentric	0.010	6.3
Adjustable cross-flow	0.010	6.5
Fixed cross-flow (MAK)	0.013	6.6
Babington (Wolcott–Sobel)	0.011	2.1
Babington (impact sphere)	0.011	3.7

[a] From [47].
[b] Result of 10 measurements.

flow nebulizer. The effect was still enhanced by the fact that a back pressure in the spray chamber occurs when a concentric nebulizer-spray chamber system is coupled to a Fassel-type plasma torch [61]. The lower pressure drop is known to produce higher mean droplet sizes and nebulizer fluctuations. The effect of different nebulizers on analytical precision is illustrated by some results for aluminum analysis published by Barnes et al. [47] (Table 6.1). The long-term stability of pneumatic nebulizers becomes rather uniform for the various nebulizers discussed when appropriate precautions are taken. In the example of ICP analyses of large series of dissolved airborne dust samples, Broekaert et al. [62] showed that even at high salt concentrations RSDs of measured intensities over 45 min periods were 1–4% (Table 6.2).

6.2.6. Perspectives

The concentric glass nebulizer or a cross-flow nebulizer is adequate for most analytical applications. The Babington nebulizer certainly has advantages for the analysis of suspensions and solutions with high salt contents. With nebulizer systems of materials other than glass, solutions containing HF can be analyzed with the same analytical performance as simple aqueous solutions. The fritted-disc nebulizer may be expected to gain interest for liquid chromatographic applications. Further efforts can be expected to improve the aerosol generation efficiency of pneumatic nebulization. These efforts include investigations of new forms of spray chambers and experiments with thermostatted spray chambers, as reported by Goulden and Anthony [105] (see also Section 4.7.8). Also, preliminary work on direct introduction of the aerosol into the ICP without a chamber has been reported [117]. Another approach for making more efficient use

Table 6.2. Analytical Precision Obtained by Recalibration with One Standard Sample after Different Periods[a]
[Reprinted with permission from J. A. C. Broekaert, B. Wopenka, and H. Puxbaum, *Anal. Chem.* **54,** 2174 (1982). Copyright (1982), American Chemical Society Washington, D.C.]

	2.5-h Measurement period	45-min Measurement Periods		
	(I–III, $n = 28$), s_r (%)[b]	I ($n = 7$), s_r (%)[b]	II ($n = 7$), s_r (%)[b]	III ($n = 6$), s_r (%)[b]
Al	0.6	0.3	0.3	0.2
Ca	4.3	0.9	1.5	1.4
Cr	1.9	1.4	1.1	1.0
Cu[c]	29.4	4.5	1.6	10.1
Fe	3.4	1.9	1.2	1.4
Mg	3.4	0.4	2.4	3.1
Mn	4.3	0.6	1.5	0.9
Pb	3.4	2.4	2.0	1.8
Sr	3.7	0.8	2.2	4.2
V*	8.4	1.9	1.7	3.1
Zn	4.0	2.1	1.4	2.8

[a] From [62].
[b] s_r = RSD in %. Standard sample: 0.08 μg/mL Al, 0.8 μg/mL Ca, 0.04 μg/mL Cr, 0.04 μg/mL Cu, 0.16 μg/mL Fe, 0.04 μg/mL Mg, 0.16 μg/mL Mn, 0.16 μg/mL Pb, 0.04 μg/mL Sr, 0.08 μg/mL V, 0.04 μg/mL Zn, 19 mg/mL H_3BO_3, 19 mg/mL HF and 310 mg/mL HNO_3.
[c] Band interferences.

of the sample was presented by Hulmston [63] who described a recirculation spray chamber for total sample consumption.

6.2.7. Discrete Sampling Using Pneumatic Nebulization

Discrete sampling by introducing small sample volumes directly into a pneumatic nebulizer was introduced by Sebastiani et al. [64] and Berndt and Jackwerth [65] in AAS. Both manual and automated systems have been applied as reviewed by Berndt and Slavin [66]. These "injection techniques" aim at

- The analysis of samples of limited volume as often encountered in biological problems
- A high power of detection by combining them with preconcentration procedures

- A reproducible nebulization of solutions with high salt contents by subjecting the nebulizer to the sample solution during short periods only
- A higher analysis throughput as compared to continuous nebulization

As discussed in a review by Cresser [67], these "injection techniques" are also useful in ICP work where the advantages can be realized in simultaneous multielement determinations. Injection techniques yield transient signals and thus may lead to some reduction of the power of detection as compared to continuous methods. Also the sample nature—especially in the case of samples with high salt content—may reduce the power of detection. However, it can be shown that the loss in power of detection is mostly smaller than the dilution factor that would be needed to introduce the solution with continuous nebulization. Avoiding such dilution is at any rate favorable in order to eliminate risks of contamination and additional labor.

The instrumentation required consists of a small sample introduction vessel connected to the nebulizer and a volume dosage system (micropipettes, automated syringe). Calibration is achieved by relating the peak heights of the amplified PMT currents to the analyte concentrations. Because of stability requirements and the choice of optimum operating conditions, the practical realization of the injection technique is slightly different when combined with either a low-power ICP or a high-power argon–nitrogen ICP.

Discrete nebulization using flow injection, as discussed, for example, by Greenfield [68], may be useful in routine analysis. Its capabilities and limitations should be kept in mind when a liquid chromatography system is coupled to an ICP–AES system.

6.2.7.1. Injection Methods Combined with a Low-Power Argon ICP

With a low-power argon ICP, a sudden introduction of air disturbs the discharge stability and output signal of the ICP. Therefore, the entry of air in the nebulization system between subsequent sample injections should be avoided. This can be achieved by

- Purging the injection funnel with argon
- Clamping the sampling tubing between subsequent injections
- Setting up a peristaltic pump between the sampling funnel and the nebulizer

Kniseley et al. [69] applied clamping of the sample tubing between subsequent injections of 25 μL sample aliquots in a cross-flow nebulizer with subsequent drying of the aerosol. They obtained detection limits for Al, Co, Cr, Cu, Mg, Mn, Ni, P, Pb, and Zn in the sub-μg/mL range and reported the direct determination of Al, Ca, Cu, Fe, Mg, and Mn in whole blood or serum using a standard addition method.

Uchida et al. [70] reported sample injection in a PTFE funnel coupled to a Babington nebulizer for the analysis of biological standard samples. The entrance of air between subsequent injections was minimal and precautions to maintain the stability of the low-power argon ICP were unnecessary. The authors also reported the nebulization of small sample aliquots (10–100 μL) with a cross-flow nebulizer [71]. In order to avoid air entrainment, they provided for a suitable alternation of blank solution uptake and aspiration of the sample aliquots which were transferred as drops on a PTFE plate. Simultaneous multielement analyses of blood and serum samples using yttrium as internal standard were reported. Results obtained for Na, K, Ca, Fe, Cu, Mg, Zn, and P with continuous nebulization after a decomposition of the organic matrix well agree with those of the injection of the biological liquids without previous decomposition of the matrix.

Broekaert and Leis [38] used a peristaltic pump placed between the injection funnel and the concentric glass nebulizer (Fig. 6.13). Sample aliquots down to

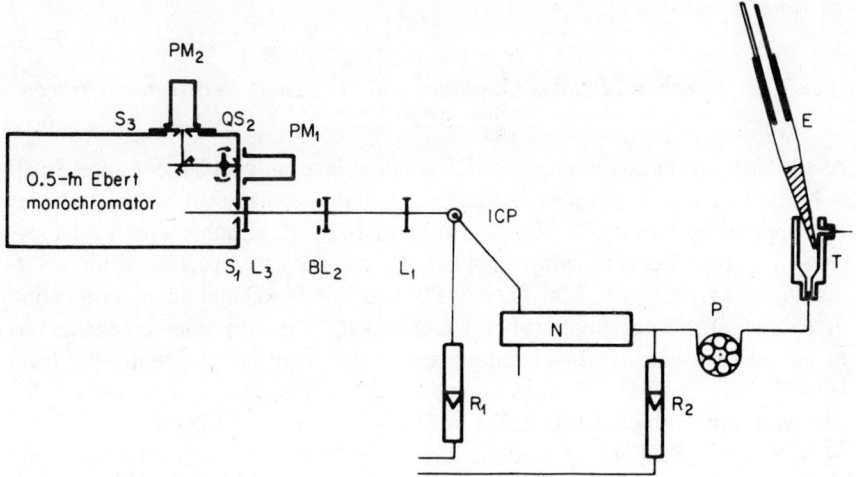

Figure 6.13. The ICP-injection setup. L_1, L_2, L_3, Spherical quartz lenses; S_1, entrance slit; S_2, fixed exit slit; S_3, moveable exit slit; Q, quartz refractor plate for alternative measurement of line and background intensity; PM_1 and PM_2: PMTs; B, diaphragm at intermediate image; R_1 and R_2, rotameters; N, spray chamber; P, peristaltic pump; T, PTFE sampling cup; E, Eppendorff micropipette [38]. [Reprinted with permission from *Anal. Chim. Acta* **109**, 73 (1979).]

20 μL are feasible. The signal rapidly increases when the front of the aerosol cloud enters the source, reaches a maximum, and tails down to the background level. After optimization of the carrier gas flow, the level of steady-state signals as found with continuous nebulization and accordingly the full power of detection is obtained at a 500-μL sample volume. When using the peak heights of the output signals as analytical signal, relative detection limits are a factor of 2–5 higher compared to continuous nebulization methods. The absolute detection limits, however, improved by a factor of 2. After carrier gas flow optimization with respect to maximum net line intensities, matrix effects caused by 20 mg/mL $NaNO_3$ are lower than 10%; this is partly due to the use of a peristaltic pump which minimized nebulization interferences. The matrix effects could be further reduced by correcting for the spectral background at an adjacent wavelength measured simultaneously in a second channel. Also an alternate line and background measurement with the aid of a vibrating quartz plate may be performed. The total analysis time of this injection method was 1 min instead of 4 min in the case of continuous nebulization comprising 10 s alternate integrations of line and background intensities and clean-out. The method was applied to the analysis of waste waters of varying composition and high salt contents after decomposition of organic particulates.

The method can be easily automated by using an automated dispenser on one side and a digital integrator for measuring the peak height or area, as proposed by Kojima and Iida [72].

6.2.7.2. Injection Methods Combined with a High-Power Argon–Nitrogen ICP

As the high-power argon–nitrogen ICP is rather insensitive to air entrance, small volume samples can be injected directly into the pneumatic nebulizer. This was first applied by Greenfield [73]. He injected 1–25-μL aliquots with a microsyringe and used a heated spray chamber. He reported the spectrographic determination of Al, Cu, Fe, Mg, Si, Ag, Pb, and P in blood and the determination of Al and Mg in organophosphorus compounds. Relative standard deviations of net signals with 10–25-μL aliquots at a 0.1–1-μg/mL concentration level were ≈ 3%. Determinations of Cr and Fe as complexes in xylene and Ba, Ca and Al in oils with detection limits of 10^{-9}–10^{-10} g were reported.

Broekaert et al. [74] reported the injection of 50-μL aliquots via a teflon cup directly connected to a Meinhard nebulizer. Detection limits for Fe, Cu, and Mg are 1–10 ng/mL. Although this is only a factor of 2–5 higher than with continuous nebulization, which required at least 0.5 mL of sample, a gain in power of detection is also realized here. Matrix effects caused by alkalis are low, and serum (dilution 1:5) and whole blood stabilized with heparine (dilution 1:10) can be nebulized reproducibly. In a subsequent paper Aziz et al. [75]

6.2. PNEUMATIC NEBULIZERS

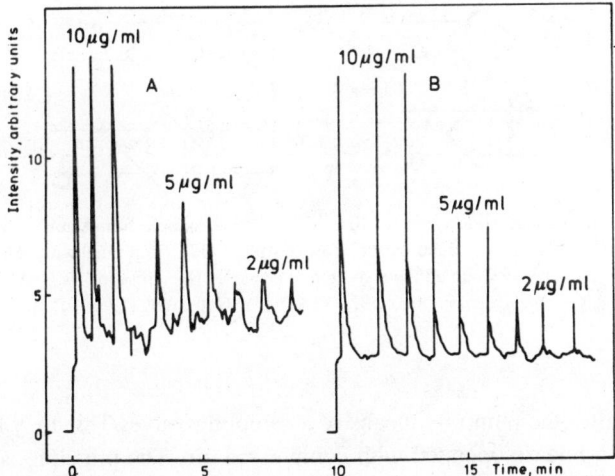

Figure 6.14. Analytical signals obtained with the ICP-injection method. *A*, Without power stabilization; *B*, with power stabilization. Line: FeII 260.0 nm. Sampling volume: 50 μl [75]. [Reprinted with permission from A. Aziz, J. A. C. Broekaert, and F. Leis, *Spectrochim. Acta* **36B**, 251 (1981). Copyright (1981), Pergamon Press.]

showed that the full analytical precision of the injection method can be only realized with a proper power stabilization (Fig. 6.14), for example, by using the rf stray field signal in the vicinity of the coil for feedback to the rf generator. The full power of detection is already obtained with sampling volumes of 50 μL. Detection limits for Ca, Cu, Fe, Mg, and Zn range from 1–50 ng/mL. In the case of Ca and Mg, they are limited by blank contributions, which originate for example from additives such as the Herrmann solution often used for the dilution of biological fluids. An RSD of 3% was achieved at a 50-μL sampling volume. By an appropriate setting of the carrier gas flow, matrix effects caused by alkali salt concentrations up to 20 mg/mL are smaller than 10%. Thus serum samples can be analyzed after a 1:5 dilution by calibrating with aqueous solutions containing only the elements to be determined.

6.2.7.3. Flow Injection Methods

By applying flow injection analysis (FIA), as proposed by Ruzicka and Hanssen [76], small sample aliquots may be analyzed with either type of ICP. As described by Greenfield [68], the carrier stream, which is fed continuously to the nebulizer, may be deionized water or whatever solvent. It should be contained in narrow-bore tubing of about 1 mm i.d. and will be driven through the system by means of a precision multiroller peristaltic pump. The sample is injected into

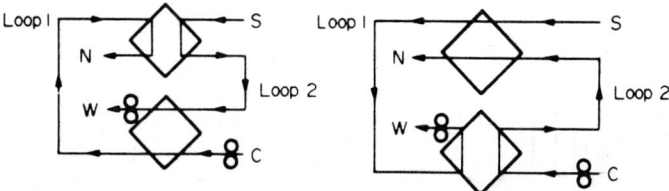

Figure 6.15. Air-operated slide valve for FIA–ICP work. S, sample from sample changer; W, to waste by peristaltic pump; N, to nebulizer; C, carrier stream [68]. [Reprinted with permission from S. Greenfield," Inductively Coupled Plasma-Atomic Emission Spectroscopy (ICP–AES) with Flow Injection Analysis," *Spectrochim. Acta* **38B**, 93 (1983). Copyright (1983), Pergamon Journals, Oxford.]

the stream after the pump by means of a sampling valve. This may be a valve with a spring release operated with compressed air. The principle is shown in Fig. 6.15. With the air pressure on, the contents of loop 1 are pumped into the nebulizer while loop 2 is being filled with sample from the sample changer. On release of the pressure, loop 1 is filled with the next sample and loop 2 is pumped into the nebulizer. Such FIA systems shorten the ICP analysis cycle and reduce the analyte consumption. However, the dilution of the sample with the carrier flow leads to a reduction of the power of detection, especially in the case of small sample aliquots.

Another possibility is the injection of the sample with the aid of a syringe through a septum into a carrier flow, as described by Alexander et al. [77]. The injector used allows a rapid flow analysis of micro samples and is shown in Fig. 6.16. The RSDs for 10-μL samples are 1–3% and decrease with the carrier flow (2–10 mL/min); detection limits with 300-μL aliquots injected into a 4 mL/min carrier flow were a factor of 4 lower as compared to those obtained with continuous nebulization. Ito et al. [78] reported that peak height and peak area measurements yielded the same analytical precision. Zagetto et al. [79] described a standard addition procedure allowing correction for interelement effects.

As shown by Ripson and de Galan [54], a sample loop between a continuous forced-fed solvent flow and a Babington nebulizer having low argon consumption (0.2 L/min) was appropriate for introducing μL samples into a low consumption argon ICP. They reported that detection limits were two times better than those obtained with a Meinhard nebulizer and an aerosol flow splitter.

6.2.7.4. Element-Specific Detection in Liquid Chromatography

ICP–AES using pneumatic nebulization since some years is applied to element-specific detection in liquid chromatography. Several approaches are followed

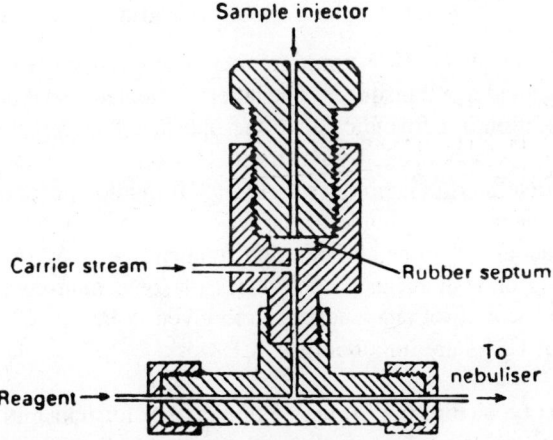

Figure 6.16. Injector for rapid flow analysis ICP-AES [77]. [Reprinted with permission from P. W. Alexander, R. I. Finlayson, L. E. Smythe, and A. Thalib, *Analyst* **107**, 1336 (1982). Copyright (1982), The Royal Society of Chemistry, London.]

to introduce the effluent of the chromatographic column into a pneumatic nebulizer.

First, a nebulizer of which the feeding flow may be externally varied over a wide range can be used. In this respect the fritted disc [31] or the Babington nebulizer [80] type are of potential interest. When the effluent flow rate of the liquid chromatograph is about 1.5 mL/min, the output may be connected to a cross-flow nebulizer directly [80–82].

Another approach is the use of a flow injector [83] or a sample loop [84, 85] for bringing the effluent into a carrier flow. This generally leads to a decrease in sensitivity as well as to a change in peak form, as discussed by Hausler and Taylor [85, 86]. The peak form was also found to depend on the form, the volume, and the location of the spray chamber (see, e.g., [84]). When organic solvents are used as effluent, which is the rule in complex separations, a change in the background intensity level of the ICP must be taken into account.

6.3. ULTRASONIC NEBULIZATION

Ultrasonic nebulization of liquid samples has been used in both AAS work [87] and in early work using microwave plasmas [88]. Also in the early ICP work, liquid samples [89–93] and even molten metals [94] were nebulized with ultrasonic devices.

6.3.1. Principles of Operation

In an ultrasonic nebulizer (USN), a longitudinal acoustic wave is produced by an oscillator coupled to a transducer. The latter is orientated so that the direction of wave propagation is perpendicular to the interface between the gas and the sample liquid to be nebulized. By the transfer of ultrasonic energy a pressure linked to a particle motion is produced. Aerosol formation occurs by the action of "geysers" when the amplitude of the waves becomes sufficiently large. In the case of aqueous solutions an operation frequency of 1 MHz was found to produce an aerosol with mean particle sizes of some micrometers. Detailed information on the aerosol characteristics are given in Part 2, Chapter 8. Two basic designs of USNs are important for ICP work.

   1. The first type (Fig. 6.17a) uses a liquid layer for transmitting the ultrasonic energy from the transducer to the analyte in a nebulization vessel. A mylar film separates the analyte solution from the coupling liquid. The transducer bears a concave focusing lens. The waveguide used and the position of the lens in the focusing chamber with respect to the sample liquid surface determine the energy transfer efficiency. This type was used in early ICP work by Wendt and Fassel [89] and Hoare and Mostyn [95]; it was again proposed by Mermet et al. [96] (Fig. 6.17b).
   2. The second type uses continuous feeding of the analyte solution onto the transducer surface. This type was used by Boumans and de Boer [92] (Fig. 6.18a) and by Olson et al. [97] (Fig. 6.18b). Here, the transducer is mounted vertically in the spray chamber itself. Floyd et al. [98] further improved this type by providing better cooling facilities [209].

6.3.2. Analytical Performance

6.3.2.1. Power of Detection

Ultrasonic nebulization of liquids enables the production of aerosols with a small mean droplet size (1–5 μm) and a narrow droplet size distribution. The efficiency of aerosol generation is high. Moreover, USNs do not require a gas for producing the aerosol but do require a gas to transport the aerosol from the spray chamber to the ICP. This implies that the carrier gas flow can be controlled independently of the aerosol production. However, the carrier gas flow does affect the rate at which the aerosol is swept from the spray chamber and thus controls the mass of analyte reaching the plasma per second. Boumans and de Boer [93] discussed this point in terms of a simple model and also showed experimentally that the flow pattern of the carrier gas in the spray chamber is a parameter affecting the mass of analyte reaching the plasma per unit of time. In general USN permits it to load the carrier gas with a substantially larger mass

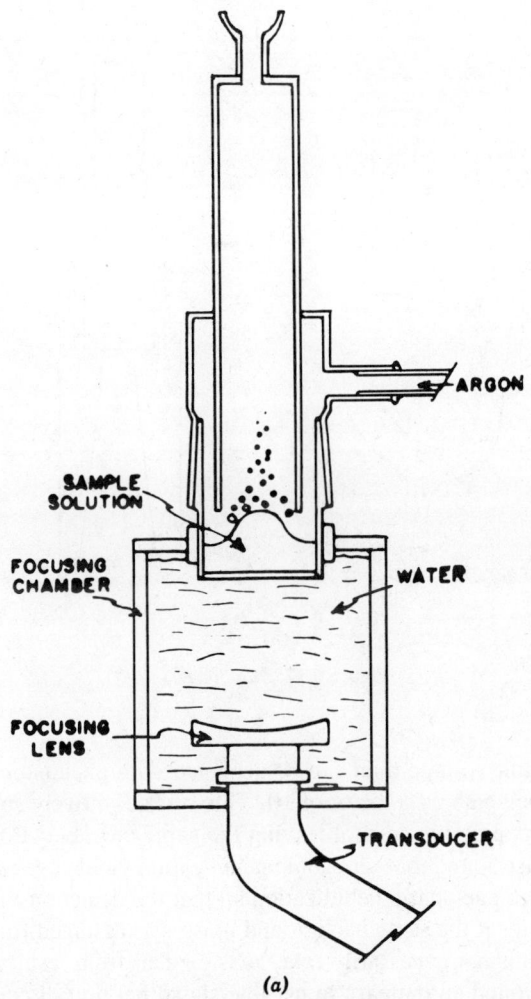

Figure 6.17. Ultrasonic nebulizers (USNs) using a transmitting bath. (*a*) USN used by Wendt and Fassel [89]; (*b*) USN described by Mermet et al.: *1*, transducer; *2*, waveguide; *3*, nebulization cell; *4*, aerosol carrier gas; *5*, drain; *6*, sheath gas; and *7*, to ICP [96]. Reprinted with permission from R. H. Wendt and V. A. Fassel, Anal. Chem. **37,** 920 (1965). Copyright (1965). American Chemical Society, Washington, DC.] [J. M. Mermet, C. Trassy, and R. Ripoche, in R. M. Barnes, ed., *Developments in Atomic Plasma Spectrochemical Analysis,* p. 245. Copyright (1981), Heyden & Son, Ltd. Reprinted by permission of John Wiley & Sons, Ltd.]

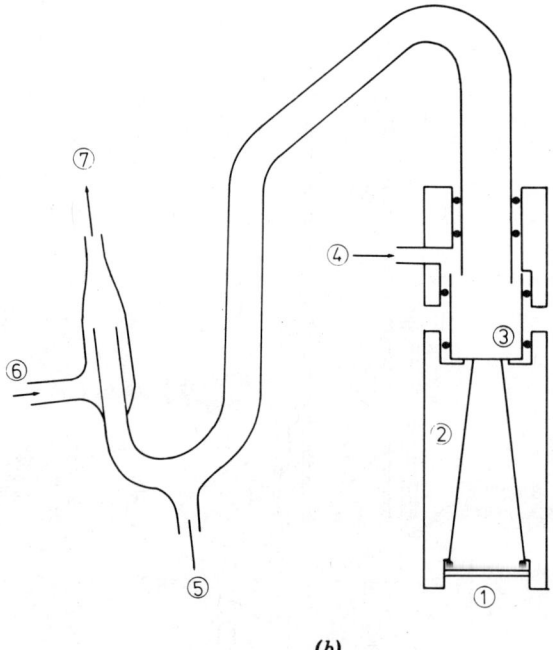

(b)

Figure 6.17. (*Continued*)

of analyte per unit volume than can be achieved with pneumatic nebulization, the advantage being about a factor of 10. This was definitively shown by Olson et al. [97] by trapping the aerosol leaving the spray chamber. Boumans and de Boer [93, 99] also found that ultrasonic nebulization yielded generally 10 times higher SBRs than pneumatic nebulization so that the detection limits improved by the same factor if the same background noise was assumed for both types of nebulizers. This is not necessarily true, as is evident from results of Berman et al. [100], who found an increase in net line signal per unit of concentration by a factor of 5–10 for the USN system and only a factor of 2–5 in power of detection. With the improved version of the USN used by Berman et al., however, Taylor and Floyd [98] generally found an improvement in the detection limits by a factor of 10 (Table 6.3) (cf. Section 4.1.7).

The high aerosol production rate of USNs calls for the solvent to be introduced into the plasma at a high rate. This may extinguish the discharge or at least drastically worsen the excitation characteristics. Here, either desolvation of the aerosol prior to feeding it into the discharge or raising the rf power is indispensable. The latter alternative meets the objection that the net signal resulting from the larger sample input be balanced by a higher background intensity so that the SBR is not essentially higher than for a pneumatic nebulizer.

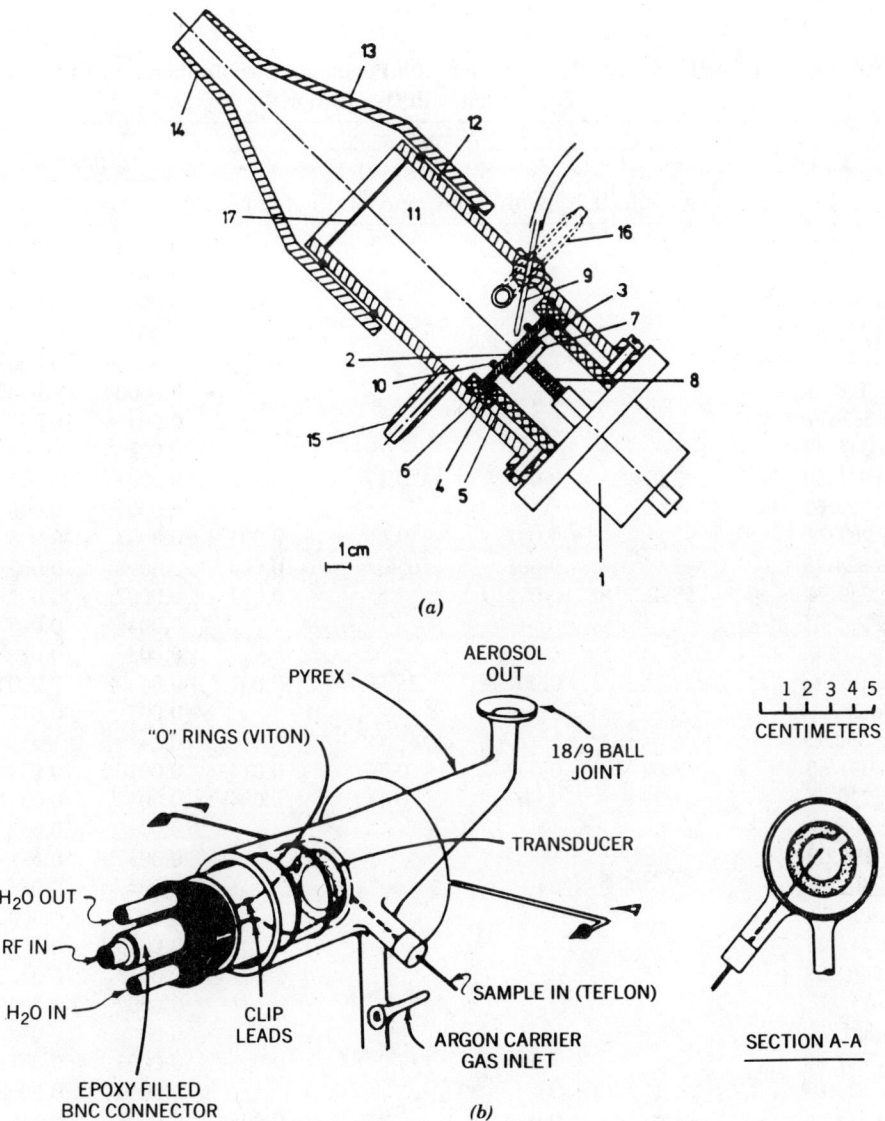

Figure 6.18. Ultrasonic nebulizers (USNs) with direct sample feed to the ultrasonic vibrator. (*a*) USN used by Boumans and de Boer: *1*, ultrasonic head; *2*, transducer plate; *3*, quartz crystal; *4*, screw ring; *5*, threaded head; *6, 10*, O-rings; *7*, mushroom electrode; *8*, spring; *9*, sample inlet capillary; *11*, fog chamber; *12*, plexiglass base; *13*, pyrex glass cap; *14*, connection to desolvation unit; *15*, drainage; *16*, carrier-gas inlet; and *17*, pierced baffle plate. From Boumans and De Boer [93]; (*b*) type described by Olson et al. [97]. [Reprinted with permission from K. W. Olson, W. J. Haas, and V. A. Fassel, "Multielement Detection Limits and Sample Nebulization Efficiencies of An Improved Ultrasonic Nebulizer and a Conventional Pneumatic Nebulizer in Inductively Coupled Plasma-Atomic Emission Spectroscopy," *Anal. Chem.* **49,** 634 (1977), Copyright (1977), American Chemical Society, Washington, DC.]

Table 6.3. Detection Limits (c_L) Obtained with Pneumatic Nebulization and Ultrasonic Nebulization with Desolvation

Element/Line (nm)	Ultrasonic Sensitivity[a] (nA/μg · mL^{-1})	c_L[a] (μg/mL)	Pneumatic Sensitivity[a] (nA/μg · mL^{-1})	c_L[a] (μg/mL)	c_L (μg/mL)[b] Ultrasonic	c_L (μg/mL)[b] Pneumatic
Ag 328.07					0.0008	0.005
Al 308.21					0.003	0.04
As 193.76					0.006	0.06
B 249.68					0.004	0.009
Ba 455.40					0.00009	0.0003
Be 313.04					0.00004	0.0002
Bi 223.06					0.009	0.05
Ca 315.89					0.002	0.006
Cd 226.50	2.02	0.003	0.17	0.035[c]	0.0003	0.005
Co 228.62					0.0004	0.006
Cr 267.72	97.2	0.002	9.6	0.005	0.0003	0.005
Cu 324.75	342	0.001	65.3	0.003	0.0008	0.003
Fe 259.94	58.4	0.002	7.5	0.004	0.0002	0.002
Hg 253.65					0.004	0.015
Mg 279.08					0.003	0.03
Mn 257.61	229	0.001	25.5	0.002	0.00008	0.00045
Mo 287.15					0.002	0.012
Na 330.23					0.24	1.5
Ni 231.60	8.02	0.002	0.79	0.014	0.0010	0.03
Pb 220.35	1.29	0.007	0.11	0.084	0.002	0.03
Sb 206.84					0.004	0.05
Se 196.09					0.003	0.6
Sn 303.41					0.004	0.06
Sr 407.77					0.00003	0.00006
Te 214.28					0.0010	0.05
Th 401.91					0.0013	0.012
Ti 351.92					0.0002	0.002
Tl 351.92					0.02	0.075
V 292.40	160	0.004	17.2	0.005	0.0003	0.002
Y 371.03					0.00008	0.0008
Zn 213.86	21.6	0.001	2.21	0.003	0.0008	0.008

[a] From [100].
[b] From [98].
[c] Cd 228.80 nm.

6.3. ULTRASONIC NEBULIZATION

Desolvation, on the contrary, makes it possible to fully exploit the higher aerosol production rate of the USN, but desolvation apparatus complicates the experimental setup and gives rise to instabilities, memory effects, and interferences [99]. In principle, desolvation is achieved by leading the aerosol through a heated tube and subsequently passing it through a reflux condensor. The heating of the aerosol and subsequent cooling separates the larger part of the solvent from the solute so that the solute together with saturated solvent vapor is fed into the plasma [92, 93, 99]. If the experimental complications including the precise control of heating and cooling temperatures are accepted, then indeed one may reach with an USN one order of magnitude better detection limits than with pneumatic nebulizers, unless the matrix of the sample emits a line-rich spectrum (cf. Section 7.3.3).

6.3.2.2. Operation Stability and Memory Effects

Ultrasonic nebulizers described in the early literature suffered from operation instability and memory effects. These problems have been overcome in recent designs described by Mermet et al. [96] and others [209]. For example, Mermet et al. lowered the memory effects by introducing a supplementary gas flow shielding the aerosol at laminar flow velocities which prevents deposition on the vessel walls.

6.3.2.3. The Sample Nature

With early types, solutions with high salt contents lead to salt deposits which blocked the system. Advanced nebulizer types are said to have overcome these difficulties and should be suitable for the routine analysis of real samples. Olson et al. [97] reported that detection limits with their USN and desolvation apparatus in the case of 1% NaCl solutions were still two times lower than with pneumatic nebulization and that cleanup times were shorter than 2 min. Floyd et al. [101] reported the application of this device to the analysis of sea water and dissolved rock samples. Berman et al. [100] showed that for the analysis of sea water good agreement was obtained between results achieved with this USN or those found with a cross-flow nebulizer. The analysis of dissolved glass samples with ICP–AES using ultrasonic nebulization was reported [102]. Mermet et al. [96] mentioned the possibility of using their nebulizer for the direct analysis of small volumes of whole blood (80 g/L of dissolved salts) and solutions of steel (150 g/L of dissolved salts).

6.3.3. Perspectives

With the present state-of-the-art equipment it appears that the original problems of USNs, viz.

- Handling solutions with high salt contents
- Instability
- Long cleanup time

have not yet been convincingly overcome to the extent that they are accepted for routine analysis of real samples. In addition, the need for desolvation of the aerosol remains a strong argument against their use for routine work.

The excellent power of detection that in principle can be achieved makes the use and further research of USNs tempting. One should realize, however, that feeding more and more sample per unit of time does not necessarily lead to better detection limits, in particular if the major elements produce a line-rich spectrum. Then the background is primarily produced by line wings. Therefore, an increase in the sample introduction rate will affect the net line signals of traces and the background to the same degree, leaving the SBRs and detection limits unchanged [103] (cf. Section 7.3.3). Years ago Fassel [104] pointed to the potentials of ultrasonic nebulization combined with an ICP for the analysis of small volumes of samples. He compared these potentials with the possibilities offered by AAS-methods using electrothermal atomization. It seems, however, that these potentials have never been further exploited.

6.4. ELECTROTHERMAL NEBULIZATION DEVICES

Electrothermal aerosol generation has been applied to liquid and solid samples in combination with FAES and FAAS [106–108] and combined with arc and spark excitation [109–110]. One has particularly aimed at the high aerosol generation efficiency that can be easily obtained here as opposed to pneumatic nebulization. In combination with ICP–AES electrothermal aerosol generation should provide for

- An increase in power of detection
- Multielement trace analysis in small volume samples

However, as selective volatilization of the different sample constituents occurs, the establishment of multielement conditions and the implications of matrix effects have to be investigated in each particular case.

6.4.1. Metal Boats and Filaments

Tantalum boats as sample bearers were introduced in FAAS by Delves [111] and allowed an increase in power of detection by a factor of 2–10. In recent work with electrically heated platinum wire loop systems, Berndt and Mes-

6.4. ELECTROTHERMAL NEBULIZATION DEVICES

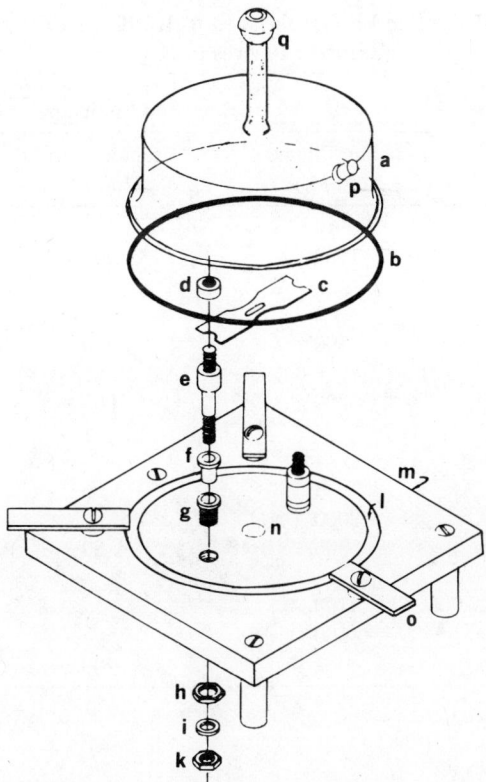

Figure 6.19. Tantalum filament vaporization apparatus. *a*, Quartz dome; *b*, O-ring; *c*, tantalum filament; *d–k*, copper port assembly; *l*, O-ring channel; *m*, aluminium base; *n*, argon gas inlet; *o*, aluminium tabs; *p*, sample injection port, *q*: to torch [113]. [Reprinted with permission from D. E. Nixon, V. A. Fassel, and R. N. Kniseley, "Inductively Coupled Plasma–Optical Emission Analytical Spectroscopy—Tantalum Filament Vaporization of Microliter Samples," *Anal. Chem.* **46,** 212, (1974). Copyright (1974), American Chemical Society, Washington, DC.]

serschmidt [112] gained a factor of 100 in the power of detection of FAAS requiring only some μL of sample. These systems cannot be used in the same form with the ICP, since it is more difficult to bring a metal sample bearer into the plasma. However, similar devices can be used as a separate atomization unit from which the dry aerosol is transported into the plasma with the aid of an argon carrier gas flow.

Nixon et al. [113] described a tantalum filament vaporization system for ICP work (Fig. 6.19). Sample volumes of 100 μL are brought onto the boat, the solvent is evaporated and then the dry analyte is evaporated by heating the boat quickly to ~ 1800°C. The dry aerosol is swept into the ICP and transient signals

Table 6.4. Detection Limits Obtained with Different Electrothermal Nebulization Devices in ICP-AES

Element/Line (nm)		Detection Limits (pg)		
		Tantalum Filament[a] (100 μL)	Graphite Yarn[b] (25 μL)	Graphite Rod[c] (10 μL)
Ag I	328.1	10	300	1
As I	228.8	1000	500	2000
Au I	267.6			10
Be I	234.9	2		1
Cd I	228.8	500	200	30
Ga I	417.2			10
Hg I	253.7	200	1000	60
In I	325.6			20
Li I	670.8			4
Mn II	257.6	3	6	1
P I	213.6	2000		200
Pb I	405.8	300	80	100
Re I	346.0			100
Sb I	259.8	100		300
Tl I	535.0	300		60
Zn I	213.8		200	20

[a] From [113].
[b] From [115, 116].
[c] From [114, 118–120].

are measured. Twenty to thirty determinations could be made per hour and one filament could be used for 300 determinations. Detection limits were in the subnanogram per milliliter range (Table 6.4).

Several authors used a tungsten loop for the vaporization of microsamples. Keilsohn et al. [132] brought 1-μL aliquots on the tungsten loop serving as the cathode of a microarc (20 mA, 1500-V dc) in pulsed operation. ICP detection limits were in the pg range. Low calcium phosphate interferences as well as moderate matrix effects arising from varying sodium content were reported. Kitazume [133] used a tungsten loop for the evaporation of one drop samples (volume 0.1 mL) with the aid of a condensed discharge and obtained detection limits ranging from 4 to 100 pg when releasing the sample vapor into an ICP. The fact that these values are higher than those given by Keilsohn et al. may be due to the difference in the devices used; however, they may also relate to the fact that the latter authors used small spray chambers whereas the former did not.

6.4.2. Graphite Yarn, Graphite Rod, and Graphite Furnace Systems

In the systems described below, graphite electrothermal atomization devices are used. These are known from AAS, as reviewed in [108], and ICP applications have often been reported with modified versions of commercially available AAS equipment. Special interest has been paid to the very high power of detection of these techniques and more recently to their capabilities for the routine analysis of real samples.

The use of a graphite yarn as atomization cell was described and patented by Dahlquist et al. [115, 116] in 1974. The sample solution is deposited on a continuous graphite yarn, dried and vaporized electrothermally, and the analyte vapor is led into the ICP. Sample volumes down to 25 μL can be used. Detection limits are in the subnanogram range (Table 6.4). Although this procedure compares favorably with pneumatic nebulization, no commercial product has been made available.

The combination of a graphite rod atomizer with the ICP was investigated especially by Kirkbright's group [114, 118–120]. With the setup shown in Part 2, Fig. 7.14 sample volumes of 1–25 μL can be used. Sample aliquots are dried on a graphite rod. By rapidly heating to ~2400°C and using a carrier gas flow of 1–2 L/min, the analyte together with the matrix is vaporized instantly. When increasing the length of the transport line between atomization cell and ICP, the height of the transient signals decreases and tailing occurs; however, the efficiency remains constant (even at 1–10-m distance). In further experiments and using a 0.5-m transport line, the influence of the injection gas flow was studied. Calibration curves are linear over five orders of magnitude of concentration. Absolute detection limits (see Table 6.4) are significantly lower than those obtained with the graphite yarn and with the tantalum filament technique. Relative detection limits are up to 10 times lower than values obtained with pneumatic nebulizers. In a subsequent paper [118], the matrix effects of the method were studied. In order to investigate their origin, experiments were performed on the vaporization, the transport efficiency, and the analyte deposition in the system. In the case of cadmium, the addition of selenium (VI) was found to effect matrix stabilization and to improve sample transport efficiency. In further work Kirkbright and Snook [119] proposed the use of a halocarbon–argon atmosphere in the sampling manifold of the graphite rod vaporization device in order to improve the capabilities of the system for refractory compound-forming elements. This improved the detection limits by one to two orders of magnitude. The absolute power of detection of the graphite rod-ICP method is high. However, its application to routine analysis is hampered by deposition of solvent vapor in the atomizer enclosure and the transport line, which might cause memory effects. Despite these limitations, Camara Rica et al. [120] used the system for the determination of manganese and nickel in blood and animal tissue calibrating by standard addition. A carbon rod device with a quartz enclosure was used by Barnes and Fodor [121] for ICP analyses of urine. They showed that peak height measurements led to lower detection limits than peak area measurements. Ng and Caruso [128, 129] reported that by using tantalum carbide pyrocoated graphite cups the detection limits for a series of elements were improved. They also analyzed samples dissolved in organic solvents [138] and found that it was then necessary to prepare the standard solutions accordingly.

Memory effects can be avoided by elimination of the solvent vapor during the drying stage as applied by Aziz et al. [122] in the case of a commercial graphite furnace atomizer. The authors eliminate the solvent vapors formed during the drying stage by introducing argon flows from both ends of the tube so that the vapors escape through the sampling hole (Fig. 6.20). After the drying stage argon flow 1 is switched off, the transport line is opened and the dry aerosol is released by heating the graphite tube and swept into the ICP. The use of a Meinhard nebulizer as aerosol injector (System A), similar to AAS work published by Kantor [107], improves the aerosol sampling efficiency and the power of detection, but is impracticable with real samples. When using a transport gas flow of 4 L/min (System B), 40% of the argon flow enters the ICP. However, the sample introduction efficiency is still higher because the dry aerosol is generated from a spot located symmetrically around the sampling hole and is pushed in the transport line by the carrier gas flow. Relative detection limits obtained with 50-μL sample aliquots and a 3-kW argon–nitrogen ICP were a factor of 3–5 lower than with pneumatic nebulization. Graphite tubes can be used for up to 50 determinations. No memory effects were found. The system was applied to the analysis of serum after 1:5 dilution with Herrmann solution and to the analysis of NBS biological standards after matrix decomposition with a $HClO_4/HNO_3$ mixture. Because of the high alkali concentrations, which cause interferences, calibrations were made by standard addition. Intermediate heating at 350°C for matrix isoformation was applied. The analysis results in Table 6.5 show that the technique is suitable for the analysis of real samples. By using a sampling syringe for powders, as proposed for furnace AAS work by Grobenski et al. [123], simultaneous determinations in microamounts of powders are possible. Crabi et al. [124] used a commercial graphite furnace atomizer with a quartz enclosure serving as aerosol outlet and analyzed biological samples. Swandan et al. [125] coupled the graphite furnace system with a simultaneous ICP spectrometer. The power of detection of the system may be raised by preconcentration of the analyte in the graphite furnace itself. This may be done by electrodeposition or aerosol deposition in the furnace; both approaches are known from AAS work [126, 127].

Recently, Smythe [130] combined "filament in furnace atomization" (FIFA) with ICP–AES relating to his former work on discrete sampling in AAS [131].

6.4. ELECTROTHERMAL NEBULIZATION DEVICES

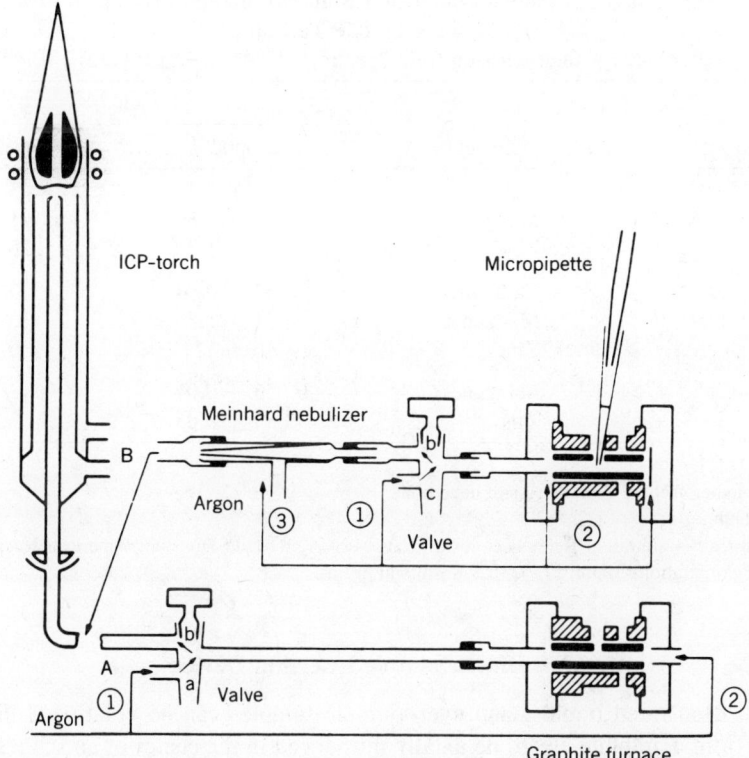

Figure 6.20. Setup for the evaporation of microsamples from a graphite furnace (HGA 74, Perkin Elmer & Co GmbH) and their introduction into an ICP. A, system using only a carrier-gas flow for aerosol transport; B, system using a Meinhard nebulizer as aerosol injector. Argon gas flows: *1*, gas flow for vapor elimination during sample drying; *2*, carrier gas for the dry aerosol; and *3*, gas supply for the aerosol injector [122]. [Reprinted with permission from A. Aziz, J. A. C. Broekaert, and F. Leis, *Spectrochim. Acta* **37B,** 369 (1982). Copyright (1982), Pergamon Press.]

Sample volumes of 1–10 μL are brought on a tungsten coil wound around a central tungsten wire (Fig. 6.21); the solvent is evaporated by electrical heating, the wire is brought into the graphite tube, and the dry aerosol is released by heating up to 2500°C. The advantages of the method lie in a freedom of solvent vapor condensation in the transport system, high speed of analysis due to easy sampling, and rapid drying of samples prior to vaporization. However, current limitations for the analysis of real samples may lie in the prefurnace vaporization stage.

Table 6.5. Analysis Results for Biological Standard Samples Obtained with the Graphite Furnace-ICP Technique[a]

[Reprinted with permission from Spectrochim. Acta. Part B [122].]

Sample	Element	Results[b] Found	Certified
SRM 1571[c]	Lead	54.5 ± 7.2	45
Orchard leaves	Manganese	87.3 ± 8.8	91
	Zinc	21.5 ± 1.8	25
SRM 1577[c]	Cadmium	0.31 ± 0.04	0.27
Bovine liver	Manganese	9.71 ± 1.36	10.3
	Zinc	146 ± 12	130
Cation-Cal[d]	Manganese	0.055 ± 0.01	
	Zinc	3.55 ± 0,35	3.01

[a] From [122].
[b] The numbers behind ± are standard deviations.
[c] Results in µg/g.
[d] Cation Cal™ Calibration Reference, lot n° CatC-109 A, B Dade Division, American Hospital Supply Corporation, Miami Fl. 33162. Results in µg/mL.

6.4.3. Direct Sample Insertion Devices

Solids, desolvated liquids, and even organic samples can be volatilized thermally from a graphite electrode axially introduced in the center of an ICP. The electrode is inserted through a central quartz tube that replaces the aerosol tube of a conventional torch. In some of these techniques samples can be changed without requiring reignition of the plasma. The principles and their applications have been described by Salin and Horlick [134], Sommer and Ohls [135], and by Kirkbright et al. [136, 137]. Selective volatilization, as known from dc arc spectroscopy, may be exploited (Fig. 6.22) to obtain the spectra of relatively volatile trace elements free from interference by the spectra of less volatile major elements as recently applied to the analysis of U_3O_8 [207]. This may substantially improve the detection limits, especially if the major elements yield line-rich spectra. Even metal speciation is possible [171]. By a slow introduction of the sample holder into the plasma, the temperature of the holder can be gradually increased. This effect can be used for in situ thermal decomposition of organic matrices, which, for example, in the case of trace element determinations in biological samples, may lower risks of contamination as compared to wet chemical decomposition. To this end, however, the sample insertion procedure must be very reproducible as it can be achieved with a microcomputer-controlled elevator [139, 140]. Further details on this technique are given in Part 2, Chapter 7 (see in particular Figs. 7.4 and 7.5) and in a recent paper by Pettit and Horlick [210].

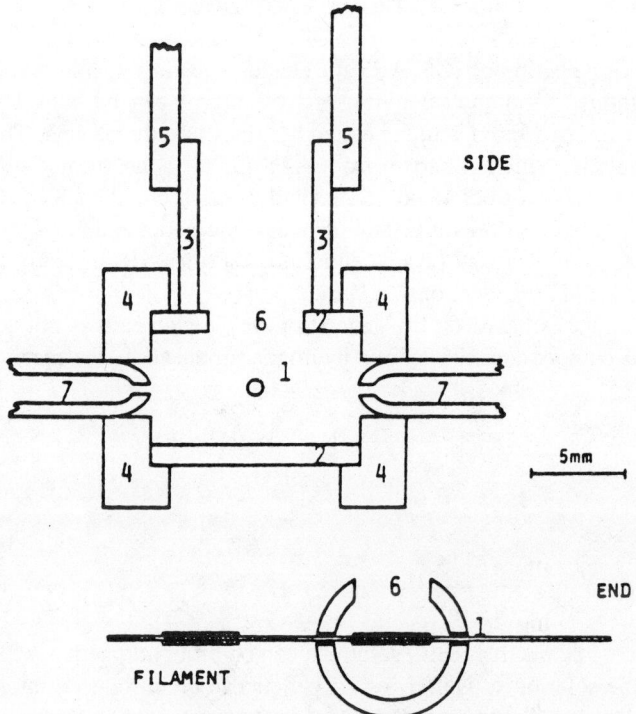

Figure 6.21. Filament in furnace atomization (FIFA). *1*, Hole for FIFA filament; *2*, furnace tube; *3*, graphite "chimney;" *4*, graphite support disc (in water-cooled electrode block); *5*, silica tube to ICP torch; *6*, exit hole; *7*, argon injector flow (0.5 L/min) [130]. [Reprinted from *ICP Information Newsletter* **6**, 224 (1980), and with permission of L. E. Smythe.]

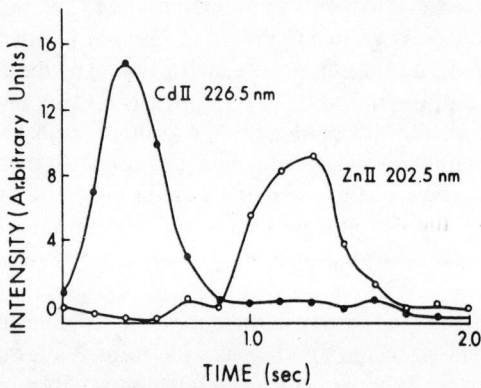

Figure 6.22. Intensity versus time curves obtained by selective volatization from a graphite rod inserted into an ICP [134]. [Reprinted with permission from E. D. Salin and G. Horlick," Direct Sample Insertion Device for Inductively Coupled Plasma Emission Spectrometry," *Anal. Chem.* **51**, 2284 (1979). Copyright (1979), American Chemical Society, Washington, DC.]

6.5. HYDRIDE TECHNIQUES

The power of detection of ICP–AES for elements having volatile hydrides (arsenic, antimony, bismuth, selenium, and tellurium) can be considerably increased by evoluting the in situ generated hydrides into the plasma. The feature of this technique, which is known from AAS [5, 6], is based on the high efficiency with which hydrides are generated and released into the plasma (virtually 100%) as compared to the efficiency of pneumatic nebulization ($\sim 1-5\%$). Its use both for AAS work and plasma emission work has been recently reviewed by Nakahara [141] (cf. Section 4.1.7.6).

The technique is based on the generation of the elements mentioned by reducing their compounds with atomic hydrogen produced by the reaction

$$NaBH_4 \xrightarrow{[H^+]} Na^+ + 2H + BH_3$$

or

$$Zn \xrightarrow{[H^+]} Zn^{2+} + H$$

Thus either stable alkali $NaBH_4$ solutions can be brought together with the acidified sample solution or $NaBH_4$ or Zn pellets can be introduced into the reaction vessel containing the acidified sample solution.

When using commercial AAS hydride generators, transient emission signals are obtained. Broekaert and Leis [142] used a similar hydride generator (Fig. 6.23) in combination with a 3-kW ICP; 40 mL of acidified sample (1 M HCl) is brought into the reaction vessel, a 5% $NaBH_4$ solution containing 50 g/L NaOH is introduced at a 0.5-mL/min pumping rate, and the generated arsine together with the excess of hydrogen is led into the plasma. When using the peak intensity of As 228.8 nm as an analytical signal, the detection limit for arsenic is 0.005 μg/mL instead of 0.4 μg/mL when using pneumatic nebulization. Sommer et al. [143] reported similar results for tin using a commercial AAS hydride generator. Thioglycolic acid was found to increase the emission signal, which might be due to the complexation of interferents. Working procedures for steel and brass were presented. As between subsequent analyses the reaction vessel must be rinsed or changed, the analysis times are at least 2–3 min.

Another technique [144] uses a syringe in which a defined aliquot of acidified sample and reagent are taken up. The hydride formed after closing the system is then injected in the ICP aerosol carrier gas flow.

Hydride techniques can be more easily coupled with a low-power ICP when the excess of generated hydrogen does not enter the plasma. This also permits

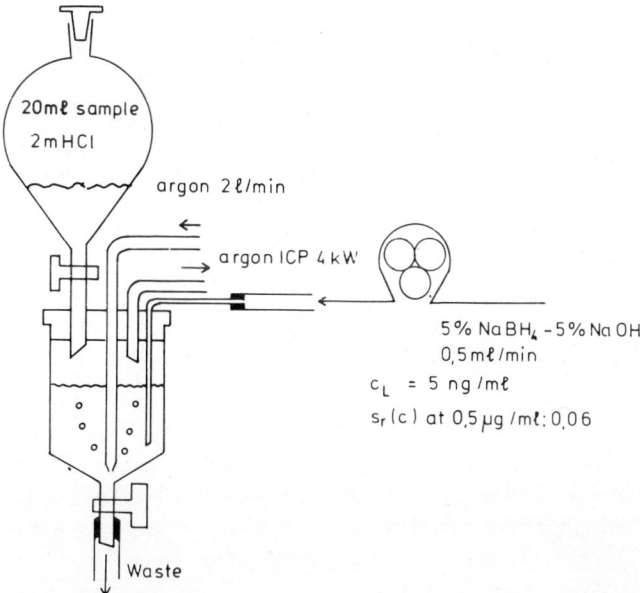

Figure 6.23. Hydride generation with transient analytical signals [142]. [Reprinted with permission from J. A. C. Broeckaert and F. Leis, *Fresenius Z. Anal. Chem.* **300**, 22 (1980).]

a further improvement in the power of detection. Fry et al. [145] described a system where the hydrides are trapped by freezing them in a vessel cooled with liquid nitrogen. When sweeping the hydrides generated from 25-mL sample volumes into a 1.2-kW argon ICP, the detection limit for arsenic is 0.05 ng/mL. Arsenic determinations in natural water were reported. Hahn et al. [146] applied such a system for simultaneous determinations of As, Bi, Ge, Sb, Se, and Sn in food. However, all systems already described are rather complex and the analysis throughput is low.

Flow-cell type hydride–ICP systems [142, 147–157] do not allow these extremely low detection limits but still give an improvement with respect to pneumatic nebulization. They enable better analytical precision because steady, instead of transient analytical signals are used. They are simple in operation and have a high sample throughput. Thompson et al. [147] described a flow-cell type hydride generator (Fig. 6.24). Detection limits obtained for arsenic, antimony, bismuth, selenium, and tellurium with a 2.7-kW argon ICP or a 5-kW argon–nitrogen ICP are listed in Table 6.6. From a comparison of various acids (HCl, H_2SO_4, $HClO_4$, H_3PO_4 and HBr) at various acidities, the use of 5 M HCl was found to be optimal. A 1% $NaBH_4$ solution was pumped at a 5-mL/min flow rate in the flow of the acidified sample; this required in the case of the

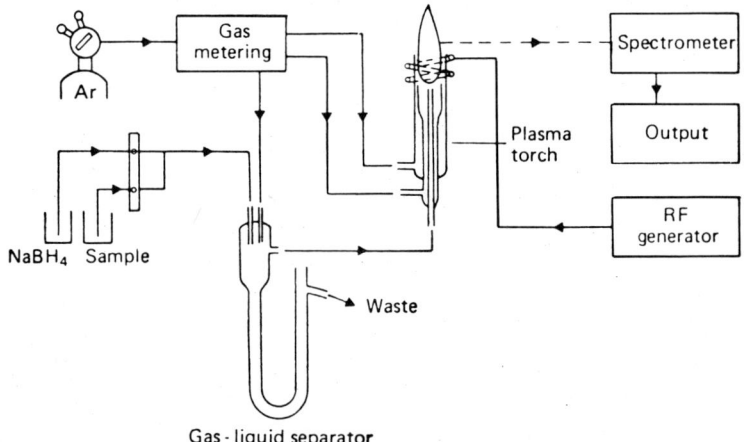

Figure 6.24. Flow-type hydride generation/ICP system [147]. [Reprinted with permission from M. Thompson, B. Pahlavanpour, and S. J. Waltor, *Analyst* **103,** 570 (1978). Copyright (1978). The Royal Society of Chemistry, London.]

Fassel torch an interruption of the sample flow (10 mL/min) between subsequent samples for preventing air entrainment in the ICP. Thompson and Pahlavanpour [148] also applied this hydride technique to tin and germanium. Detection limits are 0.02 µg/mL, and the linear concentration ranges cover nearly four orders of magnitude. Here optimal results were obtained with 0.1 M HCl. Mutual interferences reported in AAS work are probably due to insufficient

Table 6.6. Comparison of Detection Limits (in ng/mL) Obtained with an ICP Source Using Pneumatic Nebulization and Hydride Generation[a]
[Reprinted with permission from M. Thompson, B. Pahlavanpour, and T. J. Walton, *Analyst* **103,** 568 (1978). Copyright (1978), The Royal Society of Chemistry, London.]

Element	Pneumatic Nebulization		Hydride Generation	
	Fassel Torch	Greenfield Torch	Fassel Torch	Greenfield Torch
As	440	770	0.8	12.5
Sb	150	114	1.0	11
Bi	46	38	0.8	10
Se	430		0.8	
Te	17		1.0	

[a] From [147].

Table 6.7. Hydride-ICP Method: Recovery of Bismuth, Selenium, and Tellurium from Solutions Containing 1 μg/mL of the Analyte and 1000 μg/mL of Various Ions, After Separation on Lanthanum Hydroxide. Values in Parenthesis Were Obtained without Separation[a]

[Reprinted with permission from H. Thompson, B. Pahlavanpour, S. J. Walton, and G. F. Kirkbright, *Analyst* **103**, 705 (1978). Copyright (1978), The Royal Society of Chemistry, London]

Interferent	Deviation (%)		
	Bismuth	Selenium	Tellurium
V (V)	0	−5	−6
	(0)	(−11)	(−50)
Cr (VI)	0	0	0
	(−2)	(−5)	(−60)
Mo (VI)	0	0	0
	(0)	(−15)	(−10)
Co (II)	−1	0	0
	(−4)	(−4)	(−39)
Ni (II)	−3	−2	−3
	(−6)	(−12)	(−72)
Cu (II)	−1	−2	−6
	(−92)	(−99)	(−99)
Zn (II)	−2	−5	−15
	(−3)	(−7)	(−61)
Cd (II)	+2	−10	−20
	(−4)	(−12)	(−99)
Hg (II)	−5	0	0
	(−83)	(0)	(−70)
Pb (II)	−4	−20	−60
	(−5)	(−53)	(−99)

[a]From [149].

atomization in the flame as they were not found with the ICP. Transition metals cause considerable interferences which can be avoided by using tartaric acid (10 g/L) instead of hydrochloric acid. This masks the interferent and gives the same power of detection. In another paper [149] interferences for arsenic, antimony, bismuth, selenium, and tellurium in the case of the hydride-ICP method are investigated. Mutual interferences between these elements can be neglected up to concentrations of 10,000 mg/mL. Interferences from other ions (e.g., Cu, Fe, Pb) are often high, even with 5 M HCl in the sample solutions (Table 6.7). Interferences from copper could be avoided by adding potassium iodide, thus precipitating the interferent as copper (I) iodide. Coprecipitation of the hydride-forming elements with lanthanum hydroxide could be used for lowering inter-

Table 6.8. Results Obtained by the Hydride-ICP Method on Some USGS Standard Rocks Compared with Accepted Values[a]

[Reprinted with permission from B. Pahlavanpour, M. Thompson, and L. Thorne, *Analyst* **105**, 756 (1980). Copyright (1980), The Royal Society of Chemistry, London.]

Sample	Result	Concentration (μg/g)		
		As	Sb	Bi
W-1	Found	0.89	1.18	<0.04
	Accepted	1.9	1.0	0.046
G-1	Found	0.50	0.55	0.12
	Accepted	0.5	0.31	0.065
G-2	Found	0.22	0.12	0.12
	Accepted	0.25	0.1	0.043
GSP-1	Found	0.12	3.20	0.18
	Accepted	0.09	3.1	0.037

[a] From [150].

ferences from Cu, Co, Cr, Mo, V. This hydride-ICP technique was applied to the determination of arsenic, antimony, and bismuth in soils and sediments [150]. Samples are attacked with concentrated hydrochloric acid in sealed tubes at 150 °C during 2 h and subsequently, potassium iodide is added. Bismuth losses due to the presence of organic matter can be avoided by selecting milder reaction conditions in the case of samples rich in organic matter. Detection limits of 0.1 μg/g and a fair agreement of the results with certified values of standard samples were reported (Table 6.8). The hydride-ICP technique was also applied to the determination of selenium in soils after coprecipitation with lanthanum–hydroxide and reduction to selenium (IV) with potassium–bromide [151]. For the determination of arsenic, antimony and bismuth in herbage [152] samples were ashed after addition of magnesium nitrate preventing analyte losses. Broekaert and Leis [142] used a similar flow-cell hydride generator (Fig. 6.25). They modified the reagent inlet and provided an adjustable valve for the waste exit. A 3-kW argon ICP with a Greenfield burner was used. Both lead to high operation stability (RSD = 1%). A sample throughput of 2 samples/min was reported. The detection limit for arsenic was reported to be 1 ng/mL. A concentration of 5000-μg/mL Fe did not cause interferences. The method was applied to the determination of arsenic in steel and in waste water samples. Organic compounds of the waste waters are decomposed by a hydrogen peroxide/sulfuric acid decomposition giving a recovery of 90%. Calibration was done by standard addition and arsenic was determined down to a concentration of 10 μg/L. Similar flow-type hydride systems have been used by several others

6.6. THE INTRODUCTION OF GASES AND VAPORS

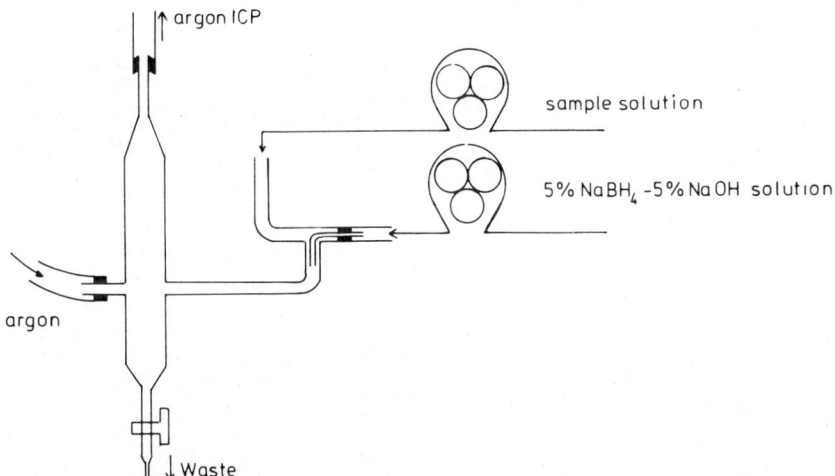

Figure 6.25. Modified flow-type hydride generator [142]. [Reprinted with permission from J. A. C. Broeckaert and F. Leis, *Fresenius Z. Anal. Chem.* **300**, 22 (1980).]

[153–157] for the analysis of real samples and are commercially available from ICP-manufacturers.

Wolnik et al. [158] proposed a tandem nebulization system for hydride generation (Fig. 6.26). The $NaBH_4$ solution is nebulized, and after separating the large droplets, the aerosol is used as nebulizer gas for the sample nebulizer. The system enables the determination of volatile elemental hydrides and other elements simultaneously. The detection limits for the hydride-forming elements are in the nanogram per milliliter range whereas the detection limits for the other elements are the same as for conventional pneumatic nebulization. The system was successfully applied to food analysis.

The hydride-ICP methods described permit the simultaneous determination of traces of the volatile hydride-forming elements at concentrations not covered with pneumatic nebulization. The application to real samples requires well elaborated sample pretreatment so as to avoid analyte loss and complete mineralization of the elements to be determined (see e.g. [159, 160]). Further, their application to real samples may be hampered by numerous interferences known from AAS, as shown, for example, in [161, 162].

6.6. THE INTRODUCTION OF GASES AND VAPORS

The introduction of gases as mixtures with argon in the ICP presents no problems. Mermet et al. [164, 165] introduced various gases into the ICP, mainly for spectroscopic purposes. Besides the earlier discussed hydride techniques,

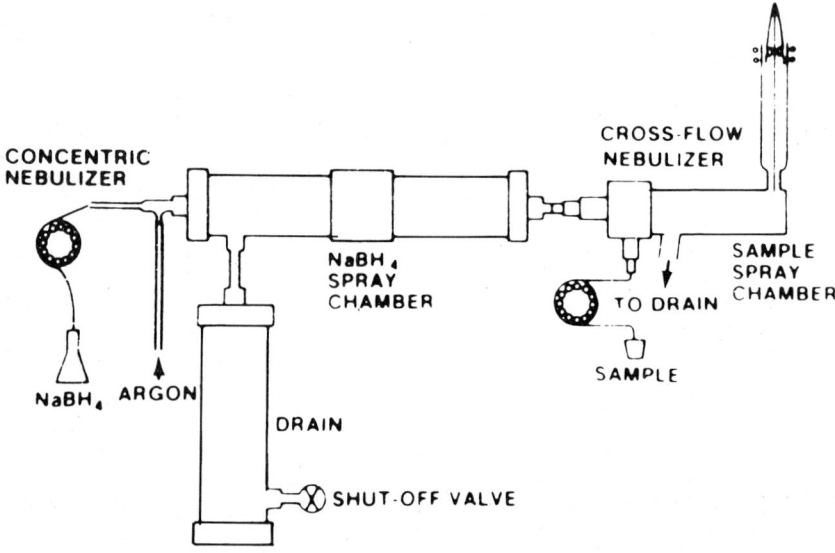

Figure 6.26. Tandem nebulization system for hydride-generation coupled to ICP–AES [158]. [Reprinted with permission from "Sample Introduction System for Simultaneous Determination of Volatile Elemental Hydrides and other Elements in Foods by Inductively Coupled Argon Emission Spectrometry," *Anal. Chem.* **53**, 1031 (1981). Copyright (1981), American Chemical Society, Washington, DC.]

other evolution techniques can be combined with ICP–AES. For mercury, for example, the detection limit obtained with pneumatic nebulization in ICP–AES was only 50 ng/mL, being insufficient for environmental applications. By using the cold-vapor technique known from AAS this value may be considerably improved. Thompson and Coles [166] combined reduction with Sn (II) chloride with pneumatic nebulization. Owing to the strong agitation during nebulization mercury vapor was released, the introduction efficiency increased and the detection limit was improved to 2 μg/L. The method described by Ellebracht et al. [167], who determined organic sulfur by a reduction to hydrogen–sulfide with hydrochloric acid and subsequent excitation in a direct current plasma, can also be applied. Alder et al. [168] determined traces of ammonium nitrogen in solutions by oxidation of the ammonium ion with sodium–hydrobromite in alkali medium and passing the evolved nitrogen in the ICP. When measuring the NH emission intensity at 336.0 nm, a detection limit of 0.1 μg/mL N for 5-mL aqueous solution samples is obtained. The method was applied to the analysis of soil samples. Brown et al. used gas chromatography coupled to ICP–AES for the determination of nitrogen-containing compounds [169] and oxygen [170] and measured the near infrared lines.

Applications of ICP-AES as chromatographic detector are also of interest for separation and speciation work. Windsor and Denton [172] used an ICP as a gas chromatographic element-specific detector for Br, C, Cl, H, O, P, and S. Sommer and Ohls [173] used a high-power argon–nitrogen ICP as element-specific detector and reported on the speciation of lead in gasolines. Here the outlet of the gas chromatograph was connected to the ICP aerosol injector by means of a heated tube. They also mentioned the possibility of separating metals from difficult samples, for example, waste waters, by using complexing agents such as dithiocarbamates or fluoroacetonates, which form thermally stable volatile compounds with a series of heavy metals [174]. Fujinaga et al. [175] injected β-diketonates into a carrier gas flow containing the ligand vapor in order to prevent its decomposition. When using a heated injection part, metal β-diketonates could be introduced with high efficiency into the ICP. The method can be used in combination with gas chromatography but also simply to determine matrix-isolated metal complexes with high power of detection. Black and Browner [176] used a loop system for the introduction of β-diketonates and determined Fe, Zn, Co, Mn, and Cr in various biological matrices. Despite the fact that ICP-AES offers good possibilities as gas chromatographic detector, less expensive devices such as microwave-induced plasmas operated at atmospheric pressure, as described by Beenakker [177, 178], are certainly more suitable for this (cf. Section 2.4).

6.7 DIRECT SOLIDS SAMPLING

The presentation of solids as powders, particulates, or metal vapors to the ICP as opposed to nebulization of liquids, has never reached the same attention from spectroscopists. Such techniques permit the omission of tedious decomposition and dissolution procedures of solids but, on the other hand, introduce severe difficulties with respect to calibration, sample conditioning, and analytical performance in general. Nevertheless, for tailored samples, direct solids sampling has been successfully applied by using powder injection, arc or spark sampling devices, or laser ablation.

6.7.1. Powder Sampling

Techniques for powder feeding into an rf plasma discharge result from the technology developed for the engineering applications of dc arc and induction plasma discharges [179–182] in the processing of solids. These systems, in Russian work often called plasmatrons, were recently reviewed by Dresvin [180]. The feeding of powders into analytical plasmas was described by Pforr and Aricot [183] and Kessler et al. [184, 185], who reported on the introduction of powders into the capacitively coupled microwave plasma (CMP). From this work it is known that the precision of powder sampling is limited by the fact that the formation of a representative cloud is hampered by segregation.

Greenfield [186] mentioned the introduction of powders into an analytical ICP in 1964 and patented the system [187]. Similar work was done by Gerasimov and Eilenkrig [188]. Analytical results were described by Hoare and Mostyn [95] and Dagnall et al. [189] (see Part 2, Chapter 7 for details). Greenfield and Sutton [190] recently proposed the use of a tape machine and an 8-kV spark for powder sampling. By using a cyclone chamber and a suitable aerosol injector they could determine 18 elements in phosphate rocks with RSDs of 0.02–0.07. Ng et al. [191] used a y-joint with an outlet diameter of 1 mm and a 1-L/min argon flow for the introduction of coal fly ash. Detection limits obtained were at the microgram per gram level and the recovery for Cr, Sr, Ti, V, Cu, and Zn in an NBS coal fly ash was 70–110% with 7–29% RSD.

It is clear that the power of detection and the dynamic range in the analysis of powders with the ICP are limited by the heat transfer processes in the plasma. It is known from calculations of Barnes and Schleicher [12] that total vaporization of the particles hardly ever occurs. This would, in the case of Al_2O_3 particles and an ICP of 6000 K, require particle diameters below 1 μm and carrier gas flows below 2 L/min. These conditions are difficult to fulfill with powders because of limitations in grinding, segregation difficulties at decreasing particle size and the necessity of substantially higher carrier gas flows for aerosol transport. With the present state of the art, the required sophisticated sample treatment and restrictions in analytical performance with powder sampling techniques do not outweigh the advantage that sample dissolution can be avoided. However, since a universal powder sampling method would be welcome in various types of applications further technological developments can be expected.

6.7.2. Arc and Spark Devices

In a survey of sampling devices in ICP–AES, Dahlquist [192] mentioned the use of arc sampling for aerosol generation from solids. He described and patented the device (Fig. 6.27) in combination with other excitation and analysis systems [193, 194]. In this commercially available Direct Solids Nebulizer (DSN) the sample acts as the cathode of a dc arc discharge. When using an open circuit voltage of 600 V and currents of 2–8 A, a broad pulse spectrum (mean frequency up to 1 MHz) is observed and a rapid movement of the cathode spot produces uniform sampling over a well-defined area. A flowing gas stream transports the aerosol particles onto the excitation source. Remote sampling at distances of up to 20 m makes the system attractive for the analysis of large-size items. On the other hand, a DSN coupled to an ICP can be used for pre-

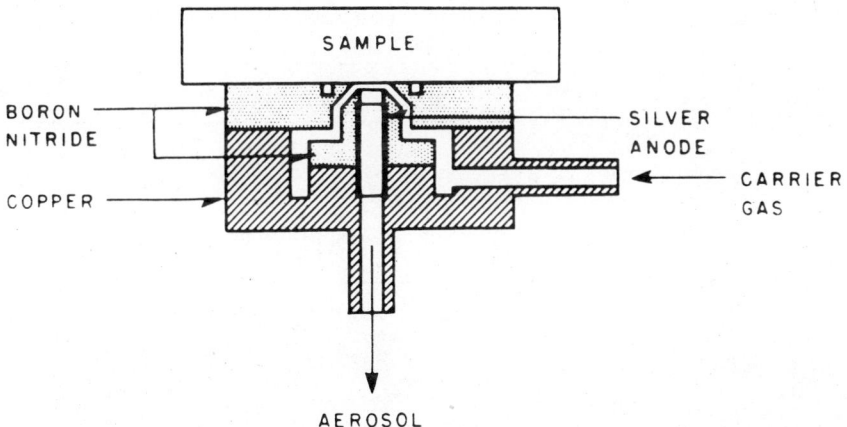

Figure 6.27. Direct solids nebulizer [193]. [Reprinted with permission from J. L. Johnes, R. L. Dahlquist, R. E. Hoyt, *Appl. Spectrosc.* **25**, 629 (1971). Copyright (1971), Society for Applied Spectroscopy, Frederick, MD.]

cision analysis of metallic samples. Indeed, linear calibration curves over several orders of magnitude of concentration and low matrix effects enabling the analysis of different types of alloys with the same calibration were reported [116]. Ohls and Sommer [195] reported on arc sampling and excitation of the vapors in a 3-kW argon–nitrogen ICP for the analysis of solids. They used a dc arc in the case of compact steel samples and an intermittent arc for oxide powders (i.e., slags and ores). The latter were mixed in a proportion of 1:5 with copper powder and briquetted into pellets as is known from glow discharge work [196]. A graphite rod was used as counter electrode and a boron nitride disc was employed for well defining the sampling surface. Remote sampling capabilities and the absence of memory effects were confirmed. Analysis results for steels, iron ores, and slags were communicated.

Spark sampling may have the advantage that risks for selective volatilization at preferred locations on the sample are lower than in arc sampling. Spark sampling was first used in ICP work by Human et al. [197] (see Part 2, Chapter 7, Fig. 7.7) and applied to compact metallic samples. Scott et al. [198] also reported the spark elutriation of powders (see Part 2, Chapter 7). Aziz et al. [199] investigated the power of detection and its limitations for the spark erosion ICP technique. In the case of aluminum samples and a medium voltage spark (repetition rate: $25\ s^{-1}$), the detection limits were considerably higher than in spark emission spectroscopy or when analyzing the dissolved samples (Table 6.9). This is related to the low analyte introduction rate (20 μg Al/min) in the ICP as compared to analyzing dissolved samples (200 μg/min when 10-g/L solutions are nebulized pneumatically). Indeed, the particle size being below 1 μm im-

Table 6.9. The Analysis of Aluminium Alloys after Dissolution and with Direct Solids Sampling

Element/Line (nm)		Detection Limits (μg/g)			
		ICP-AES of Dissolved Samples (20 g/L)[a]	Spark Optical Emission Spectroscopy[b]	Spark Erosion (25 s^{-1})[c]	Coupled to ICP-AES (400 s^{-1})[d]
Cu I	324.8	2	1	26	0.5
Fe II	259.9	0.3	3[e]	29	1
Mg II	279.6	0.05	1[e]	1	0.2
Mn II	293.3				10
Si I	288.3				2

[a] Estimated from the detection limits given in [122].
[b] From [201].
[c] From [199].
[d] From [200].
[e] Other lines used.

poses no limitations here. It was found that the power of detection can be hardly improved by increasing the condensor voltage of the spark, as this increases the erosion rate, however, also the particle size. By increasing the spark repetition rate (from 25 to 400 s^{-1}), the erosion rate may be considerably increased (to 70 μg/min Al) while the particle size remains practically unchanged. Consequently, the detection limits could be considerably improved (Table 6.9) and become similar to those in spark AES. In the example of aluminium alloys, it was shown that various types of alloys could be analyzed with the same calibration curves (Fig. 6.28). Matrix effects are much lower than in spark AES; at varying silicon contents, for example, they may be neglected up to the eutectic concentration (100–110 mg/g Si in Al). The method can also be applied to the analysis of electrically nonconducting powders (Al_2O_3, $CaCO_3$), provided that they are mixed with a metal powder (e.g., Cu) and briquetted into pellets. By using a matrix or a copper line as reference, an RSD of 5% is achieved at a concentration level of 5 mg/g. Thus the method may be of interest for the analysis of powders which are difficult to dissolve such as Al_2O_3 and TiO_2. It may be used when X-ray fluorescence (XRF) methods are not available or cannot be applied because of matrix effects.

6.7.3. Laser Evaporation of Solid Samples

As recently reviewed by Laqua [202], high-power lasers are very useful for the evaporation of both conducting and nonconducting samples prior to excitation in various sources.

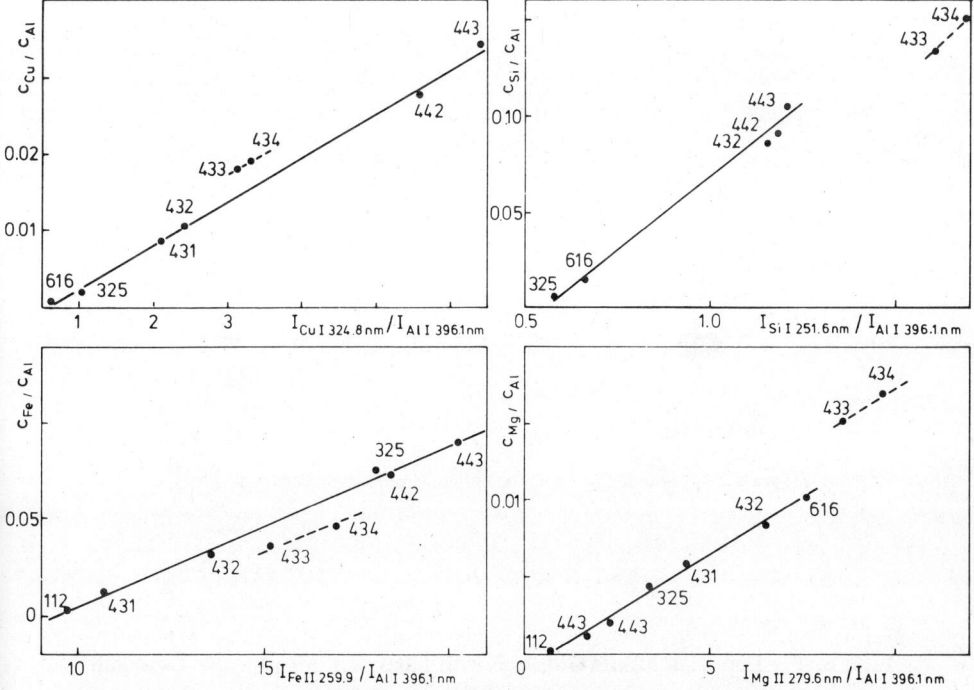

Figure 6.28. Analysis of aluminium alloys by spark erosion coupled to ICP–AES. 3 kW argon/nitrogen ICP; AlMn: 325, AlSiCuNi: 431, 432, 442-433, 434 (c_{Si} > 110 mg/g); AlSiCu: 443; AlMgSi: 616, Al: 112 [199]. [Reprinted with permission from A. Aziz, J. A. C. Broekaert, F. Leis, and K. Laqua, *Spectrochim. Acta* **39B**, 1091 (1984). Copyright (1984), Pergamon Press.]

A first combination of laser evaporation and ICP-excitation was reported by Abercrombie et al. [203]. They used a laser sampling device in air particulate analysis for geochemical exploration. A pulsed carbon dioxide laser beam was focused onto single spots of air particulate matter collected on paper-tape during aircraft flights, and the evaporated material was carried into the ICP. A fully automated instrument which could analyze a new sample for 25 elements every 10 s was described.

Thompson et al. [204] coupled laser ablation with an ICP-spectrometer using a light-activated switch for starting signal integration (Fig. 6.29). Results for steel samples and initial work on silicate rocks were reported. Absolute detection limits in the case of laser sampling are superior to those of conventional pneumatic nebulization; relative detection limits are poorer (Table 6.10). This is related to the lack of sampling efficiency, which is limited by sample deposition at the edge of the laser crater, losses in the transport system, and the physical properties of the aerosol.

Carr and Horlick [205] used a ruby laser in free running mode and with a

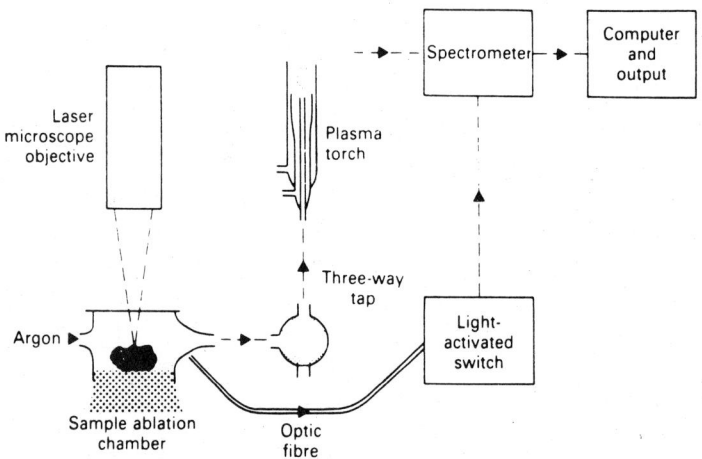

Figure 6.29. Laser-ICP microprobe system (From Thompson et al. [204].)

Table 6.10. Detection Limits Obtained with Laser ICP Microprobe Techniques and with Nebulization of Solutions

Element	Relative Detection Limits for Steel (μg/g)[a,b]		Absolute Detection Limit (pg)[a]		Relative Detection Limits for Steel (μg/g)[c]	
	Laser	Nebulizer[d]	Laser[e]	Nebulizer[f]	Q-Switched	Free Running
Al					20	2
Co					8	0.6
Cr	15	2	15	25	10	1
Cu	20	1.5	20	20	9	0.8
Mn	80	2	80	5	3	0.3
Mo	60	4.5	60	20	20	2
Ni	70	1.0	70	70	20	1
P	10	35	10	150		
S	15	45	15	250		
V	10	1.5	10	20	20	1

[a] From [204].
[b] Measured at the same ICP spectrometer.
[c] From [206].
[d] Assuming 10-g/L solutions of steel.
[e] Assuming 1 μg of sample ablated.
[f] Assuming a 2% efficiency at an uptake rate of 0.7 mL/min.

Q-switch at energies of 1–2 J. Powder samples were directly analyzed from a suitable support. Metal samples were moved between subsequent shots and the signals measured with a multichannel PD spectrometer. Best precision was obtained with 8 s between subsequent shots and signals had a duration of 4.8 s. When plotting relative intensities versus relative concentrations, calibration curves for aluminum and brass alloys were linear in a logarithmic display over two and one order of magnitude, respectively. Mean RSDs then were 5 and 3%.

Ishizuka and Uwamino [206] showed that Q-switched operation only yields a slightly better precision than free-running modes. The laser focusing considerably influences the emission intensities; at a laser beam diameter of 1.3 mm the ablated amount was 30 μg instead of 500 μg at a beam size of 0.5 mm and, nevertheless, the emission signals were higher; this was due to the occurrence of a discharge in the aerosol injector and consequent absorbance of rf power when high amounts of metal were injected. Also here remote sampling capabilities were reported. Detection limits in steel (Table 6.10) were ten times lower with normal operation as compared to Q-switched operation, which is concerned with both the ablation rate and the particle size. Differences between the results in [204] and [206] may be due to different lines used.

These works show that the ICP laser microprobe well enables the direct

Table 6.11. Applicability of the Various Sample Introduction Techniques

Problem	Pneumatic Nebulization	Ultrasonic Nebulization	Electrothermal Methods	Hydride Techniques	Direct Solids Sampling
Ultratrace analysis	+	+++	+++	+++	+
High-precision analysis	+++	+	+	++	+
Microanalysis	++	+++	+++	++	++
Universal technique	+++	+	+	−	+
Real samples	+++	+	++	++	+
Solutions with high salt contents	+++	+	+	++	−
Routine operation	+++	++	++	++	++

+++ Very good.
++ Good.
+ Moderate.
− Not at all.

analysis of solids. The analytical precision and the power of detection are similar to those of spark erosion techniques.

6.8. CONCLUSION

The analysis of liquid samples is and will remain the prime task of ICP-AES. For routine work pneumatic nebulizers, tailored to the nature of real samples are standard devices. Ultrasonic nebulization still has to be made applicable to the routine analysis of real samples. Electrothermal atomizers are of potential interest for the analysis of small-volume samples and can be expected to become useful accessories to ICP equipment in the same way as hydride generators. Direct solids sampling of both powders and compact samples may have some advantage with respect to arc and spark emission techniques but are likely to remain confined to special applications. From a summary of present aerosol generation devices and their capabilities and limitations (Table 6.11), it is clear that sample introduction is the link of the ICP system that should be the most carefully adapted to the analytical problem at hand.

REFERENCES

1. R. Herrmann and C. T. J. Alkemade, *Chemical Analysis by Flame Photometry*, Wiley, New York (1963).
2. R. Mavrodineanu and H. Boiteux, *Flame Spectroscopy*, Wiley, New York (1965).
3. E. Kranz, *Spectrochim. Acta* **27B**, 327 (1972).
4. J. M. Mermet, *ICP Inf. Newslett.* **2**, 70 (1976).
5. B. Welz, *Atomic Absorption Spectroscopy*, Verlag Chemie, Weinheim, New York (1976).
6. W. B. Robbins and J. A. Caruso, *Anal. Chem.* **51**, 889A (1979).
7. G. R. Kornblum and L. de Galan, *Spectrochim. Acta* **32B**, 71 (1977).
8. K. Visser, F. M. Hamm, and P. B. Zeeman, *Appl. Spectrosc.* **30**, 34 (1976).
9. J. Mermet, *Spectrochim. Acta* **30B**, 383 (1975).
10. D. J. Kalnicky, R. N. Kniseley, and V. A. Fassel, *Spectrochim. Acta* **30B**, 511 (1975).
11. P. W. J. M. Boumans and F. J. de Boer, *Spectrochim. Acta* **27B**, 391 (1972).
12. R. M. Barnes and R. G. Schleicher, *Spectrochim. Acta* **30B**, 109 (1975).
13. A. Syty, *CRC Crit. Rev. Anal. Chem.* **4**, 155 (1974).
14. J. E. Meinhard, *ICP Inf. Newslett.* **2**, 163 (1976).
15. C. C. Butler, R. N. Kniseley, and V. A. Fassel, *Anal. Chem.* **47**, 825 (1975).
16. R. N. Kniseley, H. Amenson, C. C. Butler, and V. A. Fassel, *Appl. Spectrosc.* **28**, 285 (1974).

17. S. E. Valente and W. G. Schrenk, *Appl. Spectrosc.* **24,** 197 (1970).
18. P. W. J. M. Boumans and M.-Ch. Lux–Steiner, *Spectrochim. Acta* **37B,** 97 (1982).
19. D. L. Donohue and J. E. Carter, *Anal. Chem.* **50,** 686 (1978).
20. V. A. Fassel, C. A. Peterson, and F. N. Abercrombie, *Anal. Chem.* **48,** 1010 (1977).
21. J. W. Novak Jr., D. E. Lillie, A. W. Boorn, and R. F. Browner, *Anal. Chem.* **52,** 576 (1980).
22. H. Anderson, H. Kaiser, and B. Meddings, in R. M. Barnes, ed. *Developments in Atomic Plasma Spectrochemical Analysis*, Heyden, London (1981), p. 251.
23. Y. Fujishiro, M. Kubota, and R. Ishida, *Spectrochim. Acta* **39B,** 617 (1984).
24. R. S. Babington, *Popular Sci.* May, p. 102 (1973).
25. R. S. Babington, U. S. Patents: 3,421,692; 3,421,699; 3,425,058; 3,425,059; and 3,504,859.
26. J. F. Wolcott and C. B. Sobel, *Appl. Spectrosc.* **32,** 591 (1978).
27. R. F. Suddendorf and K. W. Boyer, *Anal. Chem.* **50,** 1769 (1978).
28. P. W. McKinnon, K. C. Giess, and T. V. Knight, in R. M. Barnes, ed. *Developments in Atomic Plasma Spectrochemical Analysis*, Heyden, London (1981), p. 287.
29. B. Thelin, *Analyst* **106,** 54 (1981).
30. C. T. Apel, T. M. Bienewski, L. E. Cox, and D. W. Steinhaus, *ICP Inf. Newslett.* **3,** 1 (1977).
31. L. R. Layman and F. E. Lichte, *Anal. Chem.* **54,** 638 (1982).
32. R. H. Scott, V. A. Fassel, and R. N. Kniseley, *Anal. Chem.* **46,** 75 (1974).
33. P. Schutyser and E. Janssens, *Spectrochim. Acta* **34B,** 443 (1979).
34. S. Greenfield, H.McD. McGeachin, and P. B. Smith, *Anal. Chim. Acta* **84,** 67 (1976).
35. S. Nukijama and Y. Tanasawa, *Trans. Soc. Mech. Eng.* (Jpn.) **4,** 86 (1938).
36. S. Nukijama and Y. Tanasawa, *Trans. Soc. Mech. Eng.* (Jpn.) **5,** 68 (1939).
37. P. W. J. M. Boumans, *Fresenius Z. Anal. Chem.* **229,** 337 (1979).
38. J. A. C. Broekaert and F. Leis, *Anal. Chim. Acta* **109,** 73 (1979).
39. G. Ackermann and M. Münx, *Talanta* **24,** 91 (1976).
40. K. Szivos, L. Polos, and E. Pungor, *Spectrochim. Acta* **31B,** 289 (1976).
41. J. A. C. Broekaert, F. Leis, and K. Laqua, *Talanta* **28,** 745 (1981).
42. A. Miyazaki, A. Kimura, K. Ba Nsho, and Y. Umezaki, *Anal. Chim. Acta* **144,** 213 (1982).
43. V. A. Fassel, C. A. Peterson, and F. N. Abercrombie, *Anal. Chem.* **48,** 516 (1976).
44. G. L. Gouy, *Ann. Chim. Phys.* **18,** 5 (1979).
45. J. E. Meinhard, in R. M. Barnes, ed. *Applications of Plasma Emission Spectroscopy*, Heyden, London, (1979), p. 1.

46. C. C. Wohlers, *ICP Inf. Newslett.* **3,** 37 (1977).
47. R. M. Barnes, H. S. Mahanti, M. R. Cave, and L. Fernando, *ICP Inf. Newslett.* **8,** 562 (1983).
48. N. R. McQuaker, P. D. Kluckner, and G. N. Chang, *Anal. Chem.* **51,** 888 (1979).
49. D. E. Dobb and D. R. Jenke, *Appl. Spectrosc.* **37,** 379 (1983).
50. M. H. Ramsey, M. Thompson, and B. J. Coles, *Anal. Chem.* **55,** 1626 (1983).
51. G. F. Wallace, V. V. Pirc, and R. D. Ediger, *Can. J. Spectrosc.* **27,** 46 (1982).
52. R. C. Fry and M. B. Denton, *Anal. Chem.* **49,** 1413 (1977).
53. R. C. Fry and M. B. Denton, *Appl. Spectrosc.* **33,** 393 (1979).
54. P. A. M. Ripson and L. de Galan, *Spectrochim. Acta* **36B,** 71 (1981).
55. J. A. C. Broekaert, B. Radziuk, and F. Leis, *Spectrochim. Acta* **38B,** 45, Supplement 1983, 23rd Coll. Spectrosc. Int., Amsterdam, (1983).
56. M. D. Wichman, R. C. Fry, and N. Mohamed, *Appl. Spectrosc.* **37,** 254 (1983).
57. C. W. Fuller, R. C. Hutton, and B. Preston, *Analyst* **106,** 913 (1981).
58. J. R. Garbarino and H. E. Taylor, *Appl. Spectrosc.* **34,** 584 (1980).
59. S. D. Olsen and A. Strasheim, *Spectrochim. Acta* **38B,** 973 (1983).
60. L. Ebdon and M. R. Cave, *Analyst* **107,** 172 (1982).
61. R. M. Belchamber and G. Horlick, *Spectrochim. Acta* **36B,** 581 (1981).
62. J. A. C. Broekaert, B. Wopenka, and H. Puxbaum, *Anal. Chem.* **54,** 2174 (1982).
63. P. Hulmston, *Analyst* **108,** 166 (1983).
64. E. Sebastiani, K. Ohls, and G. Riemer, *Fresenius Z. Anal. Chem.* **264,** 105 (1973).
65. H. Berndt and E. Jackwerth, *Spectrochim. Acta* **30B,** 169 (1975).
66. H. Berndt and W. Slavin, *At. Absorption Newslett.* **17,** 109 (1978).
67. M. S. Cresser, *Progr. Analyt. Atom. Spectrosc.* **4,** 219 (1981).
68. S. Greenfield, *Spectrochim. Acta* **38B,** 93 (1983).
69. R. N. Kniseley, V. A. Fassel, and C. C. Butler, *Clin. Chem.* **19,** 807 (1973).
70. T. Uchida, I. Kojima, and C. Iida, *Anal. Chim. Acta* **116,** 205 (1980).
71. H. Uchida, Y. Nojiri, H. Haraguchi, and F. Fuwa, *Anal. Chim. Acta* **123,** 57 (1981).
72. I. Kojima and C. Iida, *Analyst* **107,** 1000 (1982).
73. S. Greenfield and P. B. Smith, *Anal. Chim. Acta* **59,** 341 (1972).
74. J. A. C. Broekaert, F. Leis, and K. Laqua, *Fresenius Z. Anal. Chem.* **301,** 105 (1980).
75. A. Aziz, J. A. C. Broekaert, and F. Leis, *Spectrochim. Acta* **36B,** 251 (1981).
76. J. Ruzicka and E. H. Hanssen, *Anal. Chim. Acta* **99,** 37 (1978).
77. P. W. Alexander, R. I. Finlayson, L. E. Smythe, and A. Thalib, *Analyst* **107,** 1335 (1982).
78. T. Ito, E. Nakagawa, H. Kawaguchi, and A. Mizuike, *Microchim. Acta* **7,** 423 (1982).

79. E. A. G. Zagotto, A. O. Jacintho, F. J. Krug, B. F. Reis, R. E. Bruns, and M. C. U. Arauyn, *Anal. Chim. Acta* **145**, 169 (1983).
80. M. Morita, T. Uehiro, and K. Fuwa, *Anal. Chem.* **52**, 349 (1980).
81. M. Morita and T. Uehiro, *Anal. Chem.* **53**, 1997 (1981).
82. K. Yoshida, H. Haraguchi, and K. Fuwa, *Anal. Chem.* **55**, 1009 (1983).
83. K. Jinno, H. Tsuchida, S. Nakanishi, Y. Hirata, and C. Fujimoto, *Appl. Spectrosc.* **37**, 258 (1983).
84. B. S. Whaley, K. S. Snable, and R. F. Browner, *Anal. Chem.* **54**, 162 (1982).
85. D. W. Hausler and L. T. Taylor, *Anal. Chem.* **53**, 1223 (1981).
86. D. W. Hausler and L. T. Taylor, *Anal. Chem.* **53**, 1227 (1981).
87. J. Spitz and G. Uny, *Appl. Optics* **7**, 1345 (1968).
88. H. Dunken, G. Pforr, W. Mikkeleit, and K. Geller, *Spectrochim. Acta* **20**, 1531 (1954).
89. R. H. Wendt and V. A. Fassel, *Anal. Chem.* **37**, 920 (1965).
90. G. W. Dickinson and V. A. Fassel, *Anal. Chem.* **41**, 1021 (1969).
91. J. C. Souillart and J. Robin, *Analusis* **1**, 427 (1972).
92. P. W. J. M. Boumans and F. J. de Boer, *Spectrochim. Acta* **27B**, 391 (1972).
93. P. W. J. M. Boumans and F. J. de Boer, *Spectrochim. Acta* **30B**, 309 (1975).
94. V. A. Fassel and G. W. Dickinson, *Anal. Chem.* **40**, 247 (1968).
95. H. C. Hoare and R. A. Mostyn, *Anal. Chem.* **39**, 1153 (1967).
96. J. M. Mermet, C. Trassy, and P. Ripoche, in R. M. Barnes, ed., *Developments in Atomic Plasma Spectrochemical Analysis*, Heyden, London, (1981), p. 245.
97. K. W. Olsen, W. J. Haas, and V. A. Fassel, *Anal. Chem* **49**, 632 (1977).
98. C. E. Taylor and T. L. Floyd, *Appl. Spectrosc.* **35**, 408 (1981).
99. P. W. J. M. Boumans and F. J. de Boer, *Spectrochim. Acta* **31B**, 355 (1976).
100. S. S. Berman, J. W. McLaren, and S. N. Willie, *Anal. Chem.* **52**, 488 (1980).
101. M. A. Floyd, V. A. Fassel, R. K. Winge, J. M. Katzenberger, and A. F. D'Silva, *Anal. Chem.* **52**, 431 (1980).
102. T. Catterick and D. A. Hickman, *Analyst* **104**, 516 (1979).
103. P. W. J. M. Boumans and J. J. A. M. Vrakking, *Spectrochim. Acta* **39B**, 1291 (1984).
104. V. A. Fassel, *Spec. Tech. Publ. 618*, American Society for Testing and Materials, Philadelphia, PA (1977), p. 22.
105. P. D. Goulden and H. J. Anthony, *Anal. Chem.* **54**, 1678 (1982).
106. H. M. Donega and T. E. Burgess, *Anal. Chem.* **42**, 1521 (1979).
107. T. Kantor, E. Pungor, L. Bezur, and J. Sztatisz, *Talanta* **26**, 357 (1979).
108. W. Frech, J.-A. Persson, and A. Cedergren, *Progr. Analyt. Atom. Spectrosc.* **3**, 279 (1980).
109. D. W. Shaw, O. I. Joensuu, and L. H. Ahrens, *Spectrochim. Acta* **4**, 233 (1950).
110. E. Gegus, J. Kreiter, L. Méray, and J. Inczédy, *Acta Chem. Acad. Sci. Hung.* **100**, 221 (1979).

111. H. T. Delves, *Analyst* **95**, 431 (1970).
112. H. Berndt and J. Messerschmidt, *Spectrochim. Acta* **34B**, 241 (1979).
113. D. E. Nixon, V. A. Fassel, and R. N. Kniseley, *Anal. Chem.* **46**, 210 (1974).
114. A. M. Gunn, D. L. Millard, and G. F. Kirkbright, *Analyst* **103**, 1066 (1978).
115. R. L. Dahlquist, U.S. Patent, 3,832,600, Aug. 27, 1974.
116. R. L. Dahlquist, J. W. Knoll, and R. E. Hoyt, 26th Pittsburgh Conference on Analytical Chemistry and Applied Spectroscopy, Abstr., Paper No. 341, Cleveland, 1975; Reprint 7002, Applied Research Laboratories, Goleta CA, see *ICP Inf. Newslett.* **1**, 15 (1975).
117. E. L. Kimberly, G. W. Rice, and V. A. Fassel, *Anal. Chem.* **56**, 289 (1984).
118. D. L. Millard, H. C. Shan, and G. F. Kirkbright, *Analyst* **105**, 502 (1980).
119. G. F. Kirkbright and R. D. Snook, *Anal. Chem.* **51**, 1938 (1979).
120. C. Camara Rica, G. F. Kirkbright, and R. D. Snook, *At. Spectrosc.* **2**, 172 (1981).
121. R. M. Barnes and P. Fodor, *Spectrochim. Acta* **38B**, 1191 (1983).
122. A. Aziz, J. A. C. Broekaert, and F. Leis, *Spectrochim. Acta* **37B**, 369 (1982).
123. Z. Grobenski, R. Lehmann, R. Tamm, and B. Welz, *Mikrochim. Acta* **1**, 115 (1982).
124. G. Crabi, P. Cavalli, M. Achilli, G. Rossi, and N. Omenetto, *At. Spectrosc.* **3**, 81 (1982).
125. H. M. Swaiden and G. D. Christian, *Anal. Chem.* **56**, 120 (1984).
126. G. E. Batley and J. P. Matousek, *Anal. Chem.* **52**, 1716 (1980).
127. IL 254 Fastac™, Instrumentation Laboratory.
128. K. C. Ng and J. A. Caruso, *Anal. Chim. Acta* **143**, 209 (1982).
129. K. C. Ng and J. A. Caruso, *Anal. Chem.* **55**, 1504 (1983).
130. L. E. Smythe, 12th Australian Spectroscopy Conference, Melbourne, 1980, see *ICP Inf. Newslett.* **6**, 224 (1980).
131. V. P. Garnys and L. E. Smythe, *Anal. Chem.* **51**, 62 (1979).
132. J. P. Keilsohn, R. D. Deutsch, and G. M. Hieftje, *Appl. Spectrosc.* **37**, 101 (1983).
133. E. Kitazume, *Anal. Chem.* **55**, 802 (1983).
134. E. D. Salin and G. Horlick, *Anal. Chem.* **51**, 2284 (1979).
135. D. Sommer and K. Ohls, *Fresenius Z. Anal. Chem.* **304**, 97 (1980).
136. G. F. Kirkbright and S. J. Walton, *Analyst* **107**, 276 (1982).
137. G. F. Kirkbright and Z. Li-Xing, *Analyst* **107**, 617 (1982).
138. K. C. Ng and J. A. Caruso, *Anal. Chem.* **55**, 2032 (1983).
139. Z. Li-Xing, G. F. Kirkbright, M. J. Cope, and J. M. Watson, *Appl. Spectrosc.* **37**, 250 (1983).
140. N. W. Barnett, M. J. Cope, G. F. Kirkbright, A. A. M. Taobi, *Spectrochim. Acta* **39B**, 343 (1984).
141. T. Nakahara, *Prog. Analyt. At. Spectrosc.* **6**, 163 (1983).

142. J. A. C. Broekaert and F. Leis, *Fresenius Z. Anal. Chem.* **300,** 22 (1980).
143. D. Sommer, K. Ohls, and A. Koch, *Fresenius Z. Anal. Chem.* **306,** 372 (1981).
144. C. J. Pickford, *Analyst* **106,** 464 (1981).
145. R. C. Fry, M. B. Denton, D. L. Windsor, and S. J. Northway, *Appl. Spectrosc.* **33,** 393 (1979).
146. M. H. Hahn, K. A. Wolnik, F. L. Fricke, and J. A. Caruso, *Anal. Chem.* **54,** 1048 (1982).
147. M. Thompson, B. Pahlavanpour, and S. J. Walton, *Analyst* **103,** 568 (1978).
148. M. Thompson and B. Pahlavanpour, *Anal. Chim. Acta* **109,** 251 (1979).
149. M. Thompson and B. Pahlavanpour, S. J. Walton, and G. F. Kirkbright, *Analyst* **103,** 705 (1978).
150. B. Pahlavanpour, M. Thompson, and L. Thorne, *Analyst* **105,** 756 (1980).
151. B. Pahlavanpour, J. H. Pullen, and M. Thompson, *Analyst* **105,** 274 (1980).
152. B. Pahlavanpour, M. Thompson, and L. Thorne, *Analyst* **106,** 467 (1981).
153. K. Nakahara, *Anal. Chim. Acta* **131,** 73 (1981).
154. K. Nakahara, *Appl. Spectrosc.* **37,** 539 (1983).
155. E. de Oliveira, J. W. McLaren, and S. S. Berman, *Anal. Chem.* **55,** 2047 (1983).
156. D. A. Rose, *Anal. Proc. (London)* **20,** 436 (1983).
157. E. Pruszkowska, P. Barrett, R. Ediger, and G. Wallace, *At. Spectrosc.* **4,** 94 (1983).
158. K. A. Wolnik, F. L. Fricke, M. H. Hahn, and J. A. Caruso, *Anal. Chem.* **53,** 1030 (1981).
159. J. W. Jones, S. G. Gapar, and C. T. O'Haver, *Analyst* **107,** 353 (1982).
160. D. D. Nygaard and J. H. Lowry, *Anal. Chem.* **54,** 803 (1982).
161. A. Meyer, C. Hofer, G. Tölg, S. Raptis, and G. Knapp, *Fresenius Z. Anal. Chem.* **296,** 337 (1979).
162. B. Welz and M. Melcher, *Spectrochim. Acta* **36B,** 439 (1981).
163. D. M. Flaley, D. A. Yates, S. E. Manahan, D. Stalling, and J. Petty, *Appl. Spectrosc.* **35,** 525 (1981).
164. J. F. Alder and J. M. Mermet, *Spectrochim. Acta* **28B,** 421 (1973).
165. J. Jarosz, J. M. Mermet, and J. Robin, *CR Acad. Sci. Ser. B.* **278,** 885 (1974).
166. M. Thompson and B. J. Coles, *Analyst* **109,** 529 (1984).
167. S. R. Ellebracht, P. D. Swain, and D. S. Treybig, *Anal. Chem.* **51,** 1605 (1979).
168. J. F. Alder, A. M. Gunn, and G. F. Kirkbright, *Anal. Chim. Acta* **92,** 43 (1977).
169. R. M. Brown, S. J. Northway, and R. C. Fry, *Anal. Chem.* **53,** 934 (1981).
170. R. M. Brown Jr. and R. C. Fry, *Anal. Chem.* **53,** 532 (1981).
171. E. R. Prack and G. J. Bastiaans, *Anal. Chem.* **55,** 1654 (1983).
172. D. L. Windsor and M. B. Denton, *Appl. Spectrosc.* **32,** 366 (1978).
173. D. Sommer and K. Ohls, *Fresenius Z. Anal. Chem.* **295,** 337 (1979).
174. K. Ohls and D. Sommer, in R. M. Barnes, ed. *Developments in Atomic Plasma Spectrochemical Analysis*, Heyden, London, (1981), p. 321.

175. T. Fujinaga, T. Kuwamoto, K. Isshiki, N. Matsubara and E. Nakayama, *Spectrochim. Acta* **38B**, 1011 (1983).
176. M. S. Black and R. F. Browner, *Anal. Chem.* **53**, 249 (1981).
177. C. I. M. Beenakker, *Spectrochim. Acta* **32B**, 173 (1977).
178. C. I. M. Beenakker, *Spectrochim. Acta* **33B**, 545 (1978).
179. U. Landt, *Chem. Ingen. Tech.* **42**, 617 (1970).
180. S. V. Dresvin, ed., *Physics and Technology of Low-Temperature Plasmas*, Atomizdat, Moscow (1972); H. U. Eckert, ed., Iowa State University Press, Ames (1977), Engl. Transl.
181. B. Waldie, *Chem. Eng. (N.Y.)* **259**, 92 (1972); **261**, 188 (1972).
182. A. Czernichovski and J. Jurewicz, *Pr. Nauk. Inst. Chem. Nieorg. Metal Pierwiastakow Rzadkich Politech. Wroclaw*, **24**, 3 (1975); Engl. Transl. in *ICP Inf. Newslett.* **2**, 1 (1976).
183. G. Pforr and O. Aricot, *Z. Chem.* **10**, 78 (1970).
184. W. Kessler, *Glastechn. Ber.* **44**, 479 (1971).
185. W. Kessler and F. Gebhardt, *Glastechn. Ber.* **40**, (5), 194 (1967).
186. S. Greenfield, I. L. Jones, and C. T. Berry, *Analyst* **89**, 713 (1964).
187. S. Greenfield, I. L. W. Jones, and C. T. Berry, U.S. Patent 3, 467, 471, Sept. 16, 1969.
188. R. D. Gerasimov and G. S. Eilenkrig, *Zh. Prikl. Spektrosk.* **19**, 791 (1973).
189. R. M. Dagnall, D. J. Smith, T. S. West, and S. Greenfield, *Anal. Chim. Acta* **54**, 397 (1971).
190. S. Greenfield and T. P. Sutton, *ICP Inf. Newslett.* **6**, 267 (1980).
191. K. C. Ng, M. Zerezghi, and J. A. Caruso, *Anal. Chem.* **56**, 417 (1984).
192. R. L. Dahlquist, *ICP Inf. Newslett.* **1**, 148 (1975).
193. J. L. Johnes, R. L. Dahlquist, and R. E. Hoyt, *Appl. Spectrosc.* **25**, 628 (1971).
194. R. L. Dahlquist, I. L. Jones, and K. W. Paschen, U.S. Patent 3, 602, 595, August 31, 1971.
195. K. Ohls and D. Sommer, *Fresenius Z. Anal. Chem.* **296**, 241 (1979).
196. S. El Alfy, K. Laqua, and H. Maßmann, *Fresenius Z. Anal. Chem.* **263**, 1 (1973).
197. H. G. C. Human, R. H. Scott, A. R. Oakes, and C. D. West, *Analyst* **101**, 265 (1976).
198. R. H. Scott, *Spectrochim. Acta.* **33B**, 123 (1978).
199. A. Aziz, J. A. C. Broekaert, F. Leis, and K. Laqua, *Spectrochim. Acta.* **39B**, 1091 (1984).
200. J. A. C. Broekaert, F. Leis, and K. Laqua, in B. Sansoni, ed., *Symposium on Instrumental Multi-Element-Analysis, Jülich*, Verlag Chemie GmbH, Weinheim (1985), p. 359.
201. K. Slickers, *Automatic Emission Spectroscopy*, Brühl Druck and Pressehaus, Giessen (1977).
202. K. Laqua, "Analytical Spectroscopy using Laser Atomizers," in N. Omenetto, ed., *Analytical Laser Spectroscopy*, Wiley, New York (1979), Ch. 2. p. 47.

REFERENCES

203. F. N. Abercrombie, M. D. Silvester, and G. S. Stoute, 28th Pittsburgh Conference on Analytical Chemistry and Applied Spectroscopy, Abstr., Paper No. 406, Cleveland, 1977; *ICP Inf. Newslett.* **2,** 309 (1977).
204. M. Thompson, J. E. Goulter, and F. Sieper, *Analyst* **106,** 32 (1981).
205. J. W. Carr and G. Horlick, *Spectrochim. Acta* **37B,** 1 (1982).
206. T. Ishizuka and Y. Uwamino, *Spectrochim. Acta* **38B,** 519 (1983).
207. A. G. Page, S. V. Godbole, K. H. Madraswala, M. J. Kulkarni, V. S. Mallapurkar, and B. D. Joshi, *Spectrochim. Acta* **39B,** 551 (1984).
208. R. M. Belchamber and G. Horlick, *Spectrochim. Acta.* **37B,** 71 (1982).
209. W. E. Pettit and G. Horlick, *Spectrochim. Acta* **41B,** 699 (1986).
210. V. A. Fassel and B. R. Bear, *Spectrochim. Acta*, **41B,** No. 10 (1986).

CHAPTER 7

LINE SELECTION AND SPECTRAL INTERFERENCES

P. W. J. M. BOUMANS

*Philips Research Laboratories
Eindhoven, The Netherlands*

7.1. **Prominent Lines**
7.2. **Line Selection: General**
7.3. **Types of Spectral Interference**
    7.3.1. Introduction—The ICP Background Continuum
    7.3.2. Stray Light
    7.3.3. Continua and Line Wings Contributed by the Sample
        7.3.3.1. Phenomena
        7.3.3.2. Implications of Background Due to Line Wings
    7.3.4. Spectral Lines and Molecular Bands Contributed by the Discharge Atmosphere and Water: Argon ICP
        7.3.4.1 Argon Spectrum
        7.3.4.2. Hydrogen and Oxygen Spectrum
        7.3.4.3. Carbon and Silicon Spectrum
        7.3.4.4. Molecular Bands: OH, NO, N_2^+, NH
        7.3.4.5. Absorption Bands of O_2
    7.3.5. Spectral Lines and Molecular Bands Contributed by the Discharge Atmosphere and Organic Matter: Argon ICP
    7.3.6. Spectral Lines and Molecular Bands Contributed by the Discharge Atmosphere and the Solvent: ICPs in Molecular Gases
    7.3.7. Spectral Lines and Molecular Bands Contributed by Inorganic Constituents of the Sample
7.4 **Line Selection with a View to Spectral Interferences**
    7.4.1. Direct Line Overlap and Line Coincidence
    7.4.2. Data Required for Minimizing the Risk of Errors
    7.4.3. Data Available for Checking Spectral Interferences: Classical Tabulations
    7.4.4. Data Available for Checking Spectral Interferences: Recent Approaches Specific to ICP–AES
        7.4.4.1. Line Coincidence Tables for ICP–AES Based on the Conversion of the Tables of Spectral-Line Intensities
        7.4.4.2. Measurement and Classification of ICP Lines Using a Scanning Echelle Monochromator with Crossed Dispersion

7.4.4.3. *Atlas of Spectral Information*
7.4.4.4. *ICP–AES Wavelength Tables*
7.4.4.5. *Miscellaneous Data*
7.5. **Background Correction**
7.6. **Corrections for Spectral Interferences in Polychromators**
7.7. **Spectral Interference and Resolving Power**
    7.7.1. Introduction: Analytical Performance and Resolving Power
    7.7.2. Line Profiles and Line Widths
        7.7.2.1. *Physical Profiles*
        7.7.2.2. *Instrument Function*
    7.7.3. Bandwidth Required for the Complete Physical Resolution of ICP Spectra
    7.7.4. Physical Widths of Lines Emitted from an ICP
    7.7.5. Matching of Practical Spectral Bandwidth and Physical Line Widths
    7.7.6. Resolving Power and Detection Limits
    7.7.7. Resolving Power, Selectivity, Limits of Determination, and Detection Power
        7.7.7.1. *Selectivity, Limit of Determination, and Resolving Power: Definitions and Interrelations*
        7.7.7.2. *Selectivity, Limit of Determination, and Resolving Power: An Example*
        7.7.7.3. *Selectivity and Limit of Determination: Choice of Threshold Value*
        7.7.7.4. *Limit of Detection and Limit of Determination in Dependence on the Resolving Power*
        7.7.7.5. *Detection Power Linked with the Limit of Determination*
        7.7.7.6. *Resolving Power and Analytical Performance: Dual Polychromator Arrangement*
        7.7.7.7. *Resolving Power and Analytical Performance: Slew-Scan Monochromator*
References

7.1. PROMINENT LINES

More than a century ago Hartley [1] was the first to investigate systematically the relationship between the concentration of an element and the behavior of its spectral lines using solutions that were excited in a spark. He also observed the persistence or disappearance of spectral lines when the concentration of the relevant element was gradually reduced. One of the results of these investigations was the publication of tables of "persistent lines."

De Gramont [2] made similar studies of spark spectra of fused minerals and salts, which led to a list of "raies ultimes" (R.U.) or "ultimate lines." The R.U. lines were not necessarily the most intense lines, which is understandable since it is the detection limit that governs the persistence. Lists of persistent or "most sensitive lines" for arcs and sparks appear in classical tables and in treatises on spectrochemical analysis (see Sections 1.8 and 7.4.3.).

Early investigators of ICP spectra recognized that the ranking of ICP lines in order of persistency differed from that for arcs and sparks. In particular, in argon ICPs the ionic lines were found to be far more persistent than in arc spectra [3-6], a circumstance sometimes referred to as "ionic line advantage" of argon ICPs.

Two comprehensive lists of analysis lines for the ICP, with appropriate intensity information, were published almost simultaneously at the beginning of 1979. The most extensive list, compiled by Winge, Fassel, and Peterson [7, 8], was based on scans of ICP spectra in the 189–516 nm region using logarithmic recording of ICP lines and background intensities. The lines were stated with their SBRs, the pertinent element concentrations, and estimates of the detection limits based on the assumption of a 1% background RSD for all lines. The order of prominence was associated with the SBR and thus with the detection limits (cf. Section 4.1.3). Winge et al. [7, 8] called those lines "prominent," a term which is both appropriate and convenient.[1]

Boumans and Bosveld [9] published "A tentative listing of the sensitivities and detection limits of the most sensitive ICP lines as derived from the fitting of experimental data for an argon ICP to the intensities tabulated for the National Bureau of Standards (NBS) copper arc" [10]. This list covered some 450 lines of 71 elements and was chiefly confined to the 230–680 nm region, as it was intended for spectrographic general survey analysis. The data comprised both sensitivities (i.e., intensities per unit of concentration) and estimates of detection limits.

The lists of Winge et al. and Boumans and Bosveld showed good agreement in many respects, but also discrepancies [11, 12]. The agreement between the two sets of results, the abundance of Winge's data, and the fact that the latter were entirely based on direct observations in the ICP induced Boumans [13, 14] to employ Winge's listing of wavelengths and SBRs [7] as an essential basis for converting the complete "Tables of Spectral-Line Intensities" for the NBS copper arc [10] into a table relevant to argon ICPs. This work resulted in the publication of *Line Coincidence Tables for ICP-AES* [13] covering 896 prominent lines of 67 elements. Moreover, a list of computer-predicted supplementary prominent lines was published [14].

Boumans et al. [15] showed that experimental SBRs of representative prominent lines in the 190–320-nm region measured with a 50-MHz ICP agreed well with the Winge data for a 27 MHz ICP. Subsequently Boumans [12] made an

---

[1]Prominent denotes "standing out so as to be easily seen," "standing out beyond the adjacent surface or line" [7].

assessment of some 600 prominent lines using spectrographic measurements, again for the 50-MHz ICP. Lines listed in the Line Coincidence Tables and some 90 computer predicted prominent lines were covered. The results of this assessment are reflected in the second edition of the *Line Coincidence Tables for Inductively Coupled Plasma Atomic Emission Spectrometry* [16].

Fassel's group recently published the *Atlas of Spectral of Information* [11, 17], which contains a list of 973 prominent lines (including argon lines) with estimated detection limits.

The most recent addition to the above tabulations is Wohler's *Wavelength Table for the ICP* [18], which covers the complete ICP spectra of 58 elements, excluding the rare earths with the exception of Ce. Prominent lines are marked in the main tables; for elements with line-rich spectra, they are also arranged in separate lists. These compilations will be further discussed in Section 7.4.4.

Recently, Boumans and Vrakking [107] published a list of 350 "most prominent" lines of 64 elements (at least 4 lines per element) as observed with a high-resolution spectrometer in a 50-MHz ICP. This work includes a breakdown of the detection limits using the approach described in Section 4.1.7.7, and, on this basis, a comparison of the reported detection limits with those of Winge et al. [7] and Wohlers [18]. The paper also contains a list of detection limits of the 350 "most prominent" lines normalized to a spectral bandwidth of 15 pm and a background RSD of 1%. The latter data may be considered as standards of ICP performance for 50 MHz frequency and medium resolution.

This book contains three lists of prominent lines in Section 4.1.7:

- Table 4.2 covers 67 elements and usually lists the three or four most prominent lines of each element.
- Table 4.4. covers the VUV prominent lines of As, B, I, P, S, and Sn.
- Table 4.5 covers the prominent lines of the alkalis.

Figures 4.13 and 4.14 provide an insight into the distribution of the detection capabilities of the prominent lines among the elements. Figure 4.15 shows the distributions of the prominent lines of various groups of elements over the wavelength range.

Fry and his associates [19] investigated analysis lines of nonmetals, chiefly in the infrared region.

7.2. LINE SELECTION: GENERAL

Many elements, in particular most transition elements, the rare earths, the platinum metals, and U and Th have a large number of prominent lines yielding closely spaced detection limits (cf. Figs. 4.13 and 4.14). As a consequence of the large dynamic range of the ICP, any of these lines can be used for analyses from the detection limit on to major element concentration levels. In trace analysis one will prefer the most prominent line of an element, but if this cannot be used in view of spectral interference, one will proceed to the next one and so forth. In general, it will not be too difficult to find a suitable line if the analyte has many prominent lines.

More difficulties can be encountered, however, when an element has only a few prominent lines and there is, in addition, a big gap in detection limit between the first one or two lines and the subsequent ones (cf. Figs. 4.13 and 4.14). Then an analysis at the lowest concentration level becomes impossible if the best lines suffer serious spectral interference.

On the other hand, it will hardly ever happen that a line cannot be used at the upper end of the concentration range, the only notable exceptions being Ca, Mg, and Sr. The first two prominent lines of these elements are so intense that the dynamic range of both the ICP itself and that of the detection system is exceeded at high analyte concentrations. For these elements one, therefore, prefers different prominent lines for low or high concentrations.

Generally, the sets of prominent lines discussed in Section 7.1 form the "reservoirs" from which analysis lines can be chosen.

Limitations upon the use of a particular prominent line primarily stem from (1) constraints imposed by the apparatus, and (2) spectral interferences originating from the concomitants of the sample.

Apparatus constraints, also mentioned in Section 1.4, are the spectral range of the spectroscopic instrument and—for polychromators—geometrical and mechanical constraints that limit the minimum distance between spectral lines and the maximum number of lines within a given spectral interval. The latter limitations can be reduced to some extent by the simultaneous use of various spectral orders and the application of photo-etched slits. The use of higher orders may give rise to additional spectral interferences from lines in the corresponding lower orders, unless appropriate filters are used [20, 21].

The spectral range of a spectrometer depends on the overall optical design, the blaze of the grating, the spectral response characteristic of the detector, and the atmosphere in the optical path (cf. Section 4.1.7.3).

Air-path polychromators usually operate in the 190–500-nm range, but special optics and detector assemblies may be included using zeroth order reflection from the grating to cover the alkali lines Na I 588.995, Li I 670.784, and K I 766.023 nm.

Monochromators and spectrographs provide the largest flexibility as to line choice, but access to wavelengths below 230 nm and, to a lesser extent, above 650 nm, is difficult with photographic detection.

7.3. TYPES OF SPECTRAL INTERFERENCE

7.3.1 Introduction—The ICP Background Continuum

The spectra of the sample constituents are superimposed on the ICP background spectrum which consists of a continuum, the atomic spectra of argon, hydrogen, and oxygen, and molecular spectra of various species. The continuum of the ICP fed with pure water may be considered as the natural reference level or the minimum background. Therefore, we shall treat all background contributions superimposed on this "own" ICP continuum as spectral interferences. They classify as

- Stray light
- Continua and line wings contributed by the constituents of the sample
- Spectral lines and molecular bands contributed by the discharge atmosphere and the solvent
- Spectral lines and molecular bands contributed by the constituents of the sample.

These contributions will be discussed in Sections 7.3.2 to 7.2.7. In this section we consider some characteristics of the "own" continuum. Figure 7.1 shows the absolute spectral radiance of the continuum emitted by a 27-MHz argon ICP at 1.25 kW and an observation height of 15 mm above the load coil as reported by Tracy and Myers [22]. If put on a relative scale, the trend with wavelength shown in Fig. 7.1 links up reasonably well with the results of Boumans et al. [4, 9] for a 50-MHz argon ICP. Discrepancies, in particular below 260 nm, might be attributed to the differencies in frequency and operating conditions [22].

The following features should be stressed:

- The background continuum intensity declines at an increasing rate from 400 to 200 nm (one order of magnitude over each of the intervals, 400–240 and 240–200 nm, respectively.
- The absolute radiance depends on the torch configuration, the ICP frequency, and the nebulizer assembly; it varies considerably with the ICP parameters: power, carrier gas flow, and observation height (cf. Figs. 7.2, 7.3, 4.39, 4.41, 4.51, and 4.53). However, given the system, the *relative* dependence on wavelength does not vary substantially if the operating conditions are changed (cf. Figs. 7.2, 4.39, and 4.41).

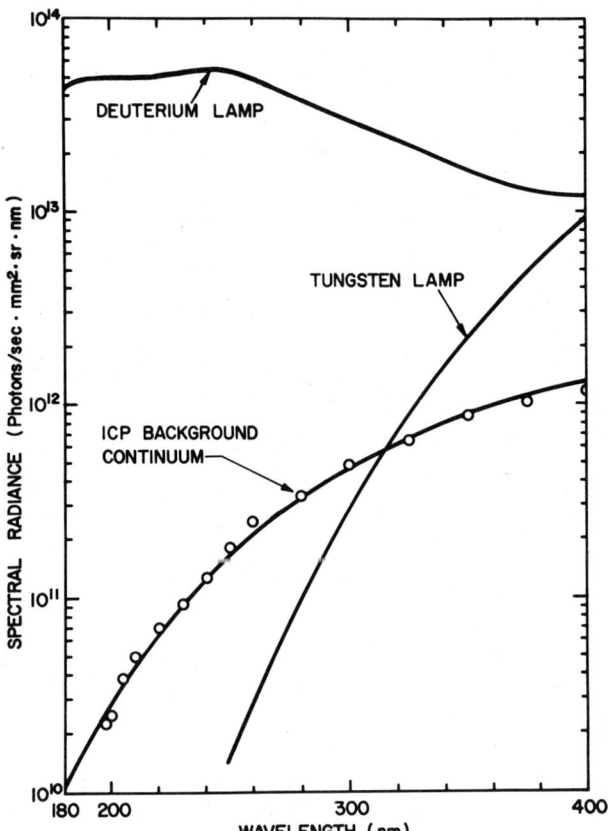

Figure 7.1. Absolute spectral radiance of background emission from argon ICP. Frequency: 27 MHz. Power: 1.25 kW. Gas flows: 12 L/min (outer), 0.5 L/min (intermediate), 0.9 L/min (carrier). Observation height: 14–16 mm above load coil. Carrier gas injected without aerosol or water vapor. Spectral radiance calibrations curves for two standard sources are shown for comparison [22]. [Reprinted with permission from D. H. Tracy and S. A. Myers, "Absolute Spectral Radiance of 27-MHz Inductively Coupled Argon Plasma Background Emission," *Spectrochim. Acta* **37B**, 1063 (1982). Copyright (1982), Pergamon Journals, Oxford.]

The literature provides hardly any data that would permit a comparison of the radiance of the background continuum of the argon ICP with those of ICPs operated with nitrogen, oxygen, or air as outer and intermediate gas. Meyer and Barnes [23] indicate that the background continuum measured at 560 nm for typical operating conditions was lowest for the air ICP (3.5 kW), three times more intense for the nitrogen ICP under identical conditions, and 50 times more intense for an argon ICP (1 kW). This statement does not permit extrapolations

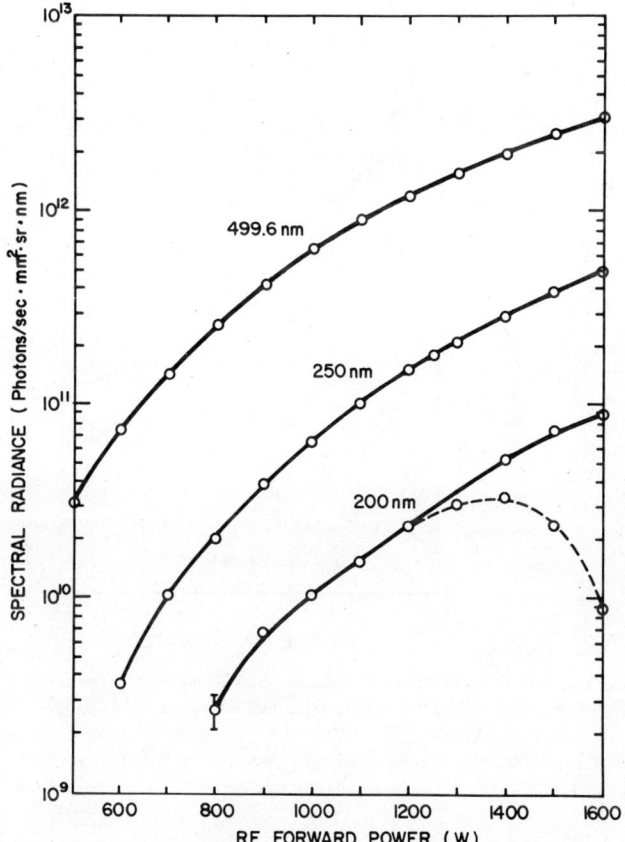

Figure 7.2. Dependence of absolute spectral radiance of ICP background on power at three wavelengths. Operating conditions as stated in caption of Fig. 7.1. The dashed line shows the effect of thermally induced absorption in the silica outer tube extension [22]. [Reprinted with permission from D. H. Tracy and S. A. Myers, "Absolute Spectral Radiance of 27 MHz Inductively Coupled Argon Plasma Background Emission," *Spectrochim. Acta* **37B,** 1064 (1982). Copyright (1982), Pergamon Journals, Oxford.

to the spectral region below 300 nm, nor does it allow whatever assessment about the consequences of the higher or lower background for analytical performance. The mere comparison of background intensities in different ICPs is not very useful for analytical assessments. Differences in background intensity should always be considered in conjunction with differences in the sensitivities of analysis lines and differences in background RSD. If one compares ICPs operated in different gaseous atmospheres, it is not only the intensity of the *continuum*, but the overall intensity of continuum plus molecular bands and the

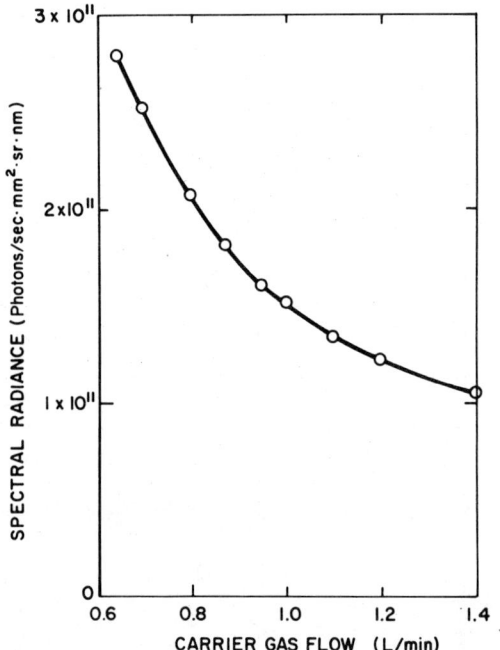

Figure 7.3. Variation of absolute spectral radiance of ICP background emission with carrier gas flow rate. The wavelength is 250 nm. Operating conditions as stated in caption of Fig. 7.1 [22]. [Reprinted with permission from D. H. Tracy and S. A. Myers, "Absolute Spectral Radiance of 27 MHz Inductively Coupled Argon Plasma Background Emission," *Spectrochim. Acta* **37B**, 1065 (1982). Copyright (1982), Pergamon Journals, Oxford.]

structure of the background that dictate the limits of detection and limits of determination. The literature provides little quantitative information about this.

7.3.2. Stray Light

Stray light is radiation at wavelengths outside of the instrumental bandpass which reaches the detector. Stray light is due to imperfections of the spectroscopic apparatus. It is a type of spectral interference both because it contributes to the background and worsens the detection limits and because the magnitude of the contribution depends on the sample composition.

In this work, stray light is discussed in Section 8.5, which also covers two important papers by Larson et al. [24] and Fassel et al. [25], who were the first to make an extensive study of stray light effects in ICP-AES. Although, since the publication of these papers, the stray light characteristics of spectroscopic instruments have greatly improved, one should remain aware that also in modern instruments the stray light problem has not been entirely eliminated and will manifest itself in particular when solutions with relatively high Ca and/or Mg contents are analyzed. Therefore, the outstanding classical papers mentioned still preserve much of their topicality.

7.3.3. Continua and Line Wings Contributed by the Sample

7.3.3.1. Phenomena

In their initial study of stray light effects Larson et al. [24] found evidence that the background enhancement between the Ca II lines at 393.366 and 396.847 nm was not only due to stray light and instrumental broadening but also to the wings of the lines themselves. The combined effects seriously interfere with the determination of low concentrations of Al in the presence of Ca using the most prominent Al lines, as is illustrated in Fig. 7.4.

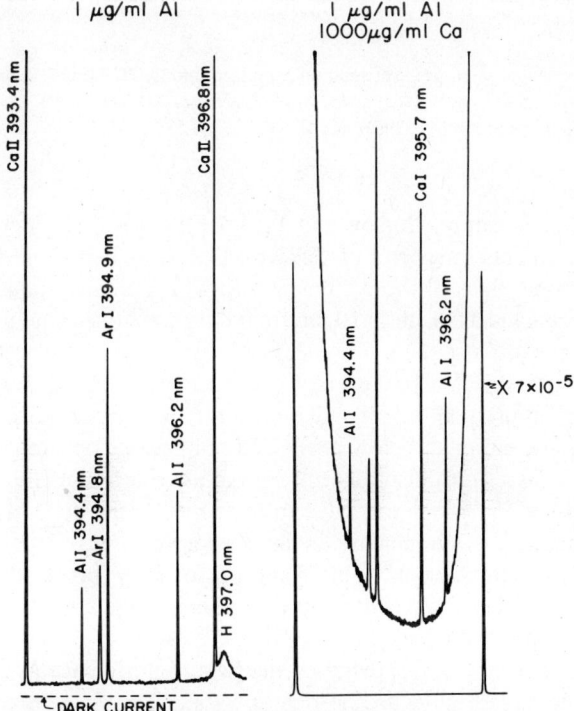

Figure 7.4. Background enhancement produced by broadening of the Ca II 393.4 and 396.8-nm lines in the plasma. The relative intensity (ordinate) scale for this figure is linear [26]. [Reprinted with permission from G. F. Larson and V. A. Fassel, *Appl. Spectrosc.* **33**, 395 (1979). Copyright (1979), Society of Applied Spectroscopy, Frederick, MD.]

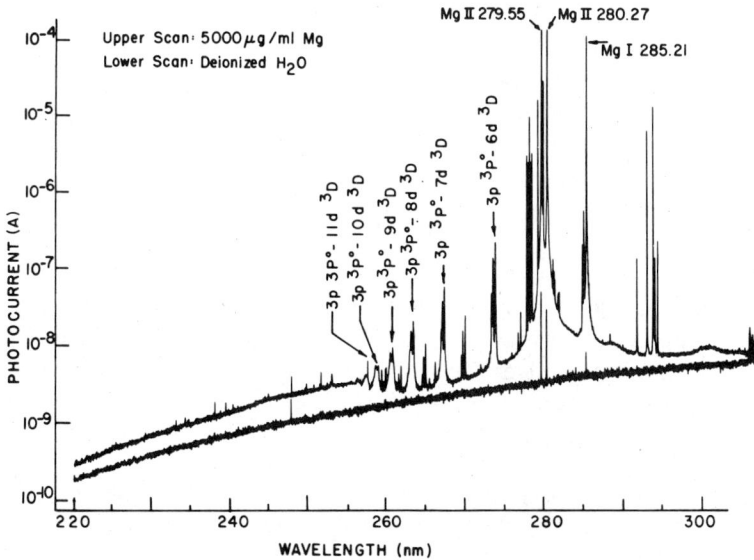

Figure 7.5. Wavelength scan for a Mg solution nebulized into the ICP [26]. [Reprinted with permission from G. F. Larson and V. A. Fassel, *Appl. Spectrosc.* **33**, 597 (1979). Copyright (1979), Society for Applied Spectroscopy, Frederick, MD.]

In a subsequent study, Larson and Fassel [26] definitely identified background enhancements in spectra of Ca, Mg, and Sr as being due to the wings of the strong ionic lines of these elements. Significant background changes at wavelengths removed as much as 10 nm from the parent line centers were found.

The spectrum of Mg (Fig. 7.5) showed not only a similar type of line broadening and associated background enhancement as found for Ca and Sr, but also exhibited unusual broadening effects for some weak lines. In addition, a recombination continuum below about 255 nm became apparent.

Larson and Fassel [26] also identified a radiative recombination continuum in the spectrum of Ca, which extended from about 302 nm upward and resulted in an approximately 50% increase in the background near 302 nm with the introduction of a solution containing 5 mg/mL of Ca. They further described line broadening effects and recombination continua in the spectrum of Al in the 193–250-nm region (Fig. 7.6).

Boumans and Vrakking [27] studied line wing interference as a major contribution to the background in line-rich spectra. They first proved that their observations of line wings—made with a high-resolution (HR) echelle monochromator—reflected truly physical characteristics of the spectra and were not produced by instrumental artifacts such as stray light. As an example Fig. 7.7 shows the behavior of the SBR in the red wing of Bi I 223.061 nm, emitted

7.3. TYPES OF SPECTRAL INTERFERENCE

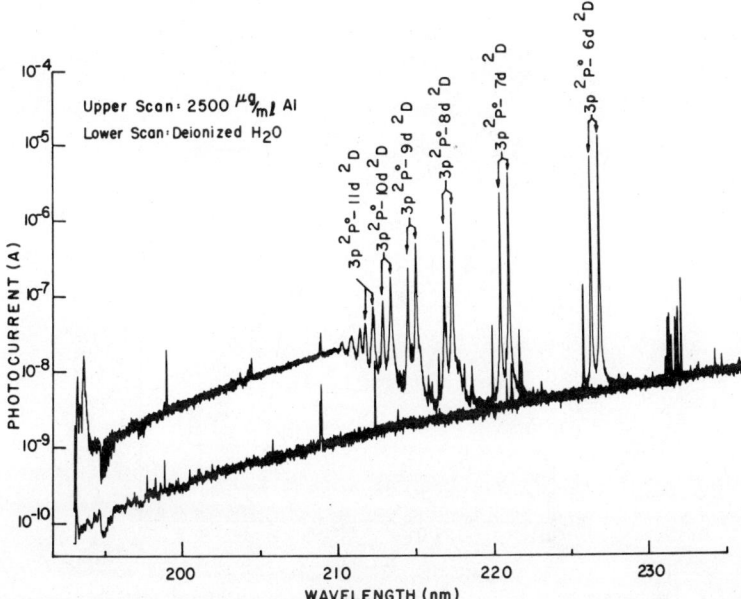

Figure 7.6. Wavelength scan for an Al solution nebulized into the ICP [26]. [Reprinted with permission from G. F. Larson and V. A. Fassel, *Appl. Spectrosc.* **33**, 599 (1979). Copyright (1979), Society for Applied Spectroscopy, Frederick, MD.]

from the ICP for a 1-mg/mL solution of Bi. For comparison the behavior of an Fe line emitted from a hollow cathode lamp is included.

Boumans and Vrakking then gave quantitative proof that the background contributed by line-rich major elements had its origin in the superposition of the wings of strong lines of these elements on the solvent background. The arguments were as follows.

1. The results of measurements of the wings of isolated lines in the spectra of Bi, Sn, Pb, Mg, and In were plotted for such concentrations of the elements that the same value of the SBR in the line peak resulted (Fig. 7.8). The curves virtually coincided with the exception of that for the In line, which was known to have an abnormally large physical width and, therefore, was not further considered.

2. The behavior of line wings in general was approximated by an average function represented by the broken curve in Fig. 7.8.

3. A mathematical expression for this curve permitted the calculation of the background at a particular wavelength (230.606 nm) in a spectrum of Ni–Co observed with a solution containing 1 mg/mL of each Ni and Co. This

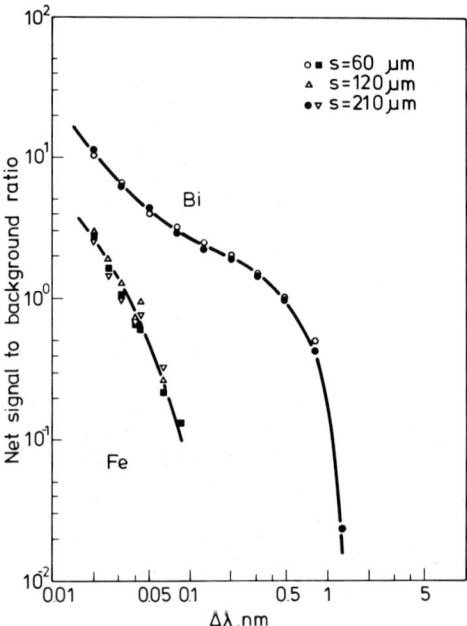

Figure 7.7. Double logarithmic plots of SBR vs. distance to peak for Bi I 223.061 nm at two slit widths as measured with an echelle monochromator with predisperser. For comparison, results for an Fe hollow cathode line have been included [27]. [Reprinted with permission from P. W. J. M. Boumans and J. J. A. M. Vrakking, *Spectrochim. Acta* **39B,** 1291 (1984). Copyright (1984), Pergamon Journals, Oxford.]

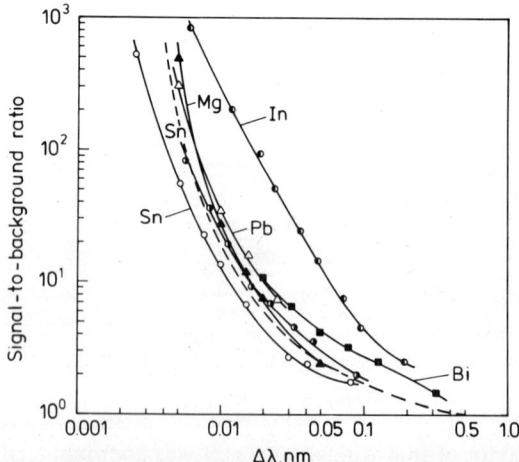

Figure 7.8. Double logarithmic plots of SBR vs. distance to peak for the ICP lines specified below. The concentrations (c) listed are for an SBR of 6000 in the line peak. ○: Sn I 235.484 nm; c = 2.5 mg/mL; ◐: Sn I 242.949 nm; c = 2.4 mg/mL; ▲: Mg II 279.553 nm; c = 0.006 mg/mL; △: Pb II 283.306 nm; c = 10 mg/mL; ■: Bi I 223.061 nm; c = 1.0 mg/mL; ◑: In I 271.026 nm; c = 35 mg/mL. [27]. [Reprinted with permission from P. W. J. M. Boumans and J. J. A. M. Vrakking, *Spectrochim. Acta* **39B,** 1291 (1984). Copyright (1984), Pergamon Journals, Oxford.]

calculation was based on the SBRs measured in the peaks of some 50 Ni or Co lines located in an interval of ± 1 nm about 230.606 nm. The background enhancement by a factor of 21.9 thus derived from the SBRs in the line peaks was identical, within the limits of experimental error, to the value found by direct measurement, that is, 22.4.

This background enhancement was attributable for 95% to the wings of 16 strong Ni and Co lines, of which six were prominent lines, the latter being responsible for 67% of the enhancement.

Boumans and Vrakking made it plausible for the results found for one wavelength in the Ni-Co spectrum to be extended to other wavelengths in this spectrum and to line-rich spectra in general; in other words, the background in line-rich spectra was made up of overlapping line wings superimposed on the solvent background. Optically the background due to line wings was found to behave as a (quasi-) continuum [28]. From the fit of the Mg ionic line in the total picture (Figure. 7.8) Boumans and Vrakking concluded that the notorious line wing interference reported in the literature [26, 29] for the ionic resonance lines of Mg, Ca, and Sr were not due to peculiarly broad wings, but to the exceptionally high intrinsic intensities of these lines.

7.3.3.2. Implications of Background Due to Line Wings

Background caused by line wings worsens the detection limits in a straightforward way via its influences on the SBRs and the background RSDs (Sections 4.1.3 and 4.1.4). The intensity of line-wing background is proportional to the concentration of the pertinent (major) element [27].

The following implications of line-wing background should be noted:

1. For a matrix yielding a line-rich spectrum the dependence of the background intensity on wavelength differs drastically from that for the pure solvent, the largest enhancements being found in regions with intense lines. Consequently, the detection limits of prominent analyte lines located in those regions will substantially differ from those in the pure solvent, even if there is no direct line overlap with lines of the matrix.

It may thus happen that relatively weak prominent analyte lines located outside the region with strong emission from the major elements are preferable to intrinsically more intense prominent lines with wavelengths in the regions with predominant wing background.

An interesting example is found in the early ICP literature, where Butler et al. [30] described the use of the relatively weak atomic lines of Cr, Mn, Nb, Ni, and W, located above 350 nm, for the determination of alloying and impurity elements in low and high alloy steels, for which the wing background in the 230–280-nm region must be appreciable. The communication of the successful use of lines above 350 nm might even have helped creating a misconception that "the presence of an iron matrix does not influence the detection limits," as stated in the abstract [30]—doubtless a statement correct in the context of that paper, but telling only part of the truth.

2. If the background contributed by line wings in the spectral region of interest exceeds that of the pure solvent by a factor of, say 5, the relative detection limits of analytes having their most prominent lines in that spectral region cannot be improved by increasing the sample concentration in the plasma by raising the solute concentration in the solution or using a nebulizer, such as USN, that produces a larger sample flow to the plasma.

As an example [27] discusses the determination of boron in iron using an HR air-path spectrometer.

3. The dependence of SBRs and detection limits on the ICP parameters (power, carrier gas flow, observation height) for samples emitting line-rich spectra may differ widely from those for pure aqueous solutions. This follows from the fact that the background is a line spectrum rather than a continuum. This point is illustrated by the example in Fig. 7.9 which shows the dependence of the net line signal (x), the background signal (x_B), the SBR, and the RSD of the background signal on ICP power for In I 230.606 nm, as observed for pure water (left) and the Ni–Co matrix already referred to (right). With pure water the net signal increases far less steeply with power than the background (Section 4.7.4); hence the SBR decreases, in this case, by a factor of 3 over the range considered. The detection limit varies less drastically because the background RSD also decreases with increasing power (Section 4.1.4). In the Ni–Co case, line and background behave closely similarly, whence the SBR varies little with power as does the background RSD, the latter because the background is high enough at all power levels to let flicker noise dictate the RSD (Section 4.1.4). Eventually, the detection limit is seen to change by a factor of 1.5 for pure water and by 15% for the Ni–Co matrix.

To conclude, although the lower power also yields the best detection limit in the Ni–Co case, the difference with respect to other power levels is marginal.

Generally the detection limits in samples emitting line-rich spectra will vary less with the ICP parameters than in samples that emit simple spectra. This leaves more freedom to optimizing performance characteristics other than detection limits, for example, precision or the magnitude of particular interferences. One should be aware, however, that for a line-rich matrix the prominent lines classify into different groups depending on whether or not they experience background enhancements from line wings. This may imply that different ICP conditions must be used for different sets of analysis lines.

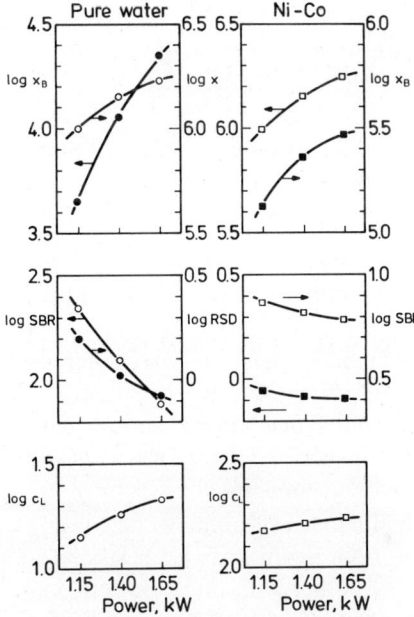

Figure 7.9. Dependence of net line (x) and background (x_B) intensities, SBR, background RSD, and detection limit (c_L) on the ICP power for pure water (left) and a Ni–Co matrix (right) containing 1 mg/mL of each Ni and Co. Spectral line: In II 230.606 nm. Note the shift of the scale for the background for Ni–Co with respect to water. The background enhancement is due to the superposition of line wings and the background behaves as a line spectrum in its dependence on the ICP power. This results in substantial differences between the curves for the SBR, the background RSD and the detection limit in the left- and right-hand frames [27]. [Reprinted with permission from P. W. J. M. Boumans and J. J. A. M. Vrakking, *Spectrochim. Acta* **39B**, 1291 (1984). Copyright (1984), Pergamon Journals, Oxford.

7.3.4 Spectral Lines and Molecular Bands Contributed by the Discharge Atmosphere and Water: Argon ICP

7.3.4.1. Argon Spectrum

The catalogs of prominent ICP lines published by Winge et al. [7, 8] contain listings of 135 and 61 argon lines, respectively, along with SBRs as measured under compromise conditions in a 27-MHz ICP.

Using a high-resolution scanning echelle monochromator, Foster et al. [20] measured the ICP background spectrum between 207.5 and 600.5 nm in a 27-MHz ICP, also under compromise operating conditions. They list 204 argon transitions with net line intensity, background, and SBR.

Wohlers [18] lists 294 argon lines between 317 and 877 nm along with relative intensities, also for a 27-MHz ICP.
The following features should be noted:

- There are no argon lines between 200 and 300 nm.
- The most intense lines are in the 420–440-nm region.
- No ionic lines are observed [31].
- Many argon lines are substantially broadened (in particularly transitions to the 4p level) as a result of Stark broadening; line widths of several 0.1 nm occur.
- The peak wavelength of argon lines may differ from the tabulated values as a result of Stark shifts [32]; the magnitude of the shifts (up to 0.004 nm) depends on the ICP operating conditions and the zone viewed; therefore, argon lines cannot be reliably used for wavelength calibration of slew-scan monochromators [33].
- A relatively small number of prominent analysis lines experiences spectral interference from argon lines [20]; second- or third-order interferences should also be borne in mind (see list in [20]). Prominent lines with serious interference from argon (or hydrogen: see Section 7.3.4.2) have been precluded from listings such as the Line Coincidence Tables for ICP–AES [16].

7.3.4.2. *Hydrogen and Oxygen Spectrum*

Hydrogen and oxygen lines originate from atomic hydrogen and oxygen formed by the dissociation of water molecules. The following transitions occur [7, 8, 18, 20]:

Line (nm)	SBR [7]
H 486.133	17.0
H 434.047	5.0
H 410.174	1.5
H 397.007	0.3
H 388.905	0.1
H 379.790	0.1

Hydrogen lines are appreciably broadened by the Stark effect (width: 0.6–1.2 nm). As an example, Fig. 7.10 shows the interference from H 397.007 nm on the prominent Ca II lines at 396.847 nm.

Wohlers [18] lists 18 atomic oxygen lines between 436 and 845 nm.

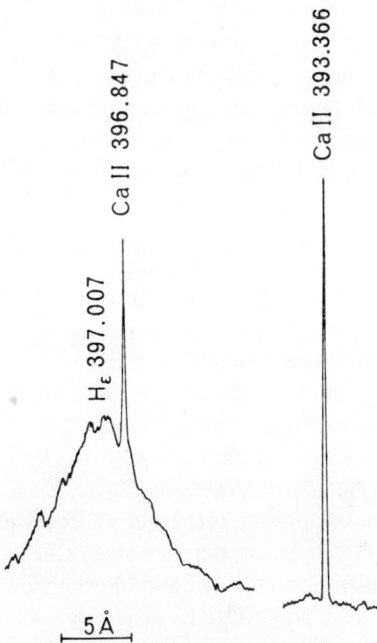

Figure 7.10. Interference from broad hydrogen line at 397.007 nm on the prominent Ca II line at 396.847 nm. The Ca II line at 393.366 nm is shown for comparison [29]. [Reprinted with permission from J. M. Mermet and C. Trassy, "A Spectrometric Study of a 40 MHz Inductively Coupled Plasma-V. Discussion of Spectral Interferences and Line Intensities," *Spectrochim. Acta* **36B**, 282 (1981). Copyright (1981), Pergamon Journals, Oxford.

7.3.4.3. Carbon and Silicon Spectrum

Usually the carbon lines C I 247.856, C I 199.362, and C I 193.091 nm will be found. The transitions result from carbon present in traces of oil in compressed argon tanks [20].

Silicon lines are commonly found as a result of erosion of silica torches, etching of glass spray chambers by hydrofluoric acid, and/or solvent blanks. Their elimination requires silicon-free materials for the carrier gas tube, nebulizer, and spray chamber and high-purity solvents.

7.3.4.4. Molecular Bands: OH, NO, N_2^+, NH

Hydroxyl (OH) radicals result from the dissociation of water and emit rotational spectra in the 281–295 and 306–325 nm wavelength regions. A detailed tabulation of 116 plus 180 major spectral features in the two regions mentioned has been published by Dieke and Crosswhite [34].

Boumans and Vrakking [35] made an analysis of the extent to which prominent ICP lines actually experience interference from OH band components at either high (HR) or medium (MR) spectral resolution (5- and 15-pm bandwidth respectively). This analysis is of general interest since it defines a rational criterion for judging the seriousness of interference from whatever band component or spectral line.

Three factors must be considered: the peak SBR (or intensity) of the interfering line, its effective width ($\triangle \lambda_{\text{eff}}$), and its distance ($\triangle \lambda$) to the analysis line. If $\triangle \lambda / \triangle \lambda_{\text{eff}} \leq 2$, line overlap occurs (cf. Section 7.4.1). Whether or not this implies serious interference should be judged from the limit of determination, c_D, rather than the limit of detection, c_L (Section 7.7). A rational criterion for an assessment is the ratio c_D/c_L, which equals 5 in the absence of line overlap. Boumans and Vrakking [35] defined $c_D/c_L \gtrsim 10$ as a reasonable threshold for designating an interference as "serious."

Over 100 prominent lines in the OH band regions were examined. Seven of them experienced very serious interference from OH band components at HR and/or MR ($c_D/c_L \gtrsim 30$). Serious OH band interference was established for 17 lines ($30 \gtrsim c_D/c_L \gtrsim 10$), while OH band interference was found to be nonexistent or negligible for 83 prominent lines. This last group comprised some 40 lines marked in the literature [7, 20] as "potentially experiencing OH band interference."

For illustration, Fig. 7.11 shows spectral scans for V II 310.230 nm, as reported by Mermet and Trassy [29], for a medium resolution monochromator with a bandwidth of about 16 pm. Fig. 7.12 contrasts spectral scans for the same line obtained with a high-resolution monochromator at spectral bandwidths of 4.6 and 13.7 pm, respectively. Further examples of OH band interference are shown in Figs. 7.41 and 7.43.

Band intensities are appreciably affected by the outer gas flow: the higher the flow the more effectively the plasma is shielded from the entrainment of air [11]. This is clearly illustrated in Fig. 7.13 for NO, OH, and N_2^+ bands.

For a low-flow, low-power ICP, Rezaaiyan and Hieftje [37] reported substantially increased OH band intensities as compared to a conventional ICP. The intensity of the continuum was somewhat lower, however, in relation to the intensity of an analyte line (Ca II 393.3 nm).

The NO bands (γ system) in principle extend from about 195 to 300 nm, and the N_2^+ bands (first negative system) from about 329 to 590 nm. Both systems are degraded to shorter wavelengths (Fig. 7.13). Not shown in Fig. 7.13, but always observable, are the rather strong and diffuse band heads of NH at 336.0 and 337.0 nm.

7.3. TYPES OF SPECTRAL INTERFERENCE

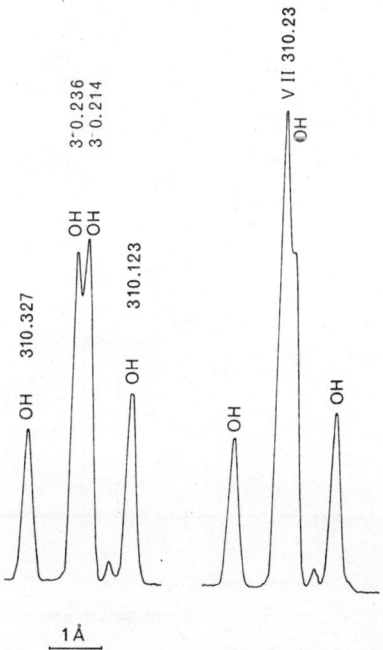

Figure 7.11. Wavelength scans illustrating OH band interference on V II 310.230 nm. Left: pure water. Right: 0.1 µg/mL V solution. Spectral bandwidth: 16 pm [29]. [Reprinted with permission from J. M. Mermet and C. Trassy, "A Spectrometric Study of a 40 MHz Inductively Coupled Plasma-V. Discussion of Spectral Interferences and Line Intensities," *Spectrochim. Acta* **36B**, 273 (1981). Copyright (1981), Pergamon Journals, Oxford.]

Bands of molecules containing nitrogen can be eliminated by the use of extended torches as specifically applied to the determination of nitrogen using atomic nitrogen lines in the VUV (cf. Sections 4.1.7.3 and 5.3).

Wavelengths of band heads and references to detailed analyses of band spectra can be found in Pearse and Gaydon's tables [38] which include spectrograms of molecular spectra.

7.3.4.5. Absorption Bands of O_2

If an air-path spectrometer is used, O_2 absorption bands produce a structured background below 200 nm as shown in Fig. 7.14. This absorption can be eliminated by using a vacuum system or purging the optical path with nitrogen or argon.

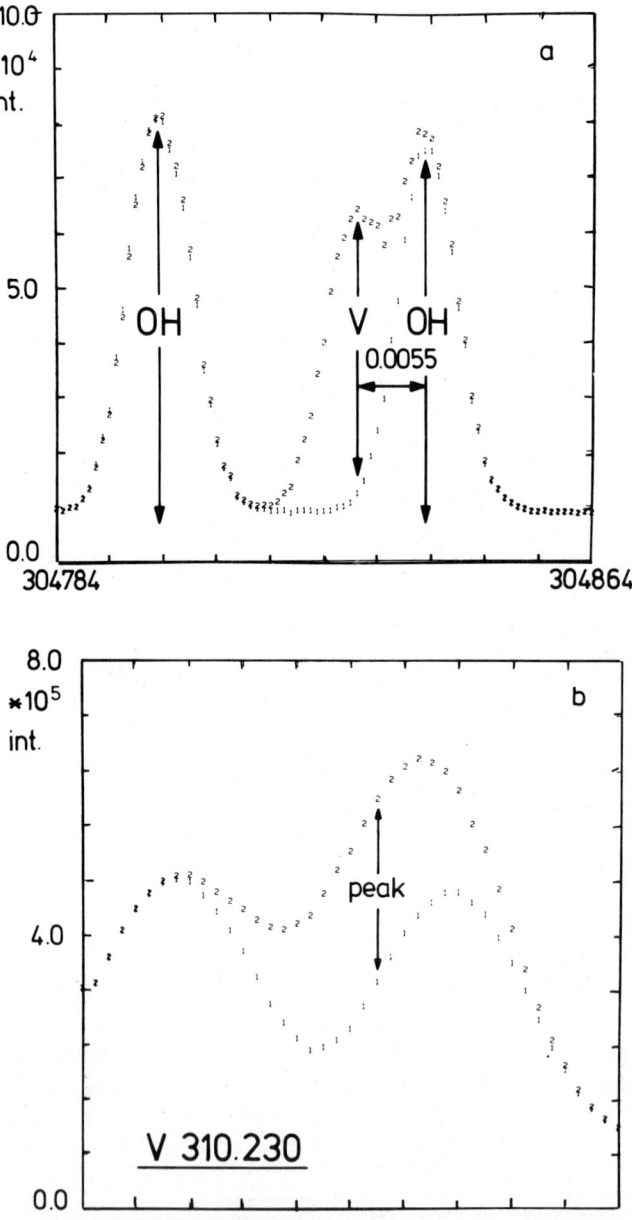

Figure 7.12. Spectral scans over a 0.0444-nm spectral window about V II 310.230 nm using a high-resolution echelle monochromator at two slit widths, s, 60 and 210 μm, respectively. Lower curves: pure water. Upper curves: vanadium concentration 70 ng/mL. (*a*) Bandwidth: 13.7 pm; $c_D/c_L \approx 90$. (*b*) Bandwidth: 4.6 pm; $c_D/c_L \approx 30$ [35, 36]. [Reprinted with permission from P. W. J. M. Boumans and J. J. A. M. Vrakking, *Spectrochim. Acta* **40B,** 1107, 1423 (1985). Copyright (1985), Pergamon Journals, Oxford.]

7.3. TYPES OF SPECTRAL INTERFERENCE

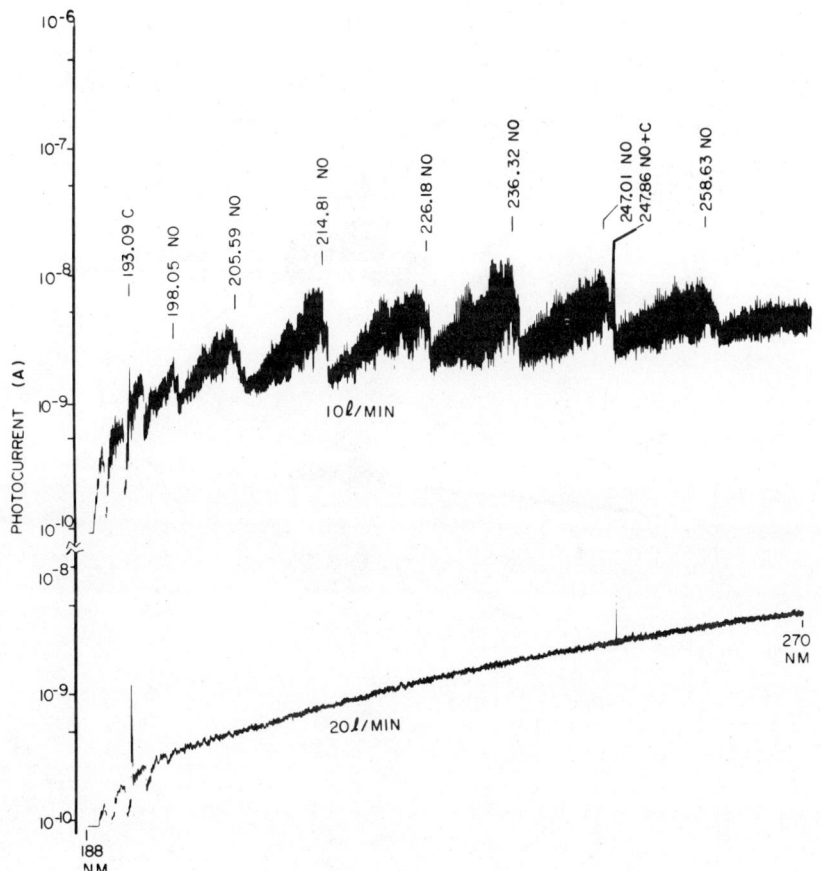

Figure 7.13. Background spectrum of the ICP under conditions favoring molecular band emission (upper spectrum) and conditions minimizing band emission (lower spectrum). The flow rates listed below each spectrum refer to the outer gas flow; other conditions were identical [11]. [Reprinted with permission from R. K. Winge, V. A. Fassel, V. J. Peterson, and M. A. Floyd, *Appl. Spectrosc.* **36**, 218 (1982). Copyright (1982), Society for Applied Spectroscopy, Frederick, MD.]

7.3.5. Spectral Lines and Molecular Bands Contributed by the Discharge Atmosphere and Organic Matter: Argon ICP

When an aqueous solution containing organic matter or an organic solvent is introduced into the ICP the C I lines at 247.856, 199.362, 193.091, and 258.288 nm will show with high intensities. In addition, the following bands are observed [39] (see Fig. 7.15):

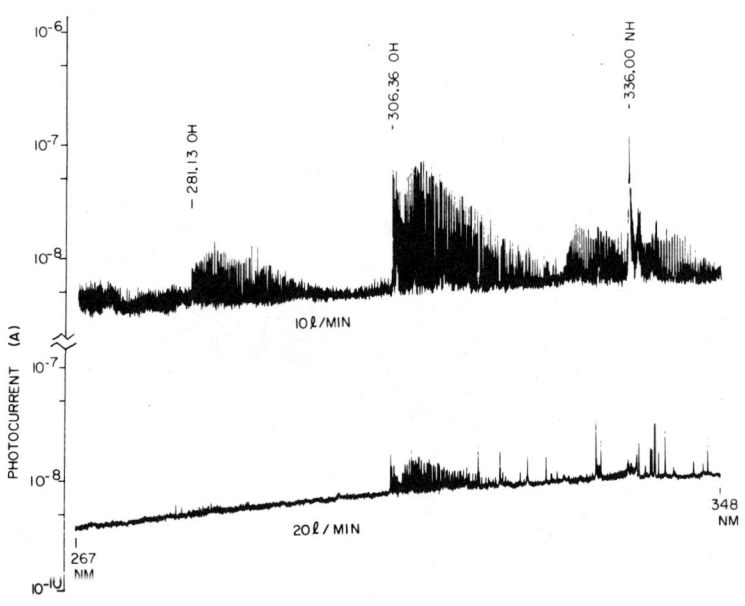
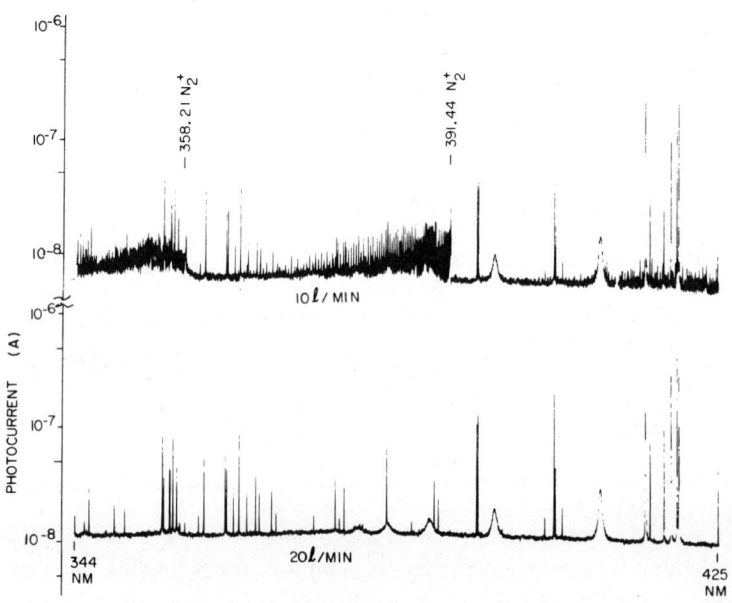

Figure 7.13. (*Continued*)

7.3. TYPES OF SPECTRAL INTERFERENCE

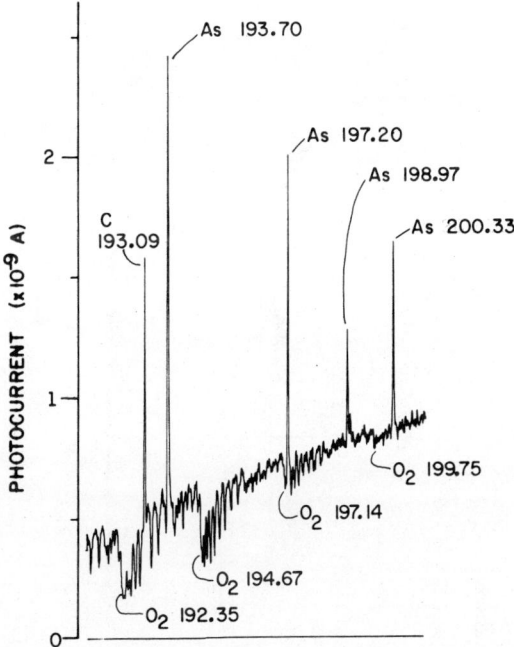

Figure 7.14. A portion of the background spectrum of the ICP showing the absorption bands arising from the presence of molecular oxygen in the optical path. The arsenic lines obtained from 5 μg/mL of arsenic are included for comparison [11]. [Reprinted with permission from R. K. Winge, V. A. Fassel, V. J. Peterson, and M. A. Floyd, *Appl. Spectrosc.* **36**, 218 (1982). Copyright (1982), Society for Applied Spectroscopy, Frederick, MD.]

- The C_2 Mullikan's system with an intensity maximum at about 232.5 nm
- The CN violet system between 358.4 and 460.6 nm with strong band heads at 421.6, 388.3, and 359.0 nm
- The CN violet system tail bands extending from 311.4 to 425.6 nm
- The C_2 Swan system between 436.5 and 667.7 nm
- The CO fourth positive system: 190–230 nm

The intensities of the various bands vary with the ICP operating conditions including observation height. This is one of the reasons why ICP optimization is more complex than for aqueous solutions with inorganic matter only (cf. Sections 4.7.4–4.7.7). Scans such as reproduced in Fig. 7.15 give an excellent overall impression of the wavelength regions where molecular bands dominate the spectrum and where interferences can be expected. However, a precise analysis for any prominent line individually is required to assess the extent of

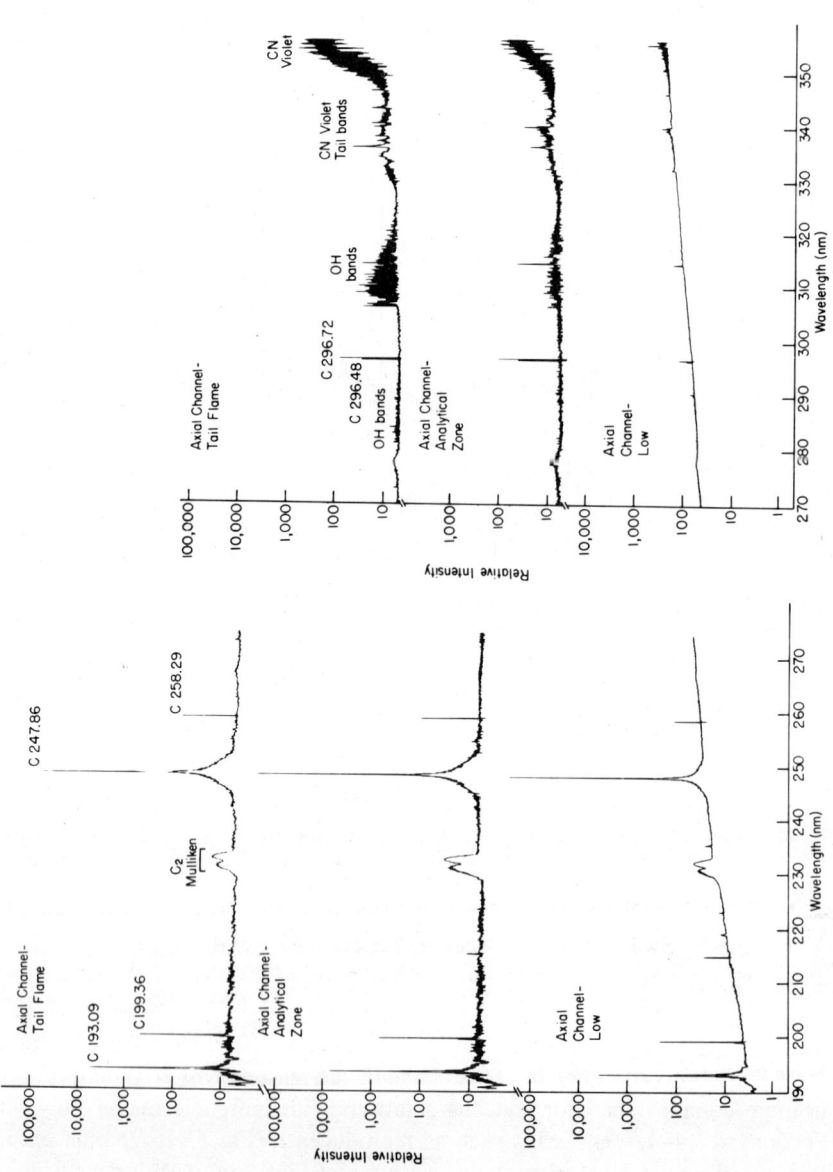

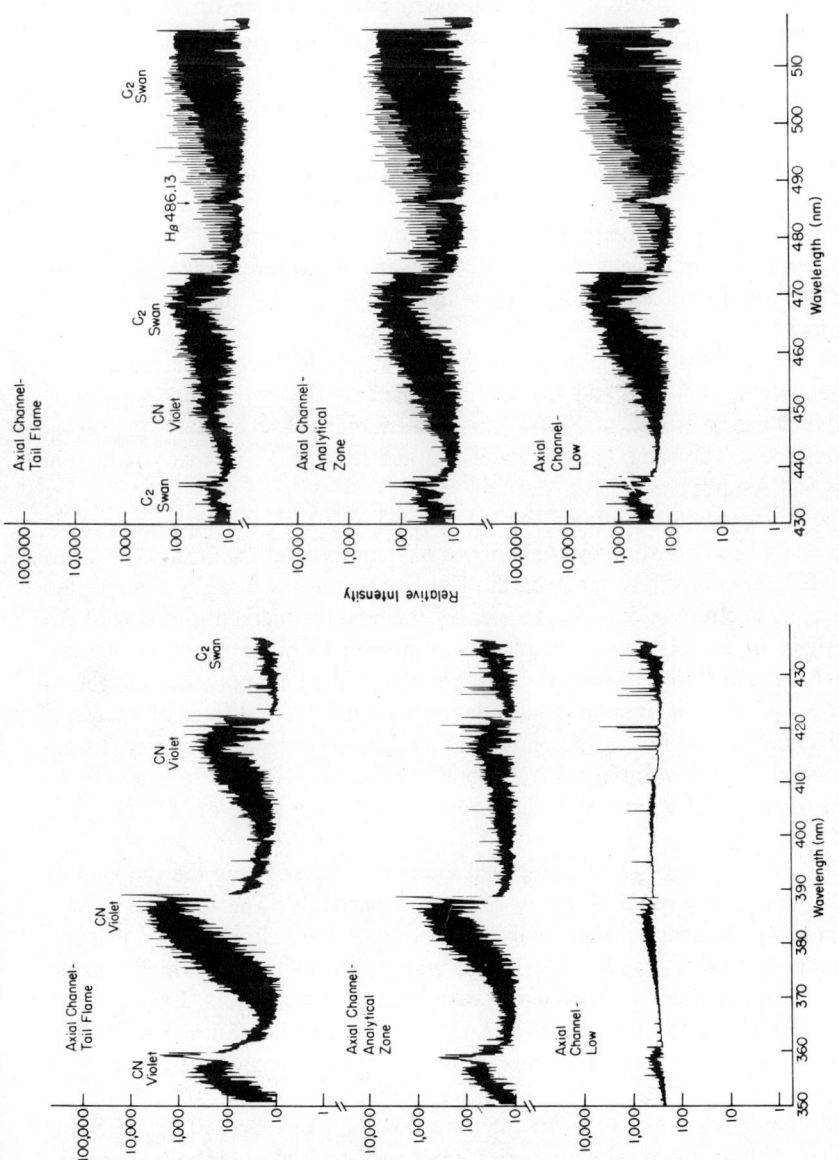

Figure 7.15. Wavelength scans from 190 to 520 nm for three observation heights in the center of the axial channel of an argon ICP while a solution of oil and MIBK (1:10 m/v) was being nebulized [39]. [Reprinted with permission from C. A. Peterson.]

the interferences concretely, as has been discussed for OH band interference in Section 7.3.4.3.

The C_2 and CN bands can be suppressed by combustion of the molecules with the aid of oxygen added to the carrier gas (see Section 4.7.7).

7.3.6. Spectral Lines and Molecular Bands Contributed by the Discharge Atmosphere and the Solvent: ICPs in Molecular Gases

Greenfield at al. [40] photographically recorded spectra from ICPs with argon as carrier and intermediate gas and air, nitrogen, or oxygen as outer gas, while methanol was nebulized. Some of these microphotometer scans are reproduced in Barnes' review [41].

The spectra illustrate the progressive removal of the C_2 Swan bands at 593–620, 547–564, 509–527, and 467–474 nm when the power of a nitrogen–argon ICP is increased from 2 to 7 kW. This is attributable to the increasing dissociation of C_2. The same happens with CO and leads to an intensity reduction of the CO Asundi bands (712–721 nm).

The CN bands are beginning to be reduced at 7 kW, but their removal is more easily accomplished by changing to oxygen as outer gas instead of nitrogen. Oxygen completely removes the CN bands but has a fairly complicated molecular spectrum of its own. In view of this and the marked lowering of the sensitivity of an oxygen–argon plasma compared to nitrogen–argon plasma, Greenfield et al. [40] indicate that the choice of using high power or oxygen to remove the bands of organic species depends on the spectral lines of interest.

Greenfield et al. [40] also contrast the spectra of argon–argon, oxygen–argon, and nitrogen–argon at 5 kW for both methanol and water spiked with 50 μg/ml of each Cr, V, and Fe. The principal bands are those of N_2^+, N_2, O_2^+, and O_2.

Ohls and Sommer [42–44] observed the effects of replacing the nitrogen of a nitrogen–argon plasma by air or oxygen respectively. The main result was that in particular spectral regions the nitrogen bands are eliminated so that important analysis lines can be observed on a smooth background, which leads to improved limits of determination and also better detection limits. Examples are Zr II 343.823, Tl 351.924, La II 379.476, and Cr I 425.430 nm. For illustration, Fig. 7.16 shows spectral scans for the Ce and Cr lines with either oxygen or nitrogen as outer gas (refer to [106] for further comments).

Meyer and Thompson [45] presented wavelength scans for a 40.68 MHz ICP operated with air or oxygen as carrier, intermediate, and outer gas. Scans are shown for wavelengths between 200 and 725 nm. Some of the band systems were identified.

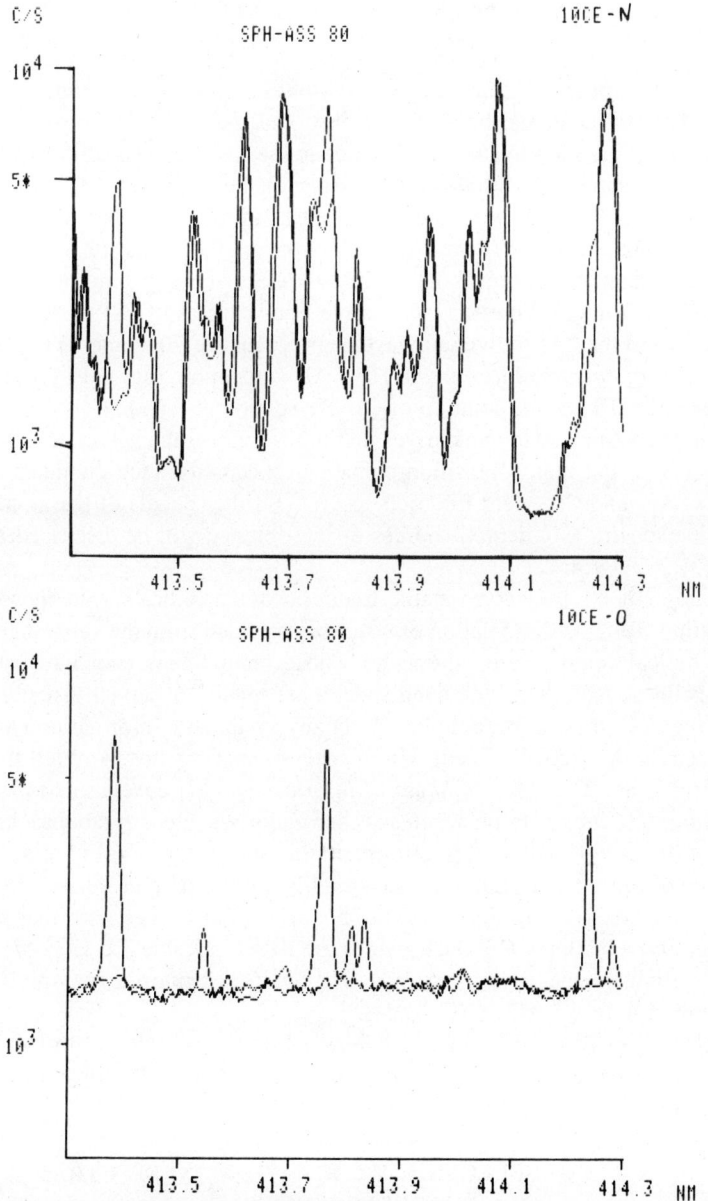

Figure 7.16. Wavelength scans in the vicinity of Ce II 413.765 nm illustrating the beneficial effect of replacing nitrogen (upper frame) by oxygen (lower frame) as outer gas [44]. [Reprinted with permission from *ICP Information Newsletter* **9,** 561 (1984). Copyright (1984), R. M. Barnes, Amherst, MA.]

7.3.7. Spectral Lines and Molecular Bands Contributed by Inorganic Constituents of the Sample

The elements present in the sample contribute atomic and/or ionic lines and sometimes also molecular bands to the observed spectrum. The types and concentrations of the sample constituents dictate the line-richness of the spectrum.

Table 7.1 shows the number of lines per element tabulated for the copper arc in the NBS Tables [10], for arcs and sparks in the MIT Wavelength Tables [46], and for the argon ICP in Wohlers' Wavelength Tables for the ICP [18]. The data give an impression about the line-richness of the spectra for the various elements. A further picture of the relative line-richness of ICP spectra can be obtained from the 232 individual wavelength scans for 70 elements reproduced in the *Atlas of Spectral Information* [17]. As an example Fig. 7.17 presents the spectrum of a 10 μg/mL solution of Cr (cf. Section 7.4.4.3).

From the various data one derives that the actinides, the lantanides, the platinum metals, and many transition elements (except for such elements as Zn, Cd, Hg, Ag, and Cu) yield line-rich spectra. Further details of the atomic and ionic line spectra as potential sources of interference will be discussed in Sections 7.4.3 and 7.4.4.

Some elements may form stable diatomic oxide radicals with the oxygen originating from the dissociation of water or entrained from the surrounding air. These radicals emit rotational spectra whose components may interfere with analysis lines. Although such band spectra are readily observed visually in the tail of the ICP, there appears to be hardly any systematic information about the interferences that actually occur with respect to analysis lines emitted from the analytical zone. The *Atlas of Spectral Information* [17] covers this type of information only incidentally, whereas other major wavelength tabulations (Section 7.4.4) do not include this information at all. In the ICP one may expect emission of those bands that have been readily observed in the dc arc when salts of the elements are introduced [38]. This applies to oxides with dissociation energies above about 4 eV. Such oxides are listed in Table 7.2 [38, 47].

For various elements (e.g., Al, Ca, and Sr) excess fluoride may give rise to band emission (e.g., AlF, CaF, and SrF).

Wavelengths of band heads and additional data can be found in Pearse and Gaydon's tables [38]. Another useful tabulation, specifically for flames, is included in the book of Alkemade and Herrmann [48].

7.4. LINE SELECTION WITH A VIEW TO SPECTRAL INTERFERENCES

7.4.1. Direct Line Overlap and Line Coincidence

To discuss spectral interferences from lines, we first define the terms "direct line overlap" and "line coincidence." Figure 7.18 shows a typical profile of a

Table 7.1. Numbers of Spectral Lines Listed for Arcs and Sparks in the *MIT Wavelength Tables* [46], for the NBS Copper Arc in the Tables of Spectral-Line Intensities [10], and for the ICP in Wohlers' Wavelength Table for ICP–AES [18][a]

Element	MIT	NBS	ICP	Element	MIT	NBS	ICP
Ag	334	15	17	Na	145	12	6
Al	84	16	39	Nb	3334	1443	711
As	263	25	24	Nd	2699	1471	
Au	345	27	37	Ni	1150	248	245
B	128	4	8	Os	1752	1036	335
Ba	424	96	43	P	45	9	11
Be	69	20	34	Pb	255	39	29
Bi	282	34	30	Pd	776	80	85
C	288	1	2	Pr	2702	1464	
Ca	727	79	79	Pt	813	177	113
Cd	256	24	31	Rb	117	13	4
Ce	5779	2654	2532	Re	2244	962	209
Co	2187	676	477	Rh	1314	512	224
Cr	2287	802	792	Ru	2720	1000	305
Cs	70	19	4	Sb	391	53	50
Cu	911	56	90	Sc	641	397	
Dy	2077	1378		Se	5	5	4
Er	1998	1211		Si	317	15	27
Eu	2407	525		Sm	3880	1562	
Fe	4741	764	461	Sn	220	59	41
Ga	97	15	18	Sr	262	76	24
Gd	1603	1431		Ta	2175	1306	659
Ge	71	43	56	Tb	2604	1657	
Hf	1507	770	582	Te	30	12	11
Hg	170	19	10	Th	2602	2231	1529
Ho	787	966		Ti	2259	964	456
In	276	22	19	Tl	133	18	11
Ir	2561	534	411	Tm	798	881	
K	55	8	3	U	5266	1966	1458
La	1380	642		V	3072	1062	648
Li	33	7	10	W	4388	1301	580
Lu	448	225		Y	798	454	
Mg	186	25	28	Yb	1256	434	
Mn	1383	353	267	Zn	127	20	25
Mo	3941	1577	655	Zr	2249	1062	575

[a] The numbers for the MIT and NBS Tables have been derived from the data stored in encoded form in the IBM computer at Philips Research Laboratories in Eindhoven [106]. Lines listed in the MIT Tables for "discharge" only have been precluded from the present tabulation.

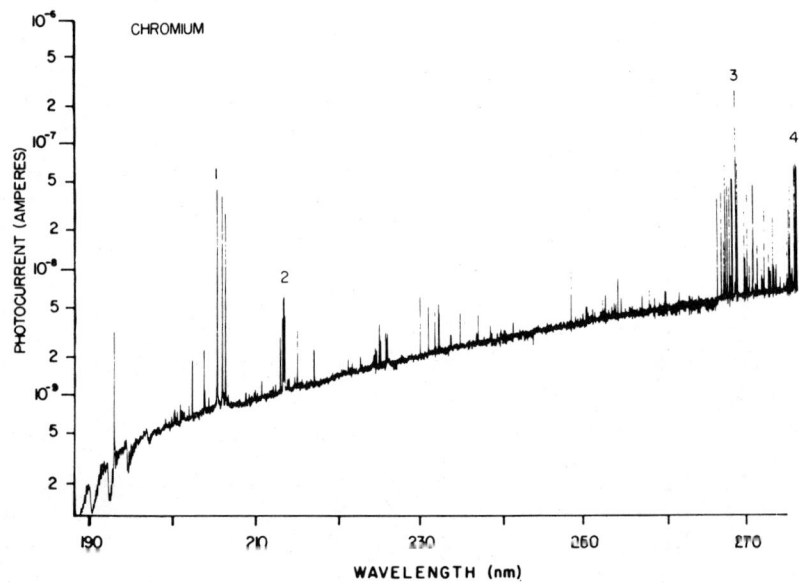

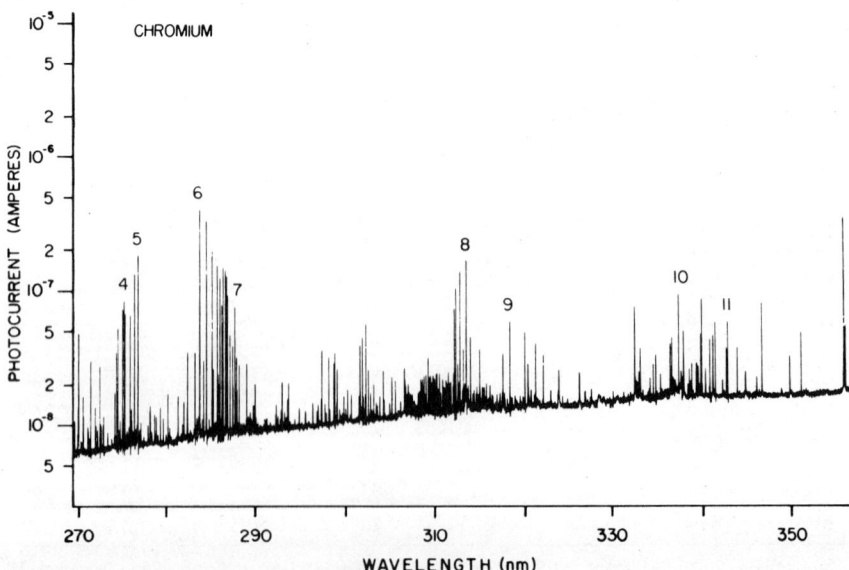

Figure 7.17. Emission spectrum of chromium, representative of the wavelength scans of 70 elements included in the *Atlas of Spectral Information* [11], [17]. [Reprinted with permission from R. K. Winge, V. A. Fassel, V. J. Peterson, and M. A. Floyd, *Appl. Spectrosc.* **36,** 212 (1982). Copyright (1982), Society for Applied Spectroscopy, Frederick, MD.]

7.4. LINE SELECTION WITH A VIEW TO SPECTRAL INTERFERENCES 389

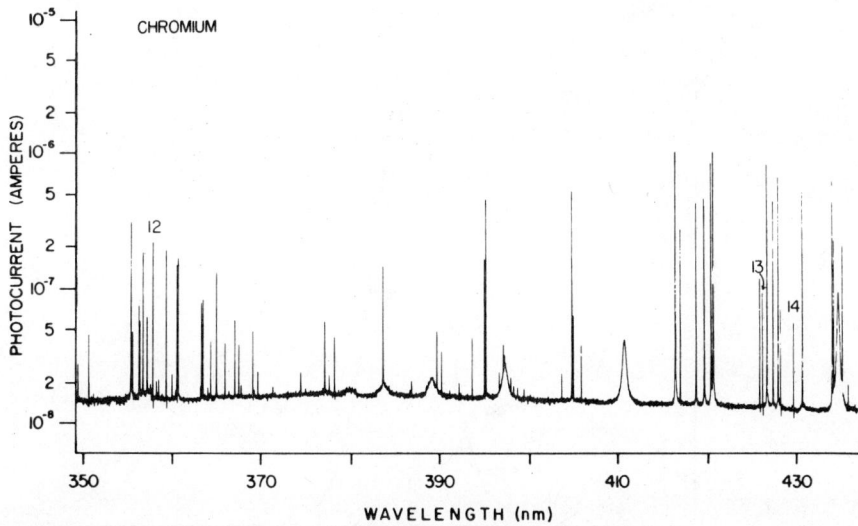

Figure 7.17. (*Continued*)

line emitted from an ICP. Such a profile is an intensity distribution as obtained by scanning a small wavelength interval about the line center. The profile is characterized by the central wavelength (λ_0), the peak intensity (I_p), the overall shape, and the effective width ($\delta\lambda = \Delta\lambda_{\text{eff}}$), the latter being expressed in terms of the full width at half maximum (FWHM). A comprehensive treatment of the complex theory of the shapes and widths of spectral lines is beyond the scope of this book; however, a few concepts and notions will be explained in Section 7.7.2. Generally the profiles found in ICP-AES are of a shape such that the bulk of the line is located within $\pm\delta\lambda$ from the line peak (Fig. 7.18). The two

Table 7.2. Oxide Radicals that Readily Emit Band Spectra from the DC Arc [38]. The Table Includes the Dissociation Energies (D) [47].

Radical	D (eV)	Radical	D (eV)	Radical	D (eV)	Radical	D (eV)	Radical	D (eV)
AlO	5.0	CeO	—	LuO	8.2	SO	5.4	VO	6.4
AsO	5.0	CrO	5.3	MgO	3.9	SbO	3.8	WO	7.2
BO	8.3	CuO	4.9	MnO	4.4	ScO	6.0	YO	9
BaO	5.0	FeO	4.3	NiO	—	SiO	8.0	ZrO	7.8
BeO	4.6	GeO	6.8	PO	6.2	SnO	4.1		
BiO	2.9	HfO	—	PbO	4.3	TaO	—		
CaO	3.9	LaO	8.2	PrO	—	TiO	6.8		

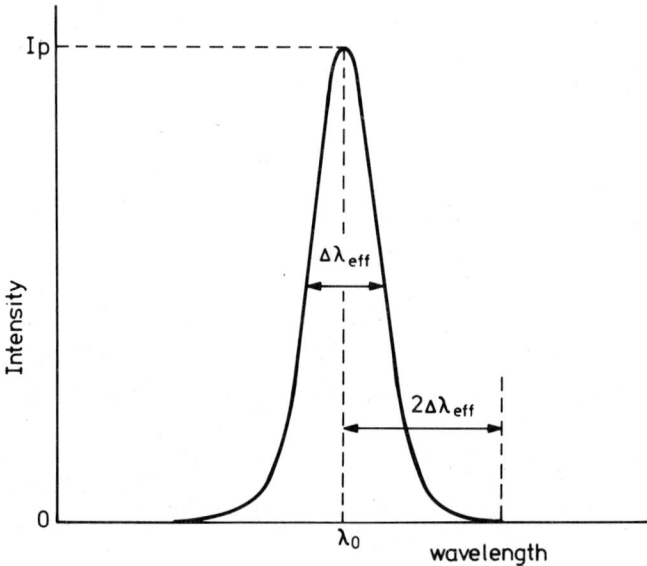

Figure 7.18. Typical profile of a spectral line emitted by an ICP. The figure shows the distribution of the net line intensity about the central wavelength λ_0 with peak intensity I_p. The quantity $\triangle\lambda_{eff}$ is the effective line width and is expressed as FWHM. For many lines the intensity has sunk virtually to zero background when the distance to the peak exceeds $2\triangle\lambda_{eff}$.

parts of the profile lying in this range are conveniently designated "shoulders," which merge into the "wings" at the points $\Delta\lambda \approx \pm 2\delta\lambda$.

A wavelength distance $\Delta\lambda = 2\delta\lambda$ between the peaks of two lines then marks the transition between line wing interference and (direct) line overlap. Two lines located at this distance are completely resolved, whatever their relative intensities (Fig. 7.19a). In the example, the analysis line hardly experiences interference from the neighboring line.

When the two lines approach each other, full resolution is maintained up to $\Delta\lambda \approx 1.5\delta\lambda$ at all three relative intensities considered (Fig. 7.19b). At more extreme intensity ratios and, in particular, if $\Delta\lambda \approx \delta\lambda$, one gets situations as shown in Fig. 7.19c, where the lines are seen to merge into each other. This may be referred to as "line coincidence," not so much because of the lack of resolution, but since from $\Delta\lambda \approx \delta\lambda$ on the peak of the analysis line begins to experience the influence of the shoulder of the interfering line. Obviously this effect becomes more serious the smaller the relative intensity of the analysis line, thus, in Fig. 7.19c from C1 to C3.

At distances $\Delta\lambda < \delta\lambda$ the interference effect from the shoulder of the interfering line grows ever worse (Fig. 7.19d), until eventually the lines completely coincide (Fig. 7.19e).

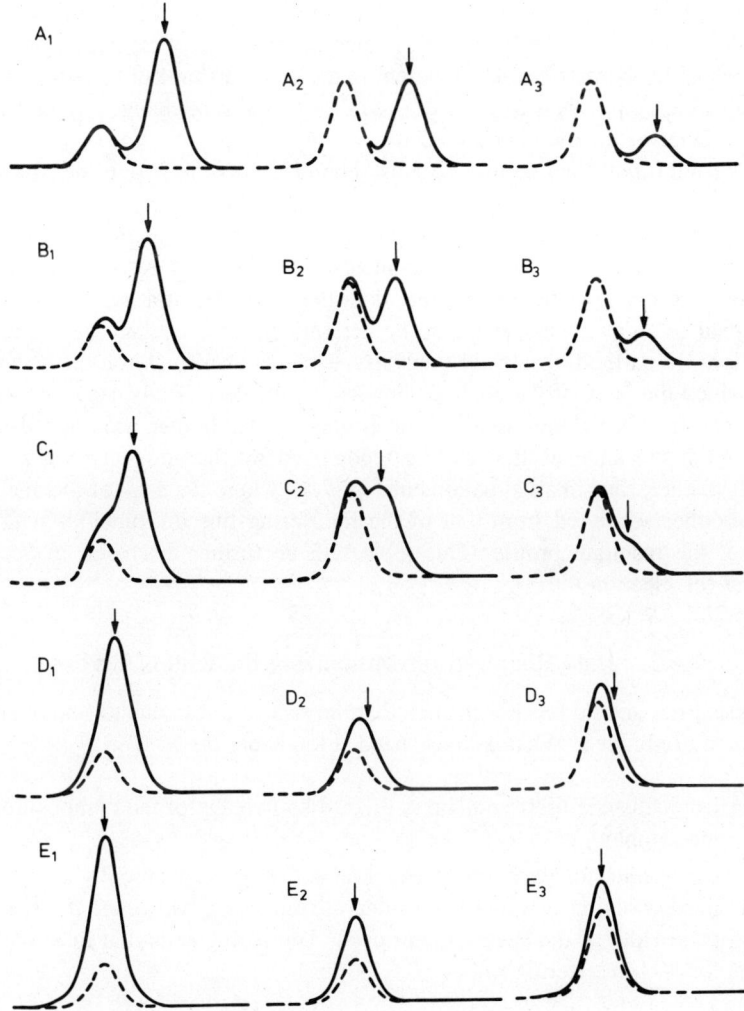

Figure 7.19. Superposition of an analysis line (peak position marked by arrow) on background with an interfering line. The broken curve represents the profile of the interfering line, the continuous curve the resultant profile. Frames *A* to *E* are for wavelength distances of 2.0, 1.5, 1.0, 0.5, and 0 FWHMs. Indexes *1*, *2*, and *3* refer to the ratio of the net peak intensity of analysis line to interfering line: 3, 1, and 1/3 respectively. The profiles shown are the results of computer manipulation of experimental profiles (cf. Section 7.4.4.1).

In summary, Fig. 7.19 defines all situations from wing interference via partial (direct) line overlap to (complete) coincidence. Clearly the transitions cannot be defined sharply, but it is inherent in the general shape of lines that in the vicinity of $\Delta\lambda/\delta\lambda = 2, 1.5, 1,$ or 0.5 more or less abrupt changes in the extent and character of the overlapping occur.

More important than defining precise boundaries is the notion that the seriousness of line interference is closely connected with (1) the wavelength distance expressed in units of halfwidth, and (2) the relative peak intensities of the two lines involved. As such it is immaterial whether the resultant profile is spectrally resolved or not. However, if in the case of partial line overlap the net signal of an analysis line has to be determined, the accuracy of this determination—in particular, in trace analysis—depends principally on the accuracy with which the peak of the analysis line can be located. If this has to be done, for example, with a slew-scan monochromator, far higher accuracy can be achieved if the analysis line can be recognized by the computer via a peak search routine, thus in all situations in Fig. 7.19 where the peak of the analysis line is either separated from that of the interfering line or coincides with the peak of the resultant profile. This topic will be further discussed in Section 7.7.7.7 (cf. Section 8.9.3).

7.4.2. Data Required for Minimizing the Risk of Errors

Analysis lines should be chosen such that the risk of errors due to line overlappings are minimized. What data are needed to ensure this?

1. Ironically the first requisite is precise knowledge of the composition of the sample.
2. One needs for each prominent line a listing of potentially interfering lines covering a wavelength interval of at least two times the spectral bandwidth of the spectrometer used. The listing should at least contain for each interfering line
    a. The element symbol and wavelength.
    b. The peak sensitivity under the conditions prevailing in the source.
    c. The physical width of the line.
3. One should know the spectral bandwidth of the spectrometer for all wavelengths of interest and the convolution of this bandwidth and the physical widths of the spectral lines.

Ideally, data set (2) as well as all data for the prominent lines should be available in the memory of a computer so that upon entering data sets (1) and (3) the computer program can provide an answer, if necessary, iteratively (cf. [106]).

Various approaches have been made to bring this ideal within reach. However, the amount of information required is so overwhelming that in spite of the efforts an entirely satisfactory procedure has not yet become available. Therefore, the results of the various efforts must still be used supplementarily when selecting analysis lines.

7.4.3. Data Available for Checking Spectral Interferences: Classical Tabulations

Classical emission spectroscopy uses various important listings of spectral lines. The most extensive tables are

1. *Massachusetts Institute of Technology (MIT) Wavelength Tables* [46] compiled by Harrison and comprising a listing of the wavelengths and intensities of about 110,000 spectral lines. The intensities chiefly refer to arcs and sparks and for a small part to discharges at reduced pressure.

2. National Bureau of Standards (NBS) *Tables of Spectral-Line Intensities* [10] compiled by Meggers, Corliss, and Scribner and containing 39,000 lines with intensities observed in a copper arc. The tables include the values of the upper and lower energy levels of the transitions.

3. *Tables of Spectral Lines* [49] compiled by Zaidel, Prokof'ev, Slavnyi, and Shreider. The lines listed have been extracted partly from the MIT tables, partly from other literature sources. Part I contains 52,000 lines between 200 and 800 nm arranged by wavelength. Part II covers 38,000 lines arranged by element. The latter includes lines in the VUV and also lists excitation energies.

4. *A Table of Emission Lines in the Vacuum Ultraviolet for All Elements* [50] and *Atomic and Ionic Emission Lines below 2000 Å, Hydrogen through Krypton* [51] compiled by Kelly and Kelly and Palumbo, respectively.

5. *Wavelengths and Transition Probabilities for Atoms and Atomic Ions* [52] compiled by Reader, Corliss, Wiese, and Martin which also appeared as *Line Spectra of the Elements,* in the *Handbook of Chemistry and Physics* [53].

These tables are invaluable for the identification of spectral lines in whatever source and will not be easily superseded. They manifest shortcomings, however, when used in ICP spectroscopy.

1. The ICP emits lines that do not appear in the classical sources.
2. The relative intensities of the spectral lines emitted from both classical sources and the ICP are different.

These are the reasons why various groups of workers have embarked upon the compilation of new data relevant to ICPs. The state of affairs and the data available at present will be reviewed in Section 7.4.4.

Finally, some other important classical tables of which the scope differs from that of tables (1)–(5) should be noted:

6. *Line Interference in Emission Spectrographic Analysis* [54] compiled by Kroonen and Vader. These tables are for a lithium-buffered arc and very closely approach an ideal in that they list on a uniform scale the sensitivities of both analysis lines and interfering lines, the latter in a wavelength interval of ± 0.06 nm about the analysis lines. The data comprise 521 analysis lines of 38 analytes.

7. *Wavelength Table of Rare-Earth and Associated Elements Including Zirconium, Thorium Hafnium, Rhenium, and Tellurium* compiled by Norris [55]. These tables cover wavelengths between 130 and 1000 nm and contain many lines not listed in the MIT Wavelength Tables. The data for the rare earths have been taken from "Spektren der Seltenen Erden" by Gatterer and Junkes [56]. The majority of the intensities listed refer to arcs.

8. *Spektren in der Glimmentladung von 1500 bis 4000 Å* compiled by Salpeter [57]. This atlas covers spectra observed in a Grimm glow discharge.

9. *An Atlas for 2 Å/mm and 4 Å/mm Grating Spectrographs* by Qiu Deren and Cheng Wan-xia [58]. The 2-Å/mm section comprises 111 charts with magnified reproductions of Fe spectra with wavelength scales and position indications of spectral lines listed in the MIT and NBS tables [10, 46], as exemplified in Fig. 7.20. Each chart contains a 28-cm-long photographic reproduction of a 2.65-nm-long portion of the Fe spectrum. The wavelength range is 233–487 nm. The 4-Å/mm section covers the range 193–660 nm in 96 charts. An inventory lists wavelengths (arranged according to element) along with intensities (NBS, MIT and as observed in ac (or dc) arcs), lower and upper energy levels (cm^{-1}), and excitation energies (eV).

10. *Tables of Spectral Lines of Neutral and Ionized Atoms* by Striganov and Sventitskii [59]. These tables are specifically intended for physical and astrophysical purposes. They include, in particular, data on highly ionized atoms. Wavelengths, intensities, energy levels, and classifications of transitions are listed.

Although not of direct interest for checking spectral interferences, a recent bibliography by Martin [60] should be noted here as an important key to spectroscopic data. This brief article presents a survey of 47 publications from the Atomic Spectroscopic Data Centers at the National Bureau of Standards. These publications deal with atomic transition probabilities, atomic energy levels, wavelengths and multiplets, and Stark widths and shifts.

7.4.4. Data Available for Checking Spectral Interferences: Recent Approaches Specific to ICP–AES

Four major approaches should be mentioned:

1. The conversion of the "Tables of Spectral-Line Intensities" into a table

7.4. LINE SELECTION WITH A VIEW TO SPECTRAL INTERFERENCES 395

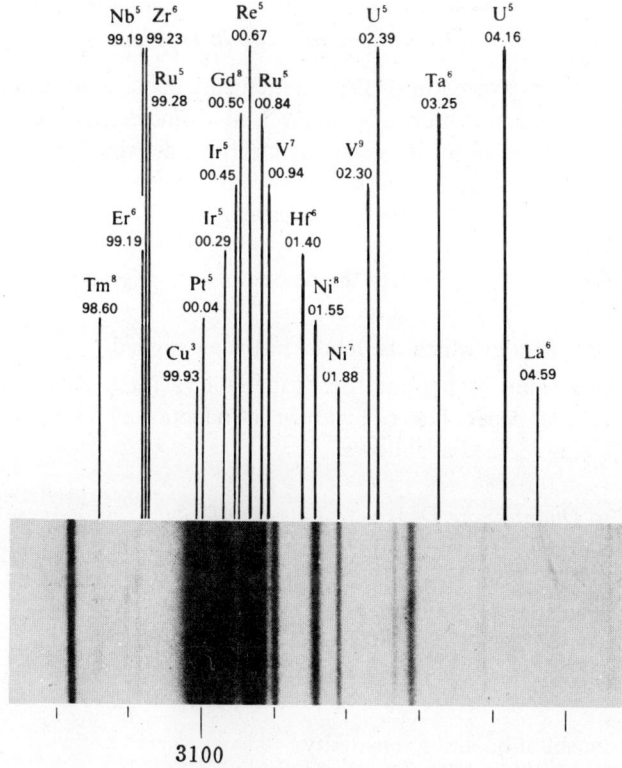

Figure 7.20. Part of a chart of the 2 Å/mm atlas of Qiu De-ren and Cheng Wan-xia [58] showing the quartet near 310 nm in the Fe spectrogram and the positions of a variety of spectral lines in this wavelength region [58]. [Reprinted with permission from Qiu De-ren and Cheng Wan-xia, *Atlas for 2 Å/mm and 4 Å/mm Grating Spectrograph*, Copyright (1984), Shanghai Scientific and Technical Publishers, P. R. China.]

relevant to the ICP [14] and the related design of *Line Coincidence Tables for ICP-AES* [13, 16] by Boumans' group in Philips Research Laboratories.

2. The measurement and tabulation of ICP spectra using a scanning echelle monochromator with crossed dispersion by Parson's group at Arizona State University, the results of which are regularly published in *Applied Spectroscopy* [20, 21, 61].

3. The compilation of wavelength scans of ICP spectra by Fassel's group at Iowa State University, published as *An Atlas of Spectral Information* [11, 17].

4. The measurement and tabulation of ICP spectra using a 0.5-m scanning monochromator by Wohlers of Allied Analytical Systems [18].

7.4.4.1. Line Coincidence Tables for ICP–AES Based on the Conversion of the Tables of Spectral-Line Intensities

In Section 7.4.2 we have outlined the requirements upon an ideal method for line selection. The approach underlying the Line Coincidence Tables is based on this outline, but the result is not yet as ideal as one would wish, owing to the lack of experimental data at the time the tables were set up. The following aspects of this approach should be distinguished:

1. The model used to quantify the degree of line interference and the assumptions underlying this model.
2. The basic data to which the model has been applied.
3. The form, aim, scope, and use of the tables. These items, covered in detail in two papers [14, 61] and the introduction of the work [13, 16], can be summarized as follows.

Model Used to Quantify the Degree of Line Interference and Assumptions Underlying This Model. Generally, the extent to which an analysis line (a) experiences interference from an interfering line (i) can be expressed by the ratio (R_{int}) of the intensities x_i and x_a received by the detector, viz.

$$R_{int} = \frac{x_i}{x_a} = \frac{s_i \times c_i}{s_a \times c_a} \tag{7.1}$$

where c is concentration and s sensitivity.

Defining intensities and sensitivities in terms of signals as received by the detector is useful only in relation to a given spectrometer for which these sensitivities can be experimentally determined in appropriate units for subsequent use in interference checks or corrections (cf. Sections 1.6 and 7.6). A tabulation for general use with a variety of spectroscopic instruments should take into account (1) the variation of the sensitivity of an interfering line with its distance to the analysis line, and (2) the variation of the spectral bandwidth with the type of apparatus. Boumans used a simplified model in which the following assumptions were involved.

1. The line profile is completely dictated by the apparatus, that is, the experimental line width is equalized to the spectral bandwidth of the instrument.
2. The profile is a Voigt profile with parameter a (cf. Section 7.7.2).
3. The sensitivity (S) of a line is identified with the peak intensity (I_p) per unit of concentration (c):

$$S = \frac{I_p}{c} \tag{7.2}$$

7.4. LINE SELECTION WITH A VIEW TO SPECTRAL INTERFERENCES

4. The intensity (I) observed at the detector is proportional to the area (A) of the line profile that overlaps a "rectangular" spectral window in the way illustrated in Fig. 7.21:

$$I = I \times A(\Delta\lambda/\Delta W, a) \tag{7.3}$$

Area A is thus formulated as a function of $\Delta\lambda/\Delta W$ and a, where a is the "a-parameter" of the Voigt profile, $\Delta\lambda$ the wavelength distance between the lines and ΔW the width of the spectral window, identified with the spectral bandwidth of the spectrometer. Figure 7.21 includes some self-explanatory numerical examples.

For conveniently connecting the degree of interference with sample composition, Boumans defined the "critical concentration ratio" (CCR) as a measure of the interference level. The CCR is the ratio of the concentrations of interferent (i) to analyte (a) at which the ratio (I_i/I_a) of the intensities of the interfering line to analysis line in the spectral window of the latter is equal to unity. In view of Eqs. (7.2) and (7.3) the CCR can be expressed as

$$\text{CCR} = \frac{S_a \times A_a(\Delta\lambda/\Delta W, a)}{S_i \times A_i(\Delta\lambda/\Delta W, a)} \tag{7.4}$$

The CCR thus is a sensitivity ratio incorporating the effects of $\Delta\lambda$ and ΔW. The

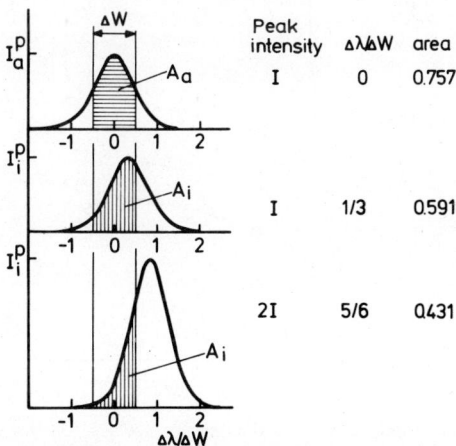

Figure 7.21. Illustration of the model used to compute the contribution of an interfering line to the total intensity in the spectral window of an analysis line [13, 16]. [Reprinted with permission from P. W. J. M. Boumans, *Line Coincidence Tables for Inductively Coupled Plasma Atomic Emission Spectrometry*, 1st and 2nd eds., Copyright (1980), and (1984), Pergamon Books, Oxford.]

quotient of the actual concentration ratio (c_i/c_a) and the CCR is the intensity ratio (I_i/I_a) to be expected. In other words, if the actual concentration ratio exceeds the CCR, the intensity of the interfering line will be higher than that of the analysis line. This represents a situation that can be hardly accepted for an accurate analysis (cf. Section 7.7.7). A correction for the interference can be attempted if the actual concentration ratio is smaller than the CCR, while the interference can be considered as nonexistent if the actual concentration ratio is of the order of $0.01 \times$ CCR or less.

In the *Line Coincidence Tables for Inductively Coupled Plasma Atomic Emission Spectrometry* [16], CCRs are tabulated for interfering lines located in an interval of ± 0.25 nm about the 892 prominent lines covered. CCRs are listed for values of the spectral bandpass ranging from 0.01 to 0.04 nm in steps of 0.005 nm. A value of $a = 0.1$ was used for the parameter of the Voigt profile. Figure 7.22 shows the effect of the numerical value of a on the critical concentration ratio: this value chiefly determines the extent to which the wing of a line contributes to the interference.

The tables further make a distinction between the "CCR for line interference" as defined above and the "CCR for background interference," the latter referring to interference from the line wing ($\Delta\lambda > 2\Delta W$). This is called "background interference" because it can be accounted for by common background correction techniques (Section 7.5). The background enhancement produced by

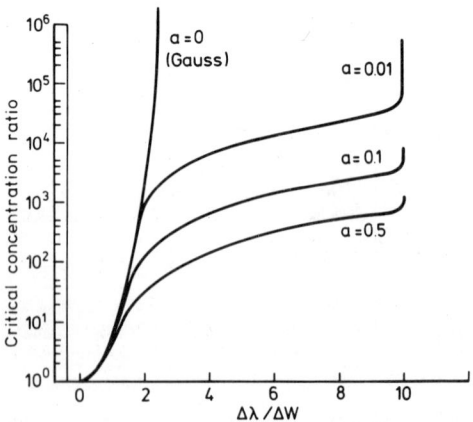

Figure 7.22. Critical concentration ratio for line interference as a function of the normalized distance ($\Delta\lambda/\Delta W$) of the interfering line to the center of the spectral window of the analysis line for various values of the "a-parameter" of the Voigt profile. The sensitivities of the analysis and interfering lines have been assumed equal [13, 16]. [Reprinted with permission from P. W. J. M. Boumans, *Line Coincidence Tables for Inductively Coupled Plasma Atomic Emission Spectrometry*, 1st and 2nd eds., Copyright (1980) and (1984), Pergamon Books, Oxford.]

7.4. LINE SELECTION WITH A VIEW TO SPECTRAL INTERFERENCES

a line wing primarily worsens the detection limit, but does not involve a risk of errors if background correction is performed. Therefore, the CCR for background interference is defined as the concentration ratio of interferent to analyte at which the detection limit is raised by a factor of 3.

It appears that the Voigt profile with $a = 0.1$ underpredicts the background enhancements produced by line wings compared to the experimental findings discussed in Section 7.3.2, in particular, at a relatively small distance from the analysis line.

Basic Data to Which the Model Has Been Applied. The peak sensitivities underlying the Line Coincidence Tables have been obtained by converting the complete "Tables of Spectral-Line Intensities" [10], hereafter referred to as "NBS Tables." To this end the detection limits of 561 prominent lines from the extensive list of Winge et al. [7] were used in a computer iteration procedure to determine the transfer factors for converting the copper arc sensitivities in the NBS Tables into values relevant to the argon ICP [13, 14, 16]. Although this approach yielded extensive data of reasonable accuracy, it could not extend the number of lines beyond the 39,000 covered by the NBS Tables. This is the most serious limitation of the Line Coincidence Tables.

Form, Aim, Scope and Use of the Tables. For illustration, Table 7.3 presents an extract from the complete coincidence table for Dy II 353.170 nm. The respective columns list the data for the interfering lines as follows: element symbol, wavelength (preceded by I or II for atom or ion), sensitivity (in relative units), and the CCRs for spectral bandwidths of 0.010, 0.020, 0.030, and 0.040 nm (those for the intermediate values of 0.015 nm, etc. have been omitted).

CCRs followed by * or # are for line interference, the remaining values for background interference. CCRs followed by # refer to wavelength distances between 1.5 and 2 times the spectral bandwidth and must be considered less reliable than those followed by *, in view of the uncertainty associated with the choice of the numerical value of the a-parameter of the Voigt profile. Symbols + and "m" following the sensitivities refer to multiple interferences, in which case the CCRs in the row labeled "m" represent the resultant CCR for the element. As an example, consider the determination of Dy in the presence of a 1000-fold excess of both Ce or La, or in the presence of a 100-fold excess of each Tb, Gd, Nd and Pr, the bandwidth of the apparatus being 0.020 nm.

The CCR for Ce is found to be 410, whereas no La line appears in the list. The Dy line may thus be further considered in the case of La as concomitant, but should be rejected in the case of Ce since the actual concentration ratio of Ce to Dy exceeds the CCR.

Similarly Dy II 353.170 is rejected if the sample contains Tb, Gd, Nd, and Pr, the CCR of 100 for Tb being decisive. If only one of these concomitants

Table 7.3. Abridged Line Coincidence Table for Dy II 353.170 nm [16].[a]

Sensitivity (S): 160000 Rank Number: 1
Detection limit: 6.7 Table: 168

Interfering line (nm)		Distance (nm)	S	Critical Concentration Ratio Spectral Bandwidth (nm)				
				0.010	0.020	0.030	0.040	
Ce II	353.095	−.075	890 +		9600000	4000000	18000 #
U II	353.111	−.059	1100 +		4700000	17000 #	5100 *
Eu II	353.115	−.055	630 +m			14000 #	3600 *
Hf I	353.123	−.047	340 +m		9000000	22000 #	5300 *
Er	353.127	−.034	n				
Ru I	353.139	−.031	200		36000 #	5300 *	2300 *
Th I	353.145	−.025	630 +	5600000	3800 *	860 *	500 *
Ta I	353.158	−.012	210 +m	9400 *	1400 *	1000 *	870 *
Ce II	353.159	−.011	640 +m	1900 *	410 *	300 *	270 *
Ce I	353.162	−.008	20 +	25000 *	11000 *	9000 *	8600 *
U I	353.164	−.006	50 +m	6000 *	3700 *	2900 *	2000 *
Tb II	353.170	.000	160	100 *	100 *	100 *	100 *
Nd	353.171	+.001	410	400 *	390 *	390 *	390 *
Mn I	353.192	+.022	4200 +m	15000 *	100 *	50 *	31 *
Th II	353.193	+.023	2300 +m		600 *	160 *	99 *
Mn I	353.212	+.042	13000 +	83000 *	180000	330 *	86 *
Ta I	353.221	+.051	60 +			190000 #	44000 *
Eu II	353.223	+.053	380 +			34000 #	8500 *

[a] The sensitivity is expressed in relative units, the detection limit in ng/ml.
n : line listed in NBS Tables but sensitivity could not be converted for lack of information on the type of species (I, II).
— : CCR larger than 10^7.
The meaning of +, m, #, and * is given in the text.

7.4. LINE SELECTION WITH A VIEW TO SPECTRAL INTERFERENCES

were present, then the Dy line would deserve further consideration in the case of Gd, Nd, and Pr. Caution would have to be exercised with Nd, since the Nd line has virtually the same wavelength as the Dy line, but in view of the CCR of 390, a determination involving a correction for the Nd blank would not be entirely impossible. Preferably one would proceed to a next Dy line and then find a more favorable situation with the third prominent line, Dy II 340.780 nm. A threefold poorer detection limit should then be accepted, but the Nd blank with Dy II 353.170 nm would also worsen the detection limit.

Let these examples serve as an illustration of the type of information in the Line Coincidence Tables and of how this information can be rationally used to speed up the process of line selection in ICP-AES. The introduction of the work [16] provides more details about the use of the tables and includes various worked examples.

The prime aim of the tables thus is to facilitate and speed up decisions about the a priori rejection of prominent lines. Decisions of this type have a reasonably high probability of being justified, although, in view of the assumptions underlying both the conversion of the NBS Tables and the coincidence model, errors resulting from overpredictions of interferences cannot be entirely precluded.

The "acceptance" of a line as being "free from interference" does not mean that the line can be safely used. It only implies that the line does not suffer from interference from relatively strong lines as tabulated in the NBS Tables. Closer inspection is thus required. This will commonly mean that scans have to be made over a spectral window of, say 0.05–0.1 nm about the center of the analysis line during the aspiration of solutions containing the analyte and each of the concomitants separately. Although it will be unnecessary to identify all details of such scans, it may be helpful to identify at least relatively strong peaks. To that end the atlases or tables discussed in Sections 7.4.4.3 to 7.4.4.5 or classical tables (Section 7.4.3) must be consulted.

In a recent paper on wavelength selection for trace analysis by ICP-AES by McLaren and Berman [63] lucidly discussed the strengths and weaknesses of wavelength tables for the prediction and identification of line overlap in ICP-AES. McLaren and Berman then illustrated their arguments by a description of the wavelength selection process for several difficult trace analyses. From the introduction of that paper is cited the following.

The choice of appropriate wavelengths for trace analysis is often difficult owing to the lack of truly comprehensive tables of ICP line intensities. Plasma spectroscopists are at present forced to use rather brief compilations of "prominent" plasma emission lines [7, 9] in combination with much more comprehensive tables for classical atomic emission sources such as the dc arc and ac spark (Section 7.4.3). Two attempts have been made to put this latter information into convenient form for use with ICP-AES [13, 64]. Notable is the work of Boumans in compiling two volumes of line coincidence tables

[13, 16], based primarily upon the U. S. National Bureau of Standards' tables of spectral intensities for the copper arc [10].

Use of the classical tables presents a number of limitations. The most severe of these is that little information about the relative intensities of ICP emission lines can be obtained, because of differences in excitation mechanisms in plasmas and those in arcs and sparks. In addition, many lines observed in plasmas are not listed in the classical tables. Furthermore, since the data for these tables were obtained almost entirely by photographic detection, many ICP lines below about 220 nm, and particularly those in the VUV, are missing.

In some respects, the Boumans tables [13, 16] are uniquely useful inasmuch as a rational attempt was made to convert line intensities observed in the copper arc to ICP emission intensities, thus permitting an estimate of the severity of a possible interference, based also on the relative concentrations of the analyte and interferent and the resolving power of the spectrometer. The primary limitation of these tables, in our experience, is that the original data base, i.e. the N.B.S. copper arc tables, lacks many lines observed in the ICP, and other sources. To overcome this, it is often useful to refer to more comprehensive listings such as the M.I.T. [46] or Zaidel [49] tables in order to predict or identify line overlap interferences.

A welcome addition to the available interference tables will be an *Atlas of Spectral Information for ICP-AES* [17], in preparation by Winge and co-workers [11] (cf. Section 7.4.4.3).

McLaren and Berman describe a three-step procedure for choosing the best wavelength for a particular trace determination and demonstrate the importance of each step. They show that, in the absence of truly comprehensive tables of ICP line intensities, an assessment of possible spectral interference based on wavelength profiling should precede the eventual line choice to ensure the quality of trace analysis. They also indicate incidentally the value of high-resolution spectrometers. In one of their papers on high-resolution spectroscopy [28], Boumans and Vrakking describe a simple procedure for the study of spectral interferences by manipulation of experimental wavelength scans stored in the computer memory.

The purpose of this manipulation is to simulate spectral scans of spiked solutions without need for preparing these solutions and acquiring accurate scans of them. The simulated scans visualize situations such as can be expected in genuine experimental scans. The assumption underlying this approach is that the contributions from the various elements and the solvent to the total intensity pattern observed in the scanned window are additive [27, 28].

The principle of scan simulation is as follows. Real scans are made over a spectral window of, say 0.05 nm (50 steps, 1 s integration time per step), centered about an analysis line. Three such scans are involved, namely, for

1. The pure main element(s) in the pure solvent: scan M ("MATRIX")

7.4. LINE SELECTION WITH A VIEW TO SPECTRAL INTERFERENCES

2. The pure analyte in the pure solvent: scan T ("TRACE")
3. The pure solvent: scan B ("BLANK")

From the stored data, the following function is constructed:

$$S = M + (T - B) \times f \tag{7.5}$$

where f is a factor entered by the operator. Function S is the simulated scan for the matrix spiked with an analyte concentration $f \times c$, where c is the concentration used in the actual scan T. Variation of f enables one to vary the ratio of analyte to main element(s) and to observe the corresponding changes in the appearance of analysis line and background.

The original scans and simulated scan S are displayed on the monitor screen and can be modified by variation of f. In addition, scan T can be shifted in the wavelength direction with respect to scans M and B. This option is included to correct for possible small changes in the wavelength calibration of the spectrometer during long periods of operation. The result on the monitor can be finally obtained in the form of hard copies of printed numerical data and/or plots.

As an example, Fig. 7.23 shows simulated scans over 0.05 nm windows about the Nd line at 396.312 nm with interference from not further identified

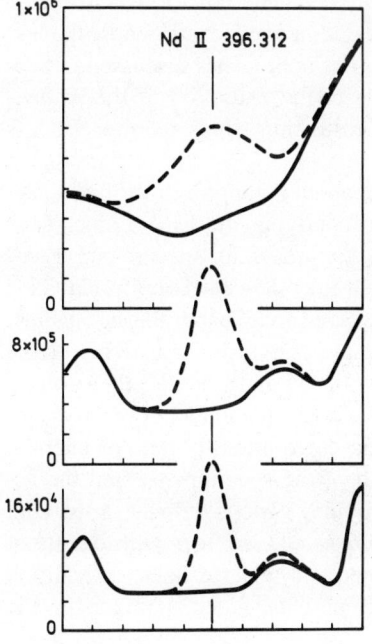

Figure 7.23. Scans over a 0.05-nm window about the Nd line at 396.312 nm with interference from not further identified Ce lines. Continuous curves: actual scans for 100-μg/mL Nd solution. Broken curves: scans for Ce solution spiked with Nd as found by the simulation technique. The vertical line indicates the position of the Nd line. All scans are for a 2-μg/mL Nd concentration. The practical spectral bandwidth was as follows: (1).8 pm, (2) 5.8 pm, (3) 2.3 pm. Ordinate: intensity (counts/s), abscissa: wavelength ($\pm$ 0.025 nm about central λ) [28]. [Reprinted with permission from P. W. J. M. Boumans and J. J. A. M. Vrakking, *Spectrochim. Acta* **39B**, 1261 (1984). Copyright (1984), Pergamon Journals, Oxford.]

Ce lines. The spectra were recorded with various spectral bandwidths. Each frame includes the actual scan for the pure matrix, that is 100 μg/mL Ce (continuous curve) and a scan for the matrix spiked with 2 mg/mL Nd (broken curve).

The scan technique is used in both methodological studies and in trace analysis of high-purity materials, for example. In the latter application, factor f is varied until the simulated scan is identical to the actual scan for the sample. Thus a value of $f \times c$ is found as a first approximation of the concentration present. If a possible inaccuracy due to multiplicative interference has to be eliminated, then the true sensitivity of the analyte signal is subsequently derived from the scan for a spiked solution of the matrix with analyte concentration $f \times c$ ("standard addition").

More important than correcting for a possible multiplicative interference effect is to ensure that the net signal for the sample derived from the gross signal at the position of the analysis line does not contain contributions from components other than the analyte itself. In other words, the pivot of the problem remains the reference point used in the background correction (see Section 7.5). Scans for the sample and for solutions of the major constituents in pure form and spiked with trace elements are invaluable and indispensable aids to achieve the highest possible accuracy.

However, a sample of complex composition will always leave the analyst with some uncertainty owing to the fact that one can only extrapolate but never cast a true look under a spectral line. The more detailed knowledge is available about the origins of the spectral patterns revealed in wavelength scans, the less the uncertainty inherent in the extrapolation. The approaches discussed in Sections 7.4.4.2–7.4.4.4 substantially contribute to the extension of this knowledge. An advanced method for background correction which requires far less knowledge will be discussed in Section 7.5.

To conclude this section are a few remarks about an approach for line selection based on the same data and model as underlying the Line Coincidence Tables. This approach has been referred to as "approach for specific purposes" [62], "software package" [65], or "computer-aided line selection" [12]. This approach relieves the user from the work involved in a search through tables. The user enters data about the expected sample composition in terms of the upper concentrations of constituents along with the analytes and their lowest concentrations of interest in the analysis. The user also enters boundary conditions, such as the tolerance, (i.e., the just accepted intensity ratio of interfering to analysis lines), the spectral bandwidth of the spectrometer and the accessible wavelength range. The computer provides various outputs: a list with one or more "free" lines per element, as requested, and lists with details of the interferences detected for both the accepted and rejected lines. In addition to the inherent appreciable gain in speed, the computer-aided approach has the

7.4. LINE SELECTION WITH A VIEW TO SPECTRAL INTERFERENCES

great advantages of permitting parameter variation (tolerance, etc.) and providing a detailed printed administration of the line search and its results.

This software is operational on the IBM VM/CMS 3083 Computer in the Philips Research Laboratories in Eindhoven. Although the present version is an update of an earlier design, the program essentially still is in Algol-60. It has not been feasible in the past years to find the required funding for appropriate translation and formatting so as to make this software available for general use on other large or medium-size computers [66, 106].

7.4.4.2. Measurement and Classification of ICP Lines Using a Scanning Echelle Monochromator with Crossed Dispersion

Parsons' group modified a commercial 0.75 m focal length echelle spectrometer to perform spectral scanning under computer control [67]. A spectrum from 207.5–600.5 nm can be acquired in approximately 6 h. Intensity data are collected over a dynamic range of 1 million by using a multirange electrometer setup.

Using this apparatus Parsons' group has embarked upon the measurement of ICP spectra using ICP compromise conditions. The results are regularly reported in *Applied Spectroscopy*. The papers published hitherto cover

1. Background emission [20]
2. The spectra of alkaline earth elements: Be, Mg, Ca, Sr, Ba [61]
3. The spectra of group IIIA elements: B, Al, Ga, In, Tl [21]

In (3) the wavelength range was extended down to 199.5 nm. Additional lines of alkaline earths with wavelengths between 199.5 and 20.7.5 nm, therefore, were included in (3).

In (1) and (2) the spectral bandwidth varied from about 3 to 7.5 pm for wavelengths from 200 to 500 nm; in (3) the bandwidth was twice as large.

The results communicated in (1) have been discussed in Section 7.3.4. The following remarks apply to (2) and (3).

For each element two full spectrum scans were taken for 10 and 100 μg/mL of the primary standards, either metal, oxide, or carbonate. From these two scans the net line intensities for the 100 μg/mL scan could be accurately determined.

The tabulated data comprise for each line identified: the element symbol, the species (I, II), wavelength (Å), lower and upper energy levels (cm^{-1}) of the transitions, background intensity (I_B), net line intensity (I_N), and ratio I_N/I_B for 100 μg/mL, and appropriate references and footnotes. The spectrum of each element is discussed and, where useful, elucidated with Grotrian diagrams.

Alkaline Earth Spectra. The spectra of the alkaline earths consist of most of the expected transitions that have been reported in this spectral range. Only a few unusual observations have been made. Several Be atomic transitions were noted that originated above the ionization potential of the element. Several transitions not listed in the *MIT Wavelength Tables* [46] were reported.

The spectra, with the exception of Ba, were mostly from atomic species with either the atomic or the ionic resonance transitions producing the most intense emission. Although the same number of transitions was observed in the ICP as in the NBS copper arc [10], namely, 209, the natures of the spectra are significantly different. Generally, the ICP emitted more ionic spectra and the copper arc more atomic spectra. Differences in the number of spectral lines varied from 10 to 40% for these elements. The authors cautioned as to this comparison. The NBS procedure provided poor sensitivity at low wavelengths and the ICP procedure followed in their work was limited to a concentration of 100 μg/mL. More lines would be observed in the ICP spectra at higher metal concentrations.

Spectra of Group IIIA Elements. Group IIIA elements' spectra were significantly different from the alkaline earth spectra in that very few ionic lines were observed. In addition, the ionic lines that were found were for the most part of very low intensity. There was a significant difference in the comparisons between the data for the NBS copper arc and those for the ICP.

In paper (3) the authors also marked lines which have not been previously used for analysis according to a literature search in 1980 [64] and yet are potentially useful analysis lines. However, virtually all these lines appear as prominent lines in the listings of Winge et al. [7, 8, 17] and Boumans [13, 14, 16].

Paper (3) also includes lists of analysis lines which experience first- or second-order interference from lines of the alkaline earth or group IIIA elements. A wavelength separation of up to 0.1 nm has been considered.

7.4.4.3. Atlas of Spectral Information

The *Atlas of Spectral Information* [17] is one of the multiple major contributions from Iowa State University to the development of ICP–AES. In an article, "On the selection of analytical lines, line coincidence tables and wavelength tables," Winge et al. [11] gave a detailed description of the contents, aim, and scope of the Atlas and the conditions under which the data were obtained. More recently, Winge et al. [68] presented a brief survey of the Atlas. From these "announcements" the following information has been extracted.
The Atlas comprises the following three sections:

1. Emission Spectra or 70 Elements and a Reference Blank. One section of the Atlas covers 232 wavelength scans with a logarithmic intensity scale, each scan extending over 80 nm. The practical wavelength resolution of the recorded

7.4. LINE SELECTION WITH A VIEW TO SPECTRAL INTERFERENCES

spectra (not the spectrometer) corresponds to somewhat less than 0.1 nm. Figure 7.17 shows an example. Marker wavelengths serving as reference points are provided at approximately regular intervals.

2. Listings of Prominent Lines. The listings of prominent lines are revised listings of the prominent lines published previously [7, 8]. This Atlas section contains 973 prominent lines arranged both in alphabetical order of the elements and wavelength. The type of information covered is illustrated by the sample page in Fig. 7.24.

3. Coincidence Profiles. The main body of the Atlas consists of so-called coincidence profiles for 281 prominent lines. A coincidence profile is a combination of wavelength profiles of individual elements (concomitants) superimposed on an analyte line profile of the same spectral region. As an example Fig. 7.25 shows the coincidence profile set for the Sc II 364.28-nm line. Each profile consists of linear and log-scaled plots of the analyte and 10 interferent test elements. These concomitants were chosen with primary emphasis on elemental relative abundancies in water, sediments, the earth's crust, and samples of bi-

APPENDIX B-1

PROMINENT LINES EMITTED BY THE INDUCTIVELY COUPLED PLASMA

Alphabetical listing with the lines of each element listed in order of increasing detection limit.

S OF I	WAVE-LENGTH (nm)		I_n / I_b	CONC (mg/L)	EST'D DET LIM (mg/L)	POSSIBLE INTERFERENCES (SEE APPENDIX C)
Ag	I	328.068	38.0	10.0	0.007	Fe,Mn,V
Ag	I	338.289	23.0	10.0	0.013	Cr,Ti
Ag	II	243.779	2.5	10.0	0.120	Fe,Mn,Ni
Ag	II	224.641	2.3	10.0	0.130	Cu,Fe,Ni
Ag	II	241.318	1.5	10.0	0.200	
Ag	II	211.383	0.9	10.0	0.333	
Ag	II	232.505	0.7	10.0	0.428	
Ag	II	224.874	0.6	10.0	0.500	
Ag	II	233.137	0.5	10.0	0.600	
Al	I	309.271	13.0	10.0	0.023	Mg,V
Al	I	309.284	13.0	10.0	0.023	Mg,V
Al	I	396.152	10.5	10.0	0.028	Ca,Ti,V
Al	I	237.335 NR	10.0	10.0	0.030	Cr,Fe,Mn
Al	I	237.312 NR	10.0	10.0	0.030	Cr,Fe,Mn
Al	I	226.922 NR	9.0	10.0		
Al	I	910 NR				

Figure 7.24. Example of the information contained in the alphabetical listing of prominent ICP lines in the *Atlas of Spectral Information* [17], [68]. [Reprinted with permission from *ICP Information Newsletter*, **10**, 446 (1984). Copyright (1984), R. M. Barnes, Amherst, MA.]

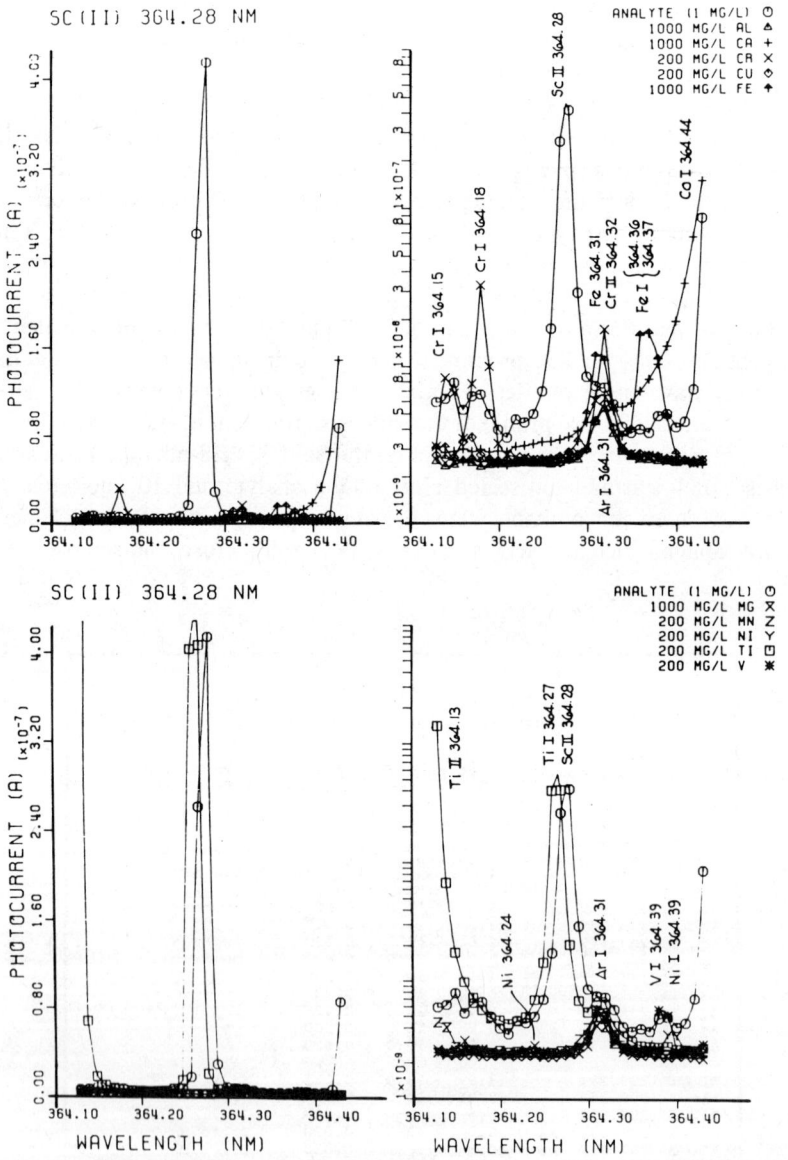

Figure 7.25. Example of a set of coincidence profiles contained in the *Atlas of Spectral Information* [17], [68]. [Reprinted with permission from *ICP Information Newsletter* **10,** 448 (1984). Copyright (1984), R. M. Barnes, Amherst, MA.]

ological origin as follows: Al, Ca, Fe, Mg Cr, Cu, Mn, Ni, Ti, and V. Solutions containing 1 mg/mL of Al, Ca, Fe, or Mg and 0.2 mg/mL of the other elements were introduced into the ICP with an USN. The type of nebulizer does not affect the intensities of analyte lines relative to interfering lines, but should be taken into account in assessing the difference in background between the profiles for the pure analytes and the pure concomitants. In this respect the individual scans in the Atlas will represent the conditions found with pneumatic nebulizers at concentrations approximately 10 times those used for the scans.

The profiles cover a wavelength range of 0.3 nm, each centered approximately at an analytical line. Each profile consists of 31 separate intensity measurements at 0.01-nm increments. The linear scaled plots allow a rapid visual estimation of the degree of interference (for the interferent/analyte concentration ratio employed) by a simple comparison of the analyte and interferent relative intensities at the analyte wavelength. The log-scale plots amplify the details of low-level interference while still allowing the full peaks to be shown for relatively intense lines. The spectral lines have been identified on the log-scaled plots when such information was available from the major compilations of wavelength data.

Winge et al. emphasize that, in contrast to tabular listings, coincidence profiles of the type provided by the Atlas have the capability of showing all types of spectral interferences. For example, coincidence profiles readily show interferences arising from recombination continua, stray light, line broadening (e.g., the wing of the Ca 364.44-nm line at the Sc 364.28-nm wavelength in Fig. 7.25), background features (e.g., the Ar 364.31-nm line in Fig. 7.25), and very significantly, previously unidentified lines.

7.4.4.4. ICP-AES Wavelength Tables

Wohlers [18] compiled a complete set of wavelength tables for the ICP using a scanning monochromator over a spectral range between 185 and 850 nm. ICP compromise conditions were used.

The tables cover 58 elements excluding the rare earths with the exception of Ce. The data are computer stored and currently comprise about 15,000 lines. The numbers of lines per element are stated in Table 7.1. The data were derived from scans of 1 mg/mL solutions of the pure elements, which were introduced into the ICP with a pneumatic nebulizer.

The tables list the wavelengths and relative intensities, along with appropriate references and notes. Background equivalent concentrations (BEC) are stated for the prominent lines. For line-rich spectra, the prominent lines are also separately listed.

As is evident from the detection limits derived from the BECs (Table 4.2), the order of prominence fits in with that in Winge's [7, 17] and Boumans' [13,

16] listings, although there exist minor and even some major discrepancies (cf. Sections 4.1.72 and 4.1.7.7, and [107]).

Since Wohlers' data were obtained from direct measurements in the ICP, they are far more relevant than classical tables or Boumans' tables [13, 16], based on the conversion of the NBS Tables [10]. Although the conversion led to reasonable results, it could not deal with unlisted lines. Currently, Wohlers' tables form the most comprehensive listing of ICP wavelengths. Ideally one would like to see these data combined with the software associated with Boumans' Line Coincidence Tables (see the end of Section 7.4.4.1) with appropriate adaptations of the coincidence model according to new insights gained by spectroscopic studies at high spectral resolution [27, 28, 33, 69, 70, 106, 107, 109].

7.4.4.5. Miscellaneous Data

Michaud and Mermet [71] studied the iron spectrum between 200 and 300 nm as emitted from an ICP. The objective was to make a comparison with available wavelength tables. More than 1000 lines were observed under compromise conditions using a 1 mg/mL Fe solution, introduced with a pneumatic nebulizer. The intensities are roughly listed on a scale of 1 to 5. Some 100 lines not reported in the classical tables [10, 46, 49] were found. The listing includes prominent lines of other elements experiencing interference from Fe lines.

Brenner and Eldad [72] compiled a "Spectral Line Atlas for Multitrace and Minor Element Analysis of Geological and Ore Mineral Samples by ICP–AES." The data were obtained with a combined polychromator-monochromator, the former being provided with a computer-controlled scanning slit facility for simultaneous wavelength profiling of all analysis lines programmed. Spectral regions of 0.08 nm on both sides of the analysis lines were simultaneously profiled in the polychromator and the monochromator during the (pneumatic) nebulization of solutions of pure elements. Concentrations (up to 1 mg/mL) corresponding to those encountered in geological and ore digests were used.

The Atlas consists of graphical representations for 45 analysis lines (Fig. 7.26) in which the potential interferences within the ±0.08-nm window are indicated by bars whose lengths indicate the extension of the interference and the thicknesses represent a qualitative measure of the intensity. The purpose of this Atlas is to aid the selection of both analysis lines and positions for background measurement in geological applications.

Botto's work [73–76], to be discussed in Sections 7.6, 7.7.7.3, and 7.7.7.6, should be mentioned here as it contains much useful information on spectral interferences in terms of "interference coefficients" for both analyte peak and background positions.

Faires [77] refers to the potentials of high resolution Fourier transform spectroscopy for compiling ICP wavelength atlases. This approach provides not

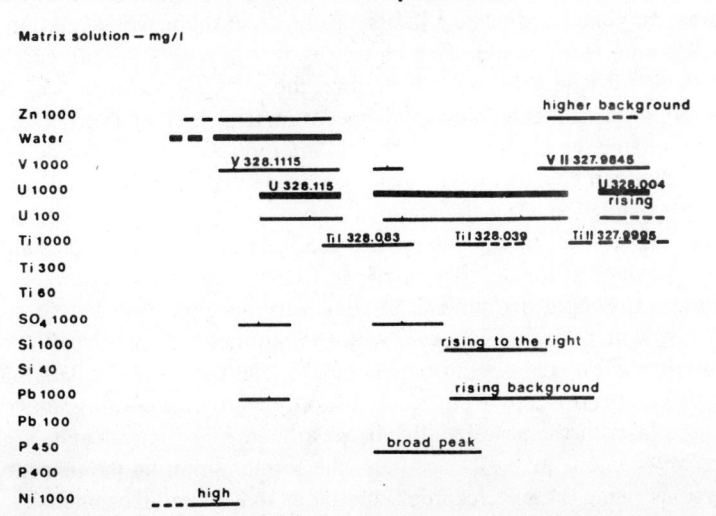

Figure 7.26. Sample page of the atlas of Brenner and Eldad [72]. [Reprinted with permission from *ICP Information Newsletter* **10**, 464 (1980). Copyright (1984), R. M. Barnes, Amherst, MA.]

only for accurate wavelength determinations and an extremely high spectral resolution (lines with a peak-to-peak separation of 0.01 nm are base-line resolved), but also permits the compilation of data on line widths and shapes and their deconvolution into Gaussian and Lorentzian components (cf. Section 7.7.2 and [78]).

As mentioned in Section 4.1.7.4, Fry's group has specifically studied the ICP spectra of nonmetals, mostly in the near infrared [19]. The emission spectra from a He MIP in this wavelength region were recently explored by Freeman and Hieftje [79].

7.5. BACKGROUND CORRECTION

An elementary calculation in Section 4.6 illustrated the importance of correct background subtraction and showed the magnitude of the errors in net signals resulting from erroneous background determinations. Background correction is one of the most difficult and ticklish problems of emission spectroscopy. This is so because the background under a spectral line cannot be directly determined and, in addition, varies with the sample composition.

It might appear from the literature that there exists a variety of approaches to background correction. This is not so! Actually the same principle is applied to a large variety of concrete situations that are only apparently different. All approaches are based on the visual inspection of wavelength scans of samples, standards, and blanks to make a judicious choice of the wavelength(s) at which the background is measured. This choice, in turn, dictates the calculation procedure to be subsequently used. Therefore, the software provided with flexible spectroscopic instruments (slew-scan spectrometers) implies interactive procedures that enable the analyst to collect and interpret data and to decide upon the approach that can be expected to yield the most reliable background correction in the given situation. The foundations of background correction are common sense and experience. In fact one applies precisely the same principles and rules that have been used for decades in spectrography. The main difference is that modern spectroscopic instruments provide very accurate wavelength scans in digital form which can be processed with an unlimited degree of sophistication. Computerized slew-scan spectrometers rapidly yield quantitative data that with photographic spectroscopy were accessible only with great efforts and even then could not claim the same reliability because neither the wavelength scale nor the intensity scale were really accurate, the former suffering the inaccuracy of registrations using a chart recorder, the latter that of indirect intensity determinations from blackening measurements.

Although, at present, background correction procedures are still governed by the rules of common sense and operator's experience, in this apparently unassailable domain, the penetration of artificial intelligence is coming within reach, as is evident from recent literature proposing background correction procedures that do not require prior knowledge about the sample or interactions by the operator [80]. However, let us first consider the conventional approaches. Discussion will be confined to the principles (also see Section 8.7).

The essential information for background correction must be derived from wavelength scans centered about the analysis line and extending over a wavelength interval of at least several times the spectral bandwidth of the apparatus.

Figure 7.27 schematically shows scans for analyte (A) and concomitants (B). The background due to the concomitants manifests as a background enhancement and can be categorized as

1. Simple ("flat") background
2. Sloping background
3. Direct line overlap
4. Complex line overlap

The complexity of the background correction procedure increases from (1) to (4) and also depends on the differences in background between the samples mutually and between samples and standards. Three situations can be distinguished, viz.

7.5. BACKGROUND CORRECTION

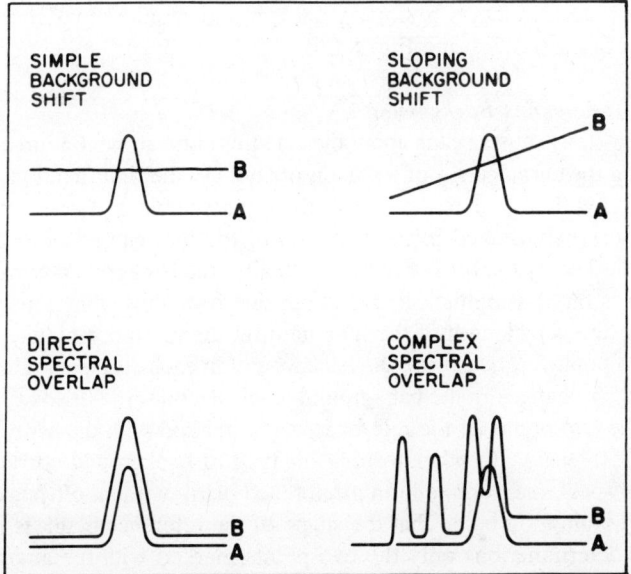

Figure 7.27. Schematic representation of the basic types of background interference. *A*, wavelength scan for the analyte; *B*, scan for the concomitants [81]. [Reprinted with permission from G. F. Wallace, *Atomic Spectroscopy*, Vol II; p. 87. Copyright (1981), Perkin-Elmer, Norwalk, CT.]

    a. Both the background level and the shape (wavelength pattern) are constant.
    b. The background level varies but the shape is constant.
    c. Both the background level and its shape are variable.

Situation I

The background correction can be based on the measurement of a reference blank at the wavelength of the analysis line. This is referred to as "on-peak background correction." In principle, this approach will yield a reliable result for all types of background 1 to 4, provided that background level and shape are entirely constant. This will happen only with simple solutions or with solutions of high-purity materials of constant composition. Caution must be exercised, however, since variations in acid concentration, for example, may give rise to changes in background level. The safest is not to rely on this constancy and to apply an "off-peak background correction" procedure, common to situations II and III.

Situation II: Simple or Sloping Background

A simple background enhancement can be corrected for by a single off-peak background measurement; a sloping background requires two such measurements (see Section 8.7.1 and Fig. 8.25).

Situation II: Complex Background

Figure 7.28 (left) shows scans about the selenium line at 196.03 nm as recorded by scanning the entrance slit of a polychromator. In the right-hand part all scans are replotted with respect to the scan of the reference blank. However, the scans are simultaneously moved in the direction of the ordinate so as to coincide at the point marked by the broken arrow. Actually, the background correction thus performed is far less sophisticated than appears from these diagrams. First, one has chosen the wavelength of the maximum in the reference blank (broken arrow) as the point for measuring the background in each scan separately in order to account for changes in the background level. Second, the intensity difference between the two points of the reference blank marked with the arrows is added to the peak intensity. In other words, this type of background correction combines an on-peak measurement on a reference blank with an off-peak measurement, the assumption being that the shape of the reference blank is invariable. In routine determinations only the two points marked with the arrows need to be measured. The scans were only required to determine the best reference point (broken arrow) such that accurate and consistent results would be obtained on all samples of this type.

Situation III

Situation III is the general situation and includes the previous ones as special cases. For conveniently discussing the background correction we consider the example shown in Fig. 7.29. A concomitant P produces a background spectrum with two lines λ_1 and λ_2, yielding the net intensities x_1 and x_2. Line λ_1 coincides with the analysis line λ_a. The various concomitants, including P are supposed to produce a simple background enhancement Δx_c at wavelength λ_c, which adds to the "initial" background x_c of solvent and discharge gas. The required background correction is

$$x_b = x_c + \Delta x_c + x_1 \tag{7.6}$$

The ratio

$$k = \frac{x_1}{x_2} \tag{7.7}$$

may be assumed to be independent of the sample composition and can be determined in a reference sample not containing the analyte. Therefore,

$$x_b = x_c + \Delta x_c + k x_2 \tag{7.8}$$

This type of background correction takes into account changes in the back-

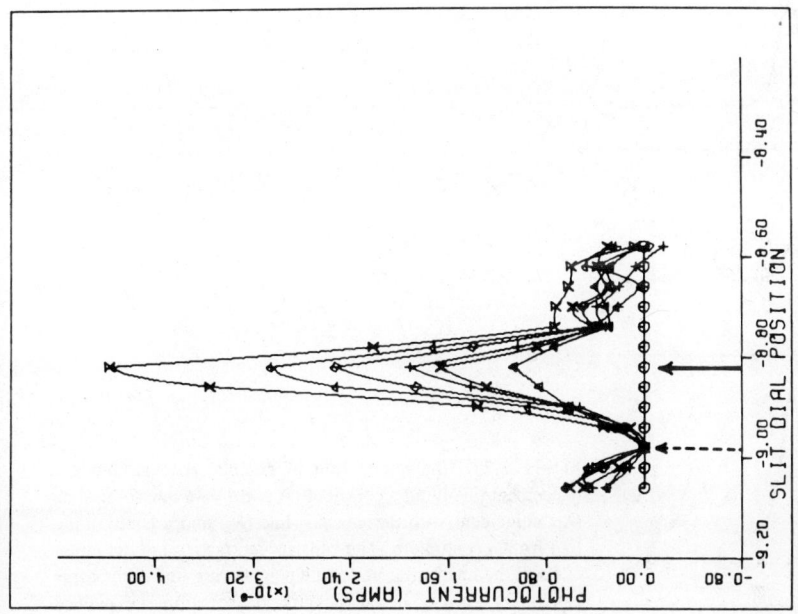
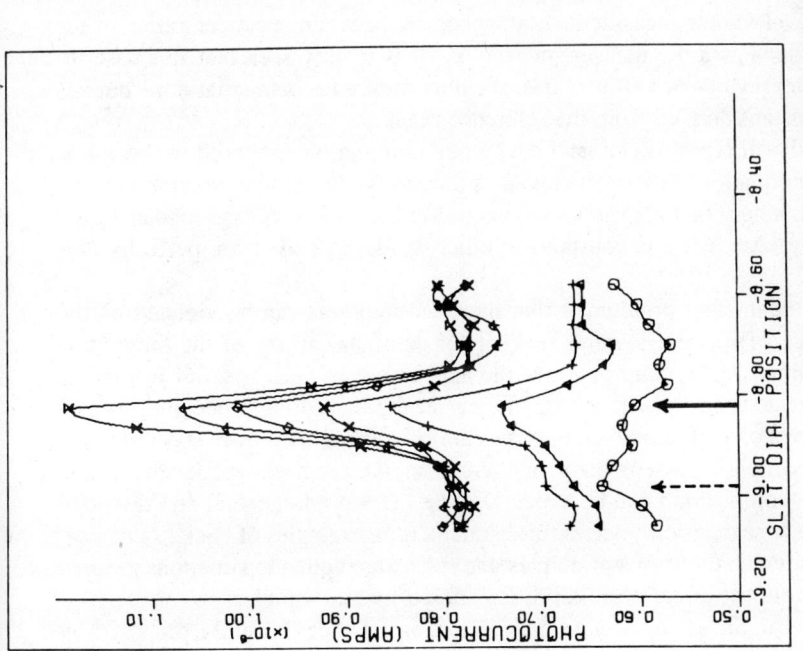

Figure 7.28. Wavelength profile data for the determination of selenium in urine. Left: as observed; right: profiles after background correction. (Note the difference in the ordinate scales.) ⊙ Reference blank (RB): 0.5 wt % NaCl and 1% (vol) HCl; △ RB + 50 ng/mL Se, + RB + 100 ng/mL Se, × normal urine sample; ◇ urine sample + 20 ng/mL Se; △ urine sample + 50 ng/mL Se; $\overline{X}$ urine sample + 100 ng/mL Se. [82]. [Reprinted with permission from W. J. Haas, Jr., V. A. Fassel, F. Grabau IV, R. N. Kniseley, and W. L. Sutherland, *Ultratrace Metal Analysis in Science and Environment*, Ch. 8, p. 91. Copyright (1979), American Chemical Society, Washington, DC.]

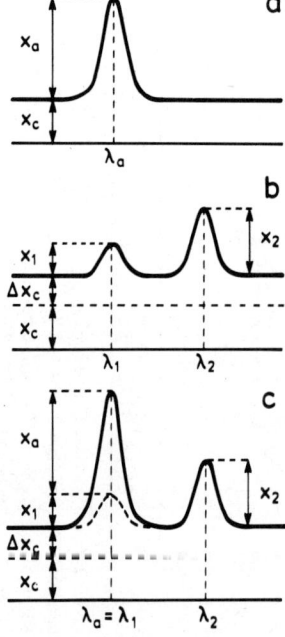

Figure 7.29. Characterization of complex background enhancement involving a contribution x_1 from an interfering line (λ_1) coincident with the analysis line (λ_a) and a contribution (x_c) from a continuum. The constant ratio (x_1/x_2) of the intensities of the interfering line and a neighboring line of the same concomitant is used in the background correction as explained in the text.

ground level via the measurement at λ_c and changes in the concentration of the concomitant P via the measurement at λ_2. It is readily seen that this case includes the previous one (Fig. 7.28), the only difference being that not k but kx_2 was a constant derived from the reference blank.

Actually all types of classical background correction are based on Eq. (7.8) or similar equations. This works well as long as the underlying assumptions are fulfilled, namely that (1) the variations of the continuous background at λ_a are identical to Δx_c; (2) k is constant, in other words, x_2 and x_1 are perfectly correlated.

The fundamental problem is that these assumptions can be violated by the appearance of spectral features (except for continua) at any of the three measuring points: λ_a, λ_c, and/or λ_2. If the appearance of such spectral features is not recognized, errors will result. Therefore, the measuring procedure, in particular the choice of λ_c and λ_2, must be carefully established from spectral scans for each sample type separately, while checks are required whenever changes in sample composition can be expected. The literature in general and the manuals of slew-scan spectrometers furnish numerous examples of background correction in which the above principles and rules are applied to situations gleaned from an infinite variety of possibilities offered by real samples.

A novel approach to background correction has been recently proposed by

7.6. SPECTRAL INTERFERENCE CORRECTIONS: POLYCHROMATORS   417

Taylor and Schutyser [80] who describe a computer program that performs quantitative spectral analysis in much the same way as the programs used in, for example, infrared spectroscopy and instrumental neutron activation analysis.

Quantitative spectral analysis here means that the information obtained during a scan (covering approximately 15 effective line widths about the analysis line) is quantitatively used and processed. This contrasts with the traditional approach outlined above where the scans have merely qualitative significance as an aid to establish the best wavelengths for background measurements. Quantitative analysis of spectral scans implies smoothing of raw data, peak detection, background correction, test and correction for spectral interference, and calculation of the final net signal and its standard deviation. The program described by Taylor and Schutyser permits accurate analyses without prior knowledge of sample composition and requires a minimum of operator-software interaction. Those authors also compared the results of the proposed method with those of the traditional "three-point method" as used in the case of sloping background of which Fig. 7.30 shows an example: the determination of Pb in silicate rocks. Figure 7.30 (top) is the scan for a Pb content of 200 ng/mL in the presence of 20 μg/ml Al (0.3% in the rock). This spectrum closely resembles that of a calibration standard and hardly shows the Al lines at 220.462 and 220.467 nm. Spectral analysis yields the continuous line as background, while visual inspection would lead to the points A and B for the conventional three-point background correction ("sloping background"); thus identical results will be obtained. If the Al concentration in the rock increases to 12%, one obtains the scan shown in Fig. 7.30 (bottom), which illustrates in a self-explanatory way the disaster if the traditional method were applied without reinspection of the scan. Quantitative spectral analysis, on the contrary, yields the continuous curve, instead of the broken one, as background spectrum and obviously leads to a correct determination of the Pb content.

7.6. CORRECTIONS FOR SPECTRAL INTERFERENCES IN POLYCHROMATORS

In the preceding section we have seen that background correction can eliminate a number of errors due to spectral interference. Generally, methods for background correction use spectral scans or at least spectral shifts that permit intensity measurements at one or more spectral positions beyond the main body of the profile of the analysis line. In what way scanning or shifting is achieved is immaterial in this context (cf. Chapter 8); essential is that off-peak spectral information is used to subtract a background signal from the gross line signal measured at the peak of the line. This approach inherently involves an extrap-

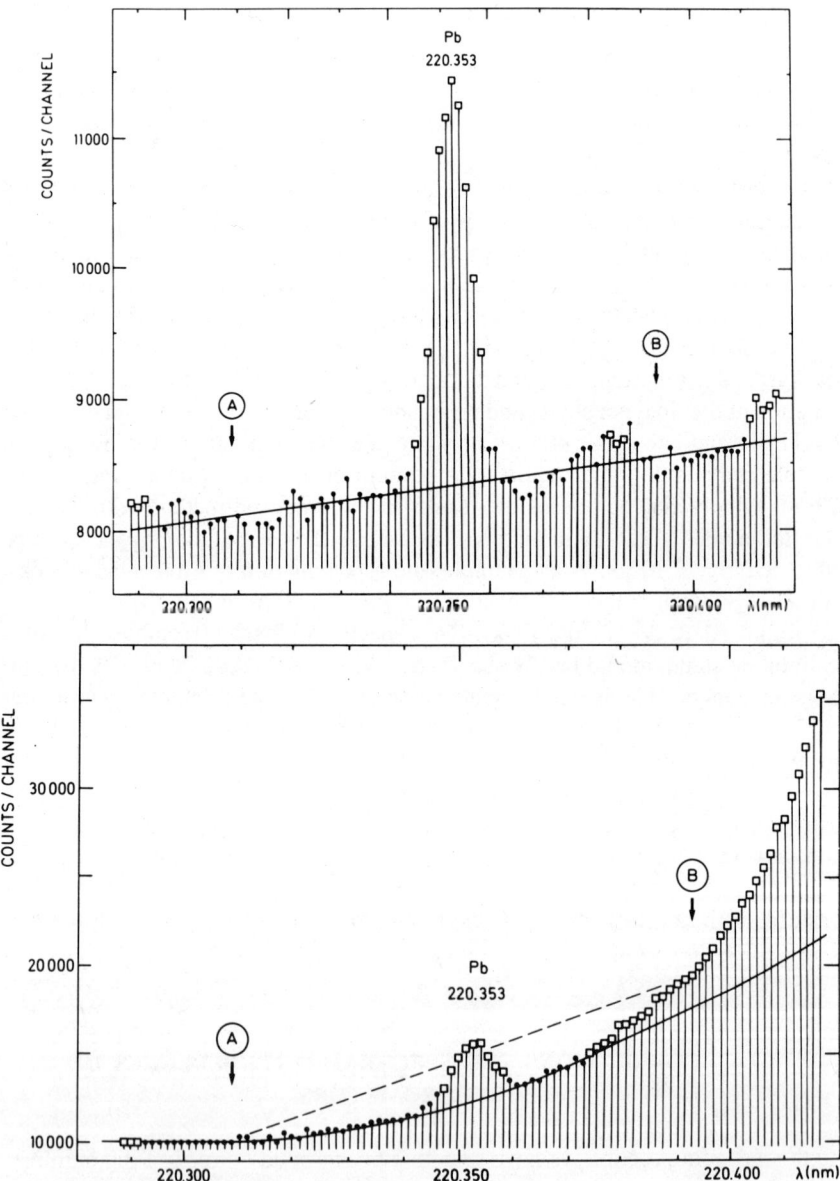

Figure 7.30. Top: spectral scan for 200 ng/mL Pb in the presence of 20 μg/mL Al. Analysis line: Pb I 220.352 nm. The dots represent the points used in the background fit; the continuous curve represents the calculated background. Points *A* and *B* are chosen for traditional background correction. Bottom: spectral scan for 200 ng/mL Pb in the presence of 600 μg/mL Al. The continuous line represents the calculated background: the broken line is the "sloping background" assumed in the traditional approach [80]. [Reprinted with permission from P. Taylor and P. Schutyser, "Description of a Computer Program for Quantitative Spectral Analysis of ICP-AES Spectra Generated a High Resolution Computer—Controlled Monochromator", *Spectrochim. Acta* **41B**, 98 (1986). Copyright (1986), Pergamon Journals.]

7.6. SPECTRAL INTERFERENCE CORRECTIONS: POLYCHROMATORS

olation of which the outcome is more reliable when better knowledge is available about the relationship between the actual background under the line and the off-peak background signal.

Background correction is traditionally associated with monochromators; this is so for various reasons. First, monochromators incorporate a scanning facility. Second, monochromators are commonly used for the determination of only a few elements of a sample which precludes the use of correction methods based on a complete analysis ("multicomponent correction methods"). Third, even if a complete analysis is made, the scanning facility deprives a monochromator to a certain extent from the stability required for the application of a multicomponent correction method. Future developments, however, might change this situation.

On the other hand, polychromators are designed for simultaneous multielement analysis and, therefore, easily permit complete analyses. This feature, added to the stable alignment of the exit slits, gives ready access to multicomponent correction methods. However, the implementation of such methods is virtually indispensable since the lack of flexibility in line choice necessitates many compromises in line choice. It will hardly ever be possible to adopt a set of lines entirely free from overlappings with lines of concomitants, in particular, if samples of a large variety have to be analyzed and low concentration levels must be covered. The problem of line overlapping becomes even more serious than with monochromators because the stability requirement tends to preclude the use of high practical resolving power (see, however, Section 7.7.7.6).

Since the advent of ICPs, polychromators are forever provided with facilities for scanning or shifting the spectrum at all exit slits so that background correction based on off-peak measurements can also be applied. This has resulted in the development of multicomponent correction methods incorporating off-peak background measurements.

Multicomponent correction methods are based on the elementary correction method for spectral line overlap described by Boumans [83], of which the principle is discussed in Section 8.7. Actually the formal basis of this method can be found in an article by Kaiser [84], which has been reprinted [85], but did not receive much attention in the literature (see, however, [86] and Section 7.7.7).

In principle the correction method is based on a set of calibration functions of the type discussed in Section 1.6.5 (Eqs. 1.14–1.16), whereby it is usually assumed that multiplicative interferences (Section 1.6.4) are absent, that is, the sensitivities of the analyte signals at their own wavelengths (S_{11}, S_{22}, . . . S_{nn}) are assumed to be independent of the concentrations of the concomitants. In fact this is also assumed for the partial sensitivities of the concomitants defined below. The starting point then is a set of calibration equations of the type

$$\left.\begin{aligned} x_1 &= S_{11}c_1 + S_{12}c_2 + \ldots S_{1n}c_n \\ x_2 &= S_{21}c_1 + S_{22}c_2 + \ldots S_{2n}c_n \\ &\quad\vdots \\ x_n &= S_{n1}c_1 + S_{n2}c_2 + \ldots S_{nn}c_n \end{aligned}\right\} \quad (7.9)$$

where S_{jk} ($j \neq k$) is the partial sensitivity of the signal of interfering component k in the spectral window in which the signal of analyte j is measured, while S_{jj} is the sensitivity of the analyte signal in its own window (cf. [83]); $x_1, x_2, \ldots x_n$ are the total signals observed in the n windows and $c_1, c_2, \ldots c_n$ are the concentrations of the n components.

Using an iteration procedure the matrix of sensitivities,

$$\left.\begin{matrix} S_{11} & S_{12} & \ldots & S_{1n} \\ S_{21} & S_{22} & \ldots & S_{2n} \\ & \vdots & & \\ S_{n1} & S_{n2} & \ldots & S_{nn} \end{matrix}\right\} \quad (7.10)$$

can be inverted to yield a set of equations of the type

$$c_j = \sum_{k=1}^{n} a_{jk} x_k \quad (7.11)$$

if in each row of the matrix the element at the diagonal is larger than the sum of all other elements in that row, that is, if

$$\frac{S_{jj}}{\sum_{k=1}^{n} |S_{jk}| - |S_{jj}|} > 1 \quad (7.12)$$

The larger the quotients (Eq. (7.12)), the better the iteration procedure converges.

Application of these equations implies the following preliminary steps:

1. The determination of $n \times n$ sensitivities using solutions of the pure components and blank solutions
2. Inversion of the matrix (Eq. 7.10) yielding the $n \times n$ coefficients a_{jk}

If this calibration and the inversion have been performed, the concentrations c_j

of the components of an unknown sample can be straightforwardly derived from the n measured signals x_k using Eq. (7.11).

Manufacturers of ICP polychromator systems usually deliver software based on Eqs. (7.9)–(7.11) or alternative versions originating from software developed for x-ray fluorescence analysis. For a mathematical treatment, refer, for example, to publications of Kowalski's group [87–89].[2]

The most comprehensive reports on the implementation of the previous method in ICP–AES have been published by Botto [73–76, 90]. In his initial works, Botto [73, 90] describes the set-up and the experience with a system for the determination of 33 analytes using 34 spectral lines and involving a square matrix of 34 × 34 interference coefficients, K_{jk}, which in terms of the formulation given above, are equivalent to ratios of sensitivities. Botto [73] defines:

$$K_{jk} = \frac{\text{(spurious concentration observed for element } n\text{)}}{\text{(actual concentration of interferent } k\text{)}} - \frac{\text{(concentration of } j \text{ observed for the blank)}}{\text{(actual concentration of interferent } k\text{)}} \quad (7.13)$$

where j refers to the analyte. The denominator in Eq. 7.13 is the spurious concentration (c_{jk}) derived from the net signal ($S_{jk}c_k$) of the interferent,

$$c_{jk} = \frac{S_{jk}c_k}{S_{jj}} \quad (7.14)$$

in the case where no off-peak background correction is applied; hence

$$K_{jk} = \frac{S_{jk}}{S_{jj}} \quad (7.15)$$

If we compare this equation to the expression for the critical concentration ratio (Eq. (7.4)) and consider that the sensitivities in Eq. (7.15) are experimental values such as observed in the window of the analysis line, we see that

$$K_{jk} = 1/\text{CCR} \quad (7.16)$$

This relationship provides a basis for a more universal interpretation of Botto's numerical results, at least on a semiquantitative basis, since the data are bound to the apparatus used, but this holds true for whatever experimental data on spectral interferences.

If off-line background measurements are employed, negative interference coefficients may occur. Then, in fact, K_{jk} becomes more complex,

---

[2] An interesting point directly related to the condition for convergence of the iteration (Eq. 7.11) is the definition of the selectivity of an analysis method as proposed by Kaiser [84, 85] and extended by Fujiwara et al. [86]. This topic will be discussed in Section 7.7.7.

$$K_{jk} = \frac{x_{jk}^G - x_{jk}^B}{S_{jj}} - \frac{x_{j0}^G - x_{j0}^B}{S_{jj}} \qquad (7.17)$$

where superscripts G and B distinguish the gross and background signals for the interferent (k) and the blank (0). If the off-line background measurement is made at an interfering line that contributes less intensity in the window of the analysis line, then $x_{jk}^B > x_{jk}^G$, which leads to a negative value of K_{jk}.

As becomes evident from Botto's work, the complete interference calibration involves a great many measurements, actually 34 × 34 measurements on standards of the pure elements and a similar number on the blanks, each with an off-line background measurement. Therefore, it is important for a calibration to remain valid over a considerable period of time.

To ensure the reproducibility of the interference calibration Botto used the intensity ratio of Cu I 324.75/Mn II 257.61 nm to monitor the excitation conditions. In a detailed analysis of the results Botto [73] showed that the accuracy of interference calibrations for ICP–AES polychromators can be adequately maintained over many months using the Cu/Mn line pair for monitoring and reproducing the plasma conditions. A period of six to eight months was considered as a maximum interval between recalibrations, in particular with a view to loss in sensitivity due to PMT fogging and the possibility of making more frequent realignments. As an example, Table 7.4 shows representative results from Botto's work [73]: interference coefficients for 29 spectral lines with uranium as interferent. The data include the results of calibrations at two dates, spanning about six months, with the percentage difference and with the grand mean of the two calibrations and 16 checks along with the grand RSD.

In a later paper Botto [75] gave a detailed account of the essential components of a quality assurance program, encompassing

1. An atom-to-ion emission intensity ratio for multielement optimization and for reproducing optimum analysis conditions
2. A concise easily applied specification for sensitivity and for precision
3. A regimen for monitoring of and correcting for calibration and background drift
4. A set of comprehensive spectral interference calibrations maintained using the atom-to-ion emission intensity ratio
5. A high resolution spectrometer for minimizing spectral interferences (Item (5) will be considered in more detail in Section 7.7.7.6)
6. A program of long-term performance monitoring and maintenance/record keeping.

7.7. SPECTRAL INTERFERENCE AND RESOLVING POWER

7.7.1. Introduction: Analytical Performance and Resolving Power

Some aspects of the influence of resolving power on analytical performance have been considered earlier. Sections 4.1.4.5 and 4.1.7.7 included results from a systematic study by Boumans and Vrakking [28] on the effect of the practical resolving power on the detection limits obtained with a 1.5-m echelle monochromator with predisperser (cf. Section 8.6.1). Section 4.1.10 referred to a comprehensive theoretical treatment of the dependence of the detection limits on the parameters of the spectroscopic apparatus and the detector by Laqua and his associates [91-95]. These sections concerned isolated spectral lines located on a smooth, structureless background.

Actually the largest effects of resolving power on analytical performance are found in the case that analysis lines suffer line overlap from concomitants. An increase in resolving power results in an increase in selectivity as a consequence of the better separation of the signals of the analyte from those of the interferents. Increased selectivity often means that better prominent lines become available for analysis. Consequently the detection power is improved. Closely related to this and even more important is that an increase in selectivity vastly lowers the limits of determination and, thus, extends an acceptable error level down to lower concentrations.

The purpose of this section is to elucidate these statements and to discuss the interrelations between the various quantities involved. The complexity of the subject is in the nature of the topic itself: selectivity is inherently linked with the sample composition and, therefore, with an infinite variety of situations. In addition, not only is the diversity of samples types infinite, also the physical profiles of spectral lines are far from identical, so the effect of the practical resolving power on analytical performance will depend on the types of line profiles involved. For this reason we shall first deal with some elementary notions on line shapes and widths.

7.7.2. Line Profiles and Line Widths

7.7.2.1. Physical Profiles

The profile of a spectral line, such as is emitted from the source, is referred to as the "physical line profile." The FWHM is called "physical line width" ($\Delta\lambda_{phys}$) or, briefly, "line width" if no confusion can arise.

The physical line profile is the resultant of various broadening effects in the source, distinguished as natural broadening, Doppler broadening, and several types of collisional broadening. A detailed discussion of this topic is irrelevant

Table 7.4 Interference Coefficients (IC) for Uranium Interference. The Table Lists the Results of Calibrations on 2 November 1980 (Cal I) and 5 May 1981 (Cal II) and the Mean Values of the Two Calibrations and 16 Monitoring Checks Performed between the Calibrations. $\triangle$ is the Percentage Difference between the Results of Cal I and Cal II. The RSD is the Grand RSD for the (16+2) determinations [73]. [Reprinted with permission from R. I. Botto, *Anal. Chem.* 54 1656 (1982). Copyright (1982), American Chemical Society, Washington, D.C.]

Spectral Line (nm)		Note	IC × 1000		$\triangle$ (%)	IC × 1000 Mean	RSD (%)
			Cal I	Cal II			
Ag I	338.289	*	0.69	0.66	−4.4	0.71	17.4
Al I	308.215		18.2	19.5	+6.9	18.6	11.6
Al I	237.312	w (x2)	1.42	1.31	−8.1	1.49	9.4
As I	193.696	* (x2)	0.98	1.11	+12.4	0.99	9.3
Ba II	455.403		0.15	0.16	+6.5	0.16	5.1
Ca II	315.887	w	13.5	21.3	+44.8	14.5	17.9
Co II	228.616	*	0.17	0.16	−6.1	0.16	7.2
Cr I	357.869	*	−1.12	−1.07	+4.6	−1.17	14.2
Cu I	324.754		0.97	0.98	+1.0	0.94	8.1
Fe II	259.940		2.03	2.04	+0.5	1.99	3.1
Fe II	238.076	w	720	779	+7.9	709	7.3
K I	766.490		0.36	0.53	+38.2	0.36	18.5

Mg I	383.231	w	7.67	8.86	+14.4	7.82	8.2
Mn II	257.610		0.48	0.57	+17.1	0.51	4.2
Na I	330.298	w	193	248	+24.9	191	14.5
Ni I	341.476	*	3.86	4.05	+4.8	3.79	7.3
P I	214.914	*(x2)	0.68	0.71	+4.3	0.70	13.0
Pb II	220.353	*	1.42	1.38	−2.9	1.41	10.9
Pb I	283.306	*	6.16	5.45	+12.2	6.42	15.2
Pt I	265.945		10.8	11.1	+2.7	9.35	9.0
Sb I	231.147	(x2)	2.22	2.33	+4.8	2.24	4.9
Se I	196.026	*	0.50	0.54	+7.7	0.56	16.3
Si I	288.158		9.93	10.7	+7.5	9.71	10.2
Si I	298.765	w	1900	2270	+17.7	1880	9.1
Sn II	189.980	*	−0.45	−0.47	−4.3	−0.49	13.3
Ti II	334.941		1.10	1.10	0	1.10	3.0
Tl I	377.572	*	28.0	35.8	−24.4	29.6	14.0
V II	292.402	*(x2)	1.88	2.24	−17.5	1.87	13.2
W II	207.911		0.88	0.85	−3.5	0.86	7.8

* Element determination made by using the spectrum shifter for background correction.
w Weak line for the determination of relatively high concentrations.
(x2) Line observed in the second spectral order.

425

in the present context and beyond the scope of this book. For a comprehensive treatment and review of literature, refer, for example, to Alkemade et al. [96]. Also the references, for example, cited by Larson and Fassel [26], Mermet and Trassy [29], and Batal and Mermet [97] may be a useful guide (cf. Part 2, Chapter 10). The following notions are of direct interest here.

In sources operating at atmospheric pressure the contribution from natural broadening is negligible in comparison with Doppler and collisional broadening. In an ICP, the prime cause of broadening is Doppler broadening [78, 97–100].

A Doppler profile is described by a Gaussian function, in normalized form [96]:

$$S_D(\lambda) = A \exp\left[\frac{-B\lambda_0 - \lambda^2}{\Delta\lambda_D^2}\right] \tag{7.18}$$

where $A = (2\sqrt{\ln 2}/\sqrt{\pi})(c/\lambda_0^2)$, $B = 4\ln 2$, c = speed of light, λ_0 = wavelength of line peak, $\Delta\lambda_D$ = Doppler width, expressed as FWHM. The Doppler width is given by

$$\Delta\lambda_D = 7.16 \times 10^{-7} \lambda_0 \sqrt{\frac{T}{M}} \tag{7.19}$$

Most types of collisional broadening yield a dispersion or Lorentzian profile, in normalized form [96]:

$$S_L(\lambda) = \frac{(2\lambda_0^2/\pi c)\Delta\lambda_L}{1 + 2^2(\lambda_0 - \lambda)^2/\Delta\lambda_L^2} \tag{7.20}$$

where $\Delta\lambda_L$ is the Lorentz width (FWHM).

The convolution [96] of a Doppler and a Lorentzian profile yields a Voigt profile characterized by the a parameter, which is defined as the ratio

$$a = \frac{\Delta\lambda_L \sqrt{\ln 2}}{\Delta\lambda_D} \tag{7.21}$$

The convolution of two Doppler profiles with halfwidths $\Delta\lambda_{D1}$ and $\Delta\lambda_{D2}$ yields a Doppler profile with halfwidth $\Delta\lambda_D$ according to

$$\Delta\lambda_D^2 = \Delta\lambda_{D1}^2 + \Delta\lambda_{D2}^2 \tag{7.22}$$

However, the resultant width of two Lorentzian profiles follows from

$$\Delta\lambda_L = \Delta\lambda_{L1} + \Delta\lambda_{L2} \tag{7.23}$$

Finally, the halfwidth ($\Delta\lambda_V$) of a Voigt profile is obtained in good approximation as [96]

$$\Delta\lambda_V = \tfrac{1}{2}\Delta\lambda_L + \sqrt{\tfrac{1}{4}\Delta\lambda_L^2 + \Delta\lambda_D^2} \qquad (7.24)$$

Figure 7.31 shows the normalized theoretical Doppler (D), Lorentzian (L), and Voigt (V) line profiles as functions of $x = 2 \ln 2 \, (\nu - \nu_0)/\Delta\nu$, where ν is frequency ($\nu = c/\lambda$ and $\Delta\nu = \Delta\lambda \cdot c/\lambda_0^2$).

7.7.2.2. Instrument Function

The interaction of a physical line profile with a spectroscopic apparatus leads to an effective line profile resulting from the convolution of the physical line profile and the instrumental line profile or instrument function. The latter is the profile the instrument yields for a purely monochromatic line. The FWHM of the instrumental profile is the practical spectral bandwidth ($\Delta\lambda_{\text{instr}}$). If both the physical and instrumental profile have a Gaussian (Doppler) shape, then

$$\Delta\lambda_{\text{eff}}^2 = \Delta\lambda_{\text{phys}}^2 + \Delta\lambda_{\text{instr}}^2 \qquad (7.25)$$

Often more complex situations in which the instrumental profile is a triangle, a trapezium, or a Voigt function and the physical profile also has a Voigt shape

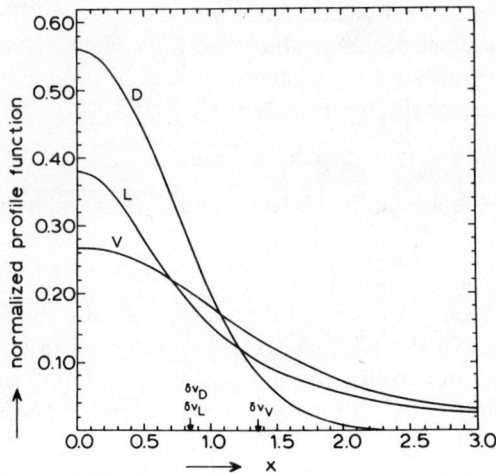

Figure 7.31. Normalized theoretical Doppler (D), Lorentzian (L), and Voigt (V) profiles are shown as functions of $x = 2\sqrt{\ln 2}\,(\nu - \nu_0)/\triangle\nu$, where $\triangle\nu = \triangle\nu_D = \triangle\nu_L$ (or $a = 0.84$). The profiles are symmetric with respect to the line center λ_0 (corresponding to $x = 0$); the areas of the half profiles shown are equal to 1/2 in accordance with the normalization applied. Arrows indicate the positions of the points at half peak intensity [96]. [Reprinted with permission from C. Th. J. Alkemade, Tj. Hollander, W. Snelleman, and P. J. Th. Zeegers, *Metal Vapours in Flames*, p. 150. Copyright (1982) Pergamon Books, Oxford.]

will be encountered. Then the quadratic addition (Eq. (7.25)) will not be necessarily valid so that corrections of Eq. (7.25) or formulas of types in Eqs. (7.30) and (7.31) must be applied (cf., e.g., [33, 100]).

Generally the instrumental width is made up of three contributions:

1. The diffraction width ($\Delta\lambda_0$) or theoretical resolution
2. The theoretical spectral bandwidth ($\Delta\lambda_s$)
3. A contribution from aberrations ($\Delta\lambda_z$)

Diffraction Width

The diffraction width is the width that results if monochromatic radiation emerging from an infinitely narrow slit is diffracted through the aperture of the spectroscopic instrument. The diffraction width is often called "theoretical resolution" and defines the theoretical resolving power

$$R_0 = \lambda/\Delta\lambda_0 \quad (7.26)$$

where $\Delta\lambda_0$ corresponds to the distance of the first minimum to the central maximum of the diffraction pattern. According to the Rayleigh criterion two lines of equal intensity will be resolved if the central wavelength of the one line coincides with the first diffraction minimum of the second line. The lines are then separated by a distance $\Delta\lambda_0$, which also is the width of each line measured at $(2/\pi)^2 = 0.405$ times the peak intensity[3].

For a grating or echelle spectrometer [33, 101, 102]

$$\Delta\lambda_0 = \lambda d/nB \quad (7.27)$$

where d = grating constant, n = spectral order, B = width of grating.

Theoretical Spectral Bandwidth

The theoretical spectral bandwidth or resultant spectral slit is the spectral width of the entrance or exit slit, whichever is the largest. The spectral width of a slit ($\Delta\lambda_s$) is the width in wavelength units computed from the angular slit width s/f (s = slit width, f = focal distance) and the angular dispersion D at that slit:

$$\Delta\lambda_s = \frac{s}{fD} \quad (7.28)$$

Chapter 8 and some recent ICP publications [33, 102, 108] contain examples of the calculation of $\Delta\lambda_s$ for grating and echelle monochromators.

---

[3] If $\Delta\lambda_0$ must be compared with widths expressed in terms of FWHM, then $\Delta\lambda_0$ should be divided by 0.885.

Aberrations

The contribution from aberrations is the most difficult to deal with because it escapes theoretical calculation and must be determined experimentally with the aid of lines whose physical widths can be neglected, such as nonresonance lines emitted from hollow cathode lamps.

Instrumental Width or Practical Spectral Bandwidth

In summary, the practical spectral bandwidth $\Delta\lambda_{instr}$ is the resultant of three contributions, for which, however, one cannot a priori assume a relationship of the form

$$\Delta\lambda_{instr}^2 = \Delta\lambda_s^2 + \Delta\lambda_0^2 + \Delta\lambda_z^2 \tag{7.29}$$

Recent literature provides two examples in which Eqs. (7.25)–(7.29) have been used.

McLaren and Mermet [102] showed for a 1-m monochromator equipped with a 3600 grooves/mm grating the validity of an equation of the type

$$\Delta\lambda_{eff} = (\Delta\lambda^2 + P)^{1/2} + Q \tag{7.30}$$

where P depends on the line and the spectral order and $Q = 0$ for the first and $Q \neq 0$ for the second order. For their purpose McLaren and Mermet considered the validity of Eq. (7.30) as a sufficient confirmation of the quadratic nature of convolution (Eq. (7.25)). This is true only, however, if $P = \Delta\lambda_{phys}^2$ and $Q = 0$, thus if $\Delta\lambda_0 = 0$ and $\Delta\lambda_z = 0$.

Actually McLaren and Mermet also show experimentally that for their system $\Delta\lambda_z$ is far from zero, but they do not further discuss how this contribution from aberrations must be precisely accounted for. Table 7.5, derived from their work, provides a good idea, however, of the magnitude of the various contributions. The resolving power computed for a slit width of 10 μm exceeds the theoretical resolving power by a factor of 3 to 5, indicating that the diffraction width is virtually negligible in comparison to the theoretical spectral bandwidth. The practical spectral bandwidth, in turn, lags a factor of 2 to 3 behind the theoretical value, as a result of aberrations.

Boumans and Vrakking [33] showed, for a 1.5-m echelle monochromator with predisperser, that at the slit widths used in their experiments $\Delta\lambda_0$ is negligible with respect to $\Delta\lambda_s$, while the effect of aberrations can be taken into account by a linear correction of the form

$$\Delta\lambda_{instr} = (1 + s_z/s)\Delta\lambda_s \tag{7.31}$$

where s is the actual slit width and s_z a correction, the magnitude of which depends on the slit height. For the slit height used in analytical work Fig. 7.32 shows the ratio of the practical-to-theoretical bandwidth for three slit widths. An approach to minimize and control s_z in a 1.5-m echelle monochromator has

Table 7.5. "Instrumental Width" and "Instrumental Resolving Power" for a 1-m Monochromator Equipped with a 3600 grooves/mm Grating Built up by Including Different Contributions [102]. [Reprinted with permission from J. W. McLaren and J. M. Mermet, *Spectrochim. Acta* **39B**, 1321 (1984). Copyright (1984), Pergamon Journals, Oxford.]

	FWHM (pm)		Resolving Power	
	Order 1	Order 2	Order 1	Order 2
Diffraction width—theoretical resolving power	0.5	0.24	500,000	1,000,000
Theoretical spectral bandwidth—resolving power, for slit width of 10 μm	2.5	0.7	90,000	330,000
Practical spectral bandwidth—resolving power, for slit width of 10 μm, with mask on grating[a]		1.6		145,000
Ditto, without mask on grating[a]	5.3	2.0	45,000	115,000

[a] The practical values are based on the measurement of the widths of narrow (nonresonance) lines emitted from a hollow cathode lamp.

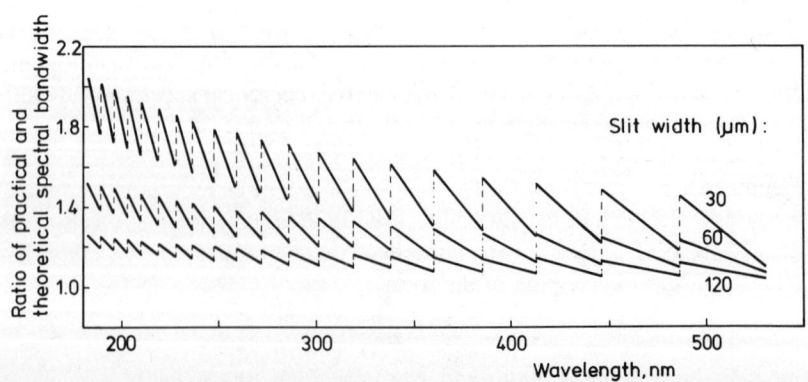

Figure 7.32. Ratio of practical and theoretical spectral bandwidth for a 1.5-m echelle monochromator with predisperser at slit widths of 30, 60, and 120 μm [33]. [Reprinted with permission from P. W. J. M. Boumans and J. J. A. M. Vrakking, *Spectrochim. Acta* **39B**, 1239 (1984). Copyright (1984), Pergamon Journals, Oxford.]

been described by Boumans and Vrakking [109] in a study aimed at the measurement of physical line widths.

7.7.3. Bandwidth Required for the Complete Physical Resolution of ICP Spectra

For judging the separative capability of a spectroscopic apparatus, criteria such as the Rayleigh criterion have been developed. In general one considers two lines of equal width ($\Delta\lambda_{\text{eff}}$) and intensity and defines the distance $\Delta\lambda_{1,2}$ between the line peaks

$$\Delta\lambda_{1,2} = k\Delta\lambda_{\text{eff}} \tag{7.32}$$

in terms of a constant k so that the intensity midway between the two peaks is a given percentage, for example, 80%, of the peak intensity. The value to be assigned to k depends on the profile shape, but in this context its precise value is irrelevant. We ask for the minimum value of r, defined as

$$r = \Delta\lambda_{\text{instr}}/\Delta\lambda_{\text{phys}} \tag{7.33}$$

yielding an optimum profit in analysis. From Eqs. (7.25), (7.32), and (7.33) one obtains

$$\Delta\lambda_{1,2} = k\Delta\lambda_{\text{eff}}\sqrt{(r^2 + 1)} \tag{7.34}$$

The step from $r = 2$ to $r = 1$ still decreases $\Delta\lambda_{1,2}$ by a factor of 1.6, whereas the step from $r = 1$ to $r = 0.5$ does not yield more than a 26% further decrease. Therefore, a practical spectral bandwidth better than the physical width does not bring an essential profit, but only leads to a further reduction of the radiant flux.

We shall see in Section 7.7.4 that the physical widths of most ICP line without hyperfine structure (HFS) are in a range between 1.4 and 5 pm. Thus bandwidths of this magnitude are needed and sufficient. The trend of line width with wavelength dictated by the Doppler effect (Eq. (7.19)) entails that the bandwidth actually needed varies with wavelength, as will be explained in the next section.

7.7.4. Physical Widths of Lines Emitted from an ICP

Line widths relevant to argon ICPs have been reported by the following authors:

1. Human and Scott [98] found a width of 3.6 pm for Ca I 422.673 nm using a Fabry–Perot interferometer.
2. Also using interferometry Kawaguchi et al. [99] measured the widths of some 15 lines of 10 elements. Their results are based on the assumption of a Lorentzian instrumental profile and Gaussian profiles for both hol-

low cathode and ICP lines, yielding a Voigt profile as experimental profile. For the deconvolution the authors used an approximation proposed by Posener [103], but Boumans and Vrakking [33] recalculated their results using the somewhat more accurate Eq. (7.24).

3. Hasegawa et al. [100] reported widths for 34 lines of 19 elements obtained with an echelle monochromator with crossed dispersion incorporating a refractor plate for wavelength scanning. Their results are based on the assumption that the experimental profile is a Voigt profile of which the Gauss (or Doppler) component ($\Delta \lambda_D$) of the halfwidth and the a parameter can be determined with a curve fitting procedure. The Gauss and Lorentz components are then separately split into an instrumental and a physical part using quadratic deconvolution for the Gauss components and linear deconvolution for the Lorentz components (Eqs. (7.22)–(7.23)).

4. Batal and Mermet [97] calculated line widths for nine atomic or ionic lines of Ca, Mg, and Sr on the basis of Doppler broadening ($T = 5000$ K) and van der Waals interaction. The total width was then computed with Eq. (7.24).

5. Broekaert et al. [104] determined the widths of 18 lines of 16 rare earths using a spectrograph provided with an order sorter. They used the same method and equipment as had been previously used by Laqua et al. [91] for the measurements of the widths of 67 lines of 40 elements emitted from four types of dc arcs and a soft spark.

6. Using high resolution Fourier transform spectrometry Faires et al. [78] determined the widths and shapes of 81 Fe I lines in the spectral range 290–390 nm.

7. Boumans and Vrakking [33] determined line widths with a 1.5-m echelle monochromator with predisperser. They covered 16 Fe I lines (364–384 nm) for comparison with the results of Faires et al. [78] and subsequently 13 lines of nine other elements. Recently, Boumans and Vrakking [109] reported the results of the measurement of the physical line widths of about 350 prominent lines of 65 elements as emitted from an ICP. The publication includes an atlas with the effective profiles of some 100 lines with HFS, measured at spectral bandwidths between 1.3 and 3.8, as dictated by characteristics of the echelle monochromator.

Table 7.6 shows representative results from the above references, while Fig. 7.33 is a plot of line width against wavelength covering a large number of results from (1) to (7), including those of Laqua et al. [91] for the dc arcs and soft spark. The latter results appear to be slightly higher than those for the argon ICP. This may be due to a difference in the source characteristics or due to an overlooked instrumental contribution to line broadening. (cf. [109]).

Table 7.6. Representative Values of Physical Line Widths Reported for the ICP and Some Types of DC Arc and a Soft Spark [33].
[Reprinted with permission from P. W. J. M. Boumans and J. J. A. M. Vrakking, *Spectrochim. Acta* 39B, 1239 (1984). Copyright (1984), Pergamon Journals, Oxford.]

Spectral Line (nm)		Physical Line Width[a]				
		[33]	[99]	[100]	[97]	[91]
Pb II	220.353	1.4	1.6	—	—	—
Cd II	226.502	2.2	1.6	—	—	—
Cd I	228.802	1.4	1.4	1.4	—	—
Mn II	257.610	3.6			—	4.6
Mn II	260.569	4.2				—
Mg II	279.553	3.4	3.9	3.3	—	4.1
Pb I	283.306	2.4			—	2.7
Mg I	285.213	3.8	3.7	3.5	3.3	4.9
Ca II	393.367	4.4	4.4	3.5	4.1	4.5
Al I	396.152	5.4	5.5	4.8	—	5.7
Sr II	407.771	3.5	3.3	2.5	—	3.7
La II	408.672	3.4	3.4		—	—
Ca I	422.673	4.6	4.6	4.2	4.4	4.8

[a] Numbers in brackets are reference numbers.

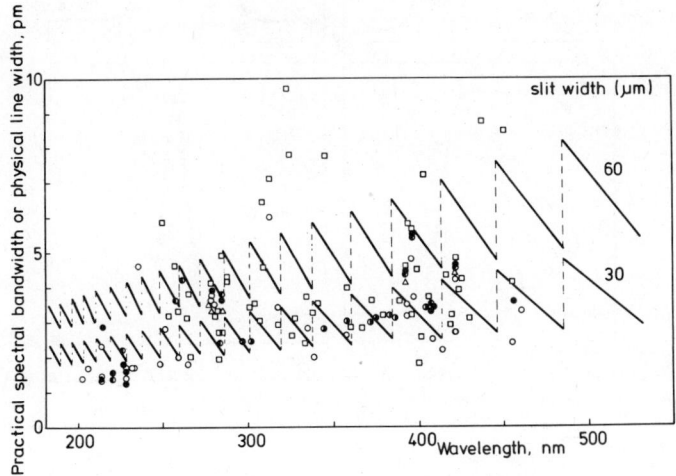

Figure 7.33. Matching of the practical spectral bandwidth attainable with the echelle monochromator and physical line widths reported in the literature. The figure shows the saw-tooth patterns of the bandwidth for 30- and 60-μm slit widths and data on line widths from Laqua et al. [91]: □, Hasegawa et al. [100]: ○, Kawaguchi et al. [99]: ●, Batal and Mermet [97]: △, Faires et al. [78]: ◑, and Boumans and Vrakking [33]: ◐. The data refer to ICPs except for those of LAQUA et al., which are mean values for four types of dc arcs and a soft spark [33]. [Reprinted with permission from P. W. J. M. Boumans and J. J. A. M. Vrakking, *Spectrochim. Acta* **39B**, 1239 (1984). Copyright (1984), Pergamon Journals, Oxford.]

Figure 7.33 also includes results for lines with unresolved HFS. These data refer to the total width of the unresolved structure. Although these widths are apparent only, they are realistic in analytical work because the HFS components are either physically unresolved or cannot be resolved under analytically usable conditions. A few examples of spectral scans of such lines are shown in Fig. 7.34, while numerical results reported by McLaren and Mermet [102] and Broekaert et al. [104] for rare earths and associated elements are shown in Table 7.7.

The bulk of the data in Fig. 7.33 (excluding results for lines with HFS) follows a trend with wavelength in agreement with the proportionality of the Doppler width with wavelength (Eq. (7.19)). The scatter is primarily caused by the dependence of the Doppler width on the atomic mass, but must also be

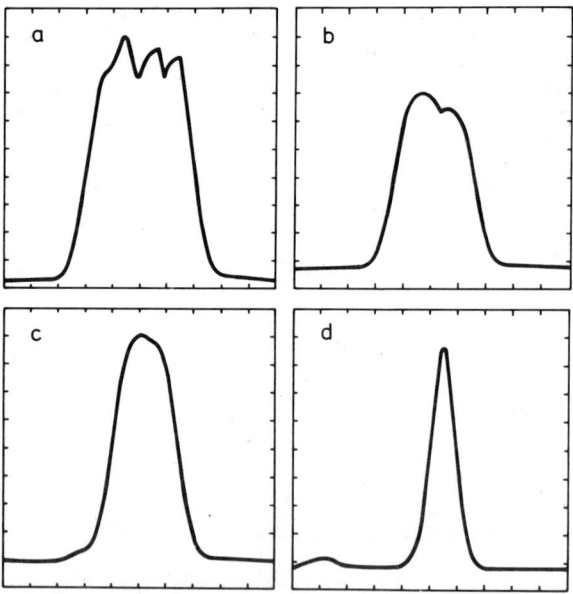

Figure 7.34. Spectral scans of lines apparently broadened by unresolved HFS (*a*, *b*, *c*) and, for comparison, a narrow line (*d*) with chiefly Doppler broadening

	Spectral Line (nm)	Effective Line Width (pm)	Spectral Bandwidth (pm)
(a)	Nb II 292.781	16.5	4.5
(b)	Lu II 289.448	12.5	4.8
(c)	Lu II 290.030	10.5	4.7
(d)	Mo II 292.339	4.5	4.5

(Based on data from Boumans and Vrakking [70].)

7.7. SPECTRAL INTERFERENCE AND RESOLVING POWER

Table 7.7. Experimental Effective Line Widths Approximating the (Apparent) Physical Widths of Lines of Some Rare Earths and Associated Elements [102].
[Reprinted with Permission from J. W. Mclaren and J. M. Mermet, *Spectrochim. Acta* **39B**, 1319 (1984). Copyright (1984), Pergamon Journals, Oxford.]

	Wavelength (nm)	Line Width (pm)[a]	
		[102]	[104]
Eu II	381.965	12.3	
II	412.973		9.5
II	420.505		14.0
II	393.051	12.2	
II	390.711	10.2	
II	372.494	12.5	
II	397.199	13.0	
Gd II	343.246	6.3	
II	364.619		5.6
Ho II	339.888	21.0	
II	345.600	21.0	19.0
La II	398.852	15.0	13.0
II	408.672	6.2	4.8
Sc II	361.384	6.5	6.0
Yb II	211.665	2.7	
II	222.445	2.6	
II	369.420		2.7

[a] Numbers in brackets are reference numbers.

partly attributed to different contributions from Lorentz broadening, experimental errors, and possibly HFS.

7.7.5. Matching of Practical Spectral Bandwidths and Physical Line Widths

Here we consider to what extent two recently discussed monochromators provide, under analytically usable conditions, spectral bandwidths matching the physical widths of ICP lines. These instruments are

1. The 1.5-m echelle monochromator with predisperser explored by Boumans and Vrakking [33] (cf. Fig. 8.23)

2. The 1-m grating monochromator considered by McLaren and Mermet [102]

For this assessment the line width data in Fig. 7.33 have been projected on two saw-tooth curves representing the practical spectral bandwidth of the echelle monochromator for two slit widths (entrance slit width = exit slit width).

The 30-μm slit width is seen to provide the bandwidth for the complete physical resolution of ICP spectra. However, in view of the required radiant flux (Section 4.1.4), a slit width of 60 μm will usually be needed. Then the practical bandwidth lags a factor of 1 to 3 behind that ideally required.

The figure also reveals that the practical bandwidth of this echelle monochromator at a given slit width reasonably follows the trend of the physical line widths with wavelength. This is an interesting property of echelle instruments not found in common grating monochromators (Fig. 7.35) for which the bandwidth at constant slit width decreases with increasing wavelength. This also happens with an echelle monochromator within an order, but the decrease in spectral order with increasing wavelength introduces such jumps in bandwidth at the order transitions that the bandwidth on average increases with wavelength. As shown in Fig. 7.35, a common grating monochromator [102] requires two gratings and frequent adaptations of the slit width for adequate matching of bandwidth and physical line widths over an extended wavelength range. In addition, when one travels to lower wavelength, the slit width must be reduced; this is not welcome, in particular, in the low UV since the background radiance and the PMT response also tend to decrease in that direction.

Note further that the curves in Fig. 7.33 refer to the theoretical bandwidth, the actual practical bandwidth being higher, as has been discussed in connection with Table 7.5 and Fig. 7.32.

McLaren and Mermet [102] show that a 1-m monochromator equipped with a 3600 or 2400 grooves/mm grating used in the second order with a 10-μm slit width and a mask on the grating can yield the bandwidth required for the complete physical resolution of the narrowest lines in the low UV. However, in contrast to Boumans and Vrakking, those authors did not systematically investigate what sacrifices are required for analytical work; in particular, they did not reveal what slit widths are needed to keep the background noise at an acceptable level (cf. Section 4.1.4).

Finally, if we consider that the required bandwidth more or less varies with wavelength from 1.4 pm at 200 nm to 5 pm at 500 nm, we can state that the practical resolving power needed and sufficient in ICP spectrometry should be in a range between 150,000 and 100,000 [33, 102]. This magnitude is identical to that stated by Laqua et al. [91] for the dc arc and soft spark; this is obviously so since the line widths in an ICP do not essentially differ from those in a dc arc or soft spark.

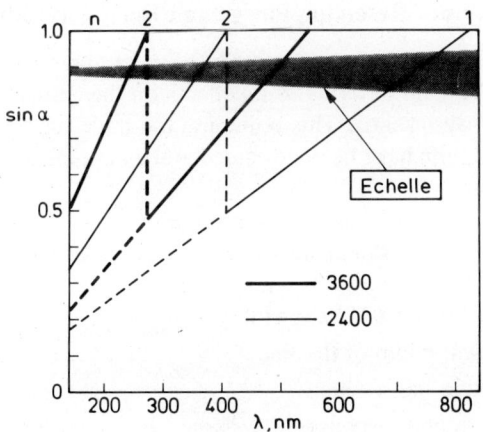

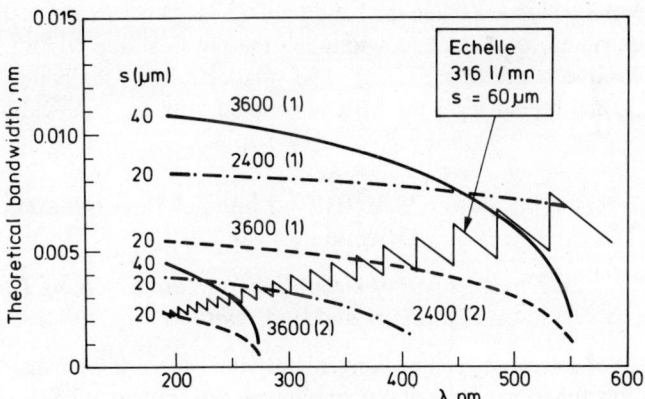

Figure 7.35. Upper diagram: dependence of the sine of the grating angle on wavelength for a 1-m monochromator equipped with a grating of 3600 or 2400 grooves/mm and used in orders 1 or 2 [102]. This picture contrasts with that for an echelle monochromator such as described in [33] (cf. Section 8.6.1) in that the echelle angle follows a saw-tooth pattern about the blaze angle and remains within the hatched area.

Lower diagram: dependence of the theoretical bandwidth on wavelength for the 1-m grating monochromator [102] and 1.5-m echelle monochromator [33]. The zigzag curve is for the echelle instrument; all other curves are for the grating apparatus. The labels refer to the groove density (3600 and 2400 groves/mm), the spectral order (1 or 2), and the slit width (s). Slit widths of 20 and 40 μm have been estimated as being required to overcome shot noise limitations. This point was not covered in the original paper [102].

7.7.6. Resolving Power and Detection Limits

The effect of the practical resolving power (or spectral resolution) on a detection limit can be understood in terms of its effects on the SBR and the background RSD (Eqs. (4.16 and 4.30)). This point and the trade-off between radiant flux and spectral resolution have been discussed and exemplified in Sections 4.1.4.5 and 4.1.7.7.

In summary, if one wishes to interpret an experimentally found change in detection limit in an individual situation, one should consider

1. The practical spectral bandwidths
2. The physical width of the line
3. The effective line widths
4. The background responses

A change in SBR is directly related to a change in effective line width ($\Delta\lambda_{eff}$). How much $\Delta\lambda_{eff}$ changes when the bandwidth ($\Delta\lambda_{instr}$) is varied depends on the relative contributions of the bandwidth and the physical line width ($\Delta\lambda_{phys}$) to the effective line width (Eq. (7.25)). The smaller $\Delta\lambda_{phys}$, the better $\Delta\lambda_{eff}$ follows $\Delta\lambda_{instr}$ and the stronger the SBR is coupled with the bandwidth [69, 70, 107].

7.7.7. Resolving Power, Selectivity, Limits of Determination, and Detection Power

7.7.7.1. Selectivity, Limit of Determination, and Resolving Power: Definitions and Interrelations

The concept of selectivity can be conveniently introduced if we consider the set of calibration functions for a multicomponent system (Eq. (7.9)), the corresponding matrix of sensitivities (Eq. (7.10)), and the condition (Eq. (7.12)) for convergence of the iteration leading to the set of inverted equations (Eq. (7.11)).

Condition (7.12) led Kaiser [84, 85] to propose as a definition of the selectivity ξ the minimum value occurring among all the quotients of type (7.12) minus 1:

$$\xi = \min_n \frac{|S_{jj}|}{\sum_{k=1}^{n} |S_{jk}| - |S_{jj}|} - 1 \quad (7.35)$$

For a fully selective procedure ξ becomes very large (formally infinite). If ξ is only a little above zero, one can hardly speak of selectivity.

Fujiwara et al. [86] extended this proposal to cover both multiplicative and additive interferences (cf. Section 1.6) and to include the concentration depen-

7.7. SPECTRAL INTERFERENCE AND RESOLVING POWER

dent selectivity, which, in the case of additive interference only, is defined as

$$\xi_c = \min \frac{|S_{jj}|c_j}{\sum_{k=1}^{n} |S_{jk}|c_k - |S_{jj}|c_j} - 1 \tag{7.36}$$

Fujiwara et al. [86] also evaluated the selectivity for samples of sea water, orange juice, and blood serum (10 analytes) for a flame atomic absorption procedure and an emission spectrometric procedure using a dc plasma with an echelle monochromator.

However interesting and rigorous this formalization, it meets the objection that the selectivity of a method is identified with the "worst case" only. It is equivalent to identifying the detection power of a method with that of the poorest available analysis line. Clearly it is far more useful to characterize the detection power of a method for a given sample type by a statement of the set of detection limits that can be reached with the analysis lines adopted.

This consideration led Boumans and Vrakking [69] to introducing the concept of "line selectivity,"

$$v_{cj} = \frac{|S_{jj}|c_j}{\sum_{k=1}^{n} |S_{jk}|c_k - |S_{jj}|c_j} \tag{7.37}$$

so that, in principle, a set of v_{cj} values for the "lowest analyte concentrations of interest" would characterize the selectivity of the method for a given sample type. However, Boumans and Vrakking then reverse the argument and do not introduce the line selectivity as an analytical figure of merit but as a criterion for defining the limit of determination. In other words, v_{cj} is equalized to a threshold value:

$$v_{cj} = v_{thr} \tag{7.38}$$

and the corresponding analyte concentration c_D then found from Eq. (7.37) is defined as the limit of determination, which is thus identified with the "lowest concentration of interest." The argument behind this proposition is that the error (Δ) in the concentration is a function of v_{cj}:

$$\Delta = f(v_{cj}) \tag{7.39}$$

Then, for defining c_D the value of v_{thr} must be chosen so that, for example, $\Delta = 10\%$.

It is rational to base c_D on a fixed value of Δ, but in view of the uncertainties inherent in error models, it is more practical to use a fixed value of v_{thr} as the basis for defining c_D. We shall distinguish the two limits of determination by the symbols c_d (fixed error of 10%) and c_D (fixed value of v_{thr}). This point will be further elucidated in Section 7.7.7.2. For the sake of continuity the definition of the limit of determination should be extended so as to include the detection limit (c_L). The eventual definition then is

$$c_D = v_{thr} \frac{\sum_{k=1}^{n} |S_{jk}| c_k}{|S_{jj}|} + 5c_L \qquad (7.40)$$

where the term with $k = j$ is precluded from the summation in the denominator. The value of 5 links up with the general custom of defining the limit of determination in a system free from spectral interference as five times the detection limit. This custom is based on the fact that under these conditions the RSD in a net line signal is 10% (Section 4.2.2). According to Eq. (7.40), generally $c_D > 5c_L$, unless the concentrations c_k of all concomitants or their partial sensitivities S_{jk} are zero.

This approach permits the full appreciation of the effect of the spectral resolution on analytical performance, in particular in the case of line overlap. Then the resolution exerts a major effect on the limit of determination via the sensitivity ratio S_{jk}/S_{jj} and a minor effect via the term $5c_L$, which, in addition, tends to be small in comparison to the first term.

The effect of the resolution on the limit of determination can be formally expressed as

$$c_D = v_{thr} \sum F_k (\Delta\lambda_{instr}) c_k + 5c_L \qquad (7.41)$$

where $k \neq j$ and c_k are the concentrations of the concomitants. The detection limit c_L, too, is a function of $\Delta\lambda_{instr}$, but c_L depends less strongly on $\Delta\lambda_{instr}$ than F_k (cf. Section 7.7.7.4); c_L also depends on the concentrations of the concomitants, but increases linearly, not proportionally with these concentrations, because it is the sum of background and interfering signals, not the sum of the net interfering signals alone, which make up the "background" in the SBR.

The introduction of the limit of determination has only apparently distracted the attention from the concept of line selectivity: the limit of determination, however, essentially covers this concept and provides a more elegant way to express the degree of selectivity for each analysis line separately by a quantitative measure with a realistic meaning. The value of this measure can be easily estimated and visualized. Should one wish to express the degree of selectivity more specifically, then the ratio of c_D to c_L, its reciprocal, or any other function of it might be proposed.

It should be noted that Eq. (7.40) appears to express an inconsistency namely, $c_D > 5c_L$. The reason is that c_D and c_L are not defined by the same experimental procedure. Detection limit c_L is determined by making repeated measurements on a blank matrix at the fixed wavelength of the analysis line. This yields the RSD of the "blank" signal instead of the RSD of the (continuous) background.

7.7. SPECTRAL INTERFERENCE AND RESOLVING POWER

The RSD of the blank signal is made up of the flicker noise contributions of background and net interfering signal (which may show correlation), the shot noise contribution of the total blank signal, and possibly a detector noise contribution. However, this "static" measurement at a fixed wavelength using a blank matrix of fixed composition does not take into account the fluctuations of the net interfering signal from sample to sample caused by (1) fluctuations in the matrix composition, (2) fluctuations in the excitation conditions, and (3) instabilities in the wavelength setting including wavelength positioning errors in slew-scan monochromators.

The limit of determination as defined by Eq. (7.40) does cover the latter types of fluctuations and, therefore, exceeds the conventional detection limit by more than a factor of 5. Truly speaking, this conventional, "static" detection limit is an irrealistic figure of merit in the case of line overlap. One should actually define the limit of detection "dynamically" as $(1/5)$ c_D, whereby c_D is defined by Eq. (7.40). This is explicitly discussed in [106].

To link up with current customs of defining and measuring detection limits we have coupled in Eq. (7.40) the static limit of detection with the "dynamic" limit of determination. It is for this reason that the common link between c_D and c_L, that is, $c_D = 5c_L$, is disrupted.

Theoretically the problem could also be approached from the other side by taking into account all blank fluctuations such as those occurring with real samples. This would lead to realistic values of the RSD of the blank signal and the detection limit. The experimental determination of this RSD would meet with the difficulty, however, that the blank signal of real samples containing the analyte cannot be measured. Therefore, Boumans and Vrakking proposed the use of the limit of determination as a starting point and then considered whether a rigorous definition based on error considerations could be replaced by an approximation, that is, Eq. (7.40), which contains quantities that can be readily determined experimentally. This approach will be discussed in the following sections.

7.7.7.2. Selectivity, Limit of Determination, and Resolving Power: An Example

For illustration, Table 7.8 shows results of a case study: Ge I 206.866 nm with interference from Ni I 206.862 nm, as measured with a matrix of 1 mg/mL of each Ni and Co using the 1.5-m echelle monochromator [28, 69]. For convenience we call here the resolution obtained with the 210 and 60-μm slit widths MR and HR respectively.

The upper part of Table 7.8 shows

1. The SBR of the analysis line, $(SBR)_a$, where the background includes the contribution from the Ni line

Table 7.8. **Results of a Case Study for Ge I 206.866 nm Suffering Interference from Ni I 206.862 nm in a Matrix Containing 1 mg/mL of each Ni and Co. [69].**
[Reprinted with permission from P. W. J. M. Boumans and J. J. A. M. Vrakking, *Spectrochim. Acta* **40B**, 1085 (1985). Copyright (1985), Pergamon Journals, Oxford.]

Slit (μm)	$(SBR)_a$	$(RSD)_B$ (%)	Detection Limit (ng/ml) (actual)	Detection Limit (ng/ml) (− Ni line)	$(SBR)_i$	v_c
210	1.0	0.63	68	29	2.3	1.4
135	1.7	0.83	52	28	1.8	2.6
90	2.7	1.2	47	29	1.3	4.8
60	4.0	1.9	51	38	0.73	9.5

Slit (μm)	c_d (ng/mL)	$\Delta = 10\%$, c_d/c_L	v_d	c_D (ng/mL)	$v = 1$, c_D/c_L	Δ (%)
210	1900	28	0.71	2700	40	11
135	1000	20	0.72	1400	27	8
90	630	13	0.79	790	17	7
60	420	8	1.04	400	8	7

  2. The background RSD
  3. The actual value of the detection limit computed from $(SBR)_a$, $(RSD)_B$, and the relevant analyte concentration of 3.8 μg/mL
  4. A hypothetical detection limit ("− Ni line") that would be obtained if the net Ni line intensity could be zeroed, the wing background from both Ni and Co remaining the same
  5. The SBR of the Ni line, $(SBR)_i$, in the window of the analysis line
  6. The selectivity v_c of the Ge line

The data show that the SBR of the analyte line, $(SBR)_a$, increases by a factor of 4 if we proceed from MR to HR. This profit is for the larger part compensated by an increase in background RSD [28], so the detection limit improves insignificantly. Even if the Ni line could be completely removed, all other conditions remaining the same, the HR profit would still be rather marginal.

The SBR of the Ni line in the window of the Ge line decreases by more than a factor of 3 and the line selectivity v_c improves by a factor of 6.8. The consequences of this improvement are revealed in the lower part of Table 7.8 where three quantities are listed under the heading "$\Delta = 10\%$":

  1. The limit of determination c_d defined here to be the concentration for which the random error in the net signal is 10%; it is calculated with the error function discussed in [69]

2. the ratio of c_d and the actual detection limit c_L
3. the line selectivity v_d at the limit of determination c_d

Clearly, the transition from MR to HR reduces the limit of determination c_d by a factor of 4.5, while the ratio c_d/c_L is improved by a factor of 3.5, covering a range from 28 to 8.

To conclude, the limit of determination c_d varies much more with the resolving power than the limit of detection, while the ratio of the limit of determination to the limit of detection is essentially larger than 5.

We now consider to what extent the limit of determination c_d can be identified with the limit of determination c_D. Table 7.8 shows that the values of the selectivity v_d are of the order of 1 at all slit widths. We now reverse the argument and compute the Ge concentrations and SBRs for which $v = 1$, and subsequently calculate the error Δ. The results are shown in the lower part of Table 7.8 under the heading "$v = 1$." In view of the uncertainties inherent in an error calculation, we may conclude that the two approaches do not yield essentially different results.

Therefore, it is considered that the convenient and practical definition of the limit of determination based on a fixed threshold value v_{thr} is justified, if v_{thr} is assigned such a value that the limiting error in the net line signal is about 10%. This value will depend on the type of errors and, therefore, on the type of apparatus.[4]

7.7.7.3. Selectivity and Limit of Determination: Choice of Threshold Value

Boumans and Vrakking [69] show spectral scans of the Ge-Ni interference (Section 7.7.7.2) for $v_c = 1$ and $v_c = 3$ at both HR and MR. Considering these scans they argue that defining c_D on the basis of $v_{thr} = 1$ appears realistic under "static" measuring conditions, that is in the absence of wavelength drift, whereas a value larger than 1 will be needed under "dynamic" conditions such as prevailing in slew-scan spectrometers, where the principal error stems from wavelength calibration drift due to optical and/or mechanical instabilities [105].

A value of $v_{thr} = 1$ also agrees with Botto's finding [73-75] for complex interferences in a polychromator system for which he established that interelement corrections could accurately compensate for spectral interferences amounting to 50% of the uncorrected intensities, but failed at larger proportions of the interfering signals. It seems likely that $v_{thr} = 1$ must be considered as the minimum acceptable threshold, larger values being often required to maintain the 10% level.

---

[4]The concept of "critical concentration ratio" (CCR) introduced in the Line Coincidence Tables for ICP-AES" [13, 16] (cf. Section 7.4.4.1) is closely linked with that of limit of determination discussed here. If a the reciprocal of the CCR is multiplied by the concentration of the interferent, one obtains the limit of determination for $v_{thr} = 1$, aside from the term $5c_L$ (Eq. (7.40)).

On the basis of their experiments with a slew-scan monochromator Boumans and Vrakking [36] suggest $v_{thr} = 2$ as a starting point for further studies of the consequences of wavelength calibration drift in slew-scan monochromators (cf. Section 7.7.7.7).

7.7.7.4. Limit of Detection and Limit of Determination in Dependence on the Resolving Power

The example discussed in Section 7.7.7.2 provided some idea about the effects of the spectral resolution on the limits of detection and determination. Boumans and Vrakking [69] studied these effects systematically with the aid of computer simulation of line overlap using experimental line profiles measured at HR and MR with their 1.5-m echelle monochromator.

Figure 7.36 shows logarithmic plots of half profiles for which mutual interactions were investigated. Essentially the following approach was made. Of two profiles one was chosen as the analysis line, the other as the interfering line. The distance between the lines was step-wise varied from zero to a few effective line widths (at HR). For each situation of line overlap the limit of detection and the limit of determination were calculated. In addition the relative intensity of the interfering line was varied.

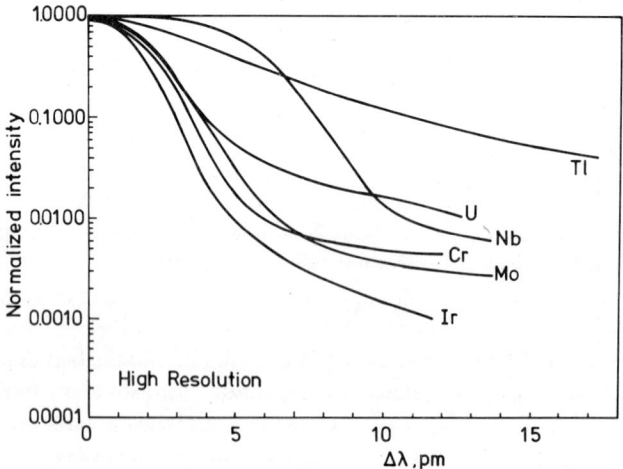

Figure 7.36. Logarithmic plots of effective half profiles observed at high spectral resolution. The curves refer to the following lines and practical spectral bandwidths (in parentheses): Ir I 284.972 nm (3.8 pm), Cr II 284.984 nm (3.8 pm), Nb II 294.192 nm (4.4 pm), U II 294.192 nm (4.4 pm), Tl I 291.832 nm (4.6 pm), and Mo II 292.339 nm (4.6 pm) [69]. [Reprinted with permission from P. W. J. M. Boumans and J. J. A. M. Vrakking, *Spectrochim. Acta* **40B**, 1085 (1985). Copyright (1985) Pergamon Journals, Oxford.]

7.7. SPECTRAL INTERFERENCE AND RESOLVING POWER

Table 7.9. Limits of detection and determination (ng/ml) in dependence on the simulated line separation [69].
[Reprinted with permission from P. W. J. M. Boumans and J. J. A. M. Vrakking, *Spectrochim. Acta* 40B, 1085 (1985). Copyright (1985), Pergamon Journals, Oxford]

	High Resolution		Medium Resolution		$(c_L)_{MR}$	$(c_D)_{MR}$
$\Delta\lambda$ (pm)	c_L (ng/mL)	c_D (ng/mL)	c_L (ng/mL)	c_D (ng/mL)	$(c_L)_{HR}$	$(c_D)_{HR}$
0	2.7	250	2.8	209	1.0	0.8
1	2.2	180	2.7	205	1.2	1.1
2	1.6	88	2.7	206	1.7	2.3
3	1.2	31	2.6	190	2.2	6.2
4	1.0	10	2.5	169	2.4	16.5
5	1.0	7.1	2.4	147	2.3	20.7
6	1.0	6.2	2.2	122	2.2	19.6
7	1.0	5.8	2.0	96	2.0	16.7
8	1.0	5.5	1.8	69	1.9	12.4
9	1.0	5.4	1.7	43	1.7	8.1
10	1.0	5.3	1.6	26	1.6	5.0
11	1.0	5.2	1.5	15	1.5	2.9
12	1.0	5.2	1.5	9.3	1.5	1.8

Note: Analysis line: Cr II 284.984 nm. Interfering line: Ir I 284.972 nm; SBR (HR) = 3; SBR (MR) = 1. Step size = 1.00 pm

As an example, Table 7.9 shows results for the interaction of Ir I 284.972 nm as interfering line with Cr 284.984 nm as analysis line. Evidently, if the distance between these lines is artificially varied from 0 to 0.012 nm, the limit of detection changes by less than a factor of 3 and 2 at HR and MR respectively. However, the change in the limit of determination is a factor of 50 and 20 respectively. From the ratios of the two limits at MR to HR shown in the last two columns of Table 7.10 we see that for the detection limit the HR profit is maximally a factor of 2.4, whereas the maximum profit for the limit of determination exceeds a factor of 20.

The six diagrams of Fig. 7.37 show MR/HR ratios of the limits of detection (top) and determination (bottom) for three profile combinations and SBRs of 30, 3, and 0.3 of the interfering line at HR. These diagrams illustrate the predominant role of the limit of determination compared to the limit of detection in an assessment of the profits of high resolving power. The diagrams also show that the magnitude of the profit varies with the types of line profiles involved.

Table 7.10 lists the maximum values of the ratios $(c_L)_{MR}/(c_L)_{HR}$ and $(c_D)_{MR}/(c_D)_{HR}$ for the six line combinations stated in column I. The magnitudes of these ratios depend on relative effective widths of the two lines, the profile shape of

Table 7.10. Maximum HR Profit in the Limit of Detection and the Limit of Determination for Six Situations Studied [69]. The Profit is Expressed as the Factor f_L or f_D Maximally Gained in c_L or $c_D{}^a$. [Reprinted with permission from P. W. J. M. Boumans and J. J. A. M. Vrakking, *Spectrochim. Acta* 40B, 1085 (1985). Copyright (1985), Pergamon Journals, Oxford.]

(SBR)$_{HR}$		30		3		0.3	
Intf/Anal [c]		f_L	f_D	f_L	f_D	f_L	f_D
Ir/Cr	9.9	9.1	68	2.4	21	1.6	4.3
Cr/Ir	12	9.9	47	2.9	20	1.9	4.9
Nb/U	23	13	54	3.6	25	2.2	6.3
U/Nb	11	4.0	8.3	1.5	6.1	1.1	2.4
Tl/Mo	20	3.4	4.1	2.3	3.7	1.8	2.9
Mo/Tl	13	6.5	33	1.9	13	1.2	3.3

[a] $f_L = (c_L)_{MR}/(c_L)_{HR}$ and $f_D = (c_D)_{MR}/(c_D)_{HR}$.
[b] SBR of interfering line at HR.
[c] The wavelengths of the lines are stated in the caption of Fig. 7.36.

the interfering line, and the SBR of the latter. Thus, for example, the relatively small HR profit for U/Nb and Tl/Mo is primarily due to the slow Lorentzian decline of the U and Tl profiles (cf. Fig. 7.36). For further details the original paper [69] should be consulted.

7.7.7.5. Detection Power Linked with the Limit of Determination

We recall here the distinction between detection limit and detection power (cf. Section 4.1). Unfortunately this distinction is not always made in the literature. However, it is rational and convenient to designate as detection limits the numerical values of the lowest detectable concentrations associated with separate elements or lines and to use the term detection power to indicate the detection capability of the method as a whole. A statement of the detection power then implies a statement of the set of detection limits of the analytes covered.

Frequently one characterizes the detection power of a method by stating the detection limits attainable in a pure solvent such as diluted HCl. Although such a characterization is useful and indispensable for an assessment of the method, it is insufficient if real samples are involved. Then the detection power may be strongly dependent on the sample type, in particular if the samples emit line-rich spectra.

In the latter case a formal statement of the detection power in terms of a set of detection limits may be misleading and useless, even if the effects of background enhancements resulting from line overlap are taken into account. Such detection limits are formally correct only to the extent that they represent con-

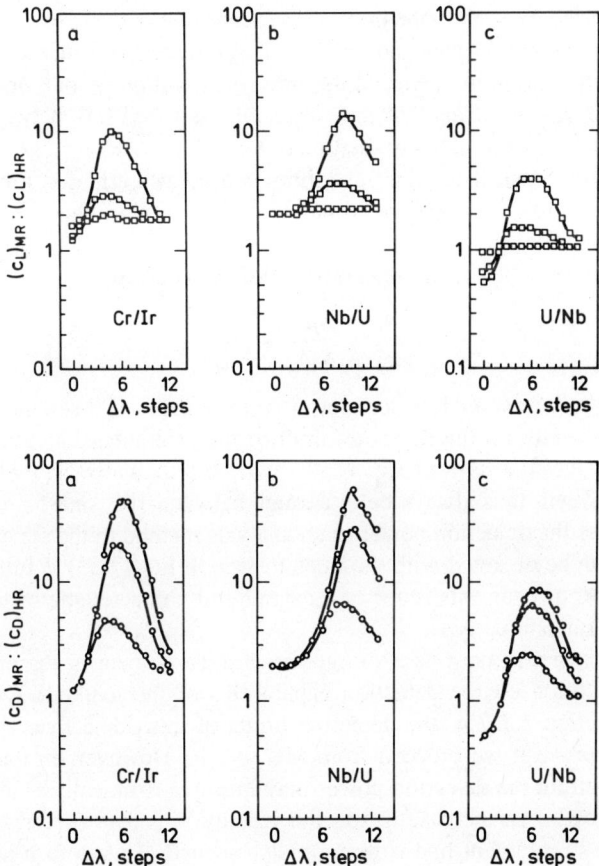

Figure 7.37. Top: ratio of detection limits (MR/HR) as a function of the line separation for three typical profile combinations (cf. Fig. 7.36) Cr/Ir, Nb/U, and U/Nb. Each frame contains three curves which, from top to bottom, are for an SBR of 30, 3, and 0.3 of the interfering line at HR. Bottom: similar representation for the ratio of the limits of determination (MR/HR) [69]. [Reprinted with permission from P. W. J. M. Boumans and J. J. A. M. Vrakking, *Spectrochim. Acta* **40B**, 1085 (1985). Copyright (1985), Pergamon Journals, Oxford.]

centrations that yield net signals equaling, say $2\sqrt{2}$ times the standard deviation of the background signal. However, those detection limits do not fulfill the condition that the analytical error in the net line signal at the detection limit is 50% (cf. Sections 4.2.2 and 7.7.7.7). Therefore, such detection limits may create the false impression that determinations at an error level less than 10% can be carried out down to concentrations equaling five times the detection limits.

Actually we have seen in the previous sections that in the case of line overlap the first term at the right-hand side of Eq. (7.39) may be considerable. Then the limit of determination may exceed the limit of detection by one or two orders of magnitude which means that the commonly assumed link between the two limits ($c_D \approx 5c_L$) is completely disrupted.

On the other hand, if the line selectivity is used as a criterion for line selection, the link between c_D and c_L is maintained to within a factor of 2. This is understood as follows.

According to the criterion, an analysis line is accepted if

$$\frac{x_I}{x_A} = \frac{S_I c_I}{S_A c_a} \leq \frac{1}{v_{\text{thr}}} \tag{7.42}$$

where c_a is the lowest analyte concentration of interest. If this concentration is equalized to five times the detection limit of the prominent line under consideration, then the first term of Eq. (7.40) will be maximally $5c_L$. The limit of determination will thus always be in a range between $10c_L$ and $5c_L$.

If we relate the detection power of an analysis method to the set of detection limits that can be obtained with the lines for which Eq. (7.33) is fulfilled, then the detection power in this sense is a meaningful characteristic of the method for the relevant sample type.

The approach also permits a straightforward assessment of the effect of the spectral resolution on the detection capabilities of the method. As has been shown in Section 7.7.7.4, the detection limits of individual lines do not substantially improve if we proceed from MR to HR. However, in the case of a line-rich spectrum the detection power may improve drastically. This is so because the effect of an increase in spectral resolution is that the threshold value for rejecting a prominent line is less rapidly reached. This in turn implies that prominent lines with lower detection limits will be accepted if the prominent lines of an analyte are searched in order of increasing detection limit. This is the basis of line selection with the aid of the Line Coincidence Tables for ICP–AES [13, 16].

Boumans and Vrakking [69] illustrated this point for a fictitious sample of rare earths and associated elements using the data underlying the *Line Coincidence Tables* as a basis. It was assumed that the sample contains the rare earths plus La, Hf, Sc, Y, and Zr, the concentrations of these elements being each 10 μg/mL except for one of them, that is, the analyte to be determined at the lowest possible concentration level. By considering each element in turn as the analyte they established for each element the best available prominent line, that is, the first prominent line that satisfies criterion (7.42) with $v_{\text{thr}} = 3$ and $c_a = 5c_L$. A threshold value of 3 was chosen because the checks were performed with respect to each interferent separately and did not take the simultaneous effect of several interferents into account.

7.7. SPECTRAL INTERFERENCE AND RESOLVING POWER

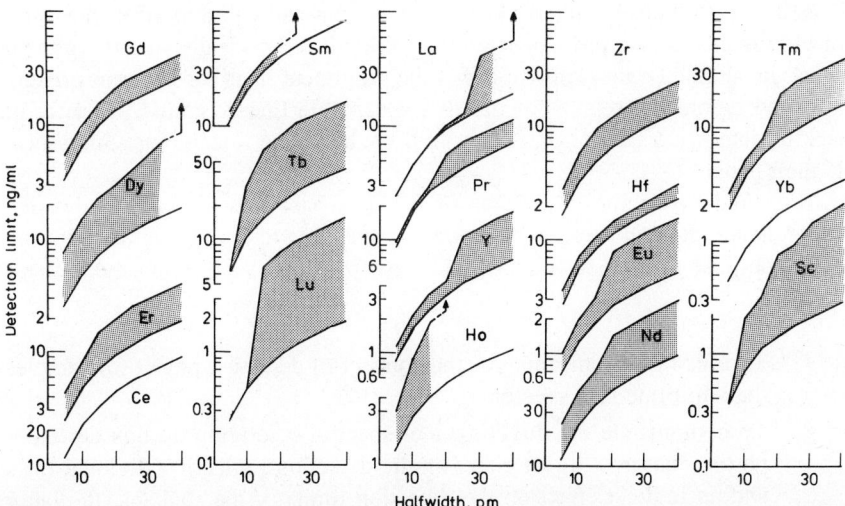

Figure 7.38. Diagram illustrating the dependence of the detection power on the spectral resolution for a fictitious sample of rare earths and associated elements as specified in the text. For compactness of presentation the frames have been arranged so as to fill the diagram effectively. The logarithmic scales shown with the frames have been interrupted but form part of one and the same continuous log scale [69]. [Reprinted with permission from P. W. J. M. Boumans and J. J. A. M. Vrakking, *Spectrochim. Acta* **40B**, 1085 (1985). Copyright (1985), Pergamon Journals, Oxford.]

The results of the search defined are collected in Fig. 7.38. It shows for each analyte a hatched area of which the boundaries represent trends of detection limits with "halfwidth" ($\approx$ spectral bandwidth). The lower boundary represents the trend of the detection limit of the best prominent line of the element. The upper boundary follows the detection limits of the lines that can be actually accepted. Arrows indicate that the stock of prominent lines is exhausted so that the detection limit is then definitely worse than that attainable with the last prominent line.

The most important feature of Fig. 7.38 is that it not only shows the trend of the detection limit with spectral resolution for individual lines (lower boundaries of the hatched areas), but also reveals what happens with the detection limit of each element when, as a consequence of decreasing resolution, better prominent lines must be rejected necessitating the acceptance of lines with poorer detection limits. The picture thus visualizes the trend of the detection power with spectral resolution in that it provides for each level of resolution a statement of the detection limits of all analytes. These detection limits are linked to the limits of determination in the way already discussed.

The approach described neglects one point: the modification of the background by the sample. Actually the wings of the lines emitted by the various constituents will change the background spectrum and will thus affect the order of prominence of the prominent lines (Section 7.3.3). Ideally the background spectrum should be first known so that the prominent lines can be appropriately arranged before the search for the first acceptable line is started. Because of lack of data Boumans and Vrakking [69] could not yet incorporate this aspect in their paper.

Both in view of the limitations of the data base and the lack of detailed experimental data on background enhancements due to line wings, it serves no useful purpose to draw detailed conclusions from Fig. 7.38. Its prime purpose is

1. To illustrate the meaning of the concept of detection power for samples that emit line-rich spectra
2. To demonstrate the dual effect of spectral resolution on this detection power, that is, a direct effect on the detection limits of individual lines and an indirect effect on the detection limits of the analytes, the latter being related to the acceptance or rejection of lines with a view to spectral interferences

For rare earths and associated elements as analytes the second effect of the spectral resolution is relatively small because the detection limits of the prominent lines of these elements do not widely differ; much larger effects can be expected, however, for analytes with a large gap between the detection limits of the first few prominent lines and those that follow. This applies to Ag, As, Au, Ba, Be, Bi, C, Ca, Cd, Ga, Hg, In, Ir, Li, Mg, Mn, Na, P, Pb, Re, Sb, Se, Sn, Sr, Te, Tl, and Zn.

7.7.7.6. Resolving Power and Analytical Performance: Dual Polychromator Arrangement

Botto [74] reports results obtained with a dual spectrometer arrangement comprising a 0.75-m Paschen–Runge polychromator and an echelle spectrometer. The latter was incorporated into the system because the Paschen–Runge spectrometer lacks sufficient spectral resolution for accurate and reliable determinations of environmentally important trace elements in complex sample matrices. If the most prominent of these elements are employed, substantial interelement corrections are needed to compensate for the spectral interferences from major constituents. This method, described in Section 7.6, can accurately correct for spectral interferences amounting to 50% of the uncorrected intensities, but fails at larger proportions of the interfering signals.

Using the high-resolution echelle monochromator Botto studied the interferences of 33 elements upon 22 analysis lines in order to judiciously select 10

Table 7.11. Summarized Results Obtained with a Dual Polychromator Arrangement [74].
[Reprinted with permission from R. I. Botto *Spectrochim. Acta* 38B, 144 (1983). Copyright (1983), Pergamon Journals, Oxford.]

Analysis line (nm)		Detection limit (ng/ml)		Reduction in Critical interferences (%)[a]
		P-R[b]	E[b]	
As	193.696	25	80	74
Be	234.861	1.4	0.3	99
Cd	214.438	8	9	>99
Mo	202.030	7	13	>99
Pb	220.353	14	35	96
Se	196.026	40	70	>99
Tl	377.572	40	70	94
Ni	341.476	16		
Ni	231.604		5	>99
Sb	231.147	30		
Sb	206.833		40	>99
U	367.007	50		
U	409.014		80	93

[a] The term "critical interferences" refers to interferents from elements often found in significant concentrations in the samples routinely analyzed. Interference from Al, B, Ba, Ca, Fe, K, Mg, Na, P, Si, and Ti is critical if the absolute value of the interference coefficient exceeds 0.00002. Interference from Co, Cr, Cu, Mn, Mo, Ni. Sr, V, and Zn is critical if the absolute value of the interference coefficient exceeds 0.0002.

[b] P−E = Paschen–Runge spectrometer, E = echelle spectrometer (multielement operation).

lines for multielement operation. Interference comparisons were made for the lines included in the Paschen–Runge polychromator (Table 7.11). The measurements were made with and without off-line background corrections.

The combination of high resolution and the ability to choose an optimum position for off-line background correction resulted in considerable reductions in the total intensity of spectral interferences. As the results in Tables 7.11 and 7.12 demonstrate, it is the increase in selectivity and not an improvement in the detection limits of individual lines that must be considered as the chief advantage of the high-resolution spectrometer. Table 7.12 shows that the Paschen–Runge polychromator provides for detection limits that are similar to or better than those for the high-resolution echelle spectrometer. The limits of determination for the Paschen–Runge instrument, however, will be at a far larger distance above the detection limits than those attainable with the echelle spectrometer.

Botto's paper [74] is an outstanding example of work in which the benefits

Table 7.12. Interference Coefficients[a] for Mo II 202.030 mn as Found for a Paschen–Runge and Echelle Spectrometer [74]. [Reprinted with permission from R. I. Botto *Spectrochim. Acta* 38B, 137 (1983). Copyright (1983), Pergamon Journals, Oxford.]

	Interference Coefficient	
Interferent	Paschen–Runge	Echelle
Al	0.00082	0.00021
Cr	0.00020	0.00008
Fe	0.00019	0.00009
Ni	0.00012	0.00002
Pt	0.00041	<0.00002
U	0.00034	0.00008
V	0.00022	0.00007
W	0.00108	0.00009

[a] See Section 7.6.

of high-resolution spectroscopy in the practical analysis of complex real samples are revealed in detail. Although the concept of limit of determination is not explicitly involved, the approach essentially covers this concept in that it amply illustrates the vast improvement in analytical precision and accuracy resulting from the increase in selectivity as a consequence of increased spectral resolution. Botto's work may be considered as the practical "precursor" of the theoretical approach made by Boumans and Vrakking [36, 69], discussed in the preceding sections and in Section 7.7.7.7. Botto's paper also provides many useful details of interferences in terms of interference coefficients.

7.7.7.7. Resolving Power and Analytical Performance: Slew-Scan Monochromator

Generally, wavelength adjustment in a slew-scan monochromator is achieved by a stepwise rotation of the grating at a slewing speed compatible with the wavelength distance to be covered. Facilities for the stepwise scanning of small wavelength intervals with selectable integration time per step are commonly incorporated. Whatever the mechanical and electronic means by which slewing and scanning are achieved, crucial is the accuracy with which the system can be set to a desired wavelength for intensity measurement.

If the errors in the mechanical and electronic control can be reduced to a minimum of better than ±1 mechanical step, there remains, however, possible drift of the optically formed spectrum with respect to the mechanical wavelength scale, that is, drift in the wavelength calibration.

7.7. SPECTRAL INTERFERENCE AND RESOLVING POWER

These errors can be reduced by improving the mechanical design of the monochromator including thermostatting [105]. Given the system, the only way to minimize errors of this type is frequent recalibration and/or the use of curve fitting or find peak routines (cf. Section 8.9.3). The latter approaches are adequate only if the peak of an analysis line is unambiguously defined. A find peak routine may lead to a wrong wavelength position, however, in the case of line overlap, in particular if the analysis line does not show to be resolved from the interfering line.

As an example Fig. 7.39 shows spectral scans of a Nd line with overlap from

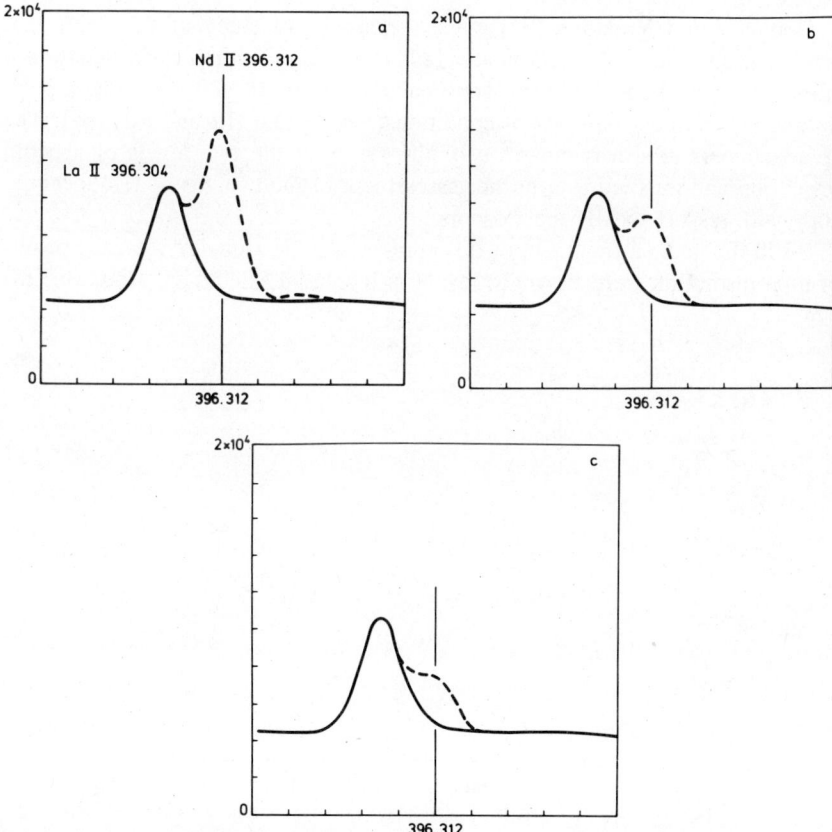

Figure 7.39. Scans over a 0.05-nm window about the Nd line at 396.312 nm with interference from La 396.304 nm. Continuous curves: scans for 100-μg/mL La solution. Broken curves: scans for La solutions spiked with Nd. The vertical line indicates the position of the Nd line. All scans are for a practical spectral bandwidth of 2.3 pm. The Nd concentrations (μg/mL) are as follows: (a) 1, (b) 0.5, and (c) 0.3. Ordinate: intensity (counts/s), abscissa: wavelength (± 0.025 nm about central λ) [28]. [Reprinted with permission from P. W. J. M. Boumans and J. J. A. M. Vrakking, *Spectrochim. Acta* **39B,** 1261 (1984). Copyright (1984), Pergamon Journals, Oxford.]

a La line at high resolution [28]. Clearly a find peak routine can be used down to 0.5 µg/mL Nd (Fig. 7.39b), but if the search window is narrow enough, the correct wavelength will also be closely approached at lower concentrations. This contrasts with the situation in Fig. 7.40 for medium resolution where the Nd line shows unresolved from the La line. If a find peak routine is used here, the measurement will drift ever more to the peak of the La line the lower the Nd concentration. This results in severe curvature of the calibration curve and in erroneous results, in particular if the La concentration is variable.

If, for lack of a better alternative, an analysis line with line overlap has to be used, the intensity measurement should be made at the correct wavelength position of the analysis line.

Boumans and Vrakking [36] made a quantitative study of the errors that result if a slew-scan monochromator fails to reach this correct wavelength position. Reasonable assumptions were made about the magnitude of these positioning errors under two sets of conditions, designated HR and MR [69]. The approach permitted a comparison of the errors at the two levels of spectral resolution and thus led to a further assessment of the benefits of HR spectroscopy with respect to MR spectroscopy.

As in the general treatment of this comparison (Section 7.7.7.4), the profits of high resolution were shown to reside in increased selectivity. However, the

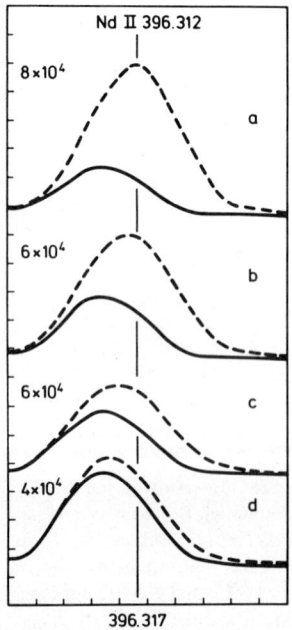

Figure 7.40. Scans for the same Nd line as shown in Fig. 7.39, but for a practical spectral bandwidth of 18 pm. As in Fig. 7.39, the La concentration is 100 µg/mL, while the Nd concentrations (µg/mL) are as follows: (a) 2, (b) 1, (c) 0.5, (d) 0.2. Ordinate: intensity (counts/s), abscissa: wavelength ($\pm$ 0025 nm about central λ). The scans have been fitted into one frame. Owing to the normalization of each scan separately the ordinate scales are different. The relevant labels indicate the number of counts/s per scale division [28]. [Reprinted with permission from P. W. J. M. Boumans and J. J. A. M. Vrakking, *Spectrochim. Acta* **39B,** 1261 (1984). Copyright (1984), Pergamon Journals, Oxford.]

7.7. SPECTRAL INTERFERENCE AND RESOLVING POWER

quantitative treatment of the errors was different in that the errors associated with defined wavelength positioning errors are systematic so that they can be predicted from the intensity distributions in spectral scans of samples and blanks.

For illustration, Fig. 7.41 shows the HR and MR scans for the Al line at 309.271 nm. This line is the strong component of a doublet 309.271–309.284 nm and suffers interference from an OH band component at 309.278 nm. The scan in frame (a) has been redrawn; the actual scans are obtained as a series of discrete points, as shown in frames (b)–(e), which also reveal the difference in step size at MR and HR.

The error treatment [36] involved the calculation of the errors associated with wavelength positioning errors equivalent to ± 1 mechanical step (Fig. 7.41). The approach led to the error diagrams shown in Fig. 7.42, which we explain by first considering the HR case (Figs. 7.41a and 7.42 (left)).

The curves labeled "H" and "L" in Fig. 7.42 (left) represent the error in the net Al signal if the wavelength is set one step above or below the correct value respectively. Starting at $c/c_L = 1000$ and proceeding to the left, we see that the high-wavelength error initially is negative; this is so because the error is dominated by the error in the emitted net signal. With decreasing c/c_L the error becomes less negative since it is ever more compensated by the positive error in the blank, until a point is reached where the two errors just balance each other and the remaining error is the random error only. The latter is represented by the lowest curve, labeled "random error (RSD)." From the point on where the H curve reaches the RSD curve the error is governed by the positive error in the blank measurement and steadily increases with decreasing c/c_L. Thus the H curve consists of two branches, labeled "H($-$)" and "H($+$)." The error associated with one error step toward lower wavelength is negative for both the net and blank signals resulting in a single smooth curve, L($-$).

Two special points have been indicated in the diagram:

1. The 10% inaccuracy limit which, in this case, is derived from the H($+$) curve
2. The find peak transition which corresponds to the situation shown in Fig. 7.41c

If c/c_L is higher than the find peak transition, a find peak routine can be used so that there will be no systematic error associated with the localization of the peak. In summary, the maximum error varies with c/c_L according to the solid curve in Fig. 7.42 (left).

The error diagram for MR (Fig. 7.42 (right)) is essentially similar to that for HR; however, both the 10% inaccuracy limit and the find peak transition have shifted to substantially higher c/c_L.

The significance of the difference between the error diagrams for HR and

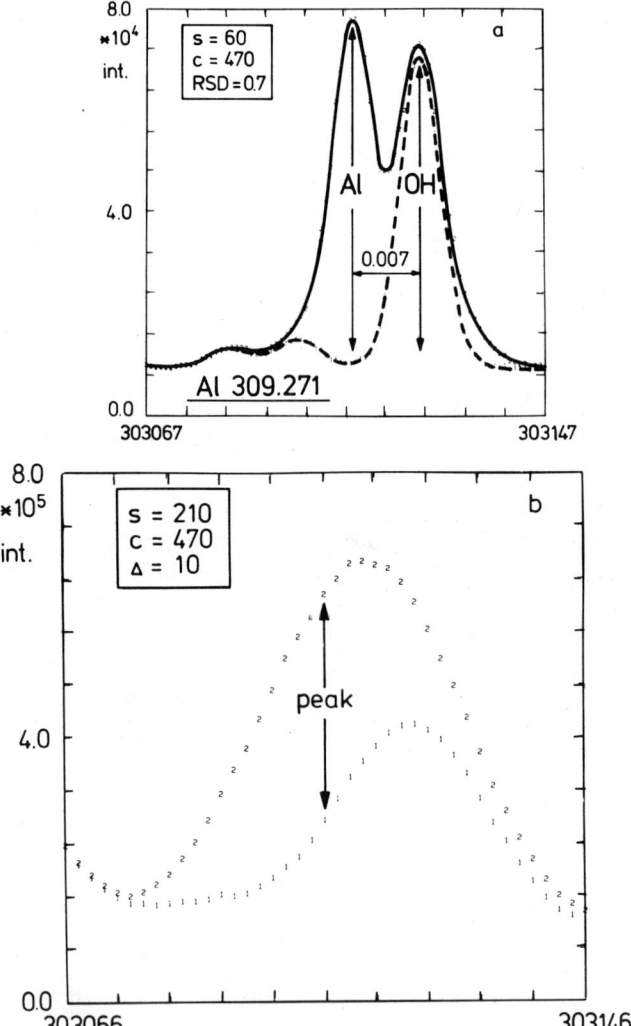

Figure 7.41. Spectral scans over a 0.0444-nm window covering Al I 309.271 nm with interference from an OH band component at 309.278 nm [36].
Left: High resolution (HR); right: medium resolution (MR). Ordinate: intensity in counts/3 s. Abscissa: wavelength in motor steps, 1 step = 0.00055 nm (HR), 1 step = 0.0011 nm (MR).
 In each frame the lower scan [broken line in (*a*)] is for the pure solvent, the upper scan [continuous line in (*a*)] for a solution spiked with Al (concentration *c*).

(*a*) c = 470 ng/mL (HR).
(*b*) c = 470 ng/mL: 10% inaccuracy limit (MR).
(*c*) c = 150 ng/mL: find peak transition (HR), random error 0.7% RSD;
(*d*) c = 150 ng/mL: inaccuracy 30% (MR).
(*e*) c = 30 ng/mL: 10% inaccuracy limit (HR).

[Reprinted with permission from P. W. J. M. Boumans and J. J. A. M. Vrakking, *Spectrochim. Acta* **40B**, 1107 (1985). Copyright (1985), Pergamon Journals, Oxford.]

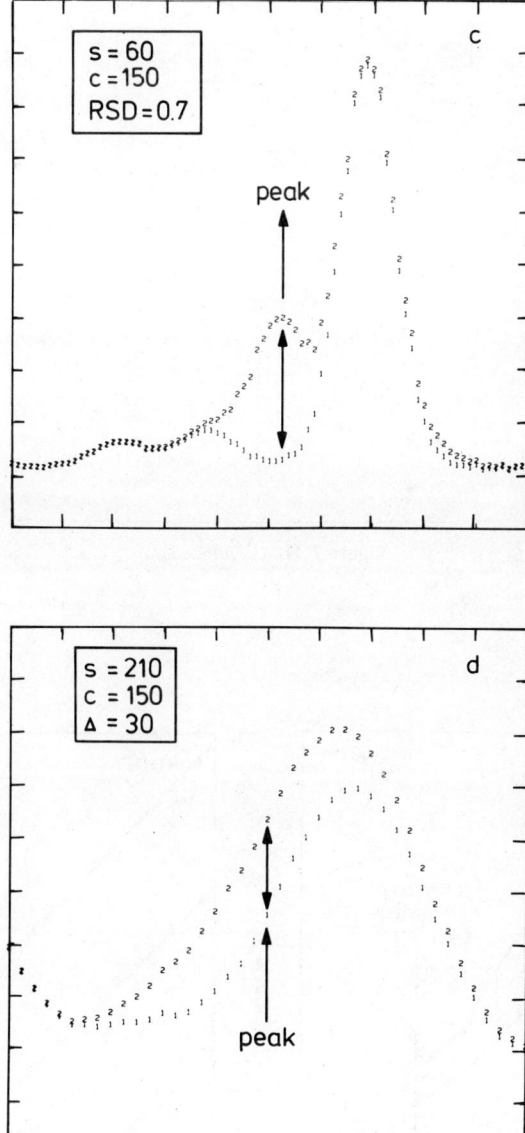

Figure 7.41. (*Continued*)

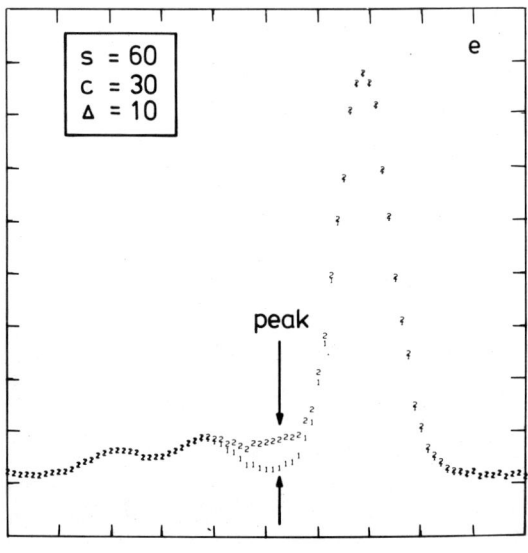

Figure 7.41. (*Continued*)

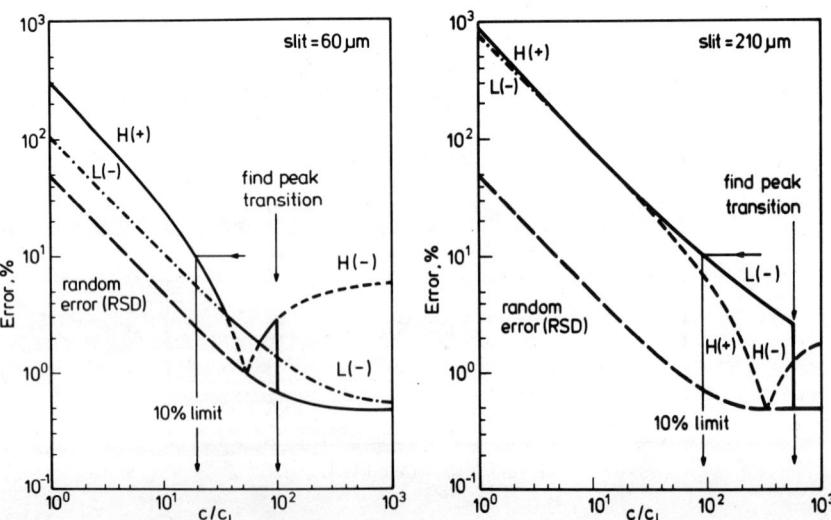

Figure 7.42. Error diagrams for high (slit = 60 μm) and medium (slit = 210 μm) resolution, as explained in the text [36]. [Reprinted with permission from P. W. J. M. Boumans and J. J. A. M. Vrakking, *Spectrochim. Acta* **40B**, 1107 (1985). Copyright (1985), Pergamon Journals, Oxford.]

Table 7.13. Al I 309.271 nm with Interference from OH 309.278 nm (SBR = 6.8 at HR): 10% Inaccuracy Limit and Associated Quantities [36]. [Reprinted with permission from P. W. J. M. Boumans and J. J. A. M. Vrakking, *Spectrochim. Acta* 40B, 1107 (1985). Copyright (1985), Pergamon Journals, Oxford.]

Quantity[a]		HR	MR	Ratio HR/MR
FWHM (nm)	OH	0.0056	0.0145	2.6
	Al	0.0058	0.0167	2.9
$\triangle \lambda$/FWHM		1.2	0.5	—
c_L		1.5	5.2	3.5
c_{10}/c_L		20	90	4.5
c_{10}		30	470	16
c_{FPT}		150	3000	20
RSD (%) at c_{FPT}		0.7	0.5	

[a] c_{10} = Concentration at 10% inaccuracy limit.
c_{FPT} = Find Peak Transition.
All concentrations are expressed in ng/mL.

MR is clearly seen from the results summarized in Table 7.13 which shows that HR improves the detection limit by a factor of 3.5, as expected. The 10% inaccuracy limit in terms of c/c_L improves by a factor of 4.5, so that in this limit in terms of the concentration itself lies a factor of 16 more favorable than for MR. The profit in the find peak transition is even a factor of 20. This means that HR enables one to determine Al with the 309.2 nm line down to 30 ng/mL with a maximum error of 10%, whereas the corresponding concentration for MR is 470 ng/mL. The find peak transitions are 150 and 3000 ng/mL for HR and MR respectively, the associated random errors being 0.7 and 0.5%, respectively.

Characteristic points are visualized by the scans shown in Fig. 7.41. Frame (*b*) is the scan at the 10% inaccuracy limit for MR (c = 470 ng/mL), while frame (*a*) represents the HR scan for the same concentration. Frame (*c*) is for the find peak transition at HR (c = 150 ng/mL) and frame (*d*) for the same concentration at MR, the associated errors being a random error (RSD) of 0.7% for the HR case and a systematic error of 30% for MR. Frame (*e*), finally, illustrates the situations at the 10% inaccuracy limit for HR.

Al 309.271 nm with OH band interference is only one example of the various case studies [36] which included complex situations such as the Bi lines, shown in Fig. 7.43, for which HR and MR provide virtually equivalent performance.

By simulating line overlap situations using experimental line profiles Boumans and Vrakking [36] made a systematic error analysis with the wavelength distance between the lines and the SBR of the interfering line as parameters.

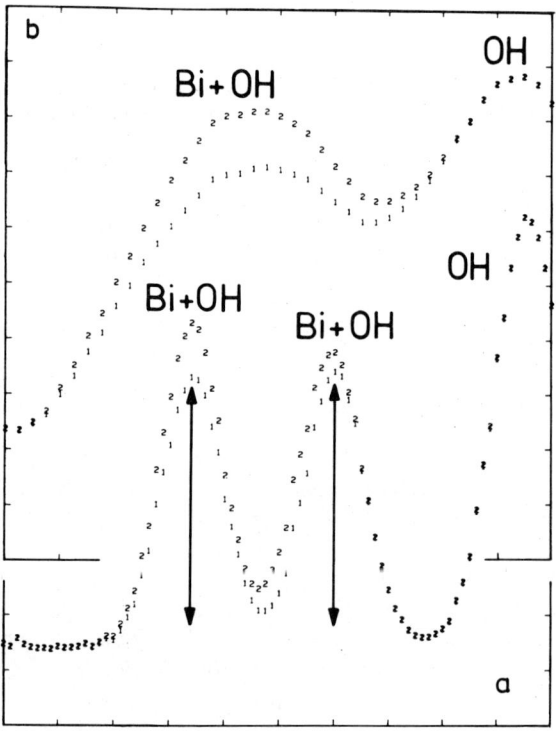

Figure 7.43. HR (*a*) and MR (*b*) scans over a 0.0444-nm window showing the doublet structure of the Bi line (306.765 and 306.774 nm) and the complex interference from three OH band components. If at HR the Bi line at 306.765 nm is used, a find peak routine remains valid down to the detection limit, which also applies to the MR situation. The scans are for a Bi concentration of 180 ng/mL, which corresponds to about 10 times the detection limit; the latter is the same at both levels of resolution [36]. [Reprinted with permission from P. W. J. M. Boumans and J. J. A. M. Vrakking, *Spectrochim. Acta* **40B**, 1107 (1985). Copyright (1985), Pergamon Journals, Oxford.]

For this simple situation of interaction of a single interfering line with an analysis line they showed that the 10% inaccuracy limit equals the limit of determination to within a factor of 2, if a threshold value of 2 is taken in Eq. (7.40). Boumans and Vrakking then suggest $v_{thr} = 2$ as a starting point for further studies in which it must be established whether or not this threshold value is generally acceptable and to what extent it covers errors due to randomized wavelength positioning errors during routine measurements.

REFERENCES

1. W. N. Hartley, *J. Chem. Soc.* **41**, 210 (1882); *Phil. Trans.* **175**, 50 (1884).
2. A. de Gramont, *Compt. Rend.* **144**, 1101 (1907); **145**, 1170 (1907); *Rev. Met.* **19**, 90 (1922).

3. P. W. J. M. Boumans and F. J. de Boer, *Spectrochim. Acta* **30B,** 309 (1975).
4. P. W. J. M. Boumans and F. J. de Boer, *Spectrochim. Acta* **32B,** 365 (1977).
5. J. C. Souilliart and J. Robin, *Analusis* **1,** 427 (1972).
6. J. M. Mermet, *C.R. Acad. Sci. Ser. B* **281,** 273 (1975).
7. R. K. Winge, V. J. Peterson, and V. A. Fassel, *Appl. Spectrosc.* **33,** 206 (1979).
8. R. K. Winge, V. J. Peterson, and V. A. Fassel, *ICP–AES: Prominent Lines,* EPA-600/4-79-017 (NTS, Springfield, VA, March 1979).
9. P. W. J. M. Boumans and M. Bosveld, *Spectrochim. Acta.* **34B,** 59 (1979).
10. W. F. Meggers, C. H. Corliss, and B. F. Scribner, "Tables of Spectral-Line Intensities," *N.B.S. Monograph* **145,** U.S. Government Printing Office, Washington, DC (1975).
11. R. K. Winge, V. A. Fassel, V. J. Peterson, and M. A. Floyd, *Appl. Spectrosc.* **36,** 210 (1982).
12. P. W. J. M. Boumans, *Spectrochim. Acta* **38B,** 747 (1983).
13. P. W. J. M. Boumans, *Line Coincidence Tables for Inductively Coupled Plasma Atomic Emission Spectrometry,* Pergamon, Oxford (1980).
14. P. W. J. M. Boumans, *Spectrochim. Acta* **36B,** 169 (1981).
15. P. W. J. M. Boumans, R. J. McKenna, and M. Bosveld, *Spectrochim. Acta* **36B,** 1031 (1981).
16. P. W. J. M. Boumans, *Line Coincidence Tables for Inductively Coupled Plasma Atomic Emission Spectrometry,* 2nd revised ed., Pergamon, Oxford (1984).
17. R. K. Winge, V. A. Fassel, V. J. Peterson, and M. A. Floyd, *Inductively Coupled Plasma Atomic Emission Spectroscopy: An Atlas of Spectral Information,* Elsevier, Amsterdam-New York (1985).
18. C. C. Wohlers, "ICP–AES Wavelength Table," *ICP Information Newslett.* **10,** 601 (1985).
19. S. K. Hughes and R. C. Fry, *Appl. Spectrosc.* **35,** 493 (1981).
20. A. R. Forster, T. A. Anderson, and M. L. Parsons, *Appl. Spectrosc.* **36,** 499 (1982).
21. T. A. Anderson and M. L. Parsons, *Appl. Spectrosc.* **38,** 625 (1984).
22. D. H. Tracy and S. A. Myers, *Spectrochim. Acta* **37B,** 1055 (1982).
23. G. A. Meyer and R. M. Barnes, *Spectrochim. Acta* **40B,** 893 (1985).
24. G. F. Larson, V. A. Fassel, R. K. Winge, and R. N. Kniseley, *Appl. Spectrosc.* **30,** 384 (1976).
25. V. A. Fassel, J. M. Katzenberger, and R. K. Winge, *Appl. Spectrosc.* **33,** 1 (1979).
26. G. F. Larson and V. A. Fassel, *Appl. Spectrosc.* **33,** 592 (1979).
27. P. W. J. M. Boumans and J. J. A. M. Vrakking, *Spectrochim. Acta* **39B,** 1291 (1984).
28. P. W. J. M. Boumans and J. J. A. M. Vrakking, *Spectrochim. Acta* **39B,** 1261 (1984).
29. J. M. Mermet and C. Trassy, *Spectrochim. Acta* **36B,** 269 (1981).

30. C. C. Butler, R. N. Kniseley, and V. A. Fassel, *Anal. Chem.* **47,** 825 (1975).
31. J. Jarosz and J. M. Mermet, *J. Quant. Spectrosc. Radiat. Transfer* **17,** 237 (1977).
32. K. Kato, H. Fukushima, and T. Nakajima, *Spectrochim. Acta* **39B,** 979 (1984)
33. P. W. J. M. Boumans and J. J. A. M. Vrakking, *Spectrochim. Acta* **39B,** 1239 (1984).
34. G. H. Dieke and H. M. Crosswhite, *J. Quant. Spectrosc. Radiat. Transfer* **2,** 97 (1962).
35. P. W. J. M. Boumans and J. J. A. M. Vrakking, *Spectrochim. Acta* **40B,** 1423 (1985).
36. P. W. J. M. Boumans and J. J. A. M. Vrakking, *Spectrochim. Acta* **40B,** 1107 (1985).
37. R. Rezaaiyan and G. M. Hieftje, *Anal. Chem.* **57,** 412 (1985).
38. R. W. B. Pearse and A. G. Gaydon, *The Identification of Molecular Spectra*, 4th ed., Chapman and Hall, London (1976).
39. C. A. Peterson, Ph.D. thesis, Iowa State University, Ames, IA (1977).
40. S. Greenfield, H. McD. McGeachin, and P. B. Smith, *ICP Information Newslett.* **2,** 167 (1976).
41. R. M. Barnes, *CRC Crit. Rev. Anal. Chem.* **7,** 203 (1978).
42. K. Ohls and D. Sommer, *ICP Information Newslett.* **4,** 532 (1979).
43. K. Ohls and D. Sommer, in R. M. Barnes, ed., *Developments in Atomic Plasma Spectrochemical Analysis*, Heyden, London (1981), p. 321.
44. K. Ohls and D. Sommer, *ICP Information Newslett.* **9,** 555 (1984).
45. G. A. Meyer and M. D. Thompson, *Spectrochim. Acta* **40B,** 195 (1985).
46. G. R. Harrison, *M.I.T. Wavelength Tables*, The M.I.T. Press, Cambridge, MA (1969).
47. P. W. J. M. Boumans, *Theory of Spectrochemical Excitation*, Adam Hilger, London/Plenum Press, New York (1966).
48. C. Th. J. Alkemade and R. Herrmann, *Fundamentals of Flame Spectroscopy*, Adam Hilger, Bristol (1979).
49. A. N. Zaidel, V. K. Prokofjev, S. M. Raiskii, V. A. Slavnyi, and E. A. Shreider, *Tables of Spectral Lines*, IFI/Plenum, New York (1970).
50. R. L. Kelly, *A Table of Emission Lines in the Vacuum Ultraviolet for All Elements*, University of California, Lawrence Radiation Lab., Livermore, CA, UCRL 5612 (1959).
51. R. L. Kelly and L. J. Palumbo, "Atomic and Ionic Emission Lines below 2000 Angstroms-Hydrogen through Krypton," NRL Report 7599. NRL, Washington, DC (1973).
52. J. Reader, C. H. Corliss, W. L. Wiese, and G. A. Martin, *Wavelengths and Transistion Probabilities for Atoms and Atomic Ions, Part I. Wavelengths, Part II. Transition Probabilities*, NSRDS–NBS Series 68. U.S. Government Printing Office, Washington, DC (1980).

53. J. Reader and C. H. Corliss, "Line Spectra of Elements," in R. C. Weast, ed., *CRC Handbook of Chemistry and Physics*, 65th ed., CRC Press, Boca Raton, FL (1982).
54. J. Kroonen and D. Vader, *Line Interference in Emission Spectrographic Analysis*, Elsevier, Amsterdam (1963).
55. J. E. Norris, *Wavelength Tables of Rare-Earth Elements and Associated Elements*, Oak Ridge National Laboratory, TN, ORNL-2774 (1960).
56. A. Gatterer and J. Junkes, *Atlas der Restlinien*, Vol. II, Spektren der seltenen Erden, Specola Vaticana, Vatican City (1945).
57. E. W. Salpeter, *Spektren in der Glimmentladung von 1500 bis 4000 Å, Parts 1–5*, Specola Vaticana, Vatican City (1973).
58. Qiu De-ren and Cheng Wan-xia, *An Atlas for 2 Å/mm and 4 Å/mm Grating Spectrographs*, Fudan University, Shanghai, P.R. China (1982).
59. A. R. Striganov and N. S. Sventitskii, *Tables of Spectral Lines of Neutral and Ionized Atoms*, IFI/Plenum, New York (1968).
60. G. A. Martin, *ICP Information Newslett.* **10**, 534 (1984).
61. T. A. Anderson, A. R. Forster, and M. L. Parsons, *Appl. Spectrosc.* **36**, 504 (1982).
62. P. W. J. M. Boumans, *Proc. 21th Coll. Spectr. Int. and 8th Int. Conf. Atomic Spectr.*, Cambridge 1979, Keynote Lectures, Heyden, London (1979); p. 49.; *Spectrochim. Acta* **35B**, 57 (1980).
63. J. W. McLaren and S. S. Berman, *Spectrochim. Acta* **40B**, 217 (1985).
64. M. L. Parsons, A. Forster, and D. Anderson, *An Atlas of Spectral Interferences in ICP Spectroscopy*, Plenum, New York (1980).
65. P. W. J. M. Boumans, in K. Fuwa, ed., *Recent Advances in Analytical Spectrometry*, Pergamon, Oxford (1982), p. 61.
66. P. W. J. M. Boumans, *ICP Information Newslett.* **10**, 430 (1984).
67. D. L. Anderson, A. R. Forster, and M. L. Parsons, *Anal. Chem.* **53**, 770 (1981).
68. R. K. Winge, V. A. Fassel, V. J. Peterson, and M. A. Floyd, *ICP Information Newslett.* **10**, 444 (1984).
69. P. W. J. M. Boumans and J. J. A. M. Vrakking, *Spectrochim. Acta* **40B**, 1085 (1985).
70. P. W. J. M. Boumans and J. J. A. M. Vrakking, *Spectrochim. Acta* **40B**, 1437 (1985).
71. E. Michaud and J. M. Mermet, *Spectrochim. Acta* **37B**, 145 (1982).
72. I. B. Brenner and H. Eldad, *ICP Information Newslett.* **10**, 451 (1984).
73. R. I. Botto, *Anal. Chem.* **54**, 1654 (1982).
74. R. I. Botto, *Spectrochim. Acta* **38B**, 129 (1983).
75. R. I. Botto, *Spectrochim. Acta* **39B**, 95 (1984).
76. R. I. Botto, *Spectrochim. Acta* **40B**, 397 (1985).
77. L. M. Faires, *ICP Information Newslett.* **10**, 449 (1984).
78. L. M. Faires, B. A. Palmer, and J. W. Brault, in R. M. Barnes, ed., *Proc. 1984*

Winter Conference on Plasma Spectrochemistry, San Diego, Spectrochim. Acta **40B,** 135 (1985).
79. J. E. Freeman and G. M. Hieftje, *Spectrochim. Acta* **40B,** 475 (1985).
80. P. Taylor and P. Schutyser, in P. W. J. M. Boumans, ed., *Proc. 1985 European Winter Conference on Plasma Spectrochemistry, Leysin, Spectrochim. Acta* **41B,** 81 (1986).
81. G. F. Wallace, *Atomic Spectrosc.* **2,** 87 (1981).
82. W. J. Haas, Jr., V. A. Fassel, F. Grabau IV, R. N. Kniseley, and W. L. Sutherland, "Simultaneous Determination of Trace Elements in Urine by Inductively Coupled Plasma—Atomic Emission Spectrometry," in *Ultratrace Metal Analysis in Science and Environment*, Ch. 8. American Chemical Society, Washington, DC (1979), p. 91.
83. P. W. J. M. Boumans, *Spectrochim. Acta* **31B,** 147 (1976).
84. H. Kaiser, in F. Korte, ed., *Methodicum Chimicum*, Vol. 1, *Analytical Methods, Part A*. Academic, New York/Georg Thieme, Stuttgart (1974).
85. H. Kaiser, *Spectrochim. Acta* **33B,** 551 (1978).
86. K. Fujiwara, J. M. McHard, S. J. Foulk, S. Bayer, and J. D. Winefordner, *Can. J. Spectrosc.* **25,** 18 (1980).
87. B. E. H. Saxberg and B. R. Kowalski, *Anal. Chem.* **51,** 1031 (1979).
88. J. H. Kalivas and B. R. Kowalski, *Anal. Chem.* **53,** 2207 (1981).
89. C. Jochum, P. Jochum, and B. R. Kowalski, *Anal. Chem.* **53,** 85 (1981).
90. R. I. Botto, in R. M. Barnes, ed., *Developments in Atomic Plasma Spectrochemical Analysis*, Heyden, London (1981), p. 141.
91. K. Laqua, W.-D. Hagenah, and H. Waechter, *Z. Anal. Chem.* **225,** 142 (1967).

92. U. Haisch, *Spectrochim. Acta* **25B,** 597 (1970).
93. U. Haisch, K. Laqua, and W.-D. Hagenah, *Spectrochim. Acta* **26B,** 651 (1971).
94. U. Haisch, *Z. Anal. Chem.* **259,** 1 (1972).
95. U. Haisch, K. Laqua, W.-D. Hagenah, and H. Waechter, *Fresenius Z. Anal. Chem.* **316,** 157 (1983).
96. C. Th. J. Alkemade, Tj. Hollander, W. Snelleman, and P. J. Th. Zeegers, *Metal Vapours in Flames*. Pergamon, Oxford (1982), p. 157.
97. A. Batal and J. M. Mermet, *Spectrochim. Acta* **36B,** 993 (1981).
98. H. G. C. Human and R. H. Scott, *Spectrochim. Acta* **31B,** 459 (1976).
99. H. Kawaguchi, Y. Oshio, and A. Mizuike, *Spectrochim. Acta* **37B,** 809 (1982).
100. T. Hasegawa and H. Haraguchi, in R. M. Barnes, ed., *Proc. 1984 Winter Conference Plasma Spectrochemistry, San Diego, Spectrochim. Acta* **40B,** 123 (1985).
101. P. Bousquet, *Spectroscopy and its Instrumentation*, Adam Hilger, London (1971), p. 101.
102. J. W. McLaren and J. M. Mermet, *Spectrochim. Acta.* **39B,** 1307 (1984).

103. D. W. Posener, *Austral. J. Phys.* **12,** 184 (1959).
104. J. A. C. Broekaert, F. Leis, and K. Laqua, *Spectrochim. Acta* **34B,** 73 (1979).
105. P. Taylor, E. Janssens, R. Dams, and J. Hoste, *Spectrochim. Acta* **39B,** 867 (1984).
106. P. W. J. M. Boumans, *Fresenius Z. Anal. Chem.* **324,** 397 (1986).
107. P. W. J. M. Boumans and J. J. A. M. Vrakking, *Spectrochim. Acta, Part B*, in press for Vol. **42B** (1987).
108. F. J. M. J. Maessen and J. A. Tielrooij, *Fresenius Z. Anal. Chem.* **323,** 490 (1986).
109. P. W. J. M. Boumans and J. J. A. M. Vrakking, *Spectrochim. Acta* **41B,** No. 12 (1986).

CHAPTER

8

SPECTROMETERS

J. W. OLESIK

Department of Chemistry
Venable and Kenan Laboratories
University of North Carolina
Chapel Hill, North Carolina

Partial support for the author was provided by the National Science Foundation (under grants CHE-82-14121 and CHE-83-20053) and the Office of Naval Research (under grant N14-76-C-0838) while at Indiana University (when the manuscript was begun) and the Department of Chemistry and the University of North Carolina at Chapel Hill (where the manuscript was completed).

The author thanks Drs. Gary Hieftje, Gary Rayson, and Susan Olesik, Mr. Eric Williamsen, Ms. Su Jane Den and Mr. Dan Wilson for useful comments about this manuscript.

8.1. Introduction
8.2. Fundamental Equations and Figures of Merit for Grating Spectrometers
  8.2.1. The Grating Equation
  8.2.2. Dispersion
  8.2.3. Free Spectral Range
  8.2.4. Theoretical Resolving Power
  8.2.5. Experimental Resolving Power
  8.2.6. The Distribution of Light into Multiple Orders—Blazed Gratings
8.3. Grating Spectrometer Mountings
  8.3.1. Classical Concave Grating Mounts
    8.3.1.1. Rowland Circle Based Mountings
    8.3.1.2. Paschen–Runge Mounting
    8.3.1.3. Seya–Namioka Mounting
  8.3.2. Classical Plane Grating Mountings
    8.3.2.1. Czerny–Turner and Other Ebert-Type Mountings
    8.3.2.2. Optical Characteristics of the Ebert, Fastie–Ebert, and Czerny–Turner-Type Mountings
8.4. Characteristics of Inductively Coupled Plasmas Important in Choosing a Spectrometer
8.5. Stray Light
  8.5.1. Ruled Grating Imperfections
  8.5.2. Improvements in Ruled Grating Production
  8.5.3. Holographic Gratings

- 8.5.4. Stray Light Produced by Perfect Gratings
- 8.5.5. Reentry Spectra and Their Elimination
- 8.5.6. Approaches to Reduce Stray Light Effects
- 8.6. **High Resolution Spectrometers**
  - 8.6.1. Obtaining High Resolution with Grating Spectrometers
  - 8.6.2. Advantages and Limitations of the Use of High-Resolution Spectrometers to Reduce Spectral Interferences in ICP–AES
- 8.7. **Background and Spectral Overlap Correction**
  - 8.7.1. Off-Peak Subtraction
    - *8.7.1.1. Wavelength Modulation*
  - 8.7.2. On-Peak Subtraction
- 8.8. **Direct Reading Spectrometers**
  - 8.8.1. Direct Reading Spectrometers Based on Concave Grating Mounts
  - 8.8.2. Direct Reading Spectrometers Based on the Echelle Mounting
  - 8.8.3. Mechanical and Optical Stability
  - 8.8.4. Wavelength Scanning Capability
  - 8.8.5. Choice of Lines Monitored by the Direct Reading Spectrometer
  - 8.8.6. The "$n + 1$" Channel
  - 8.8.7. Polychromators with Photographic Detection
- 8.9. **Slew-Scan Spectrometers**
  - 8.9.1. Mechanics for Slewing and Scanning by Grating Rotation
    - *8.9.1.1. Sine Bar Drive*
    - *8.9.1.2. Direct Drive*
    - *8.9.1.3. Magnetically Driven Grating Rotation*
    - *8.9.1.4. Encoders*
  - 8.9.2. Wavelength Calibration
  - 8.9.3. Peak Finding and Signal Acquisition
    - *8.9.3.1. Single Peak-Finding Methods*
    - *8.9.3.2. Moving Window Peak-Finding Method*
    - *8.9.3.3. Peak Area Fitting Routine*
    - *8.9.3.4. Side Line Indexing Method*
  - 8.9.4. Moving Detector Slew-Scan Spectrometers
- 8.10. **Choosing a Spectrometer for ICP–AES**
- 8.11. **New Directions in Spectrometers for ICP–AES**
  - 8.11.1. New Systems for Simultaneous, Multichannel Detection
  - 8.11.2. Intelligent Instruments
- **References**

8.1. INTRODUCTION

The function of a spectrometer is twofold. First, the light emitted from a source, such as the ICP, must be separated by wavelength. A characteristic spectrum, a measure of energy versus frequency, is obtained. The observed spectral lines are used to identify the elements present in the sample. Second, the relative

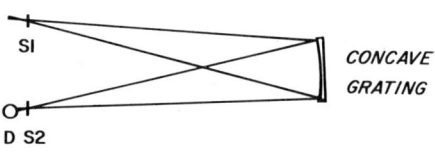

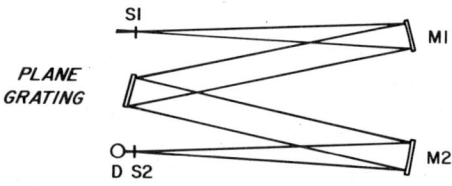

Figure 8.1. Spectrometer optical systems. Top: Concave grating based system; Bottom: plane grating based spectrometer.

intensities of the elemental lines must be measured in order to ascertain their concentration in the original sample.

Wavelength dispersion or discrimination can be provided by a number of devices including prisms, interferometers, or grating spectrometers. It is difficult to make large, transparent, homogeneous prisms with high dispersion. Therefore, prism-based spectrometers are not often used for ICP-AES. Interferometers have a free spectral range which is several orders of magnitude less than grating spectrometers. Thus, the grating spectrometer is by far the most prevalently used wavelength dispersive device for ICP-AES at the present time.

The typical grating spectrometer consists of an entrance (primary) slit, grating, exit (secondary) slit, and a detector. The entrance slit acts as a narrow object for the spectrometer's optical system. Light falling on the grating is dispersed spatially, perpendicular to the grooves of the grating. Optics in the spectrometer transfer many images of the entrance slit, one for each wavelength, to positions along a focal plane. One or more exit slits are placed on the focal plane so that only light over a small wavelength region passes through the slit and falls onto the detector. The image transfer optics consist of collimating and focusing mirrors used with a plane grating or the grating itself, which is concave, as shown in Fig. 8.1.

Requirements of the spectrometer depend on the nature of the source and its emitted spectra. Over the last 10 to 15 years commercially available spectrometers have evolved to match the unique properties of the ICP. In this chapter some of the fundamental characteristics of spectrometers are reviewed (Sections 8.2 and 8.3). Most of this chapter (Sections 8.4-8.9) deals with the requirements of spectrometers used for ICP-AES together with recent developments

8.2. FUNDAMENTALS OF GRATING SPECTROMETERS

which have improved the performance of these spectrometers and, therefore, ICP–AES as an analytical technique. Finally, the factors which should be considered in choosing an ICP–AES spectrometer (Section 8.10) and possible spectrometers of the future (Section 8.11) are discussed.

8.2. FUNDAMENTAL EQUATIONS AND FIGURES OF MERIT FOR GRATING SPECTROMETERS

The diffraction grating is made up of a number of grooves (typically 600–3600 per millimeter) in a reflecting material, usually aluminum coated glass or quartz. Ideally, the groove shape is well controlled and the grooves are parallel, straight, and equally spaced. A magnesium fluoride overcoat is often used to improve the reflective properties of the grating in the UV region.

The theory of the diffraction grating was first described by Rowland [1]. A number of good references are available concerning grating operation [2–9] including most elementary optics texts; only a summary of important points will be presented here.

8.2.1. The Grating Equation

A schematic diagram of a portion of a grating is shown in Fig. 8.2. The *grating normal* is defined as a line perpendicular to the flat surface of the grating blank before ruling. When the grating is illuminated by light incoming at an angle α, some of the light will be reflected at an equal angle on the opposite side of the grating normal. At this angle of diffraction the grating acts as a mirror without separation of the different wavelengths. This undispersed light is called the "zero order."

The separation of wavelengths by the grating is based on multiple slit dif-

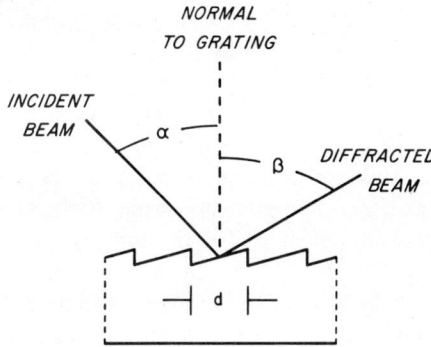

Figure 8.2. Definition of diffraction grating angles: α is the angle of incidence; β is the angle of diffraction; d is the distance between grooves.

fraction [2–5]. Constructive interference occurs at a particular spatial position when the path difference between rays reflected from neighboring grooves is an integral number of wavelengths. Equation 8.1 describes the constructive interference condition mathematically.

$$n\lambda = d(\sin \alpha \pm \sin \beta) \tag{8.1}$$

The angles α and β are the angles of incidence and diffraction, respectively, relative to the grating normal, as shown in Fig. 8.2. The spacing between grooves is represented by d, while n is an integer called the "order." When the incoming and diffracted rays are on the same side of the grating normal, Eq. 8.1 is satisfied by the addition ($+$) of the two terms. The order is then positive. Conversely, negative orders occur when the incident and diffracted rays are on opposite sides of the grating normal. The grating equation consists of a subtraction ($-$) of the two terms for negative orders.

For example, consider a spectrometer having a grating with 1200 grooves per millimeter. If the angle of incidence is 5° and the intensity of the Na line at 589 nm is to be monitored, the angle of diffraction can be calculated from Eq. 8.1.

For $n = 1$: $n\lambda = 589$ nm

$$\beta = \sin^{-1}\left[\left(\frac{n\lambda}{d}\right) - \sin \alpha\right]$$

$$= \sin^{-1}\left[\frac{598 \times 10^{-9} \text{m}}{\frac{10^{-3}}{1200}\text{m}} - \sin 5°\right]$$

$$= 38°$$

For $n = 2$: $n\lambda = 589$ nm; $\lambda = 589/2$ nm

Therefore, the exit slit and detector should be placed at an angle of 38° relative to the grating normal. Note that second-order light of wavelength 589/2 or 295 nm will also be present at this angle.

The wavelength of light falling on the detector can be changed by moving the detector to different angles of diffraction along the focal plane. This can be accomplished in a number of ways. The detector and exit slit could be moved along the focal plane, although this is not common in commercial instruments. A number of exit slits and detectors can be placed along the plane to simultaneously monitor multiple wavelengths. This arrangement is called a "polychromator" or "direct reading spectrometer." Alternatively, a photographic plate or film could be used as a multichannel detector. The resulting *spectrograph* provides a huge amount of information due to the nature of the photographic emulsion, which is comprised of a large number of small, discrete detectors. *Polychromators* can also be configured with optoelectronic imaging detectors such as linear photodiode (PD) arrays or vidicons. Detectors for ICP–AES will be discussed in detail in Chapter 9.

Alternatively, the angles of incidence and diffraction could be changed by rotating the grating while using fixed entrance and exit slits. The *scanning monochromator* provides flexibility in wavelength selection while using a single detector, most often a PMT.

8.2.2. Dispersion

The separation in angle or space per unit wavelength is called "dispersion." The *angular dispersion* can be determined by differentiation of the grating equation (8.1):

$$\frac{d\beta}{d\lambda} = \frac{n}{d \cos \beta} \quad (8.2)$$

The *linear dispersion* describes the distance along the focal plane per unit wavelength. The angular dispersion and the focal length of the spectrometer, F, determine the *reciprocal linear dispersion* as described by Eq. 8.3.

$$\frac{d\lambda}{dx} = \frac{d \cos \beta}{nF} \quad (8.3)$$

If Eq. 8.1 is rearranged, we obtain

$$d = \frac{n\lambda}{\sin \alpha + \sin \beta} \quad (8.4)$$

Substituting for d in Eq. 8.3,

$$\frac{d\lambda}{dx} = \frac{\lambda (\cos \beta)}{(\sin \alpha + \sin \beta) F} \quad (8.5)$$

The angular dispersion depends on the angles of incidence and diffraction, as shown in Fig. 8.3. This fact has two important consequences. First, the dispersion is *not constant* at all points along the focal plane or as the grating is rotated. However, at angles near 0° the dispersion is fairly constant since the cosine function changes slowly. For example, as the angle is changed from 0° to 30° the relative change in reciprocal linear dispersion is 1 to 0.87 for a fixed angle of incidence. Second, to obtain high dispersion, large angles and long focal lengths should be used.

As pointed out by Harrison [10, 11], a grating gives the same dispersion at a given angle whether it has 100 or 100,000 grooves per millimeter. The number of grooves per millimeter will determine the order observed at that angle, and affect free spectral range, as will be discussed.

If the dispersion provided by a spectrometer in first order is compared to that

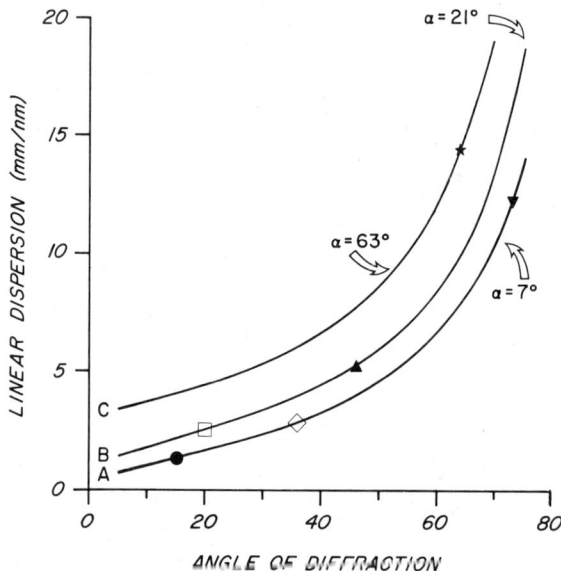

Figure 8.3. Dispersion as a function of grating angle for a 1-m monochromator. A, Angle of incidence is 7°; B, Angle of incidence is 21°; C, Angle of incidence is 63°; ●, 1200-grooves per millimeter grating used in first order; ◊, □, 2400-grooves per millimeter grating used in first order or 1200-grooves per millimeter grating used in second order; ▲, ▼, 3600-grooves per millimeter grating used in first order or 1200-grooves per millimeter grating used in third order; ★, 76th order with a 79-grooves per millimeter grating.

from the same spectrometer in second order, the dispersion will not simply be doubled. The angle of diffraction (and possibly incidence) also changes. Table 8.1 shows the calculated dispersion for different orders.

Some monochromators, including Ebert and Czerny–Turner mounts (described in Section 8.3) are arranged with a constant angle ϕ between the incident and diffracted beams. The angles of incidence and diffraction can be defined relative to the grating normal when at the central image position, as shown in Fig. 8.4:

$$\alpha = \theta - \frac{\phi}{2} \quad \beta = \theta + \frac{\phi}{2}$$

When the grating is rotated to change wavelength, both the angles of incidence and diffraction are changed. Table 8.1 (spectrometer B) shows the relationship between dispersion and the order used for a Czerny–Turner or Ebert type monochromator. For a one meter Czerny–Turner monochromator with a 3600 g/mm grating used in first order and $\phi = 14.74°$, the reciprocal linear dispersion at 500 mm is 0.083 nm/mm and at 180 nm is 0.248 nm/mm [12].

8.2. FUNDAMENTALS OF GRATING SPECTROMETERS

Table 8.1. Dispersion Provided by Various Spectrometers

Order	α (°)	β (°)	λ (nm)	Reciprocal Linear Dispersion (nm/mm)	Spectrometer[a]
1	10	17.8	400	0.79	A
2	10	51.8	400	0.26	A
1	7	10.2	250	0.82	A'
2	7	28.6	250	0.37	A'
3	7	51.1	250	0.17	A'
1	6.3	21.3	400	0.78	B
2	20.9	35.9	400	0.33	B
3	38	53	400	0.17	B

[a] Spectrometer A has a fixed angle of incidence and a 1200 grooves per millimeter grating. Spectrometer A' is similar to A except with a smaller angle of incidence. Spectrometer B is a Czerny-Turner monochromator with a 1200 g/mm grating. The grating was rotated as indicated for each order. All spectrometers have a one meter focal length.

8.2.3. Free Spectral Range

More than one wavelength can satisfy the grating equation. Therefore, light from different wavelengths can reach the detector for given angles of incidence and diffraction. The wavelength range between overlapping orders, called the "free spectral range," is described by Eqs. 8.6 and 8.7.

$$F_{SR} = \lambda_{n+1} - \lambda_n \tag{8.6}$$

$$F_{SR} = \frac{\lambda_{n+1} \lambda_n}{d (\sin \alpha + \sin \beta)} \tag{8.7}$$

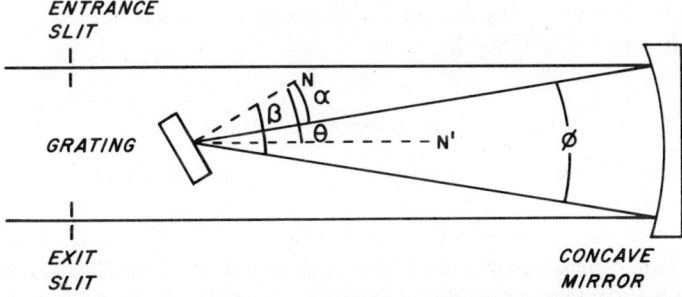

Figure 8.4. Monochromator with a constant angle between incident and diffracted beams. ϕ is the angle between the incident and diffracted beams, α is the angle of incidence, β is the angle of diffraction, N is the grating normal, N' is the normal when the grating is in the zero position, θ is the angle between N and N'.

The free spectral range for given angles of incidence and diffraction will be dependent on the number of grooves per millimeter since d, α, and β will determine the order n.

8.2.4. Theoretical Resolving Power

The *resolving power* is a measure of the wavelength separation between two signals, such as emission lines, necessary to distinguish the two. The Rayleigh criterion states that two lines of equal intensities are resolved if the maximum of one falls on the minimum of the other [3, 13]. Diffraction results in an intensity pattern with an angular distance between the principal maximum and the first minimum of

$$\Delta\beta = \frac{\lambda}{W \cos \beta} \tag{8.8}$$

where W is the width of the grating which is illuminated.

The *theoretical resolving power* is often defined by

$$R = \frac{\lambda}{\Delta\lambda} = nN = n\left(\frac{W}{d}\right) \tag{8.9}$$

where n is the spectral order, N is the total number of grating lines which are illuminated, and W is the ruled width of the illuminated grating. However, the resolving power is *independent* of the number of lines per millimeter for a given angle [10, 11], as was dispersion. The resolving power can also be expressed by [9],

$$R = \frac{W (\sin \alpha + \sin \beta)}{\lambda} \tag{8.10}$$

Therefore, a fixed size grating will provide highest resolution, and dispersion, at large angles. The Rayleigh criterion must be modified slightly to be applied to concave gratings [3, 13].

8.2.5. Experimental Resolving Power

The theoretical resolving power is often not observed in practice. A number of other factors may limit the experimental resolution. The optical fidelity of the spectrometer will always be limited by aberrations [3, 6] such as coma and astigmatism, which become more important as the slit width is decreased and the light collection angle of the optics is increased. Since the observed resolution is critically dependent on the image of the entrance slit at the focal plane, aberrations will degrade the spectral resolution.

The observed resolving power will also be limited by the bandpass of the

8.2. FUNDAMENTALS OF GRATING SPECTROMETERS

Table 8.2. Bandpass of Spectrometers as a Function of Slit Width

Czerny–Turner Monochromator with 0.83-nm/mm Dispersion[a,b]			
Slit Width (μm)	Calculated Bandpass (nm)	$R = \dfrac{\lambda}{\Delta\lambda}$[a]	Minimum Grating Width (mm)[d]
10	0.008	50,000	42
25	0.021	19,050	16
50	0.042	9,520	8
100	0.083	4,820	4
Echelle Spectrometer with 0.26-nm/mm Dispersion.[a,c]			
Slit Width (μm)	Calculated Bandpass (nm)	$R = \dfrac{\lambda}{\Delta\lambda}$[a]	Minimum Grating Width (mm)[d]
10	0.0026	154,000	34
25	0.0065	61,500	14
50	0.013	30,800	7
100	0.026	15,400	4

[a] At 400 nm.
[b] 1200 g/mm grating.
[c] 79 g/mm grating.
[d] Illuminated width of grating where bandpass limited resolution is equal to the theoretical resolving power.

spectrometer, as determined by the slit width and reciprocal linear dispersion. If the exit slit is similar to or wider than the entrance slit, the bandpass may be expressed as

$$\text{bandpass} = \text{exit slit width} \times \text{reciprocal linear dispersion} \quad (8.11)$$

Table 8.2 lists the calculated bandpass for two typical spectrometers used for ICP–AES as a function of slit width. The width of the grating which must be illuminated to provide a theoretical resolving power equal to the bandpass limited resolution is also listed.

8.2.6. The Distribution of Light into Multiple Orders—Blazed Gratings

The shape and angle of the grating grooves control the amount of light in a given order. While the theory describing grating efficiency is complex and dependent on the electromagnetic nature of light [14], useful approximations, which will be described, can be made when λ/d is less than about 0.2.

The greatest percentage of incident light can be concentrated in an order when the angles of incidence and diffraction measured relative to the normal of the groove surface are equal. Under these conditions the angles are equal to those for specular reflection, that is, each groove is acting as a mirror. Wood [15] first successfully controlled the face angle of grooves in order to produce a *blazed* grating. Using a blazed grating, up to 90% of the incident light can be directed into a single order.

The wavelength at which the efficiency (amount of light in one order relative to the total reflected light) is maximum is called the "blaze wavelength." If the grating equation is solved under the conditions for specular reflection from the groove surface, the blaze wavelength can be calculated. Referring to Fig. 8.5, $\alpha - \theta = -\beta + \theta$ at the blaze angle. Solving for θ one obtains

$$\theta = \frac{(\alpha + \beta)}{2} \tag{8.12}$$

Substituting into the grating equation (8.2) one obtains

$$n\lambda_{\text{BLAZE}} = 2d \sin \theta \cos (\alpha - \theta) \tag{8.13}$$

The sign of θ is positive when on the same side of the normal as α. The blaze wavelength is usually specified for the Littrow configuration where $\alpha = \beta$. Then

$$n\lambda_{\text{BLAZE}} = 2d \sin \theta \tag{8.14}$$

For Ebert and Czerny–Turner mounts, where there is a constant angle between the incident and diffracted light, the blaze wavelength can be calculated from (Fig. 8.4):

$$n\lambda_{\text{BLAZE}} = 2d \sin \theta \cos \frac{\phi}{2} \tag{8.15}$$

The intensities or efficiencies at wavelengths removed from the blaze wave-

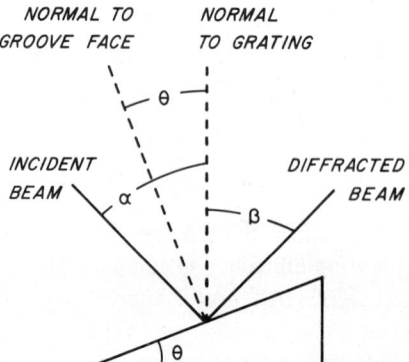

Figure 8.5. Grating angles defining the blaze. α is the angle of incidence, β is the angle of diffraction, and θ is the blaze angle.

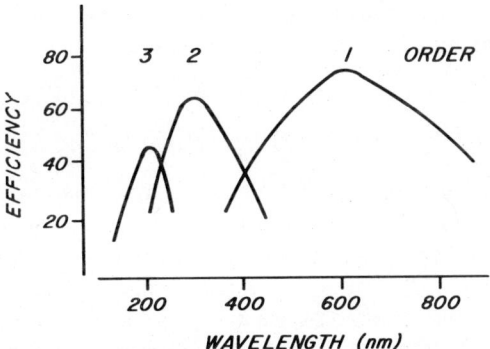

Figure 8.6. Efficiency in first, second, and third order for a grating blazed at 600 nm [17].

length can be estimated using the equations describing diffraction from a single groove [3, 9]. The simplified expression for the intensity is

$$I = \frac{A_0 \sin^2 b}{b^2} \tag{8.16}$$

where $b = \pi d \sin \beta / \lambda$. As a general guideline [16], the efficiency of the grating drops to 70% of that at the blaze wavelength for wavelengths

$$\lambda = \lambda_{BLAZE} \pm \frac{\lambda_{BLAZE}}{2n} \tag{8.17}$$

Figure 8.6 shows the efficiency for a grating with a blaze wavelength of 600 nm.

8.3. GRATING SPECTROMETER MOUNTINGS

Many different optical arrangements are used for grating spectrometers. A number of good reviews have been published [2, 6-9]. The mountings described in the following sections are those which are among the most popular for ICP-AES. The characteristics of each mounting determine its use. For example, most direct reading spectrometers for ICP-AES are based on concave grating mounts while most of the scanning monochromators use a plane grating system.

8.3.1. Classical Concave Grating Mounts

Spectrometers with a concave grating use the grating for both imaging the entrance slit at the focal plane and dispersing the various wavelengths. The grating grooves are ruled on a concave spherical mirror blank. The distance between grooves is equal along a chord of the grating surface.

8.3.1.1. Rowland Circle Based Mountings

Rowland [18] first described a spectrometer using a concave grating for both imaging and dispersion. The positions of the entrance slit, grating, and exit slit or focal plane, where focused images of the slit are formed, is described by the Rowland circle (Fig. 8.7). The grating is placed tangent to the Rowland circle. When the entrance slit is placed on the circle the images of the slit making up a spectrum are in focus on the Rowland circle. The diameter of the Rowland circle is equal to the radius of curvature of the spherical grating blank, which is twice its on-axis focal length.

A number of different spectrometers are based on the Rowland circle including the Rowland, Paschen–Runge, Abney, Eagle, Seya–Namioka, and grazing incidence mountings. Since the Paschen–Runge system is the only Rowland circle-based mounting commonly used in commercial spectrometers for ICP–AES, refer to the literature [2, 6–9] for descriptions of others. The Seya–Namioka mounting is also briefly discussed (Section 8.3.1.3) because it has been employed for ICP–AES in the VUV spectral region [19, 20].

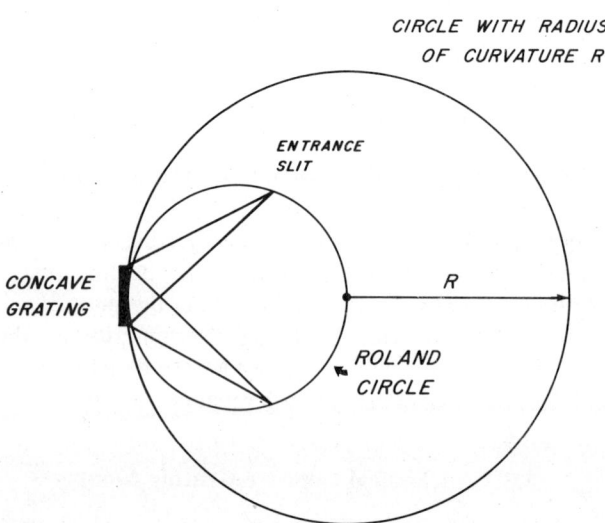

Figure 8.7. Rowland circle arrangement. R is the radius of curvature of the concave grating blank.

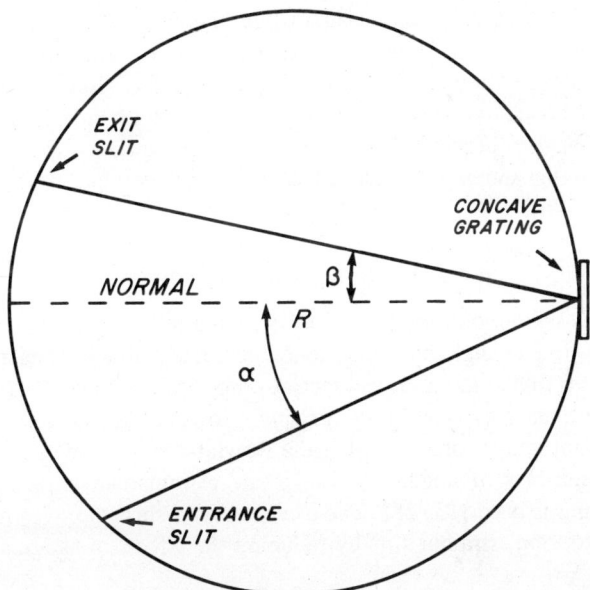

Figure 8.8. Paschen–Runge spectrometer mounting. R is the radius of curvature, α is the angle of incidence, and β is the angle of diffraction.

8.3.1.2. Paschen–Runge Mounting

The Paschen–Runge [21] mounting (Fig. 8.8) is the most popular concave grating mounting for AES because of the wide wavelength coverage along its focal plane. The large focal plane allows a number of photographic plates or exit slit-PMT modules to be placed along the focal plane. As a result, many commercial direct reading spectrometers for ICP–AES are based on the Paschen–Runge mount. However, because the angles of incidence and diffraction change substantially, the dispersion also varies along the focal plane.

The mechanically ruled concave gratings show efficiency or blaze behavior which is somewhat broader than the 70% "rule of thumb" value given by $\pm 1/2n$. The peak efficiency is somewhat less than for plane gratings. This is due to a change in the groove shape across the face of the grating. Because the blank is concave and the ruling diamond which cuts the grooves is in a fixed orientation, the shape of the grooves changes as the slope of the blank changes. The change in shape is often reduced by ruling the concave grating in up to seven separate zones [22].

Astigmatism in the concave grating mountings can be large so that the horizontal and vertical foci are at different distances from the grating [2]. As a result, the exit slit height must be larger than the entrance slit height to collect the entire image of the entrance slit. Astigmatism and field curvature can be reduced by ruling with a nonparallel, complex series of grooves [23]. Aberration corrected concave gratings are now produced holographically [22, 24] with impressive imaging properties. For example, gratings which are stigmatic at three wavelengths and exhibit reduced abberrations over the entire spectrum are available [22, 25].

8.3.1.3. Seya–Namioka Mounting

The Seya–Namioka mounting (Fig. 8.9) is used almost exclusively in the VUV spectral region. The angle between the incident and diffracted beams is fixed at 70° because at this angle a large spectral range is well focused [26, 27]. The spectrum is scanned by rotating the grating. However, as the angle is changed from the central image, or zero order, the Rowland circle moves away from the slits with some loss of image fidelity. Large astigmatism is present unless a toroidal grating is used [13, 25]. The use of curved slits improves resolution in comparison to long, straight slits by reducing the effects of astigmatism.

8.3.2. Classical Plane Grating Mountings

Plane grating spectrometers use separate optics for focusing and dispersion. As a result, it is possible to design spectrometers with few aberrations. A concave

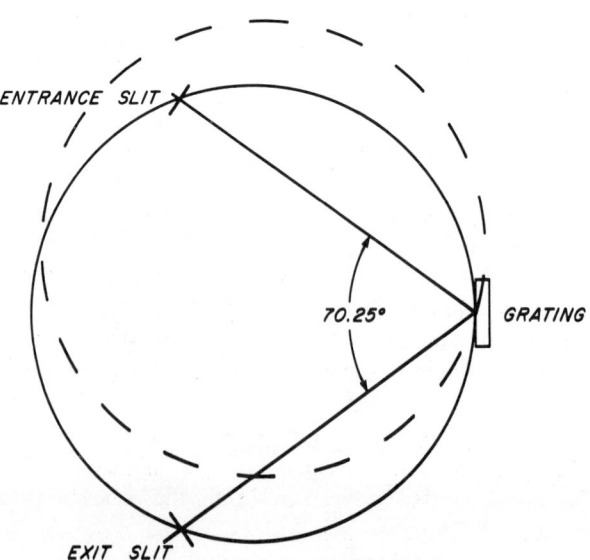

Figure 8.9. Seya–Namioka spectrometer mounting.

mirror is placed at its focal length from the entrance slit, illuminating the grating with collimated light. Light diffracted from the grating is collected by a second mirror which forms images of the entrance slit on the focal plane.

8.3.2.1. Czerny–Turner and Other Ebert-Type Mountings

The Ebert mounting [28] shown in Fig. 8.10, consists of an entrance slit, a single large concave mirror, grating, and exit slit or imaging detector mount. The concave mirror serves a dual purpose, both collimating light which is then directed toward the grating and focusing the diffracted light onto the focal plane.

The symmetrical system results in some cancellation of coma, as first described by Czerny and Turner [29]. However, the main limitation of the Ebert mounting is the need for a very large mirror. Further, the system has stray light problems apparently due to light from the entrance slit which is reflected directly by the mirror to the focal plane [30]. The size of the focal plane is limited because of the on-line arrangement of the entrance slit, grating, and focal plane.

Fastie [31] independently developed a spectrometer similar in design to that of Ebert, but with improvements which addressed many of the limitations de-

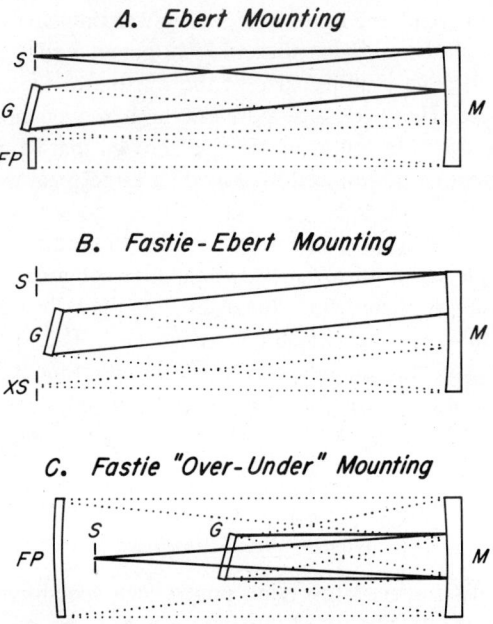

Figure 8.10. Ebert-type spectrometer mountings. (a) Ebert mounting. (b) Ebert–Fastie mounting. (c) Fastie "over-under" optical arrangement. M is a concave mirror, G is the grating, S is the entrance slit, XS is the exit slit, and FP is the focal plane.

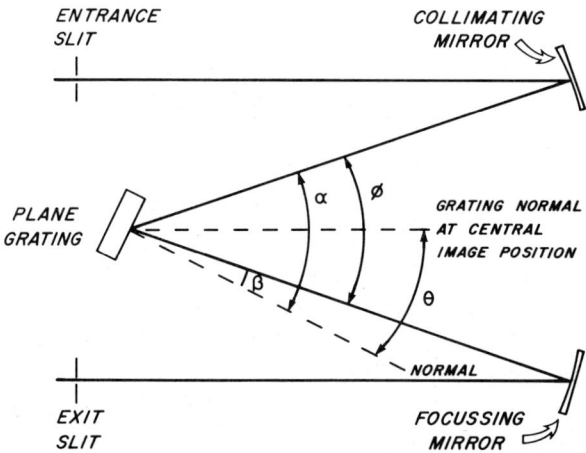

Figure 8.11. Czerny–Turner mounting. α is the angle of incidence, β is the angle of diffraction, θ is the grating angle measured relative to the zero order normal, and ϕ is the fixed angle between the incident and diffracted rays.

scribed. The arrangement is shown in Fig. 8.10. Baffles and higher quality gratings reduced the scattered light. The grating was moved toward the concave mirror, off-line from the entrance slit and focal plane. As the grating is moved closer to the mirror, the required size of the mirror for the same wavelength coverage is reduced. However, when the grating is moved closer to the mirror the off-axis angle becomes large. Fastie also showed that an off-plane, "over-under" arrangement could be used to provide a large focal plane as shown in Fig. 8.10.

Czerny and Turner [29] further improved the Ebert and Fastie designs by replacing the single large mirror with two smaller spherical concave mirrors. The Czerny–Turner spectrometer, shown in Fig. 8.11, has become the most widely used scanning monochromator for ICP-AES. The main advantages of the Czerny–Turner mounting relative to the Fastie–Ebert mounting are the lower cost of two smaller mirrors and the ability to independently position and choose the focal lengths of the mirrors, which allow good aberration correction.

8.3.2.2. Optical Characteristics of the Ebert, Fastie–Ebert, and Czerny–Turner-Type Mountings

The experimentally observed resolving power of a spectrometer depends on, among other factors, the fidelity of the optical system. Extensive work [6] has been done to minimize the aberrations present in the Ebert and modified Ebert-type spectrometers.

Fastic [31] examined a number of the aberrations present. When an entrance slit with a large height is used, astigmatism results in a curved image of the slit at the focal plane. The effect of astigmatism on resolution can be minimized by using properly curved entrance and exit slits. However, if the slit is used as a mask to observe emission from a selected portion of the source, curved slits are inconvenient.

Although coma is not present to a first approximation in the Ebert system, it is observed. Coma manifests itself to produce an asymmetric image of the entrance slit. This results from the effect of the diffraction grating on the wavefront. The coma in Czerny–Turner spectrometers can be reduced by an asymmetric tilt of the focusing mirror or by using a different focal length focusing mirror [32]. The optical arrangement conditions for coma-free operation have been described mathematically [6, 33, 34]. When the off-axis angles are small this condition is

$$\frac{\beta}{\alpha} = \frac{\cos^3 \alpha_g}{\cos^3 \beta_g} \qquad (8.18)$$

The angles are defined in Fig. 8.12a.

Note that the coma correction point changes as the angles of incidence and diffraction are changed. However, when equal focal length mirrors are used the image will be almost coma-free along the entire focal plane at a grating angle of 46° [35].

The spectra provided by Ebert or modified Ebert mountings show field curvature introduced by the collimating and focusing mirrors. It has been shown that a nearly flat focal plane can be obtained when the grating is a distance $(1 - 1/\sqrt{3})R_2$ from the focusing mirror [33, 36]. R_2 is the radius of curvature of the focusing mirror. When the grating is moved along the x-axis shown in Fig. 8.12.1, corrected flat field and super flat field behavior is observed as indicated in Fig. 8.12b [37].

Many of the commercially available Czerny–Turner spectrometers use one or more of the corrections previously described to prevent loss of resolution due to aberrations.

8.4. CHARACTERISTICS OF INDUCTIVELY COUPLED PLASMAS IMPORTANT IN CHOOSING A SPECTROMETER

The inductively coupled plasma has become a popular source for atomic spectroscopy because of its unique analytical characteristics as discussed in this book and previously published reviews [38, 39]. These characteristics are due to the physical properties of the ICP and the spectra it emits. However, the ICP

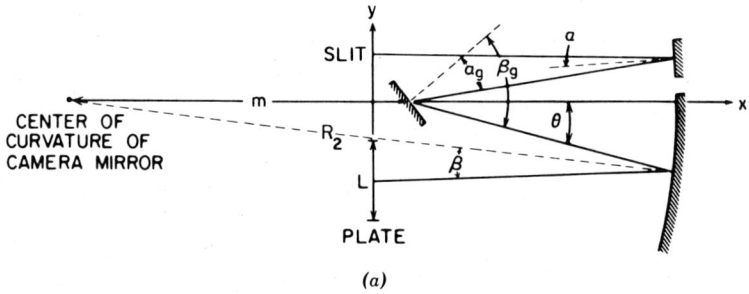

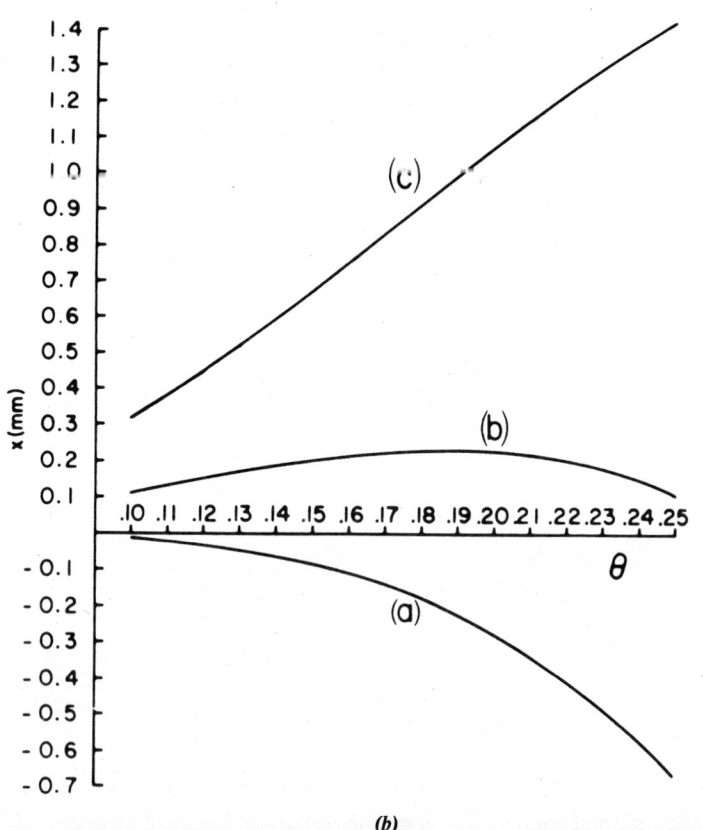

Figure 8.12. Positioning of the grating for a flat field focal plane. (*a*) Definition of angles important in coma minimization and field curvature. (*b*) Focus position versus grating angle. (a) Flat field, (b) corrected flat field, (c) super flat field [37]. [Reprinted with permission from J. Reader, J. Opt. Soc. Am. **59**, 1191 (1969). Copyright (1969), Optical Society of America.]

8.4. CHOOSING A SPECTROMETER FOR ICP-AES

characteristics also result in unique requirements for the spectrometers and measurement procedures which are used.

The ICP possesses a linear dynamic range of three to six orders of magnitude, allowing a wide range of element concentrations to be determined simultaneously without preconcentration or dilution. ICP detection limits are much lower then dc arc or flame emission methods, and are often similar to those of atomic absorption. As a result, it is possible to determine elemental concentrations from major to ultratrace. The ICP spectral lines can have a wide range of intensities, from up to 10^6 times the background level to a fraction of the background. This requires a high contrast (high resolution, low stray light) spectrometer to separate the weak spectral signals from intense emission signals. Also, the detectors and measurement electronics must provide a large linear dynamic range.

Chemical interferences such as those due to differing vaporization rates are small in the ICP in comparison to other atomic sources such as flames. The reduction in interferences is due to the highly excited plasma (temperatures of 4,000 to 10,000 K have been measured) and relatively long residence time (ms) of the analyte in the source. ICP emission from up to 70 elements can be observed simultaneously providing rapid, multielement analysis.

Because the plasma is so "hot," a large number of excited states are populated for most elements. Therefore, a huge number of elemental lines, both atom and ion, are observed [40–43]. As a result, spectral interferences can be prevalent. Many of the spectral interferences observed are due to insufficient resolving power of the spectrometer.

Thus, high-resolution spectrometers can reduce the number of spectral interferences. However, once the bandpass becomes equal to or smaller than the physical linewidths [12, 44–47] of the ICP emission, no further reduction in interferences will be gained by using higher resolution. The necessary spectrometer resolution to provide a bandpass equal to the physical linewidths is listed in Table 8.3 (cf. Section 7.7.5).

The use of high-resolution spectrometers does not preclude the occurrence of spectral interferences. If the wavelength difference between two lines is less than approximately 0.01 nm, separation is usually not possible, whatever the spectrometer's resolving power [12]. Two different steps can then be taken. A line may be chosen which is relatively free of spectral interferences. However, this will involve a trade-off between spectral interferences and sensitivity if the most intense line suffers from interference. Alternatively, it may be possible to correct or compensate for spectral interferences if the interferences can be identified and quantified. The latter approach is possible because of the stability of the ICP emission. However, the identification and quantitation of all significant interferences may be a difficult task. Spectral interferences will be most severe in samples with matrices such as iron or rare earth elements, which have complex emission spectra.

Table 8.3. **Spectrometer Resolution Necessary to Provide Resolution Equal to the Physical Linewidth of ICP Emitted Radiation**

Element	λ (nm)	Ref.	Physical Linewidth	Resolution $R = \dfrac{\lambda}{\Delta\lambda}$
As I	228.81	47	0.0017	135,000
Ba I	553.55	44	0.0037	150,000
Ba II	455.40	44	0.0034	134,000
Ca I	422.70	44	0.0042	101,000
Ca II	393.37	44	0.0039	101,000
Ho II	345.60	46	0.0190[a]	18,000
La II	398.80	46	0.0130[a]	31,000
Pb I	283.31	44	0.0016	177,000
Mg II	279.55	44	0.0036	78,000
Sc II	361.30	46	0.0060	60,000
Y II	371.00	46	0.0028	133,000
Zn I	213.86	44	0.0015	143,000

[a] Large physical linewidth may be due to unresolved hyperfine structure.

Background emission other than direct spectral overlap is also important. For analysis near the detection limits, the background is of the order of 100 times the net analyte signal (see Section 4.1). The ability to perform quantitative analysis when the background is larger than the analyte signal is possible only if the ICP background is reproducible (RSD near 1%). When working at concentrations where the background is a significant portion of the signal, accurate and precise background subtraction is imperative. Further, background emission intensity is dependent on the sample composition. Therefore, background enhancement (sometimes called "background shift") must be measured while sample or blank containing all of the species responsible for the background enhancement is being introduced into the plasma, rather than a simple blank such as distilled water.

Background emission may have a number of sources. Continuum emission is due to recombination of ions and electrons in the plasma. Much of the energy mismatch in this reaction is released as light. Radiative recombination of aluminum ions and electrons can double the continuum background from approximately 210–190 nm when the aluminum concentration is just 250 µg/mL [48]. Calcium and magnesium can also be important sources of background emission.

Spectra due to the plasma support gas (usually argon; nitrogen and helium have also been used) are also observed. In the case of argon, narrow atomic lines are observed. Molecular band emission may also be present [49] due to air entrainment (OH, NH, NO), sample solvent (OH from aqueous samples, C_2

Table 8.4. Characteristics of the Ideal Spectrometer System for ICP-AES

1. Record all spectral information simultaneously.
2. Rapid signal acquisition and recovery.
3. Provide high contrast (high resolution, low stray light).
4. Possess a wide dynamic range—at least 10^6.
5. Provide accurate, precise wavelength identification and selection for analysis.
6. Highly stable, insensitive to environmental changes (temperature, humidity, vibration).
7. Provide means to identify and correct for interferences including displaying spectra for operator.
8. Measure and subtract background.
9. Provide a permanent record of spectra and analysis results.
10. Computerized operation: control, readout, storage, data manipulation, statistical analysis, report generation.

---

from organic solvents), and formation of stable metal oxides (such as with Y, Sc, rare earths). However, the extent of stable monoxide formation in the analytical zone of the ICP is much less than in flame sources.

Emission from collisionally broadened elemental lines may also appear as background. The FWHM intensity measure of line width is in most cases due to Doppler broadening [50]. However, the wings of the line may produce significant background enhancements at wavelengths as far as 10 nm from the line center [48] (also see Section 7.3). The broadening is a function of the fifth power of wavelength and the elemental concentration of that element in the plasma [48]. Stark broadening may also produce background shifts at large displacements from the peak wavelength for some species such as Mg [50] and H.

Considering the characteristics of ICP-AES described, a "wish list" for the ideal spectrometer can be made (Table 8.4). The ideal spectrometer is not available at present, but significant progress toward it has been made over the last ten years.

8.5. STRAY LIGHT

Light at wavelengths outside of the instrumental bandpass which reaches the detector is called "stray light." Stray light often follows an optical path other than the intended one. Sources of stray light include reflections or scattering from optical components, dust, interior surfaces of the spectrometer, slit jaw imperfections, and grating imperfections. Other sources include diffraction at the entrance or exit slits, diffraction due to mirror apertures, and multiple diffraction.

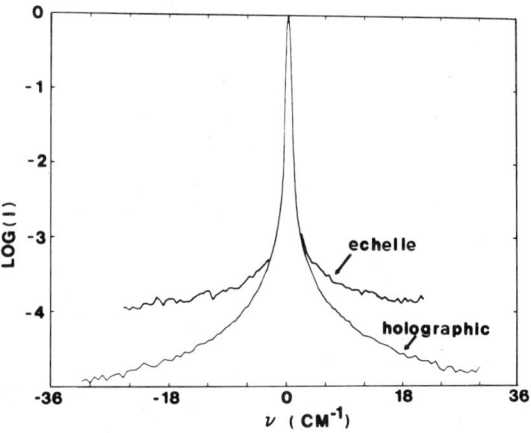

Figure 8.13. He 388.8-nm spectra obtained using spectrometers with a ruled echelle and holographic gratings. The higher apparent background observed with the echelle grating is due to stray light [51]. [Reprinted with permission from J. Kielkopf, *J. Opt. Soc. Am.* **20**, 3330 (1981). Copyright (1981), Optical Society of America.]

Experimentally, stray light may appear as an increase in the continuum background, such as when light is scattered from optics or spectrometer housing surfaces. For example, Fig. 8.13 shows [51] a comparison of stray light from a mechanically ruled echelle grating and a holographically ruled grating near the He 388.8-nm line emitted from a low-pressure lamp. The apparently larger background when the mechanically ruled grating was used is due to higher stray light. Line-like stray light can be produced by periodic ruling imperfections in the grating or multiple diffraction. Figure 8.14 shows [52] line-like stray light

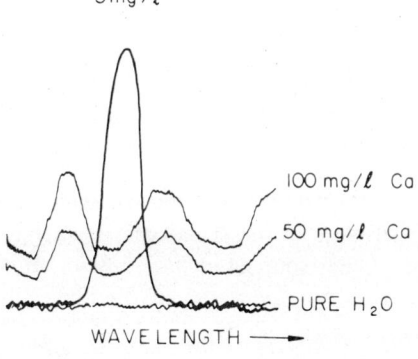

Figure 8.14. Line-like stray light due to (50 mg/L and 100 mg/L calcium) Ca II 396.85 nm near the Al 396.1-nm line [52]. [Reprinted with permission from G. F. Larson, V. A. Fassel, R. K. Winge, and R. N. Kniseley, *Appl. Spectrosc.* **30**, 386 (1976). Copyright (1976), Society for Applied Spectroscopy, Frederick, MD.]

due to the Ca II 396.85-nm line which appears near an Al line at 396.1 nm. Stray light intensities which are several percent of the source line intensity can result.

Enhancements in the background are observed when the concentration of concomitants in the sample changes, often due to changes in the stray light intensity [53], as shown in Fig. 8.15. As a result, it is important that the sources

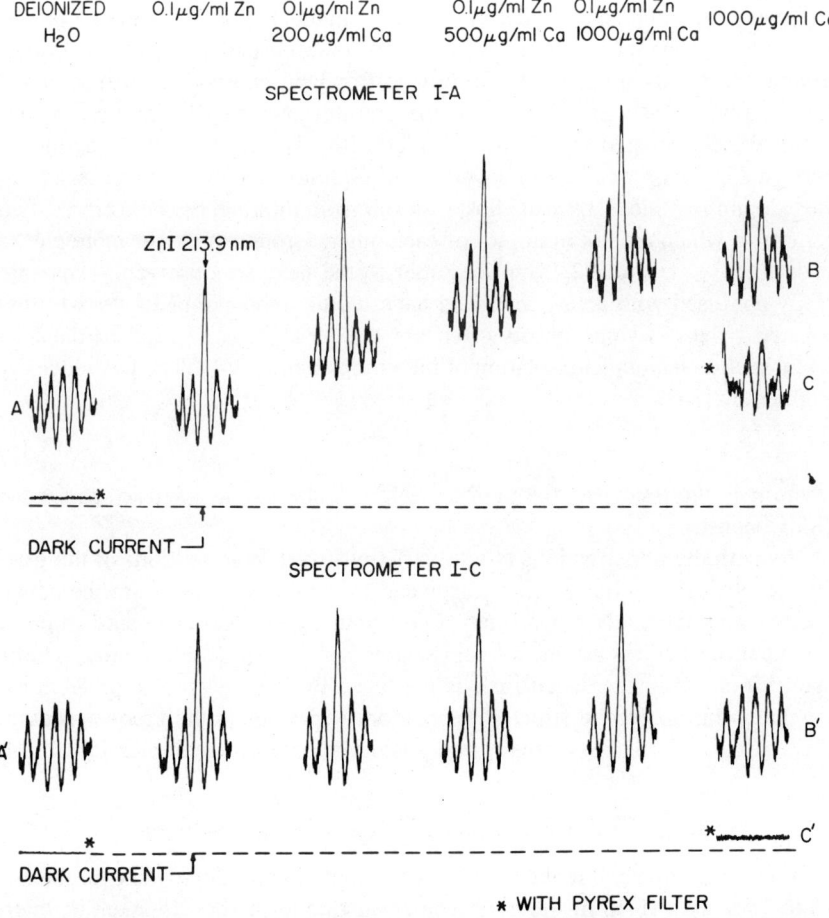

Figure 8.15. Background enhancements in ICP spectra due to changes in concomitant (Ca) concentration. Spectrometer IA has a mechanically ruled grating with high stray light. Spectrometer IC has a holographically ruled grating to reduce stray light [54]. [Reprinted with permission from G. F. Larson, V. A. Fassel, R. K. Winge, and R. N. Kniseley, *Appl. Spectrosc.* **30,** 387 (1976). Copyright (1976), Society for Applied Spectroscopy, Frederick MD.]

of stray light be identified and minimized. Stray light effects are most important when attempting to measure weak signals (small concentrations) in the presence of strong emission lines due to intense transitions and large concomitant concentrations [52].

8.5.1. Ruled Grating Imperfections

Periodic imperfections in the ruling of gratings can result in false lines called "ghosts" [6]. Ghosts will appear at apparent orders n', defined by $n' = n \pm 1/N$, where n is the actual order and N is the number of grooves in one period of the spacing error. The periodic ruling errors are usually traceable to imperfections in the ruling engine lead screw. "Rowland ghosts" result from alternate narrower and wider spacing of the grating grooves. Rowland ghosts are symmetrically located from parent wavelength. The stray light appearing as lines in Fig. 8.14 was due to seventh and eighth order Rowland ghosts from the calcium ion line. "Lyman ghosts" result from multiple periodic errors, two periods of which are not multiples of each other. Groups of two or more ghosts at distance $\pm 1/5$ to $1/2$ from one order to the next are observed. These are easily confused with actual lines and have led to some claims of "newly discovered" lines. Lyman ghosts often are due to sources external to the lead screw such as incomplete isolation of the ruling engine [24]. The ghost intensity is given by [24]

$$I_g = n^2 \pi^2 E_0^2 a^2 I_p \tag{8.19}$$

where n is the order, E_0 the periodic error, I_p the parent intensity and a the groove density.

Nonperiodic imperfections (slight variations over large portions of the grating) result in "satellites," lines displaced by small distances from the actual lines by approximately 2.5–5.0 nm. The cumulative effect of the production of a number of satellites results in "near scatter" or "grass." "Far scatter," more than 5.0 nm from the actual line, is produced by local groove imperfections, roughness in the groove structure or random variations in the groove spacing. The rules governing the intensity and position of ghosts is complex [53].

8.5.2. Improvements in Ruled Grating Production

Grating production and replication has been significantly improved over the last 10 to 15 years. Much of the improvement is due to the development of interferometrically controlled ruling engines. First order Rowland ghost intensity has been reduced from as much as 2% of the parent line intensity to less than 0.001% [54, 55]. Figure 8.16 shows a comparison of spectra [56] obtained with gratings which were mechanically ruled with and without interferometrically controlled engines.

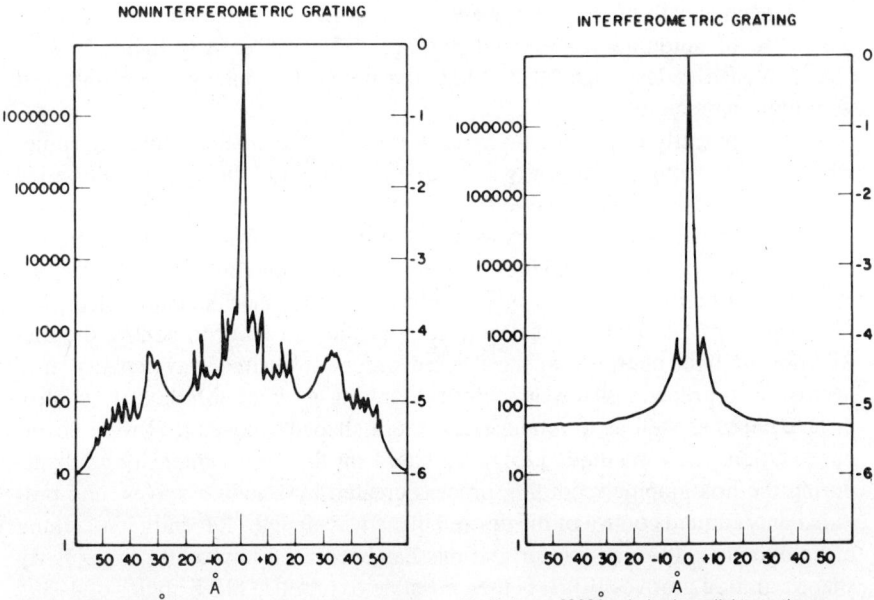

Figure 8.16. Comparison of observed He–Ne laser spectra from gratings ruled with a noninterferometrically and interferometrically controlled ruling engines. Note that significant improvements in interferometrically controlled ruling engines since the gratings used here were produced have resulted in gratings with less stray light than indicated in the spectrum on the right. [56]. [Reprinted with permission from *Spex Speaker* **11**(2), 4 (1966). Copyright (1966), Spex Industries, Metuchen, NJ.]

8.5.3. Holographic Gratings

One of the most exciting recent developments in spectrometers is the production and continued evolution of holographic gratings. These are produced [16, 24, 57] using an interference pattern formed by crossing two laser beams. A photoresist with a very small grain size deposited on an optically flat surface (within one-tenth of a wavelength) is exposed to the light. The exposed photopolymer is then removed by a solvent. Master gratings produced in this manner have groove spacing control at distances less than the wavelength of the laser light used. As a result, virtually ghost-free gratings with low scattered light characteristics can be produced.

The ghost-free nature of the holographically ruled grating results in excellent stray light properties. A spectrometer's stray light is often specified as the intensity measured 0.5 nm from a parent line relative to the parent line intensity. A He–Ne laser is often used as an intense, narrow line profile source [58]. Measured stray light intensities for spectrometers with interferometrically controlled mechanically ruled gratings are as low as 10^{-6} of the parent line intensity while use of holographically ruled gratings can provide stray light of 10^{-7}. Ghost intensities less than 10^{-12} times the parent line intensity are observed with holographic gratings.

Holographically ruled gratings possess other characteristics which are somewhat different from mechanically ruled gratings. Initially, holographically ruled gratings had a sinusoidal groove shape with incomplete control of the groove depth. As a result, these gratings were not blazed as mechanically ruled grating could be and, therefore, efficiency was low (30–55% maximum). However, the efficiency at peripheral wavelengths is higher than for mechanically ruled gratings resulting in a relatively flat efficiency profile. In order to control the distribution of light energy into the desired orders at desired wavelengths, it is necessary to properly shape or sculpt the grating grooves. Methods to control groove depth as well as to produce saw-tooth-shaped grooves [57] were developed. Often, these methods [25] were based on the use of intensity gradients during the holographic recording process created by standing waves, and provided only limited control of the desired blaze wavelength. Recently ion-etching of holographically ruled master gratings has been used to produce triangularly shaped grating grooves [9]. It is then possible to produce blaze angles of 4–39° [9], depending on the groove density. Holographically ruled gratings with large blaze angles, as required for echelle spectrometers, are yet to be developed.

8.5.4. Stray Light Produced by Perfect Gratings

Even perfectly ruled gratings will exhibit some ghosts due to the diffraction pattern produced by an aperture width equal to the width of the grating projected onto the optical axis. The "feet" of this diffracted image can be reduced through apodization [59] such as the use of a diamond-shaped mask on the grating when narrow slits are used. However, in almost all cases the theoretically predicted far scattering is several orders of magnitude lower than that experimentally observed due to other sources.

8.5.5. Reentry Spectra and Their Elimination

Reentry spectra due to multiple diffraction have been observed with Czerny–Turner and Ebert mounts. As shown in Fig. 8.17, once-diffracted light can be returned to the grating by the focusing mirror. The returned light is then diffracted again, and a wavelength determined by the grating equation directed toward the exit slit. Since the incidence angles of the singly and doubly refracted light are different, two wavelengths will pass through the exit slit and fall on the detector. Reentry spectra can be eliminated by proper choice of the grating groove spacing, grating or mirror masks, and geometric arrangement of

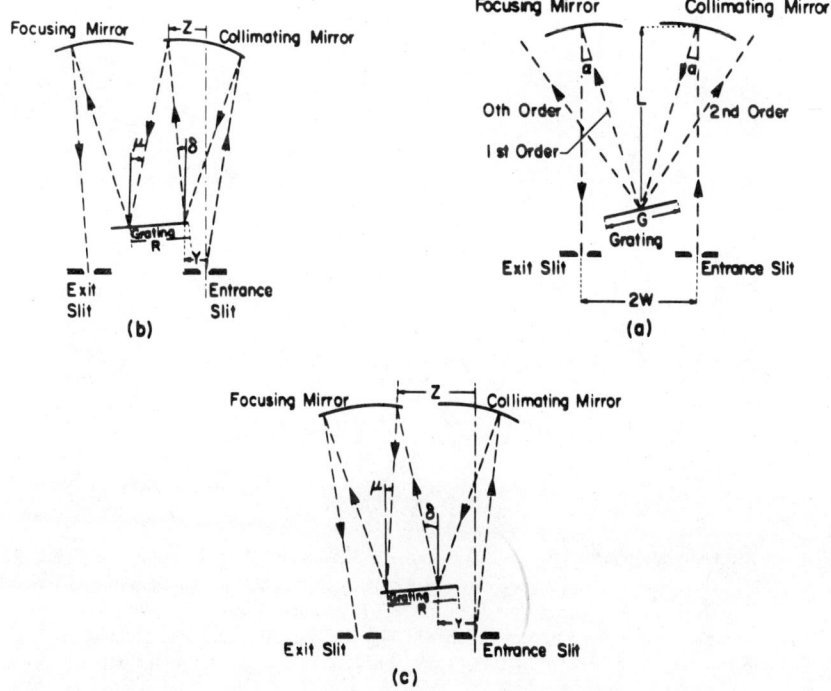

Figure 8.17. Diffracted stray light in Ebert or Czerny–Turner mountings. (*a*) Top view of a Czerny–Turner monochromator showing path of first-order diffracted light under normal conditions. (*b*) Schematic showing possible path of doubly diffracted light where the light after first diffraction hits the collimating mirror. (*c*) Schematic showing a possible path of doubly diffracted light where the light after first diffraction hit the focusing mirror [Reprinted with permission from J. K. Pribram and C. M. Penchina, *Appl. Opt.* **7,** 2006 (1968). Copyright (1968), Optical Society of America.]

the spectrometer. For example, a crossed Czerny–Turner mount, shown in Fig. 8.30 can be used to preclude reentry spectra.

8.5.6. Approaches to Reduce Stray Light Effects

The use of holographically ruled or interferometrically controlled, mechanically ruled gratings provides great reduction in stray light as discussed in the previous section. Many commercial ICP-AES spectrometers employ holographically ruled diffraction gratings. The echelle spectrometers are an exception since holographically ruled gratings with high blaze angles are not yet available. However, interferometrically controlled ruling engines are able to produce echelle gratings with few ghosts and relatively low stray light [51, 60]. Figure 8.18 shows a comparison of calibration curves acquired with mechanically ruled and

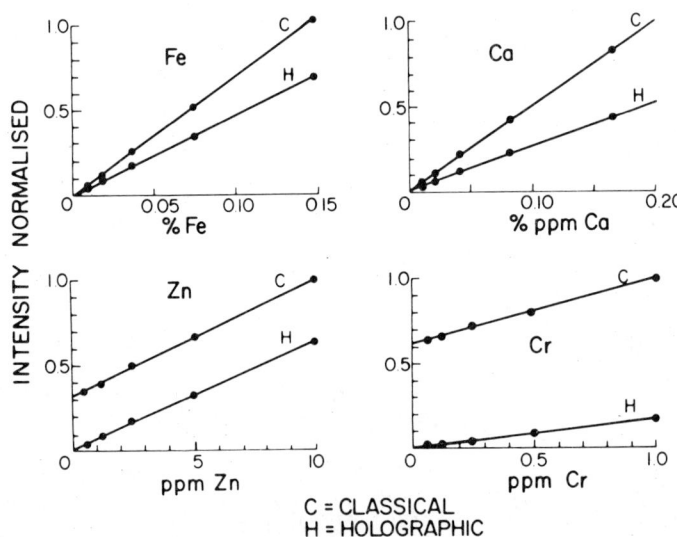

Figure 8.18. Calibration curves for Fe 259.95 nm, Ca 315.88 nm, Zn 213.86 nm, and Cr 425.43 nm with a 1440 groove per millimeter classically ruled grating (*C*) and a holographically ruled grating (*H*) [61]. [Reprinted with permission from N. M. Walters, A. Strasheim, and A. R. Oakes, "The Influence of Dispersion and Stray Light On The Analysis of Geological Samples By ICP-AES," *Spectrochim. Acta* **38B,** 963 (1983). Copyright (1983), Pergamon Press, Oxford.]

holographically ruled gratings [61]. The offset observed for Cr and Zn when the mechanically ruled grating is used is the result of stray light. The higher sensitivity for the Cr, Ca, and Fe signals when the mechanically ruled grating is used is due to the higher efficiency compared to the holographically ruled grating.

The interior of the spectrometer housing should be made as nonreflecting as possible to minimize spurious reflections or scattering. This can be done by using flat black surfaces except for optics.

Baffles can be used to reduce stray light following optical paths which deviate from the intended ones. Radiation from orders other than the one of interest are often trapped through use of baffles or light traps. The exit slit, any secondary optics, and the PMT in direct reading spectrometers should be protected from stray light by light-tight baffles where possible. This will reduce "cross talk" effects due to light scattered from one detection channel assembly to its neighbors.

Interference filters may also be used to reduce stray light. Narrow band filters placed in front of each exit slit or PMT tube can be used as "order sorters," allowing only the desired wavelength to pass. Stray light outside of the bandpass of the filter will also be reduced. Narrow band filters are used when a

8.5. STRAY LIGHT

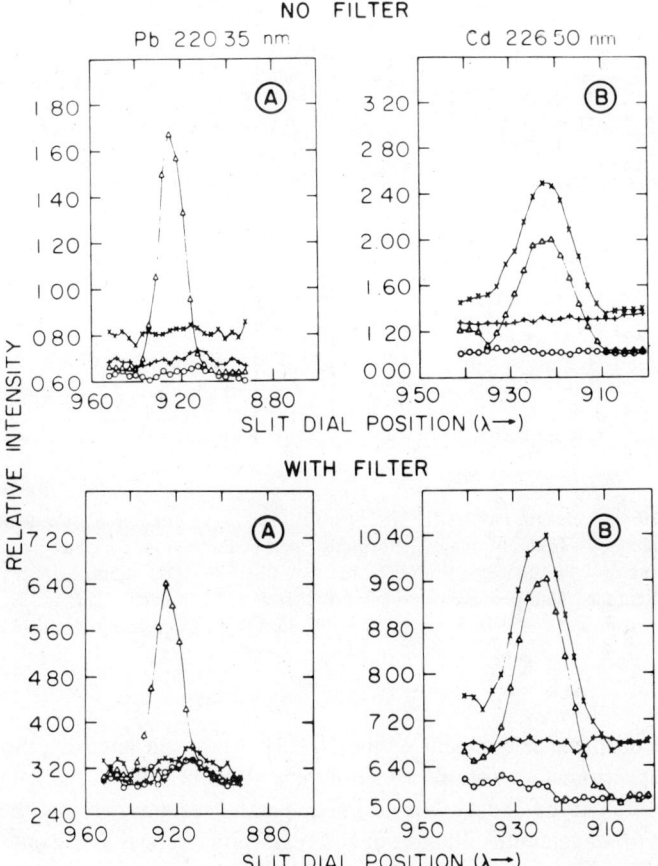

Figure 8.19. Wavelength profiles of Pb 220.35 and Cd 226.50 nm. (*a*) ○, 1% HNO₃ reference blank; △, 1 μg/mL Pb; +, 200 μg/mL Ca; x, 1000 μg/mL Ca; (*b*) ○, 1% HNO₃ reference blank; △, 1 μg/mL Cd; +, 200 μg/mL Mg; x, 100 μg/mL Mg and 0.1 μg/mL Cd [54]. [Reprinted with permission from V. A. Fassel, J. M. Katzenberger, and R. K. Winge, *Appl. Spectrosc.* **33**, 3 (1979). Copyright (1979), Society for Applied Spectroscopy, Fredrick, MD.

number of lines contribute to the stray light. The filters can reduce stray light from Ca to background levels, as shown in Fig. 8.19. However, when the background enhancement is due to a true increase in background rather than stray light, filters will not be effective (Fig. 8.19). Band blocking filters may also be used, before or immediately after the entrance slit, to reduce the light which falls on the grating from intense lines which are not of interest. This strategy is most effective when the stray light is due to one or two intense lines of known origin. Figure 8.20 shows the improvement in detected stray light due to Ca

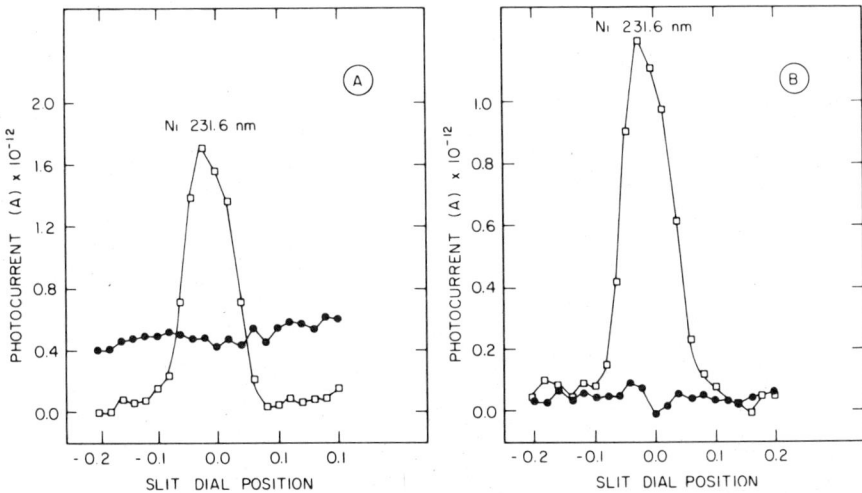

Figure 8.20. Wavelength profiles for Ni 231.6 nm. The background profile for deionized water has been subtracted from all profiles. All profiles were redrawn from computer plotted data. (a) Without band rejection filter (b) With band rejection filter. ●, 1000 μg/mL Ca; ☐, 0.67 μg/mL Ni [54]. [Reprinted with permission from V. A. Fassel, J. M. Katzenberger, and R. K. Winge, *Appl. Spectrosc.* **33**, 4 (1979), Copyright (1979), Society for Applied Spectroscopy, Frederick, MD.]

emission on the measurement of the Ni 231.6 nm line intensity. Some commercial spectrometers use up to 19 different filters in front of the exit slits.

Detectors can be chosen with limited spectral response so that they are insensitive to wavelengths outside of the region of interest. For example, solar blind PMT tubes can be used to detect UV radiation from elements with sensitive lines below about 300 nm, such as zinc, while discriminating against wavelengths longer than 400 nm, such as calcium at 422 nm.

Double monochromators may be used to increase the spectrometer's stray light rejection capabilities, since the stray light rejection will be the square of the two individual monochromators. Detection systems for Raman spectroscopy, where stray light from the incident laser light is extremely important, have successfully used double monochromator systems for a number of years.

Stray light as low as 10^{-10} times the intensity of the parent line at 0.5 nm removed from that line have been measured with double monochromator systems [54]. A commercially available double monochromator for ICP–AES is shown in Fig. 8.21.

Spectrometers for ICP–AES have evolved over the last ten years to include most of the methods discussed to reduce stray light. Modifications of early commercially available ICP–AES spectrometers can reduce the stray light levels by almost two orders of magnitude [62, 63].

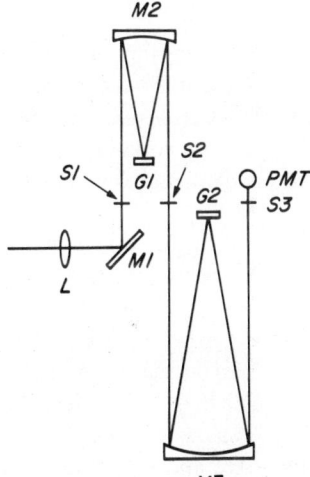

Figure 8.21. Schematic drawing of a double monochromator for ICP-AES. L, lens; $M1$, plane mirror; $M2$, $M3$, concave mirrors; $G1$, $G2$, gratings; $S1$, $S2$, $S3$, slits; PMT, photomultiplier tube.

8.6. HIGH RESOLUTION SPECTROMETERS

Spectral interferences can be reduced by using high resolution spectrometers. Also, an increase in the ratio of analyte line to continuum background intensities may be gained, possibly improving the minimum detectable concentration of analyte. (These topics are discussed in detail in Sections 4.1.3 to 4.1.7 and 7.7.)

Both interferometers and grating spectrometers can be configured to provide high resolution. For example, Fabry-Perot interferometers have been used to study the physical linewidths of ICP emission [44-45]. However, the small free spectral range of Fabry-Perot interferometers requires a second spectral discriminator, such as a monochromator. Michaelson interferometers can also provide high resolution [64]. Experimental limitations such as the difficulty of measuring weak lines in the presence of strong emission signals (as a consequence of the multiplex disadvantage) will require further work [65, 66] before becoming commercially viable for ICP-AES. As a result, choices for most ICP spectroscopists at present lie with grating spectrometers.

8.6.1. Obtaining High Resolution with Grating Spectrometers

High spectral resolution can be attained by increasing the angles of grating illumination and diffraction. For a given wavelength, the angles of illumination can be increased by using gratings with a larger number of grooves per millimeter or higher orders.

8.6.1.1. Echelle Spectrometers

Echelle spectrometers [67–70] use large angles of incidence and diffraction (45° and greater) while detecting high orders (20–170) to obtain good spectral resolution. The echelle grating is coarsely ruled, with 79–300 grooves per millimeter, typically. Only the steep, or short, side of the grooves is illuminated as shown in Fig. 8.22. A theoretical resolution of 500,000 can be obtained in the UV region with an echelle spectrometer. However, because ghost intensity increases with the square of the order, very precise control of the groove shape is required. This is now possible with interferometrically controlled ruling engines.

The echelle grating is often used in the Littrow mode, where the angles of incidence and diffraction are equal. The grating is blazed at the same angle. Since many orders of diffraction are viewed over a small range of angles, the echelle grating is always used near the blaze angle. As a result, grating efficiency is high despite the use of large angles of incidence and diffraction. As wavelength is increased in an echelle spectrometer, successively lower orders provide the highest efficiency. Choosing an order other than that of the peak efficiency for the particular wavelength of interest can result in lower intensities than expected and some irregularity in the decrease in efficiency [71]. The grating equation for the Littrow configuration can be written as

$$n\lambda = 2d \sin \beta = 2t \quad (8.20)$$

where t is the width of a step as shown in Fig. 8.22. The angular dispersion is then

$$\frac{d\beta}{d\lambda} = \frac{2 \tan \beta}{\lambda} \quad (8.21)$$

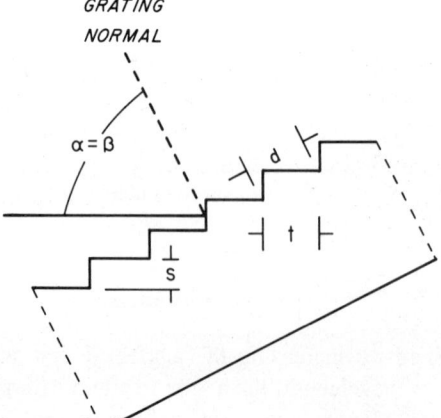

Figure 8.22. Echelle grating. α and β are the angles of incidence and diffraction, respectively; d, groove spacing; s, groove height; t, groove length.

8.6. HIGH RESOLUTION SPECTROMETERS

At the blaze angle, $\tan \beta = t/s$, where s is the step height. When the groove ratio, t/s, is equal to 2 the blaze angle is 63° 26'. This is the most commonly used blaze angle for ICP-AES echelle systems. The linear dispersion is given by

$$\frac{dx}{d\lambda} = \frac{2F \tan \beta}{\lambda} \tag{8.22}$$

The theoretical resolving power can be expressed as

$$R = W (\sin \alpha \pm \sin \beta)/\lambda \tag{8.23}$$

When $\alpha = \beta$,

$$R = 1.79 \, W/\lambda \tag{8.24}$$

The free spectral range, given by λ/n, will be much smaller than for spectrometers with a conventional grating. For example, the free spectral range at 400 nm for a spectrometer with a conventional grating used in first order is 400 nm. A typical echelle spectrometer views 400-nm light in order 56, with a free spectral range of 1.4 nm. Therefore, order overlap is a problem. Many orders must be observed in order to cover a broad wavelength range.

To overcome the order overlap problem, a predispersing monochromator [13, 72, 73, 145] can be used to limit the wavelength range passed into the echelle spectrometer, as shown in Fig. 8.23a. Such a system, with computer control of the monochromators and automatic wavelength calibration has been described and characterized in detail by Boumans and Vrakking [145]. These researchers used an angle approximately 1.5° removed from the blaze angle in order to reduce the occurrence of sharp intensity jumps at the transition from one order to the next.

Alternatively, cross dispersion [74, 75] can be used. A prism or second grating used in low order is used to disperse the various wavelengths of light in a direction perpendicular to that of the echelle grating, as shown in Figure 8.23b. This results in a compact, two dimensional display of the spectrum.

Since the slit height for the spectrometer will act as the slit width for the cross dispersing system, it must be relatively small (100 μm is typical). The small slit height and the use of high angles of incidence and diffraction reduce the luminosity. A number of steps can be taken to compensate for the reduction. A wider grating is normally used to increase light throughput. Shorter focal lengths (typically 0.5–0.75 m) can be used than for conventional grating spectrometers while continuing to provide high resolution, resulting in higher light throughput for similarly sized optics. Finally, the luminosity is a function of the grating efficiency, which is high since the angles of incidence and diffraction are always close to the blaze angle. A limitation of the echelle system is the need to adjust both the echelle grating and prism or cross-dispersing grating in

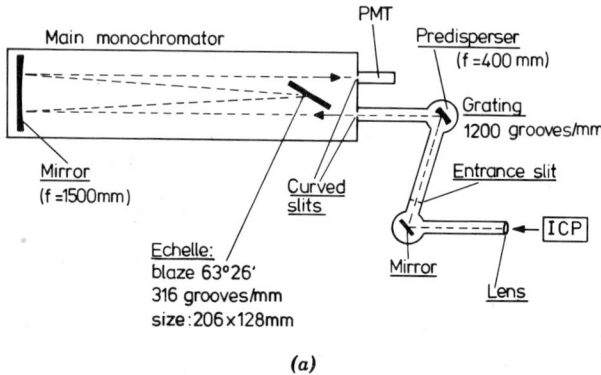

(a)

Figure 8.23. Echelle spectrometer arrangements. (a) A single channel echelle spectrometer system using a predispersing monochromator [145]. [Reprinted with permission from P. W. J. M. Boumans and J. J. A. M. Vrakking, *Spectrochim. Acta* **39B**, 1243 (1984). Copyright (1984), Pergamon Press.] (b) A prism is used for cross dispersion with a modified Czerny–Turner spectrometer, resulting in a two-dimensional display of the spectrum [76]. [Reprinted with permission from M. B. Denton, H. A. Lewis, and G. R. Sims, "Charge-Injection and Charge-Coupled Devices," in *Multichannel Image Detectors*, Vol. 2 (1983), Ch. 6, p. 152. Copyright (1983), American Chemical Society.]

order to scan wavelength. However, this can be done under computer control [58].

8.6.1.2. Long Focal Length Spectrometers

As discussed in Section 8.2, the observed resolution provided by a spectrometer is often limited by the bandpass, which is determined by the slit width and dispersion. One way to increase the linear dispersion is to increase the focal length of the spectrometer. For example, a 3.4-m Ebert spectrograph with a 1180 or 1800 groove per millimeter grating can provide sufficient resolution to study the physical linewidths of ICP emission [46] (also see Section 7.7). However, as the focal length is increased, the optics including mirrors and the grating must be enlarged in order to maintain a similar light throughput, often described by the f/number, greatly increasing the cost. Also, the mechanical stability must be higher as the focal length is increased since the effective "lever arm" also increases. Spectrometers used for ICP–AES typically have focal lengths of 0.6–1 m.

8.6.1.3. Spectrometers Using High Groove Density Gratings in Low Orders

When gratings with a large number of grooves per millimeter (3600 or greater) are used, a large free spectral range can be maintained while working at large

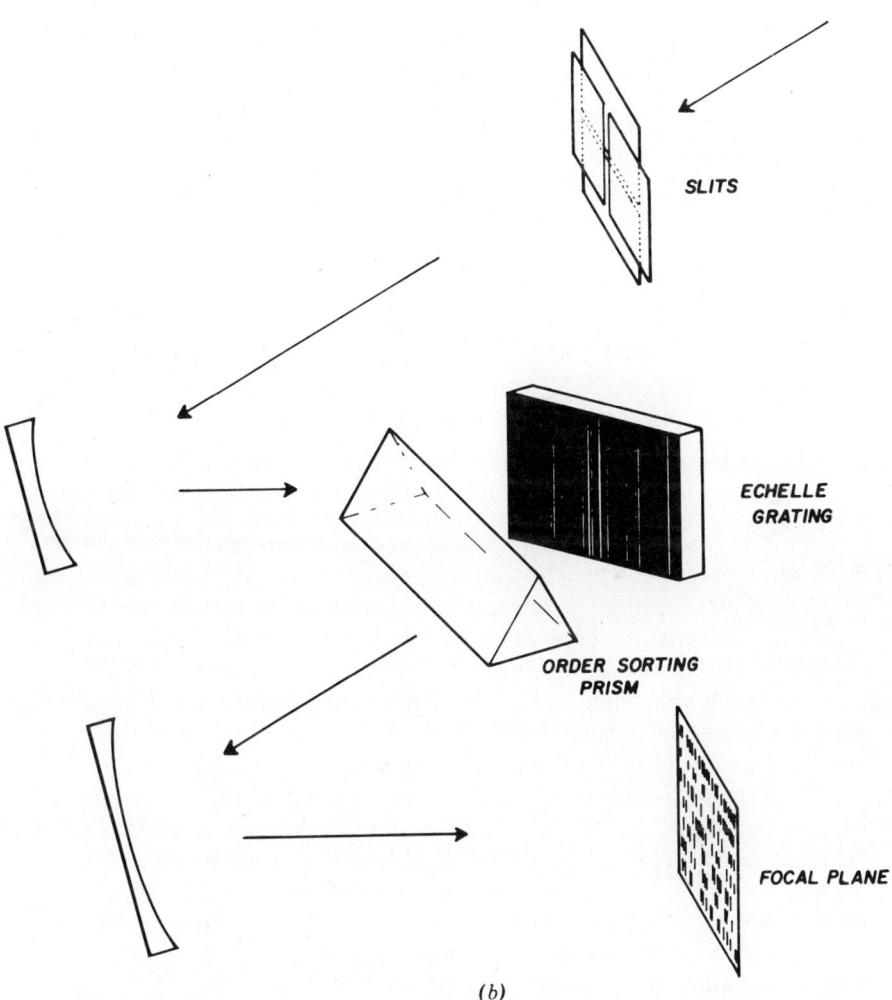

(b)

Figure 8.23. (*Continued*)

incidence and diffraction angles necessary for high resolution. In order to provide a resolution of 150,000 (bandpass of 0.0013 nm at 200 nm) which is near ideal for ICP-AES [12, 145], using a one meter spectrometer with a grating having 3600 grooves per millimeter, orders of two or greater must be used.

Gratings with a larger number of grooves per millimeter could be used, but at the expense of the range of wavelengths which can be used. For example, consider a Czerny–Turner monochromator with an angle of 15° between the incident and diffracted beams and a 3600 groove per millimeter grating. The largest wavelength which can be observed will be limited by the maximum grating angle that can be used

$$\lambda = 2d \sin \theta (\cos \phi/2) \quad (8.25)$$

If the largest grating angle, θ, is 60°, the maximum wavelength is 477 nm. As the grating angle is increased, the light throughput is reduced since the projected area of the grating decreases. This effect can be minimized by using a wider grating. Various commercial monochromators will have different maximum grating angles determined by their scanning mechanism.

Table 8.5 lists characteristics of some spectrometers which might be considered for ICP-AES. The data show how the observed resolution can be increased, and the resulting trade-offs in the maximum observable wavelength of free spectral range.

8.6.2. Advantages and Limitations of the Use of High-Resolution Spectrometers to Reduce Spectral Interferences in ICP-AES

A number of sample types produce complex spectra giving rise to numerous spectral interferences. High resolution spectrometers have demonstrated the ability to reduce spectral interferences in a number of samples including those containing rare earth elements [46], environmental samples containing major amounts of iron and titanium [77], steels [74], and marine samples [78]. Figure 8.24 shows the separation of As I 228.812 nm and Cd I 228.802 nm as the spectrometer resolution is increased. However, this is a "best case" since both lines have narrow physical linewidths of approximately 0.014 nm.

The determination of elements present at trace concentrations (including As, Be, Cd, Pb, Sb, Se, and Tl) in environmental samples such as coal fly ash in the presence of major concentrations of elements with intense lines (Ca and Mg) or complex spectra (such as Fe and Ti) is difficult or impossible with medium resolution (0.03 nm bandpass) spectrometers. The spectral interferences at the analyte wavelength or at the nearby wavelengths normally used for background correction often account for more than 50% of the detected emission intensity. As a consequence, correction procedures may be imprecise and inaccurate [77]. Table 8.6 shows the reduction in the magnitude of interferences using a high resolution (bandpass of 0.005–0.02 nm) echelle spectrometer in comparison to a medium-resolution spectrometer.

The analysis of phosphorous in steel is also difficult because of spectral interferences. Phosphorus has few intense lines above 200 nm and all suffer from interferences from iron when medium-resolution spectrometers are used. The most intense phosphorus line, at 213.618 nm, can be separated from the iron lines at 213.596 nm and 213.652 nm by using a high-resolution spectrometer [74]. As a result, the detection limit of phosphorus in steel can be improved from 0.002 to 0.0006% [74].

Table 8.5. Characteristics of Commercial Spectrometers for ICP–AES

Spectrometer	Grooves/mm	Order	Reciprocal Linear Dispersion[a] (nm/mm)	Bandpass[a] (nm)	Free Spectral Range[a]	Maximum Wavelength[b] (nm)
1.0-m Paschen–Runge	1080	1	0.93	0.023	400	820
	1080	2	0.46	0.012	200	
	1440	1	0.69	0.017	400	620
	1667	1	0.60	0.015	400	530
	2160	1	0.46	0.012	400	410
1.0-m Czerny–Turner	1200	1	0.83	0.021	400	1300
	2400	1	0.42	0.011	400	650
	3600	1	0.21	0.005	400	430
		2	0.10	0.003	200	220
0.75-m Echelle	79	56[a]	0.14	0.007	1.4	>800

[a] At approximately 400 nm. These will vary for other wavelengths.
[b] These will vary somewhat depending on the maximum grating angle which can be used in each particular spectrometer and scanning mechanism.

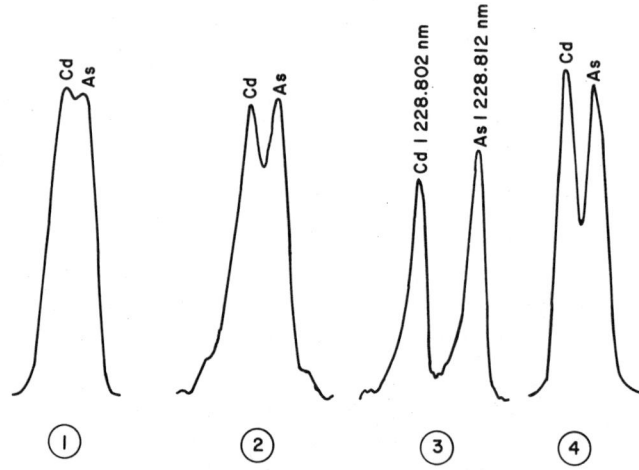

Figure 8.24. Influence of the spectrometer on the separation of the As I 228.812-nm and Cd I 228.802-nm lines *1*, ARL 35000 with 2160 grooves per millimeter and 20-μm slits; *2*, JY HR 1000 with 3600 grooves/mm and 10-μm slits; *3*, same as *(2)* but working in second order; *4*, spectrametrics echelle monochromator with 25-μm slits. [12]. [Reprinted with permission from J. M. McLaren and J. M. Mermet, "Influence of the Dispersive System in Inductively Coupled Plasma Atomic Emission Spectrometry, Spectrochimica Acta **39B**, 1307 (1984). Copyright (1984), Pergamon Press, Oxford.]

Table 8.6. Comparison of Interference Corrections for Medium- versus High-Resolution Spectrometer[a]

Analyte Line (nm)		Interferent (nm)		Interference Reduction Factor for High- versus Medium-Resolution
Be I	234.861	Fe II	234.830	2285
Cd II	214.438	Pt	214.422	297
		Fe	214.443	
		Fe	214.454	197
As I	193.696	Al II	193.693	
		Al II	193.628	4
Tl I	377.572	Ni I	377.533	67

[a]From [77].

Spectrometers with a narrow spectral bandpass may also result in improvements in the SBRs and SNRs. For simplicity, consider two spectrometers with equal light throughput but different resolution. If the background is due to continuum emission, it can be reduced by decreasing the spectral bandwidth. However, the analyte line signal is not reduced as long as the instrumental bandwidth is wider than the physical linewidth. Therefore, the SBR will be increased by using the higher-resolution spectrometer (see Sections 4.1.3 to 4.1.7).

High-resolution spectrometers require high mechanical and electronic stability and control. As the angular dispersion is increased, a smaller distance perturbation is required for a similar wavelength displacement. Therefore, care must be taken to minimize the effects of ambient temperature changes and vibration. Scanning systems will require a large number of steps to scan wide wavelength regions. As a result, costs will be higher and the analysis speed may be reduced if the spectrometer must scan a number of lines widely separated in wavelength. Detectors and associated electronics may require improvement in order to take advantage of the higher resolution. (See Sections 4.1.3. to 4.1.7.)

Finally, the analysis and peak search algorithms may require careful reconsideration and modification. When the bandpass of the spectrometer is much greater than the physical linewidth, the instrumental slit function will determine the observed line shape. However, if the bandpass of the instrument becomes similar to the physical linewidth, the observed line shapes will be dependent on the particular line used. Neither the relative physical nor instrumental linewidths will be constant over different analyte lines and wavelengths. Also, the observed line shape will be affected when resonance broadening or HFS is important. Therefore, it may be necessary to use different intensity measurement approaches depending on the line of interest and degree of broadening.

Peak search routines are often used to precisely set the spectrometer wavelength to that of the analyte line peak intensity. Low- or medium-resolution spectrometers can normally use the same peak finding algorithm (these are discussed in Section 8.9.3) for all lines. In contrast, the peak finding routines used with high-resolution spectrometers may require adaptation to the observed shape of the line profile [12].

8.7. BACKGROUND AND SPECTRAL OVERLAP CORRECTION

Background and spectral interferences can be significant even when spectrometers with high resolution and low stray light are used. Therefore, dynamic background and spectral overlap correction are often imperative for accurate results.

In general, there are two different types of correction methods which can be used separately or together. ff-peak or on-peak subtraction. The relative effectiveness of the two methods depends on the source of the background near the analyte line. Off-peak subtraction is useful to correct for featureless, continuum background. On-peak subtraction is necessary to compensate for direct spectral overlaps.

8.7.1. Off-Peak Subtraction

Using the off-peak subtraction procedure, one or two wavelengths close to the analyte line, but removed from it by 0.01 to 0.10 nm, are chosen to represent the background. If the background is flat as in Fig. 8.25a, a 1-point (one wavelength) correction is sufficient. The corrected analyte intensity is simply the difference between the intensity measured at the analyte wavelength, λ_A, and the background wavelength. Often the background is not flat in the vicinity of the analyte line. If a sloping, but featureless background is present, a two-point subtraction can be used. This might be the case when the analyte line is near the broadened wings of a stronger emitter, such as Mg^+. Two wavelengths are chosen for the background measurement, one below the analyte line wavelength and one above the analyte line wavelength. A simple average of I_{B1} and I_{B2} is

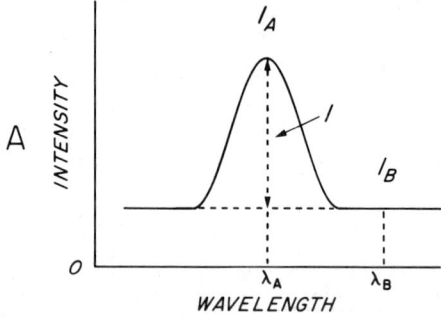

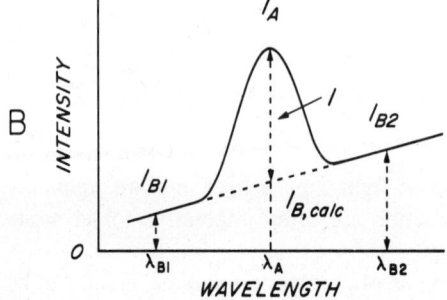

Figure 8.25. (Top) Spectral line on a flat background. One-point background subtraction using background measured at λ_B is sufficient. The net line intensity, I, is calculated from $I = I_A - I_B$. (Bottom) Spectral line on a sloping background. Two points are used to estimate the background at the analyte wavelength. The net line intensity, I, is calculated from $I = I_A - [(I_{B1} + I_{B2})/2]$ when points $B1$ and $B2$ are equal spaced from the wavelength of peak intensity.

8.7. BACKGROUND AND SPECTRAL OVERLAP CORRECTION

sometimes used to estimate the background at λ_A. A more effective method uses a weighted average or finds $I_{B,\text{calc}}$ by defining a line through I_{B1} and I_{B2} and then determining its value at λ_A, as shown in Fig. 8.25b.

The off-peak subtraction procedure can correct for continuum background and partial wing overlap from atomic or molecular emission. However, this method cannot correct for direct spectral overlap. The potential for spectral interferences biasing results is increased somewhat since measurements are made at two or three different wavelengths. If the background signal has a significant contribution from elemental line or molecular band emission, error will be introduced. Therefore, it is important that the wavelengths for background correction be carefully chosen for each analyte line, where no other line or spectral structure is present. Display of wavelength scans in the vicinity of the analyte line before choosing the background wavelength(s) is helpful and often necessary. Most commercially available ICP spectrometer systems provide display of wavelengths scans, to aid in detecting interferences and choosing wavelengths for background correction. Knowledge about the sample matrix and possible spectral interferents is also helpful in background wavelength selection. These concerns become more crucial as the analyte concentration approaches the limit of detection and the complexity of the sample matrix increases.

Algorithms have been developed to automatically find the best wavelengths for background subtraction [79]. One method finds all minima in a region 0.125 nm on either side of the analyte line wavelength. The intensities of each successive group of three channels or steps (0.0025 nm each) are summed. A frequency distribution is determined from a group of standards, a group of blanks, and a group of different sample types. The most frequently occurring minima (66%) are considered the best choices for two-point background subtraction.

One advantage of the off-peak correction method is that the species causing the featureless background enhancements need not be identified or quantitated. Only the analyte concentrations of interest need to be determined. Also, the off-peak method can be used with a multielement standard addition calibration scheme.

8.7.1.1. Wavelength Modulation

Rapid, automatic off-peak background correction can be performed by modulating the detected wavelength. This is most often done by placing a quartz refractor plate behind the entrance slit of the spectrometer. The refractor plate introduces a displacement of the beam falling on the grating, effectively changing the angle of incidence. The displacement, as depicted in Fig. 8.26, is approximately proportional to the refractor plate angle relative to the incident beam normal and the thickness of the plate.

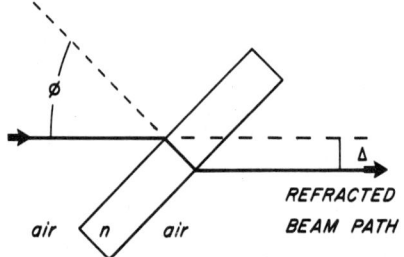

Figure 8.26. Use of a refractor plate to scan within a small wavelength region. $\triangle$ is the distance by which the beam is offset by passing through the refractor plate; n indicates the index of refraction of the plate; ϕ is the refractor plate angle.

The refractor plate is vibrated or rotated resulting in series of sinusoidal [80] wavelength scans. A flat background produces no ac component. A linearly sloping background generates an ac signal at the modulation frequency. Peaks or lines within the wavelength region scanned result in an ac signal with a frequency twice that of the scanning frequency. A frequency selective detection system, such as a lock-in amplifier, is then used to measure the analyte line signal which will occur at the second harmonic of the refractor oscillation or scanning frequency.

A two-wavelength scan (analyte peak and background) may also be obtained by moving a refractor plate in and out of the optical path [81]. In this case the measurement duty factor (percentage of time at a measurement wavelength) will be higher than in the sinusoidal scan technique. Intensities measured at the background wavelength are simply subtracted from those at the analyte wavelength by use of a multiplexed analog-to-digital converter and computer manipulation of the data. This provides an automatic, one-point (wavelength) background correction. A stepped-wave modulation scheme can also be used for automatic 2-point background subtraction [82].

Derivative spectra acquired by wavelength modulation can improve the ability to separate the analyte line from a nearby line or band [83, 84]. Wavelength modulation coupled with a signal averaging—peak determination scheme [85]—can also be used to correct for spectrometer misalignment. A selective line modulation technique [86, 87] can be used to modulate the emission signal from a narrow wavelength region.

The effectiveness of wavelength modulation to correct for scattered light, recombination continuum, and partial wing overlap in ICP–AES has been demonstrated by the analysis of phosphorous in steel [74] and the determination of trace quantities of cadmium and lead in biological tissues and sediments [78]. In both examples, high-resolution echelle spectrometers were used. Without the narrow bandpass provided by the echelle spectrometer, direct spectral overlaps would have prevented accurate results when only the off-peak background correction method was employed.

While wavelength modulation is effective in background correction when the background is smooth, a number of limitations are inherent. The introduction of the refractor plate into the spectrometer may cause some defocusing and loss of spectral resolution. More importantly, the control over the wavelengths used for background measurement is limited. This will be particularly important in cases where the spectral region near the analyte line includes other atomic lines or molecular bands.

8.7.2. On-Peak Subtraction

On-peak subtraction can be used to correct for background enhancements including direct spectral overlaps which are caused by specific species. All species causing background enhancements at a particular analyte line must be identified. The relative contribution of each of the interferent species is then experimentally determined so that correction factors for each analyte line, which are dependent on the concentrations of concomitants in the sample, can be calculated. The "correction factors" or "sensitivity factors" are then related to the observed signal and each species concentration in a linear fashion as described by Boumans [88] and summarized below. Nonlinear descriptions of the interference effects have also been used [89].

If the relationship between the analyte line intensity and concentration is linear, the analyte concentration can be expressed as

$$C = I/S \tag{8.26}$$

where C is the analyte concentration, I is the background corrected intensity, and S is the sensitivity of the analyte line. If an interferent is present which emits light within the spectrometer bandpass of the analyte line, the measured light intensity will be $I = I_{analyte} + I_{inter}$, where I_{inter} is the contribution from the interferent to the peak intensity at the analyte wavelength. If the measured signal is not corrected for the presence of a spectral interference the calculated concentration will be

$$C' = (I_{analyte} + I_{inter})/S_{analyte} \tag{8.27}$$

However, the correct analyte concentration is

$$C = (I_{analyte}/S_{analyte}) - (I_{inter}/S_{inter}) \tag{8.28}$$

or

$$C = C' - C_{inter}/v_{interf} \tag{8.29}$$

where v_{interf} is the ratio of sensitivities of the analyte and interferent at the analyte wavelength, $S_{analyte}/S_{inter}$. If more than one interferent contributes to the intensity measured at the analyte wavelength, then sensitivity factors for each interferent must be determined to obtain the correct analyte concentration

$$C = C' - (C_{\text{inter1}}/v_{\text{inter1}} + C_{\text{inter2}}/v_{\text{inter2}} + \ldots) \quad (8.30)$$

Expressing Eq. 8.30 in terms of the sensitivity ratio (v) and the concentration ratio of the interferent to analyte is convenient to assess the importance of each correction:

$$C = C'/(1 + (1/v)(C_{\text{inter}}/C_{\text{analyte}}) \quad (8.31)$$

Sensitivity factors are determined by using ultrahigh purity standard solutions of known concentrations (usually fairly large—1000 μg/mL) of each possible interferent. Extreme care must be exercised to be sure that the standard solutions are not contaminated. Sources of contamination include the solid standard, solvent or acids, and leaching from the container used to hold the standard solution. The use of alternate methods of analysis to confirm purity of the solutions is recommended.

The determination of sensitivity factors is a time-consuming process (in some cases 100 to 200 correction factors are determined [90, 91]). However, the sensitivity factors remain useful over a number of months [90] if plasma and instrumental conditions are properly maintained. The Cu I 324.75–Mn II 257.61 nm intensity ratio has been used by Botto [90] as an indicator of plasma conditions. This ratio can be maintained by making adjustments in the aerosol carrier gas mass flow rate. Usual adjustment was ± 1–2 mL/min at a nominal flow rate of 700 mL/min. For systems where plasma conditions are not held constant, it is important to choose lines for determination of the interferent species concentration which respond to changes in plasma conditions similarly to the interferent line [92].

The experimentally determined sensitivity factors are sometimes found to be negative [90] when off-line background correction is used. Negative factors indicate that the interferent contributes a higher intensity at the wavelength(s) used for measurement of the background than at the analyte line.

Often, many of the correction factors are due to a few elements. Botto [90] determined correction factors for 33 elements at a total of 34 lines resulting in a 34 $\times$ 34 matrix of correction factors. Of the 220 nonzero factors, 20 were due mainly to Cr, 22 to Mo, 29 to U, and 17 to V (also see Section 7.6).

For analysis of analyte concentrations in matrices with complex spectra, such as iron or rare earth elements, a large number of corrections will be necessary. The error in the corrected concentration will increase as the relative contribution of the interferents becomes large. If the uncertainty introduced by the interferent corrections becomes larger than the correction itself, the method is not effective. Table 8.7 shows the results [88] for the correction of Ca interference on the Pb I 283.3-nm line in surface water samples. In this case the correction is effective even when the uncorrected Pb concentration was four times the actual Pb concentration.

Once the correction factors have been determined, the analysis requires only

Table 8.7. Results for Correction of Calcium Spectral Interference on Pb I 283.3 nm in Surface Water Samples[a]

	Sample 1	Sample 2	Sample 3
Uncorrected lead concentration (ng/mL)	54	30	24
% Relative standard deviation	3.8	5.5	6.5
2 × Standard deviation (ng/mL)	4.1	3.3	3.1
Calcium concentration (µg/mL)	80	80	80
% RSD	2	2	2
$v (= S_{Pb}/S_{Ca})$	4350	4350	4350
%RSD	2	2	2
Corrected Pb concentration (ng/mL)	36	12	6
% RSD	6	14	27
2 × Standard deviation (ng/mL)	4.3	3.4	3.2

[a] Reproduced with permission from *Spectrochim. Acta, Part B* [88].

one measurement per analyte line. This method has most often been used with direct reading spectrometers so that all of the important analyte channel intensities can be measured simultaneously. While the on-peak method can correct for spectral overlaps, unlike off-peak subtraction, it cannot correct for nonspecific background enhancements unless the sample matrix is constant.

Both the off-peak and on-peak methods correct only for the source of the light measured. Neither method corrects for changes in the sensitivity, or light intensity per unit concentration, which may be due to changes in the plasma itself, or for changes in the processes converting sample solution to emitting atoms and ions. Many matrix effects result in changes in the spatial distribution of emission in the ICP source. Therefore, changing the observation region or experimental conditions may minimize the effects. Correction for changes in sensitivity are further complicated by their nonadditivity [93] (see multiplicative interferences, Section 1.6). Also, two factor interactions were found to be important [93].

8.8. DIRECT READING SPECTROMETERS

One of the advantages of AES is the ability to perform simultaneous, multielement analysis. By detecting emission from all species of interest simultaneously, a large number of samples can be analyzed in a short time. Multichannel detection is also important where the amount of sample available is small or where transient signals are produced. For example, laser ablation [94–

96], graphite rod direct sample insertion [97, 98], electrothermal atomizer [99], and rf arc [100] ICP sample introduction methods produce transient signals, often from small amounts of material. Commercial ICP-AES systems for simultaneous, multielement analysis consist of direct reading spectrometers with a number (up to 90) of individual exit slits and PMT tube detectors.

8.8.1. Direct Reading Spectrometers Based on Concave Grating Mounts

The majority of the first commercially available ICP-AES systems employed direct reading spectrometers based on the Rowland circle, most often the Paschen-Runge mounting. This trend continues today although the use of rapid scanning, sequential spectrometers has grown dramatically, as discussed in Section 8.9. A number (20-90) of exit slits and PMTs are placed on the focal plane of the spectrometer, as shown in Fig. 8.27. The signal from each PMT tube is integrated, with a separate analog integrator card for each PMT. The integrators are multiplexed to an analog-to-digital converter followed by storage of the digital values in computer memory. At least one channel, consisting of an exit slit, photomultiplier tube and electronics is needed for each element to be determined. A bandpass, or order sorting, filter may also be placed in front of each exit slit to reduce stray light and order overlap effects.

The concave grating based spectrometers have only one optical component,

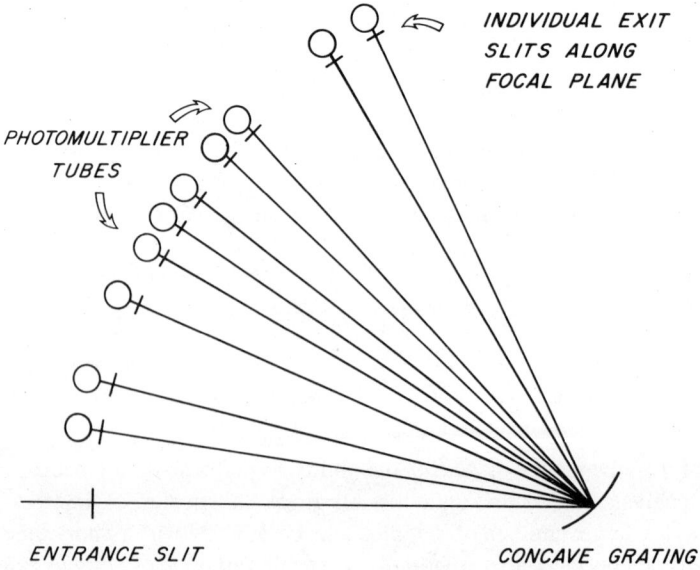

Figure 8.27. Direct reading spectrometer.

the grating, between the entrance and exit slits (except when refractor plates are also used for fine wavelength adjustment or scanning). This arrangement provides high efficiency and low scattered light because the number of optical surfaces is minimized.

Two different exit slit arrangements have been used: individual slit mounts and slit strips. The slit strip is comprised of a single sheet of steel with up to 130 individual slits machined by photoerosion. Alignment and changing exit slit wavelengths to view the appropriate lines of interest for a particular sample is somewhat simpler with the slit strip approach. The alignment of the slits relative to each other is fixed so that only one slit needs to be experimentally aligned on the focal plane. This places a larger burden on the manufacturing of the slit strip since highly precise and accurate positioning of the slits on the strip is required. The individual slit arrangement requires separate alignment of each exit slit assembly. However, if only one or two channels need to be changed, the other channels do not need to be realigned. Most often, predrilled holes are available in the spectrometer for each exit slit—PMT channel assembly. A refractor plate in front of the exit slit is used to fine tune or align the exit slit exactly on the proper line.

8.8.2. Direct Reading Spectrometers Based on the Echelle Mounting

The echelle spectrometer has gained popularity due to its inherently high spectral resolution. A number of PMTs can be placed behind exit slits on the focal plane of an echelle spectrometer to make a direct reading spectrometer. However, because of the compact, two-dimensional focal plane, it is difficult to mount a large number of PMT tubes on the focal plane. Also the choice of analyte lines is often limited by how closely the PMTs can be packed. Commercial echelle direct reading spectrometers with up to 20 [101, 102] or 48 [103, 104] channels are available. Exit slits may be arranged on removable cartridges with different cartridges available [101] to analyze various sample types (geological, environmental, metallurgical, biological, etc.). The 20 PMTs are fixed so that changing the wavelengths used requires only that the cartridge be changed.

8.8.3. Mechanical and Optical Stability

Stable optical alignment of the direct reading spectrometer is imperative for reproducible operation. The relative position of the entrance slit, grating, and exit slit must be maintained with high precision. For example, if a 25-μm entrance slit and 50-μm exit slit were used, movement of the exit slit along the focal plane relative to the image of the entrance slit of more than 13 μm would result in loss of signal from the analyte line.

A number of steps are taken to maintain and check the stability of the spectrometer. The optical components including entrance slit, scanning mechanism, grating, and exit slits are often mounted on an A-frame of a rigid material with a low thermal coefficient of expansion such as invar. Further, the spectrometer is usually thermostated to ± 0.1 °C at a temperature above room temperature (30–40 °C) to reduce susceptibility to thermal expansion or contraction. The spectrometer typically has some type of vibration isolation. The exit slits are normally wider than the entrance slit (20 vs. 30–75 μm) to reduce the effects of small changes in position. An auxiliary light source, such as a Hg lamp, is often provided to check spectrometer alignment. The spectrometer is then adjusted (by movement of the entrance slit or refractor plate) to compensate for drift in the alignment.

8.8.4. Wavelength Scanning Capability

As discussed in Section 8.7, background correction and identification of spectral interferences can be very important. As a result, commercially available direct reading spectrometers have evolved to provide limited scanning capability in the vicinity of each analyte line so that off-peak as well as on-peak background subtraction can be performed. Scanning is most often accomplished in one of two ways. The entrance slit may be moved tangentially to the Rowland circle short distances under stepper-motor driven computer control. By moving the position of the entrance slit the incidence angle to the grating is changed, resulting in a change in the wavelength of light reaching each of the fixed exit slits. Another popular method involves use of a refractor plate immediately behind the entrance slit, as described in Section 8.7.1. As the angle of the refractor plate with respect to the optical axis from the slit to the grating is changed, the incoming path is offset, thereby changing the incidence angle to the grating.

8.8.5. Choice of Lines Monitored by the Direct Reading Spectrometer

The choice of analyte lines on the direct reading spectrometer is limited. This is due to space limitations (how closely exit slit—PMT assemblies can be placed) and cost. Commercial systems will usually have only one or two lines per element. Compromises must be made in the selection of lines used for analysis as dictated by the presence of interferent lines, the distance between adjacent channel assemblies, the number of elements to be determined and the sensitivity of the lines. Since line selection will depend somewhat on the sample type and matrix, it is important to know what sample types will be analyzed before purchase.

8.8.6. The "$n + 1$" Channel

In order to alleviate the limited line selection somewhat, many manufacturers also offer a scanning monochromator which can be used simultaneously with the direct reading spectrometer. The monochromator is often called an "$n + 1$" channel and treated as a separate exit slit–PMT channel by the direct reading spectrometer control and readout electronics. The monochromator may be used to select a line for analysis not available on the direct reader, to scan near analyte lines in order to identify possible interferents, or as a relative background measurement.

The "$n + 1$" channel capability may range from a fairly low resolution monochromator mainly to provide detection at wavelengths higher than the direct reading spectrometer (in particular, for alkali element emission—Na, K) to a fully automated medium to high-resolution slew-scan spectrometer. As expected, the cost of the "$n + 1$" is related to the capabilities of the scanning monochromator.

Alternatively, Philips provides on optional "roving slit" which is employed as the "$n + 1$" channel. A periscope on the roving slit assembly deflects light from the grating to a plane below the focal plane of the direct reading spectrometer before reaching the normal, fixed exit slits. A carriage carrying the periscope mirrors, exit slit, and PMT tube can be precisely positioned along the focal plane (from 190 to 700 nm) on a track which follows the Rowland circle. Only a small portion of the normal direct reading spectrometer focal plane is obstructed by the roving slit.

8.8.7. Polychromators with Photographic Detection

Multichannel detection systems with photographic detection provide the ability to simultaneously detect a large wavelength range with an enormous number of detection elements, limited only by the photographic grain size. However, spectrographic systems for quantitative analysis are hampered by the time required to process the photographic plates, limited dynamic range, fairly low sensitivity, and nonlinear response. The use of computerized densitometers [105, 106] makes photographic detection more competitive with PMT-based direct reading spectrometers. However, the extra time required for signal recovery prohibits the use of spectrographic detection for routine quantitative analysis. The huge amount of spectral information provided is as yet much greater than obtainable with any other detector. Therefore, spectrographs are especially useful for qualitative and some semiquantitative analysis [107, 108]. Some manufacturers provide an option for photographic detection, particularly with the echelle systems which have compact focal planes. Other detection schemes which may provide a large number of resolution elements and rapid signal recovery are discussed in Section 8.10 and Chapter 9.

8.9. SLEW-SCAN SPECTROMETERS

Scanning monochromators have become increasingly popular in response to the limited line selection and high cost of direct reading spectrometers. Slew-scan spectrometers use a two-speed wavelength movement. The grating is rapidly moved, or "slewed" to a wavelength near that of the analyte line of interest. Wavelength scans near each analyte line are then obtained by slowly changing the wavelength in a number of small (0.01–0.001 nm) steps. High slew rates are desirable since slewing from one analyte line wavelength to the next is often the time-limiting step in sample analysis throughput. Significant improvements in mechanical design allowing more rapid slewing together with the continuing development of computer-controlled spectrometers have led to further acceptance of the slew-scan spectrometer for ICP–AES.

Because the monochromator can be set to any particular wavelength within its mechanical limits, as determined by the range of grating angles and the grating groove density, the best lines for analysis can be chosen, in contrast to direct reading spectrometers. The choice may depend on the element to be determined, the analyte concentration, the sample matrix including the resulting spectral interferences, and plasma conditions. Signals from a number of lines due to the same element may be monitored to check for unsuspected interferences [114].

The cost of the slew-scan instrument is 50–70% that of a direct reading spectrometer since only one exit slit, detector, and set of signal processing electronics are required. Further, if a scanning mode is used rather than integrating on-peak for analysis, the mechanical stability of the system can be somewhat less than the direct reading spectrometer.

A number of unique experimental difficulties can arise in attempting to construct a slew-scan spectrometer. Signals must be detected over a wide range of wavelengths (190–780 nm) and intensities (a range of up to 10^7) by a single detection channel (including exit slit, PMT, and electronics). Therefore, a PMT with a wide wavelength response is often employed. Alternatively, two or three PMTs can be used. The appropriate PMT is then automatically positioned behind the exit slit depending on the wavelength. Different PMT voltages and signal integration times are often utilized to accommodate the wide range of intensities. High-resolution analog-to-digital converters are needed to take advantage of the wide dynamic range of the ICP. If bandpass or order sorting filters are to be used to reduce stray light, the appropriate filter must be rapidly placed in the optical path for each different wavelength region.

The combination of rapid slewing and slow, high-resolution scanning presents mechanical problems which can limit the accuracy and precision of the spectrometer. In particular, positioning the grating to select the proper wavelength for analysis and, therefore, peak intensity measurements, are more complex than in direct reading spectrometers where the grating position is fixed. High-resolution stepper motors and position-sensing transducers are needed. Software and hardware to perform wavelength calibration and peak searching are also important.

8.9.1. Mechanics for Slewing and Scanning by Grating Rotation

The mechanical requirements of the ideal slew-scan spectrometer are stringent. The slew rate must be as high as possible to minimize "dead time" when no signal is being acquired. The scan step size must be smaller than the spectrometer bandpass by at least a factor of 2 so that resolution is not lost. The wavelength must be known and controlled accurately and precisely (within a single bandpass—0.01–0.001 nm) to be sure that the proper line is observed and that the peak intensity is measured.

8.9.1.1. Sine Bar Drive

The sine bar drive provides a mechanical means of scanning wavelength linearly with rotation of a lead screw. Figure 8.28 shows a simple sine bar drive. The sine bar is affixed to the grating holder and held in contact with the drive plate or rod. The sine bar is the hypotenuse of angle θ. When the grating is perpendicular to the bisector of angle ϕ, $\alpha = \beta = \phi/2$. If the grating is rotated away from this position $\alpha = \theta - \phi/2$ and $\beta = \theta + \phi/2$. When α and β are on the same side of the grating normal,

$$n\lambda = d\left[\sin\left(\theta - \frac{\phi}{2}\right) + \sin\left(\theta + \frac{\phi}{2}\right)\right] \quad (8.32)$$

This can be rewritten

$$n\lambda = 2d \sin\theta \cos\frac{\phi}{2} \quad (8.33)$$

As a result, wavelength is linearly related to the $\sin \theta$. Note that $\sin \theta = X/R$. R is fixed by the length of the sine bar. Therefore, as X changes linearly with rotation of the lead screw, $\sin \theta$ also changes linearly.

The lead screw–nut drive arrangement experiences some backlash. As a result, scans from opposite directions will show an offset. Also, fine pitch and small angles of rotation of the lead screw are needed to provide small wavelength steps. It is difficult to mechanically drive the sine bar mechanism rapidly.

8.9.1.2. Direct Drive

The use of the sine bar drive came from the need to provide linear wavelength scans mechanically. The advent of microprocessor-controlled stepper motors

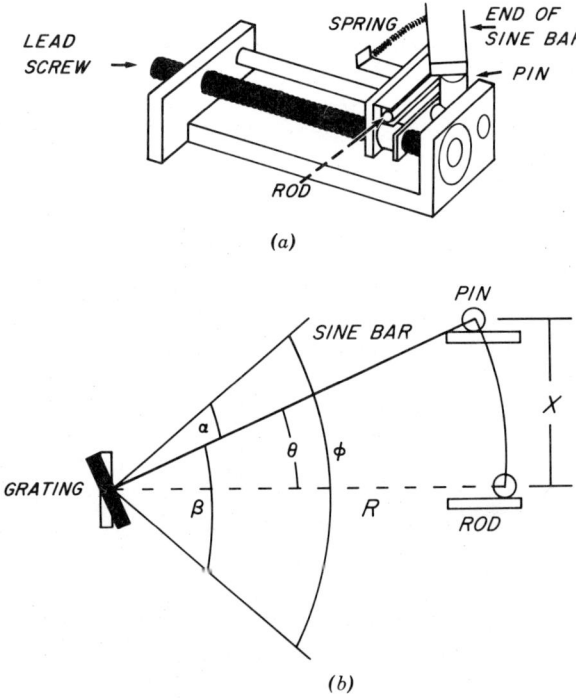

Figure 8.28. Sine bar grating drive. (*a*): Mechanics of the sine bar drive. (*b*): Relationship between the grating angle, θ, and linear movement of the sine bar. The sine of the grating angle, θ, is simply equal to X/R, where X is the sine bar displacement from zero position and R is the length of the sine bar.

with high resolution and relatively low cost removed the need for a linear mechanical scanning mechanism. For example, the grating can be mounted on a turntable which is driven by a stepper motor through a worm gear system, as shown in Fig. 8.29. Direct drive systems such as this may allow more rapid slew speeds. The computer is easily able to relate the number of steps to wavelength through the grating equation relationship and the number of steps per unit angle. This system may also allow a wide range of grating angles and, therefore, wavelength, with a simpler mechanical system than the sine bar drive. A high-resolution stepper motor is required. For example, if the step size at 400 nm is to be 0.01 nm, a typical 1 m Czerny–Turner monochromator with a 1200 groove per millimeter grating would require a grating rotation of 8×10^{-3}°. However, unlike the sine bar drive arrangement, this system will not have a constant step size over all wavelengths. At 250 nm, 10 steps would result in a wavelength movement of 0.102 nm while at 700 nm, 10 steps would cause a 0.093-nm wavelength change.

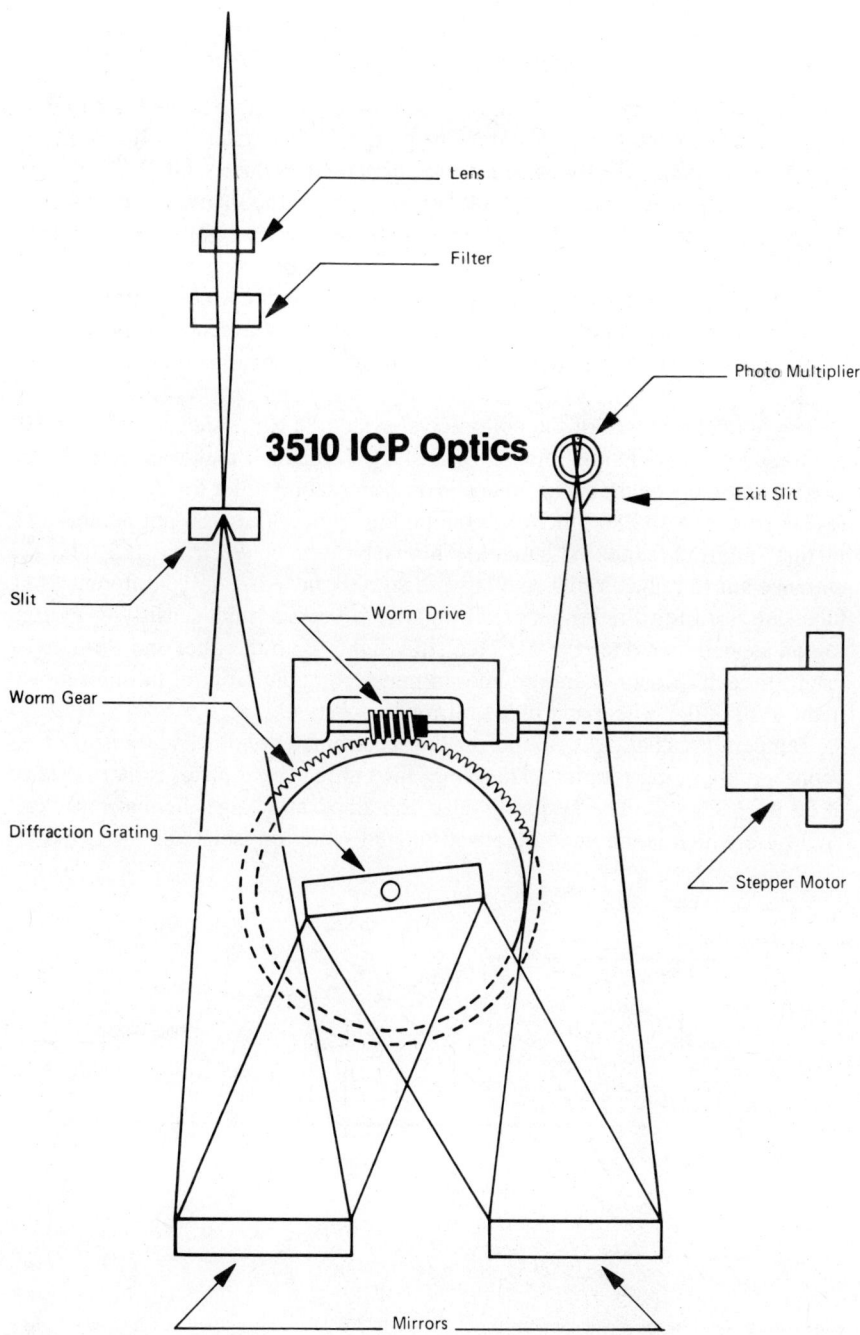

Figure 8.29. Worm gear driven monochromator [109]. [Reprinted with permission from ARL Model 3510 ICP Brochure. Copyright (1983), Bausch & Lomb, Sunland, CA.]

8.9.1.3. Magnetically Driven Grating Rotation

The fastest slewing monochromator commercially available to date is based on a galvanometer-like grating rotation. The grating in this system [110] is mounted directly to a "magnetically driven torsion bar and transducer unit." The torsion bar can be driven at speeds up to 2000 nm/s. As a result, any wavelength within the spectrometer's range (178–780 nm) can be reached in significantly less than one second. Further, since no mechanical linkages or drives are used, there is no backlash so that scanning can be bidirectional without loss of precision or accuracy. The position of the grating is detected by a transducer which is used to provide feedback to the servo amplifier which controls the torsion bar movement.

The optical diagram of the commercial system [110] is shown in Fig. 8.30. A crossed Czerny–Turner mount is used to eliminate "reentry spectra" (see Section 8.5.6). The monochromator is used in second order for the wavelength region from 178 to 380 nm, while the region from 380 to 780 nm is measured in first order. A series of bandpass filters is used between the ICP and the entrance slit to reduce order overlap and stray light. Also, three different PMT tubes are used for different spectral regions to provide high sensitivity in each region as well as reduce the detected stray light. Both the filter and PMT to be used for each measurement are chosen under computer control through movement of the filter wheel and plane mirror, respectively.

Temperature changes could affect the spectrometer geometry, transducer response, and grating rotation. Therefore, the entire spectrometer is thermostated at 35 to ±0.05°C. The system is also contained in a large thermal mass cast enclosure which is mounted on air-cushioned vibration isolators.

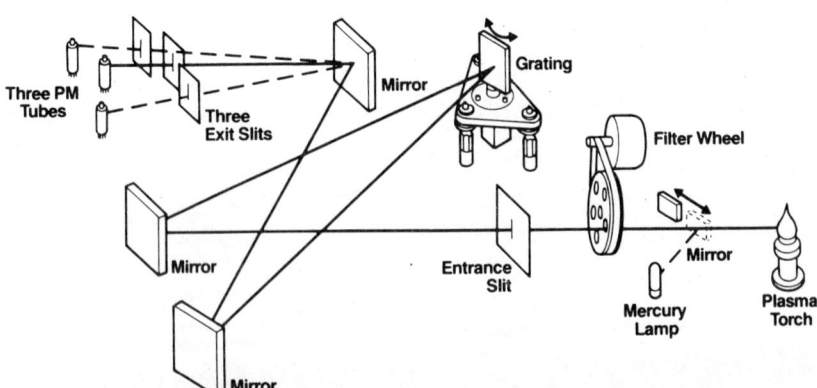

Figure 8.30. Optical schematic drawing of magnetically driven grating system in a crossed Czerny–Turner mounting [110]. [Reprinted from *American Laboratory*, volume 15, number 3, page 63, 1983. Copyright 1983 by International Scientific Communications, Inc.]

8.9. SLEW-SCAN SPECTROMETERS

The analog voltage to the torsion bar on which the grating is mounted is controlled by a microcomputer through an 18-bit digital-to-analog (D/A) converter. The full range of the D/A is used for the 178–380-nm region, providing a bit resolution of approximately 0.0009 nm. When the full range of the D/A is used from 380 to 780 nm the bit resolution is approximately 0.0017 nm.

A 1-s total integration time is typical for most elements. Thus the determination of up to 30 elements per minute is possible.

8.9.1.4. Encoders

Optoelectric encoders can provide feedback information on the angle by which a grating has been rotated. As the step size is decreased and scan or slewing speeds are increased, it becomes more important to have grating position feedback. Even with high-resolution stepper motors, mechanical parts limit the grating angle reproducibility. Angle encoders [111–113] can provide accurate feedback for high angle reproducibility ($2 \times 10^{-4}°$). For a 0.6-m Czerny–Turner monochromator used in first order with a 2400 groove per millimeter grating, 0.003-nm wavelength resolution accuracy was provided by the encoder [111]. Rotation of the grating by a single step must then be less than 0.003 nm to fully use the encoder capability. Also the encoder and stepper motor should be chosen to provide resolution better than the spectral bandpass of the spectrometer. A predetermined reference position (reported by the encoder) is used as a starting point. The displacement angle from the reference point is measured by counting the number of pulses emitted by the encoder from the reference position to the current position.

8.9.2. Wavelength Calibration

A number of methods to calibrate the spectrometer wavelength are used. In virtually all of these, the number of steps required to move the grating to provide a known wavelength change is determined experimentally. In the simplest case, with a sine bar drive, the system can be calibrated by a two-point calibration [114]. The zero order image and the 407.201-nm argon line are used in one commercial spectrometer [114] to obtain the linear relationship between number of steps and wavelength to 0.01 nm.

Mercury lamps are often used as a calibration source. For example [62], a number of mercury lines may be used to experimentally determine the coefficients in a cubic equation, relating number of stepper motor steps from a reference point to wavelength.

For an analysis, the monochromator is first scanned to a position near the line of interest. A solution of the element of interest is then introduced into the ICP and the position of maximum intensity found by scanning in a continuous fashion over a 0.50 nm (first order) or 0.25 nm (second order) region. Corrections in the wavelength calibration near each analyte line of interest are determined in this way before analyzing the sample. Changing from one line to the next during analysis of the sample then requires approximately 3 s.

Recalibrations can be performed to check for wavelength drift. The frequency at which recalibration is necessary is directly related to the wavelength stability of the spectrometer. Some instruments [115] perform recalibration before each analysis, while others may require recalibration only once per day.

Possible causes of drift and calibration errors include temperature variations (resulting in geometric changes due to expansion or contraction), vibration, refractive index changes (in the gas inside of the spectrometer), changing gears from slew to scan, and spurious electronic signals or noise taken as a step or encoder pulse. Changes in wavelength of up to 0.05 nm have been observed with a 1-m spectrometer due to a change in the room temperature of $1.0°C$ [116]. The spectrometer did not become stable until the room temperature was stable for several hours. Short-term drift was reported [116] to be related to refractive index gradients in the atmosphere within the spectrometer itself.

8.9.3. Peak Finding and Signal Acquisition

Either static or dynamic methods of peak intensity measurement may be used. The spectrometer can be tuned to the analyte line peak and then the signal integrated for a given sampling time, in an on-peak, static measurement. In order to do this, both short- and long-term wavelength reproducibility and drift must be excellent. As a result, few commercial slew-scan systems employ this approach. Alternatively, the line peak may be found dynamically by scanning a short wavelength region (0.01–0.03 nm) around the expected (calibrated) peak position. The signal may then be acquired by repetitively scanning across the peak or positioning the grating so that the peak intensity is monitored. The short scans can be done with small grating movements or with refractor plate rotation. The size of the search window is a tradeoff between a large window (e.g., 0.06 nm), where more than one intense line may appear and be falsely identified as the analyte line, and a window which is too small, such that calibration errors or drift would often cause the analyte line to be outside of the window.

8.9.3.1. Simple Peak-Finding Methods

Peak finding procedures can be simple when the analyte line intensity is well above the background and without nearby spectral interferences. After slewing to near the wavelength of interest a short wavelength region is scanned. The highest or three highest intensities within the window can be used to define the peak position. The experimentally determined peak position should be close to the calculated position, for example, within 0.02 nm. If not, a line other than the analyte line may have been found. Large drift or calibration errors could also be responsible, so the spectrometer should be recalibrated if this situation occurs. If the calibration is correct, then observation of a shift in the peak intensity could be due to a spectral interference [117].

A somewhat more complex peak search routine [11] uses 0.01-nm steps across a 0.3-nm wavelength region while introducing a pure, single element reference solution. A peak is sensed when at least five consecutive points with increasing intensity are followed by at least five points of decreasing intensity. However, this routine is used only for initial calibration and not for the final peak location with the sample.

As the concentration of analyte approaches within an order of magnitude of the detection limit, such as in the analysis of metals at the nanograms per milliliter level in water samples [79], simple peak finding methods become unreliable. Interferences due to other elemental or molecular emission can also perturb the simple peak finding routines.

8.9.3.2. Moving Window Peak-Finding Method

This method involves passing a five-point window over a 0.05 nm region centered on the calculated analyte peak position in nine steps. A number of tests are applied to the intensity data [118] to determine if the peak is real. The tests are listed in Table 8.8. All of the tests will be positive only at the peak wavelength. Highly structured background or small analyte line intensities provide difficulty for this peak finding method. Therefore, the final step of the test rejects assignment of the peak position if the peak intensity is less than 20% more than the background as measured by the first or last points in the entire 0.05 nm wavelength scan. Once the peak position is determined, the peak intensity is measured by fitting a Gaussian function to the three most intense points in the window surrounding the peak as shown in Fig. 8.31. In this way the maximum intensity can be calculated with high precision, even if it lies between two wavelength steps [119].

8.9.3.3. Peak Area Fitting Routine

A partial peak area fitting routine [120] has been applied to peak finding and net intensity measurement [79]. The presence of the peak is tested at five central wavelengths from 0.005 nm before to 0.005 nm after the calculated peak position using a partial peak area method. At each of the test wavelengths the area under the curve, as defined by five points and the background at points A and

Table 8.8. Tests Applied to Each Point in Window of Five Points for Identification of the Peak Location

1. Is $I_x > (I_{x-1} + I_{x+1})/2$? — Is the intensity at the center point greater than the average of its two neighboring points?

2. Is $(I_{x-1} + I_{x+1})/2 > (I_{x-2} + I_{x+2})/2$? — Is the average of the neighboring two points greater than the average of the intensities two points away?

3. Is $I_x >= I_{x-1}$? — Is the intensity greater than or equal to that of the next point to be tested?

4. Is $I_x >= I_{x+1}$? — Is the intensity greater than or equal to that of the previous point tested?

5. Is $I_{x+1} >= I_{x+2}$? — Is the intensity of the next point greater than or equal to its following point?

6. Is $I_{x-1} >= I_{x-2}$? — Is the intensity of the previous point greater than or equal to its previous point?

7. Is $I_x >= 1.2 (I_{x-4} + I_{x+4})$? — Is the intensity at least 1.2 times the background of determined by the first and last points of the entire 9-point scan?

B, is calculated. The resulting partial areas are used to define the peak position. For a pure single peak the five partial areas can be used to define a parabola, numerically fitted by the method of Savitsky and Golay [121]. The position of the top of the parabola is used to estimate the peak centroid, as shown in Fig. 8.32. The net signal intensity is determined by integrating the nine steps around the experimentally identified centroid.

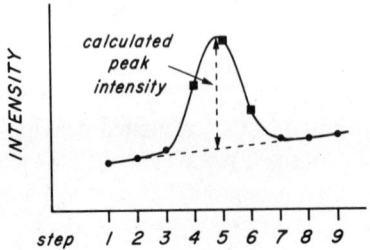

Figure 8.31. The procedure for identifying and quantifying a spectral line in a scanning monochromator. The tests listed in Table 8.8 are used to estimate the background at the analyte wavelength. If the points within the window are identified as a peak by passing all of the tests, the peak intensity is calculated by fitting a Gaussian function to the center three points.

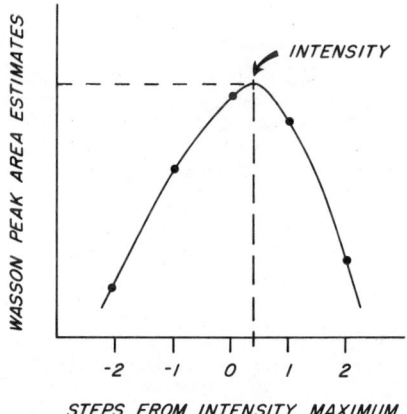

Figure 8.32. Determination of peak centroid by fitting a parabola to five partial areas.

8.9.3.4. Side Line Indexing Method

An alternative peak search method, the side line indexing method [122], uses an intense line near the region of the analyte line as a reference position. The small movement from the reference position to the analyte line wavelength can then be calculated with higher accuracy and precision than from a previous, wide wavelength range calibration. The intense line used as a reference can be one already present in the plasma or sample matrix or added to the sample prior to analysis.

8.9.4. Moving Detector Slew-Scan Spectrometers

As an alternative to single exit slit, rotating grating slew-scan spectrometers, moving detector systems have been recently developed [123, 124]. A number of fixed exit slits together with moving a single PMT tube detector are used to perform slew-scan functions without rotating the grating (and measuring its rotation with an encoder) or using peak finding routines. Further, on-peak signal acquisition and background or spectral interference correction is made much more facile than if the grating was rotated.

8.9.4.1. Paschen–Runge Moving Detector Spectrometer

One commercial spectrometer [123] is based on a Paschen–Runge mount with 255 fixed exit slits with 2-mm slit-to-slit spacing. A carriage holding the PMT tube, as shown in Fig. 8.33, is positioned at the exit slit nearest the analyte line of interest. The entrance slit is then translated to present light of the desired wavelength through the exit slit onto the PMT. The entrance slit position is

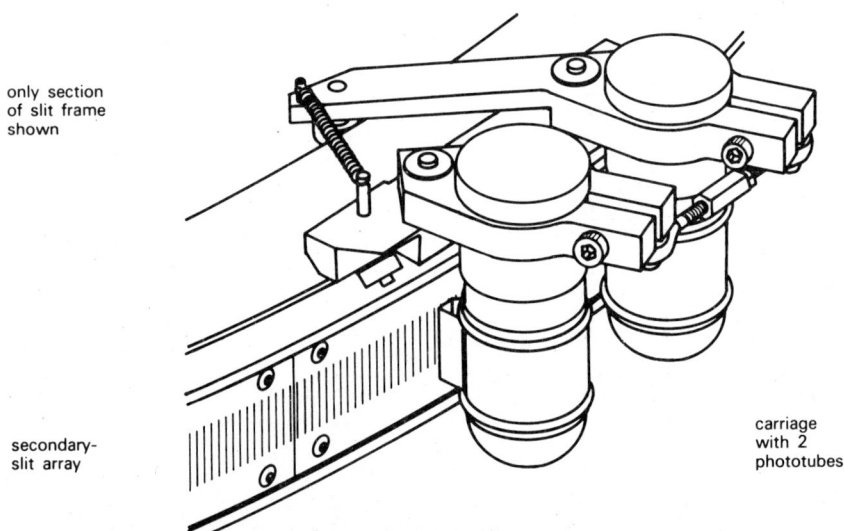

Figure 8.33. Carriage, with two PMT tubes, which moves along the focal plane of a Paschen–Runge spectrometer to observe different wavelengths [123]. [Reprinted with permission from ARL Model 3500 AES Series Emission Spectrometer Systems. Copyright (1982), Bausch & Lomb, Sunland, CA.]

controlled to 1.25-μm per step (0.0005 nm per step in second order with a 1080 groove per millimeter grating in the one-meter spectrometer). Long-term mechanical stability of the system is specified [123] to maintain better than 1-step precision over 18 months.

One of two PMT tubes is used, depending on the wavelength. Seven different filters are used to reduce the effects of stray light. Selection of both are under computer control, as is the PMT carriage movement, entrance slit movement, signal integration time, and PMT voltage. Analysis speeds of 6–12 elements per minute is possible.

8.9.4.2. Echelle Moving Detector Spectrometer

A moving detector system based on an echelle spectrometer is also available commercially [124]. The PMT is mounted on a carriage with movement in both x and y directions. An aperture plate with up to 300 photo-etched exit slits is mounted on the focal plane of the spectrometer. The PMT is then moved in a sequential manner to each of the desired wavelengths or exit slits. The time to move from one line to another is typically less than 1 s. Three 2-s integrations are used for each line so that five elements can be determined in 45 s.

8.10. CHOOSING A SPECTROMETER FOR ICP-AES

The choice of an ICP-AES spectrometer will depend on a number of factors. Most importantly, the sample types and the number of analyses that will be done must be known to make a choice. If rapid analysis of similar, routine samples is most important, a direct reading spectrometer system is preferable to a scanning instrument. However, if a smaller number of dissimilar samples are to be analyzed, the flexibility of line selection with the slew-scan spectrometer is desirable. If sufficient funding is available, both a direct reading spectrometer and slew-scan system can be purchased. The cost of the dual systems is less than purchasing each separately. Also, the cost of an instrument with a scanning spectrometer is approximately 60% of the price of an ICP system with a 35-channel direct reading spectrometer.

Knowledge of the samples to be analyzed is of the utmost importance in choosing the lines available. If a direct reading spectrometer is to be acquired, compromise decisions of line sensitivity versus interferences must be made before purchase and delivery. In any case, the range of wavelengths necessary for analyses of the elements of interest must be available. For example, some systems do not observe wavelengths above 500 nm. If alkali metals such as Na or K are to be determined, the spectrometer must be able to observe lines at 598 and 690 nm, respectively. Other elements [19, 125], such as S, P, and B, are best determined with lines below 190 nm. As a result, a vacuum or purged spectrometer system may be necessary.

As the spectral complexity of the samples to be analyzed increases, the importance of high resolution and flexibility in choosing the spectral lines to be used for analysis also rises. For example, analyses of samples with high concentrations of rare earth elements, Fe or Ti, will be best performed by high-resolution spectrometers. Also, more complex samples will require better interference (on-peak) correction.

As the analyte concentration approaches the detection limits or concomitant concentrations become large, background correction as well as interference correction become increasingly important. This will also be true for samples with intensely emitting species such as calcium or magnesium. High resolution and low stray light are also beneficial.

The software which comes with the system can vary from simple control of wavelength selection or signal acquisition to sophisticated control of the instrument, data display, manipulation, and reporting. Again, the number of samples to be analyzed is an important factor. If the system is to be used for a high sample throughput, statistical data treatment, report generation, permanent record keeping, and communications with other computers, then the software becomes more valuable.

Finally, manufacturers are normally willing to accept samples for analysis to test their particular instrument with the sample types to be used. Guarantees of detection limits, precision, accuracy, stability, linear working range, and other analytical characteristics are normally made. Tests of desired specifications should be performed before purchase.

8.11. NEW DIRECTIONS IN SPECTROMETERS FOR ICP-AES

Future spectrometers will closely approach the ideal spectrometer described in Table 8.4. Multichannel or multiplex detection will be used to detect the entire spectrum simultaneously with high resolution. As a result, the flexibility in line selection characteristic of slew-scan spectrometers will be merged with the rapid analysis provided by direct reading spectrometers.

Beyond improvement in the optical and electronic hardware, the spectrometers and complete ICP-AES instruments will become "intelligent." That is, the instrument itself will be able to optimize the sample introduction, excitation, line selection, and detection conditions dependent upon the particular sample being analyzed and its emission signals.

8.11.1. New Systems for Simultaneous, Multichannel Detection

The "electronic spectrograph." For many years, photographic detection was used for AES to provide simultaneous detection over a wide wavelength range with good resolution. However, because the readout process is slow and photographic emulsions have a limited dynamic range, many spectroscopists sought an "electronic emulsion" [126].

Attempts to construct an "electronic spectrograph" [127, 128] have been limited by the detectors or optical systems employed. Because optoelectronic imaging detectors with more than 20,000 detector elements are two dimensional, echelle spectrometers have been used. Image dissectors [129] and vidicons [130] have been used with echelle spectrometers. However, these detectors suffered from one or more limitations including poor SNRs, blooming or cross talk, and lag, incomplete signal readout.

Linear diode arrays [131-134] provide good SNRs when cooled, and the signal is allowed to integrate on the detector for periods up to 1 second. However, the number of detector elements is limited to approximately 2048. The diode arrays could be used to sequentially detect portions of the spectrum. This arrangement would provide complete analyte line and background detection simultaneously, but in a number of 5-20-nm pieces. Alternatively, a number of linear PD arrays could be placed along the focal plane of a spectrometer.

New solid state detectors with low-noise characteristics (when cooled to dry ice or liquid nitrogen temperatures) may lead to a viable electronic spectrograph. Denton [76, 135] has described a system based on charge injection device (CID) or charge coupled device (CCD) detectors with an echelle spectrometer. An aberration corrected cassegrain mirror system is used to provide optical reduction of the focal plane to match the size of the detectors (typically on the order of $\frac{1}{2} \times \frac{1}{2}$ in.). Solid state detectors with up to 300,000 elements will soon be available. When used with the echelle spectrometer, it would be possible to detect the spectrum from 200 to 800 nm with a single detector element viewing an average of 0.002 nm. The resolution provided by this system would be better than many commercially available ICP–AES systems now. For a more complete discussion of detectors for ICP–AES and electronic spectrograph systems see Chapter 9.

Multiplex Detection. The alternative to the electronic spectrograph is multiplex detection, and particularly Fourier transform (FT) interferometry. The Michaelson interferometer can provide multichannel detection with good resolution, as has been shown in the infrared region. Horlick and co-workers [66, 136, 137] have pioneered the use of FT–UV–Visible spectrometry for atomic spectrometry, with promising results. Other researchers, using the Kitt Peak one-meter interferometer [64], have obtained high-resolution spectra from an ICP.

A number of problems must be overcome for FT interferometry to become accepted. The most severe is the multiplex disadvantage. The detected signal (an interferogram) is a superposition of sine waves of different frequencies rather than a spectrally separated signal as with a grating spectrometer. Therefore, the weak lines must be measured from a small contribution to a large signal. However, to fully exploit the large linear dynamic range of ICP–AES and its even wider range of line intensities, weak lines must be detectable in the presence of other intense emission signals. Horlick's group [65, 66] has proposed and explored some possible solutions. A low-resolution monochromator could be used to pass a portion of the spectrum at a time through the interferometer. Alternatively, collimated light exiting the interferometer is fed directly to a plane grating before detection. In this arrangement the high-light throughput characteristic of the interferometer is maintained. Signal acquisition can be done in two ways. Fourier transform procedures could be used to obtain the spectrum or a correlation treatment [138] could be used to detect the elements present directly from the observed interferograms.

8.11.2. Intelligent Instruments

The dramatic improvement in the computer performance to cost ratio has had a large impact on ICP –AES. Extensive computer control of the grating angle (wavelength selection), filters placed in the optical path, PMT tube to be used, the PMT voltage, and signal acquisition time have made slew-scan spectrometers competitive with direct reading spectrometers. A large effort has gone into automation of the ICP-AES systems [139, 140]. The on-line exchange of information between the operator and the spectrometer has increased. Wavelength scans are displayed so that the operator can detect possible interferences or background problems.

Much of the computerization has concentrated on the collection and processing of data rather than adjustment of the instrument. Report generation and statistics are provided with many instruments. Some computers in ICP-AES instruments make wavelength tables and interference lists available.

However, decisions such as line selection, background wavelength, optimum ICP operating conditions, position within the plasma from which light is collected, and signal acquisition time are either selected by the operator or fixed. In the future, these decisions will be made more efficiently, quickly, an effectively by the instrument itself through experimental "learning." If the entire spectrum is acquired with good resolution, the instrument can choose a number of lines for each element and determine which is best. The choice will include compromises such as interferences versus sensitivity.

The intelligent instrument may also provide improved performance when matrix effects occur. The intelligent instrument may be able to determine if matrix interferences are significant, and if so, change the experimental conditions or data treatment to compensate for them. For example, the presence of easily ionizable elements causes a shift in the observed emission spatial profile [141, 142]. The profile might be used as an indicator of plasma conditions and matrix interferences. Adjustments in plasma flow rates or power, as suggested by some researchers [143], could then be used to reduce matrix interferences. The intelligent instrument could use this or other feedback routes together with chemometric experiment design and data interpretation to improve analysis accuracy, precision, and reduce the susceptibility to matrix effects.

REFERENCES

1. H. A. Rowland, *Phil. Mag.* **35,** 397 (1893); *Astronomy and Astrophys.* **12,** 129 (1893).
2. S. P. Davis, *Diffraction Grating Spectrographs*, Holt, Rinehart and Winston, New York (1970).
3. F. A. Jenkins and H. E. White, *Fundamentals of Optics*, 4th ed. McGraw-Hill, New York (1976).
4. E. Hecht and A. Zajac, *Optics*, Addison-Wesley, Reading, MA (1974).
5. G. S. Monk, *Light: Principles and Experiments*, Dover, New York (1963).
6. R. M. Barnes and R. F. Jarrell, "Gratings and Grating Instruments," in E. L. Grove, ed., *Analytical Emission Spectroscopy,* Part I, Analytical Spectroscopy Series, Vol. 1, Dekker, New York (1971).

7. G. R. Harrison, R. C. Lord, and J. R. Loofbourow, *Practical Spectroscopy*, Prentice-Hall, New York (1948).
8. R. A. Sawyer, *Experimental Spectroscopy*, Prentice-Hall, New York (1944).
9. A. P. Thorne, *Spectrophysics*, Chapman and Hall, London (1974).
10. G. R. Harrison, *Appl. Optics* **12**, 2039 (1973).
11. G. R. Harrison, *J. Opt. Soc. Am.* **39**, 522 (1949).
12. J. W. McLaren and J. M. Mermet, *Spectrochim. Acta* **39B**, 1307 (1984).
13. J. A. R. Samson, *Techniques of Vacuum Ultraviolet Spectroscopy*, Wiley, New York (1967).
14. E. G. Loewen, M. Neviere, and D. Maystre, *Appl. Optics* **16**, 2711 (1977).
15. R. W. Wood, *J. Opt. Soc. Am.* **34**, 509 (1944).
16. *Jobin-Yvon Handbook of Diffraction Gratings—Ruled and Holographic*, J and Y *Diffraction Gratings*, Jobin-Yvon, Longjumeaux, France.
17. "ICP Spectrometry: Instrumentation and Operating Parameters," Bausch & Lomb/Applied Research Laboratories, publication ICP-01-252-0, Bausch & Lomb, Sunland, CA.
18. H. A. Rowland, *Phil. Mag.* **16**, 197 (1883).
19. D. R. Heine, J. S. Babis, and M. B. Denton, *Appl. Spectrosc.* **34**, 595 (1980).
20. J. W. Carr and M. W. Blades, *Spectrochim. Acta* **39B**, 507 (1984).
21. C. R. Runge and G. Paschen, *Abhandl. Kaiserl. Akad. Wiss. Berlin, Anhang* **1** (1902).
22. J. M. Lerner, J. Flamand, J. P. Laude, G. Passereau, and A. Thevenon, "Aberration corrected holographically recorded diffraction gratings," in *Periodic Structures, Gratings, Moire Patterns and Diffraction Phenomena*, vol. 240, Society for Photo-instrumentation Engineers (1980), p. 72.
23. G. W. Stroke, "Diffraction Gratings," in *Handbuch der Physik*, Vol. XXIX, Springer-Verlag, Berlin (1967), pp. 426–754.
24. J. M. Lerner, J. Flamand, J. P. Laude, G. Passereau, and A. Thevenon, "Diffraction Gratings Ruled and Holographic—A Review," in *Periodic Structures, Gratings, Moire Patterns and Diffraction Phenomena*, Vol. 240, Society for Photo-instrumentation Engineers (1980), p. 82.
25. J. M. Lerner, "The Blazing of Holographic Gratings Using Ion Etching," National Conference on Spectrochemical Excitation, Abstr. published in *ICP Information Newslett.* **9**, 431 (1983).
26. M. Seya, *Sci. Light (Tokyo)* **2**, 8 (1952).
27. T. Namioka, *J. Opt. Soc. Am.* **49**, 951 (1959).
28. H. Ebert, *Wied. Ann.* **38**, 498 (1889).
29. M. Czerny and A. F. Turner, *Z. Physik* **61**, 792 (1930).
30. H. Kayser, in S. Hirzel, ed., *Handbuch der Spektroskopie*, Vol. I, Leipzig (1900).
31. W. G. Fastie, *J. Opt. Soc. Am*, **42**, 641 (1967).
32. A. B. Shafer, L. R. Megill, and I. Droppleman, *J. Opt. Soc. Am.* **54**, 879 (1964).
33. W. G. Fastie, U.S. Patent 3, 011, 391 (1961).

34. G. R. Rosendahl, *J. Opt. Soc. Am.* **52,** 412 (1962).
35. C. D. Allemand, *J. Opt. Soc. Am.* **58,** 159 (1968).
36. S. A. Khrshanovskii, *Opt. Spectr. (USSR)* **9,** 207 (1960).
37. J. Reader, *J. Opt. Soc. Am.* **59,** 513A, 1189 (1969).
38. R. M. Barnes, *CRC Crit. Rev. Anal. Chem.* **7,** 203 (1978).
39. V. A. Fassel, *Science* **202,** 183 (1978).
40. P. W. J. M. Boumans, *Line Coincidence Tables for Inductively Coupled Plasma Atomic Emission Spectrometry*, Pergamon, Oxford (1980); 2nd revised ed. (1984).
41. M. L. Parsons, A. Foster, and D. Anderson, *An Atlas of Spectral Interferences in ICP Spectroscopy*, Plenum, New York (1980).
42. R. K. Winge, V. A. Fassel, V. J. Peterson, and M. A. Floyd, *Appl. Spectrosc.* **36,** 210 (1982).
43. P. W. J. M. Boumans, *Spectrochim. Acta* **38B,** 747 (1983).
44. H. Kawaguchi, Y. Oshio, and A. Misuike, *Spectrochim. Acta* **37B,** 809 (1982).
45. H. G. C. Human and R. H. Scott, *Spectrochim. Acta* **31B,** 459 (1976).
46. J. A. C. Broekaert, F. Leis, and K. Laqua, *Spectrochim. Acta* **34B,** 73 (1979).
47. T. Hasegawa and H. Haraguchi, in R. M. Barnes, ed., *Proc. 1984 Winter Conf. Plasma Spectrochem.* (San Diego, CA, 1984), *Spectrochim. Acta* **40B,** 123 (1985).
48. G. F. Larson and V. A. Fassel, *Appl. Spectrosc.* **33,** 592 (1979).
49. R. D. Reeves, S. Nikdel, and J. D. Winefordner, *Appl. Spectrosc.* **34,** 477 (1980).
50. A. Batal and J. M. Mermet, *Spectrochim. Acta* **36B,** 993 (1981).
51. J. Kielkopf, *Appl. Optics* **20,** 3327 (1981).
52. G. F. Larson, V. A. Fassel, R. K. Winge, and R. N. Kniseley, *Appl. Spectrosc.* **30,** 384 (1976).
53. W. Kaye, *Anal. Chem.* **55,** 2022 (1983).
54. V. A. Fassel, J. M. Katzenburger, and R. K. Winge, *Appl. Spectrosc.* **33,** 1 (1979).
55. J. W. McLaren, S. S. Berman, V. J. Boyko, and D. S. Russel, *Anal. Chem.* **53,** 1802 (1981).
56. Anon., *Spex Speaker* **11**(2), 4 (July 1966).
57. G. Schmahl and D. Rudolph, "Holographic Diffraction Gratings," in E. Wolf, ed., *Progress in Optics*, Volume XIV, North-Holland, New York (1976).
58. D. L. Anderson, A. R. Foster, and M. L. Parsons, *Anal. Chem.* **53,** 770 (1981).
59. P. Jacquinot and B. Roizen-Dossier, "Apodisation," in E. Wolf, ed., *Progress in Optics*, Volume III, North-Holland, New York (1964).
60. A. R. Forster and M. L. Parsons, 1982 Pittsburgh Conference on Analytical Chemistry and Applied Spectroscopy, Abstr. 427, Atlantic City, NJ, Mar. 1982.
61. N. M. Walters, A. Strasheim, and A. R. Oakes, *Spectrochim. Acta* **38B,** 959 (1983).
62. Instrumentation Laboratories IL-100 described in "Plasma Emission Spectros-

copy: The Present Status," H. L. Kahn, S. B. Smith, Jr., and R. G. Schleicher, IL Report #113, 1979.
63. C. E. Taylor and T. L. Floyd, *Appl. Spectrosc.* **34,** 472 (1980).
64. L. M. Faires, B. A. Palmer and J. W. Brault, in R. M. Barnes, ed., *Proc. 1984 Winter Conf. Plasma Spectrochem.*, San Diego, CA,; *Spectrochim. Acta* **40B,** 135 (1985).
65. G. Horlick, 1984 Winter Conf. Plasma Spectrochemistry, Paper No. 29, San Diego, CA, 1984.
66. E. A. Stubley and G. Horlick, *Appl. Spectrosc.* **39,** 753, 805, 811 (1985).
67. G. R. Harrison, *J. Opt. Soc. Am.* **39,** 522 (1949).
68. G. R. Harrison, J. E. Archer, and J. Camus, *J. Opt. Soc. Am.* **42,** 706 (1952).
69. A. T. Zander, *Plasmaline* **2**(1), 4 (1981). (Publication of Spectrametrics, Inc.).
70. P. N. Keliher and C. C. Wohlers, *Anal. Chem.* **48,** 333A (1976).
71. A. T. Zander and P. N. Keliher, *Appl. Spectrosc.* **33,** 499 (1979).
72. J. Reader, L. C. Marquet, and S. P. Davis, *Appl. Opt.* **2,** 963 (1963).
73. R. F. Jarrell, *J. Opt. Soc. Amer.* **45,** 259 (1955).
74. J. Xu, H. Kawaguchi, and A. Mizuike, *Appl. Spectrosc.* **37,** 123 (1983).
75. R. Gerharz, *J. Quant. Spectrosc. Radiat. Transfer* **6,** 59 (1966).
76. M. B. Denton, H. A. Lewis, and G. R. Sims, "Charge-Injection and Charge-Coupled Devices in Practical Chemical Analysis: Operation, Characteristics and Considerations," in *Multichannel Image Detectors*, Vol. 2, ACS Symp. Series 236, American Chemical Society, Washington, DC (1983).
77. R. I. Botto, *Spectrochim. Acta* **38B,** 129 (1983).
78. J. W. McLaren and S. S. Berman, *Appl. Spectrosc.* **35,** 403 (1981).
79. E. Janssens, P. Schutyser, R. Dams, and J. Hoste, *Spectrochim. Acta* **38B,** 337 (1983).
80. W. Snelleman, T. C. Rains, K. W. Yee, H. D. Cook, and O. Menis, *Anal. Chem.* **42,** 394 (1970).
81. R. K. Skogerboe, P. J. Lamothe, G. J. Bastiaans, S. J. Freeland, and G. N. Coleman, *Appl. Spectrosc.* **30,** 495 (1976).
82. S. R. Koirtyohann, S. D. Glass, D. A. Yates, E. J. Hinderberger, and F. E. Lichte, *Anal. Chem.* **49,** 1121 (1977).
83. T. O'Haver, *Anal. Chem.* **51,** 91A (1979).
84. H. Ishii and K. Satoh, *Talanta* **29,** 243 (1982).
85. G. M. Hieftje and R. J. Sydor, *Appl. Spectrosc.* **33,** 499 (1979).
86. R. L. Cochran and G. M. Hieftje, *Anal. Chem.* **49,** 98 (1977).
87. S. W. Downey, J. G. Shabushnig, and G. M. Hieftje, *Anal. Chim. Acta* **121,** 165 (1980).
88. P. W. J. M. Boumans, *Spectrochim. Acta* **31B,** 147 (1976).
89. P. L. Kempster, J. D. R. Malloch, and M. V. D. S. de Klerk, *Spectrochim. Acta* **38B,** 967 (1983).

90. R. I. Botto, *Anal. Chem.* **54,** 1654 (1982).
91. R. I. Botto, *Spectrochim. Acta* **38B,** 129 (1983).
92. J. M. Mermet and C. Trassy, *Spectrochim. Acta* **36B,** 292 (1981).
93. F. J. M. J. Maessen, J. Balke, and J. L. M. de Boer, *Spectrochim. Acta* **37B,** 517 (1982).
94. J. W. Carr and G. Horlick, *Spectrochim. Acta* **37B,** 1 (1982).
95. M. Thompson, J. E. Goulter, and F. Sieper, *Analyst* **106,** 32 (1981).
96. T. Ishizuka and Y. Uwamino, *Spectrochim. Acta* **38B,** 519 (1983).
97. G. F. Kirkbright, A. M. Gunn, and D. L. Millard, *Analyst* **103,** 1066 (1978).
98. D. L. Millard, H. C. Shan, and G. F. Kirkbright, *Analyst* **105,** 502 (1980).
99. K. C. Ng and J. A. Caruso, *Anal. Chim. Acta* **143,** 209 (1980).
100. P. B. Farnsworth and G. M. Hieftje, *Anal. Chem.* **55,** 1414 (1983).
101. Spectrametrics, Inc., Haverhill, MA.
102. L. A. Fernando, *Spectrochim. Acta* **37B,** 859 (1982).
103. Leeman Labs Plasma-Spec, see [104].
104. J. C. MacDonald, *Amer. Lab.* **15**(9), 90 (1983).
105. A. W. Witmer, J. A. Jansen, G. H. Van Gool, and G. Brouwer, *Philips Tech. Rev.* **34,** 322 (1974).
106. M. E. Grandy, M. A. Sainz, and D. M. Coleman, *Appl. Spectrosc.* **36,** 643 (1982).
107. P. W. J. M. Boumans, F. J. de Boer, A. W. Witmer, and M. Bosveld, *Spectrochim. Acta* **37B,** 535 (1978).
108. P. W. J. M. Boumans, *Spectrochim. Acta* **38B,** 747 (1983).
109. Bausch and Lomb/ARL 3510 Spectrometer, described in Applied Research Laboratories publication 86-12, 1983, Bausch & Lomb, Sunland, CA.
110. K. Scharicz, *Amer. Lab.* **15**(3), 62 (1983).
111. M. Nippus, *Labor Praxis*, **26** (1980); translated into English and published in *ICP Information Newslett.* **6,** 297 (1980).
112. Heidenhain R800 and EXE 710 Encoder. See [111.]
113. A. Strasheim, M. E. Thain, N. M. Walters, C. Claase, H. G. C. Human, and N. P. Ferreira, *Spectrochim. Acta* **38B,** 921 (1983).
114. J. Burman, B. Johansson, B. Morefalt, K. Narfeldt, *Anal. Chim. Acta Computer Techniques and Optimization* **133,** 379 (1981).
115. P. Barrett, C. G. Fisher, III, and T. W. Barnard, *At. Spectrosc.* **3,** 43 (1982).
116. B. A. R. Tait, *Appl. Spectrosc.* **36,** 693 (1982).
117. E. Janssens, J. De Donder, P. Taylor, R. Dams, and J. Hoste, *Spectrochim. Acta* **38B,** 865 (1983).
118. M. Thompson and J. N. Walsh, *A Handbook of Inductively Coupled Plasma Spectrometry*, Blackie and Son, Bishopbriggs, Glasgow (1983).
119. M. A. Floyd, V. A. Fassel, R. K. Winge, J. M. Katzenberger, and A. P. D'Silva, *Anal. Chem.* **52,** 431 (1980).
120. P. A. Baedecker, *Anal. Chem.* **43,** 405 (1971).

121. A. Savitsky and M. J. E. Golay, *Anal. Chem.* **36,** 1627 (1964).
122. D. A. Leighty, D. D. Nygaard, D. S. Chase, and S. B. Smith, Jr., 1984 Winter Conference on Plasma Spectrochemistry, Abstr. P21, San Diego, CA, 1984.
123. ARL 3520 Sequential Spectrometer, described in ARL publication 87-18, 1083, "3500 AES Series Optical Emission Spectrometer Systems," Applied Research Laboratories, Division of Bausch and Lomb, Sunland, CA (1982).
124. J. C. MacDonald, *Amer. Lab.* **15**(9), 90 (1983).
125. D. D. Nygaard, D. S. Chase, and D. A. Leighty, *Research and Development*, Feb. 1984, p. 172.
126. W. D. Metz, *Science* **175,** 1448 (1972).
127. P. Gloersen, *J. Opt. Soc. Am.* **48,** 712 (1958).
128. D. L. Wood, A. B. Dargis, and D. L. Nash, *Appl. Spectrosc.* **29,** 310 (1975).
129. A. Danielson and P. Lindblom, *Appl. Spectrosc.* **30,** 151 (1976).
130. H. L. Felkel and H. L. Pardue, *Anal. Chem.* **49,** 1112 (1977).
131. G. Horlick and E. G. Codding, *Anal. Chem.* **45,** 1490 (1973).
132. G. Horlick and E. G. Codding, "Photodiode Arrays for Spectrochemical Measurements," in D. M. Hercules, G. M. Hieftje, L. R. Snyder, and M. A. Evenson, eds., *Contemporary Topics in Analytical and Clinical Chemistry*, Vol. 1, Plenum, New York (1977).
133. M. Kubota, Y. Fujishiro, and R. Ishida, *Spectrochim. Acta* **37B,** 849 (1982).
134. F. Grabau and Y. Talmi, "Inductively Coupled Plasma—Atomic Emission Spectroscopy with Multichannel Array Detection," in *Multichannel Image Detectors*, Vol. 2, ACS Symposium Series 236, American Chemical Society, Washington, DC (1983).
135. G. R. Sims and M. B. Denton, "Multielement Emission Spectrometry Using a Charge-Injection Device Detector," in *Multichannel Image Detectors*, Vol. 2, ACS Symposium Series 236, American Chemical Society, Washington, DC (1983).
136. G. Horlick and W. K. Yuen, *Anal. Chem.* **47,** 775A (1975).
137. W. K. Yuen and G. Horlick, *Anal. Chem.* **49,** 1446 (1977).
138. R. C. L. Ng and G. Horlick, *Spectrochim. Acta* **36B,** 543 (1981).
139. C. Allemand, *ICP Information Newslett.* **2,** 1 (1975).
140. J. R. Garbarino and H. E. Taylor, *Spectrochim. Acta* **38B,** 323 (1983).
141. M. W. Blades and G. Horlick, *Spectrochim. Acta* **36B,** 881 (1981).
142. R. Rezaaiyaan, J. W. Olesik, and G. M. Hieftje, in R. M. Barnes, ed., *Proc. 1984 Winter Conf. Plasma Spectrochem.*, San Diego, CA, 1984; *Spectrochim. Acta* **40B,** 73 (1985).
143. S. R. Koirtyohann, J. S. Jones, and D. A. Yates, *Spectrochim. Acta* **35B,** 49 (1980).
144. K. Laqua, W. D. Hagenah, and H. Waechter, *Z. Anal. Chem.* **225,** 142 (1967).
145. P. W. J. M. Boumans and J. J. A. M. Vrakking, *Spectrochim. Acta* **39B,** 1239 (1984).

CHAPTER

9

DETECTION AND MEASUREMENT

H. BUBERT and W.-D. HAGENAH

Institut für Spektrochemie und angewandte Spektroskopie
Dortmund, Federal Republic of Germany

9.1. **Introduction**
9.2. **Detection of Radiation**
   9.2.1. Single-Channel Detectors
      9.2.1.1. Photomultiplier Tube (PMT)
      9.2.1.2. Single Photodiode (PD)
   9.2.2. Optoelectronic Image Detectors (OIDs)
      9.2.2.1. Self-scanning Photodiode Arrays (SPDs)
      9.2.2.2. Silicon Vidicon Tube (SV)
9.3. **Electronic Measurement Systems**
   9.3.1. Introduction
   9.3.2. Modes of Measurement and their Characteristics
      9.3.2.1. DC Measurement
      9.3.2.2. AC Measurement ("Lock-In" Measurement)
      9.3.2.3. Pulse Measurement ("Photon Counting")
   9.3.3. Some Electronic Measurement Devices
      9.3.3.1. DC Amplifier with Low-Pass Filter
      9.3.3.2. DC Integrator
      9.3.3.3. Boxcar Integrator
      9.3.3.4. Multichannel Averager
      9.3.3.5. AC Amplifier ("Lock-In")
9.4. **Data Processing and Control of Electronic Circuits**
9.5. **Error and Noise Considerations in Detection and Measurement Devices**
   9.5.1. Introduction
   9.5.2. Systematic Errors
   9.5.3. Random Errors
      9.5.3.1. Detector Noise
         9.5.3.1.1. Noise of Photomultiplier
         9.5.3.1.2. Noise of Photodiode
      9.5.3.2. Noise of Measuring Systems
         9.5.3.2.1. Noise of Conventional Measuring Systems
         9.5.3.2.2. Noise of Optoelectronic Image Detector Systems

References

9.1. INTRODUCTION

The requirements that have to be met by a detection and measuring system (DMS) for ICP–AES depend on the analytical problems to be solved. These problems are manifold and diverse; therefore, different emphasis may be put on each of the essential properties by which the DMS can be characterized: accuracy, precision, sensitivity, linearity, dynamic range, speed, cost, and perhaps, reliability and compactness.

Accuracy. A DMS that does not yield accurate results seems to be of little use in AES. However, inaccuracy based on unknown systematic errors will be compensated for by relative measurements, for example, the determination of concentrations by means of analytical calibration curves obtained by the same experimental procedure.

Precision. High precision of analytical results requires low noise and long-term stability of the DMS. These can be achieved by the use of selected and expensive types of PMT (see Section 9.2.2.1), the use of high precision field-effect-transistor (FET)-input operational amplifier (see Section 9.3.2.1), and other high-precision electronic devices and, eventually, by cooling of the PMT. Good long-term stability of detector and electronic devices is needed to exploit long measuring times for increasing the SNR. Thus, the requirement of high precision determines other characteristics, such as cost or speed.

Sensitivity. The sensitivity of a detector should be as high as possible in the wavelength region of interest and as low as possible in other regions to avoid or to reduce the detection of undesired but inevitable radiation (e.g., stray light). This can be reached in some cases by choosing detectors with suitable spectral response (see Section 9.3.2.1) or by using optical filters.

Linearity and Dynamic Range. Linearity and dynamic range concern the detector as well as the measuring system, and will be given for various detectors (see Sections 9.2.1 and 9.2.2) and modes of measurements (Section 9.3.1). The dynamic range should be as wide as possible so that samples can be analyzed for major, minor, and trace constituents without accommodation by dilution or preconcentration of the constituents. Only if the same type of samples, the concentrations of which do not vary greatly, have to be analyzed, one can waive the requirement of wide dynamic range.

Speed. Especially in industrial laboratories, where a great number of analyses have to be carried out, is speed a dominant factor. Dilution or preconcentration of the constituents of samples necessitated by the lack of a sufficiently wide dynamic range, or frequent recalibration of the instrument imposed by insufficient long-term stability, reduce the desired and often required speed of an analytical procedure. A much higher speed can be achieved if the instrument is computerized. In relation to the DMS, this means that the accommodation of the amplifier gain or the integration time of the DMS to the analytical requirements ensues automatically under computer control (see Section 9.4).

Cost. If a system is designed in such a way that the best performance is achieved in every respect, the cost may be prohibitive. Therefore, one must always agree to a compromise between cost and performance.

Finally, we should point to the two different approaches that might be considered for obtaining results for multielement analysis: the multiplex technique and the dispersive technique. AES measurements are shot-noise limited since the spectral lines are located in the UV–VIS region; therefore, the Fellgett advantage [1] does not apply here. Thus, dispersive systems [2] yield a higher SNR than multiplex systems do. Consequently, ICP–AES only uses dispersive systems. The systems most frequently used are the multichannel spectrometer and the sequential-reading polychromator, the behavior of which is identical to that of the sequential slew-scan monochromator. These systems all use PMTs as detectors. Other systems, such as echelle systems with image dissectors or TV cameras, and systems incorporating linear PD arrays, have not yet reached the stage of development that makes them generally useful and acceptable in ICP–AES.

9.2. DETECTION OF RADIATION

The detectors used for spectrochemical analysis can be divided into two groups: single-channel and multichannel detectors. PM tubes (in connection with exit slits) have emerged as the commonly used single-channel detectors. Because of their internal amplification and large linear dynamic range, they are capable of converting radiation intensities directly into electronic signals of sufficient high levels for further handling. The most widely used multichannel detector is the photographic plate. The number of channels is determined by the grain size of the emulsion, and the plate is capable of recording thousands of spectral lines in a single exposure. The greatest disadvantages of the photographic plate are the low sensitivity, the nonlinear response characteristic, and the tedious and time-consuming readout.

However, in the last two decades, optoelectronic image detectors have become available which, to a certain extent, combine the major advantages of multichannel detection with the desired properties of a radiation detector.

9.2.1. Single-Channel Detectors

9.2.1.1. Photomultiplier Tube (PMT)

The PM tube is a vacuum tube with a photocathode from which photoelectrons are generated by incident photons. A dynode system multiplies the number of photoelectrons by a factor of up to 10^8, and an anode collects these electrons. The quantum efficiency of the photocathode (i.e., the number of photoelectrons ejected by one incident photon) depends on the photocathode material and the energy of the single photon, that is, the wavelength of the incident radiation. There is a long-wave cutoff due to the work function of the photocathode and a short-wave cutoff caused by the absorption of incident photons by the entrance window. Unfortunately, even in the total absence of incident radiation, emission of "dark current electrons" from the photocathode occurs. These electrons get sufficient energy to leave the photocathode thermally or from cosmic or radioactive rays. The thermal dark electrons can be reduced by cooling or by using a type of PMT of which the sensitivity is restricted to the wavelengths needed for the analytical or measuring problem. Selected modern PMTs with bialkali photocathodes can be purchased such that the dark current is very low without tedious cooling. Low-detection limits can be achieved with them. With high dc currents at the anode of a PMT (but still far below the maximum anode current allowed by the manufacturers to avoid destruction), the amplification of most types of PMTs will decrease slowly with time (fatigue effect [3]). High pulse currents will result in nonlinear amplification since the voltage between the last dynodes varies with the pulse current. To restore linear amplification, capacitors are required to stabilize the voltage between the last dynodes.

In Table 9.1, the vital properties of some types of PMTs used as radiation detectors in analytical equipment for AES are listed. Side-on (circular) types, the mounting of which is more convenient, show lower effective quantum efficiency than end-on (linear) types due to losses of photoelectrons occurring on the way from the photocathode to the anode. Line 10 (square root of the number of dark electrons) gives a fairly good estimation of the fluctuations of the dark current to which the detection limits are proportional. The linearity of PMTs is sufficient, while good stability in dc conditions and linearity are obtained if anodic currents higher than 1 or 2 μA are avoided and the current in the divider chain ("bleeder current") is at least 100 times higher than the maximum anode current. Normally, the ratio of maximum signal current to dark current noise is 10^5. This dynamic range cannot be expanded by changing the voltage supplied to the PMT since the maximum signal current will be increased (or decreased) in the same way as the dark current. This was also found by Boumans et al. [4] for a selected PMT. In principle, two possibilities exist to enlarge the dynamic range: (1) reduction of the dark current, and (2) extension of the range to higher signal currents.

Table 9.1. Characteristics of Six Photomultiplier Tubes (PMT) Used in AES

Type: Manufacturer:	1 P28/V1 RCA	4837 RCA	R 212 Hamamatsu	R 166 Hamamatsu	9783 B EMI	9789B EMI
Spectral response designation	20 ERX (103)	35 ER (118)	S-5	Similar to (121)	Modified S-5	—
Number of dynodes	9	9	9	9	9	13
Structure of dynodes	Circular	Circular	Circular	Circular	Circular	Linear
Photocathode material	Cs–Sb	Bialkali	Cs–Sb	Cs–Te	Cs–Sb	Bialkali
Typical cathodic sensitivity in mA/W	48	59	50	60	60	64
Amplification	3.3×10^6	2.3×10^6	5×10^6	1.7×10^5	3×10^7	4×10^7
Maximal quantum efficiency in electrons/photon eff. quantum efficiency in brackets	0.20 (0.07)	0.24 (0.08)	0.22 (0.07)	0.37 (0.12)	0.22 (0.07)	0.20 (0.18)
Wavelength of maximum quantum efficiency in nm	~300	~300	270	200	340	400
Cathodic dark current I_{DK} in A	6×10^{-16}	6.5×10^{-16}	4×10^{-16}	6×10^{-16}	1.7×10^{-16}	5×10^{-18}
Cathodic dark current noise in electrons $\cdot s^{-1/2}$	87	90	71	87	46	8

9.2. DETECTION OF RADIATION

A reduction of the thermal dark current can be achieved by cooling the tube as previously mentioned. In the case of a Cs–Sb photocathode, the dark current can be suppressed by a factor of 10 by cooling from room temperature to 0° C; in this case of trialkali photocathodes, a reduction by a factor of 16 may be obtained under similar conditions (see, e.g., [5, 6]). Another well-known, but rarely applied method to reduce the dark current and hence, to expand the dynamic range, is shielding the PMT by a metallic coating biased at or near cathode potential. This method may reduce the dark current by as much as three orders of magnitude [7]. Sometimes it is necessary to impede cosmic or radioactive rays from reaching the tube or to reduce the influence of magnetic and/or electrostatic fields. This may be achieved by means of a μ-metal shielding which generally reduces the dark current.

The extension of the range to higher signal currents is normally accompanied by a loss of linearity and of the high stability in dc measurements. Only in ac measurements, may higher anodic signal currents be permitted dependent on the frequency, the upsensing time of the PMT, or light-on time, respectively.

9.2.1.2. Single Photodiode (PD)

PDs are used to a great extent mainly in the IR spectral region (e.g., molecular spectroscopy, laser diagnostics) because they have a good sensitivity in that region. However, in the UV region the PDs have to compete with PMTs; both have a high sensitivity in this spectral region and beyond it, but PMTs have a high internal, low-noise amplification of the generated cathodic photocurrent. Only if properties other than sensitivity dominate in a particular application, can the use of PDs be considered.

Figure 9.1 represents the current-voltage characteristic of a junction PD. Except for the reverse voltage breakdown, the curves obey the expression for the total PD current,

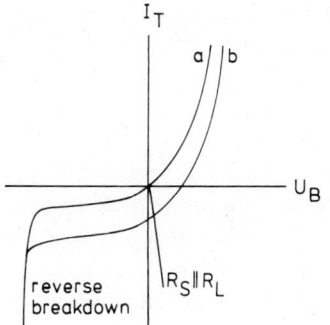

Figure 9.1. Current–voltage characteristic of a junction PD: a, unilluminated and b, illuminated diode; I_T, total PD current; U_B, voltage across the PD; R_S, series resistor; R_L, load resistor.

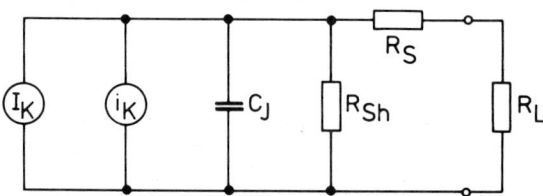

Figure 9.2. Electrical equivalent circuit for a photovoltaic PD. I_K, photoelectron current; i_K, noise current; C_J, pn-junction capacitance; R_{Sh}, shunt resistor; R_S, series resistor; R_L, load resistor.

$$I_T = I_0 \left[\exp\left(\frac{eU_B}{kT}\right) - 1 \right] - I_K \quad (9.1)$$

which follows from a simple solid-state theory [8]. I_0 is the leakage current, I_K the photoelectron current, kT the thermal energy ($kT = 4.08 \times 10^{-21}$ J at $T = 296$ K), e the elementary charge, and U_B the voltage across the PD.

In Fig. 9.2 the electrical equivalent circuit of a PD is shown. In analytical spectroscopy, PDs can operate in either the photoconductive or the photovoltaic mode. In the photoconductive mode, a reverse bias voltage is applied to the PD by which the junction capacitance C_J is reduced. Thus, the time constant of junction capacitance and shunt resistor R_{Sh} is lowered so that high-frequency signals can be accommodated. But the reverse bias voltage causes an increase of the reverse leakage current, which enhances the noise of the PD. If signals with frequencies not larger than approximately 1 kHz have to be measured, the PD should operate in the photovoltaic mode, which means that no reverse bias voltage is applied to the PD. In this case the total PD current I_T is nearly $-I_K$. Since I_K is proportional to the photon flux Φ (photon per second),

$$I_T \approx \eta e \Phi \quad (9.2)$$

where η is the quantum efficiency (electron per photon). Thus I_T itself is proportional to the photon flux so that a good linearity is achieved over about five orders of magnitude. State-of-the-art PDs have a high quantum efficiency in the UV region. In Fig. 9.3 the spectral response is given for the UV series manufactured by EG&G [9].

Although PDs show good linearity and high sensitivity over a wide spectral range, they have hardly been introduced in AES because they're lacking in internal amplification of the generated photocurrent. The amplification must be produced by an operational amplifier and this results in a lower SNR than is obtainable with a PMT. Only when an array of single PDs, isolated from each other, is fabricated on one wafer can the other properties such as the possibility of simultaneously detecting the radiation at closely spaced wavelengths make it worthwhile to exploit this possibility [10, 11].

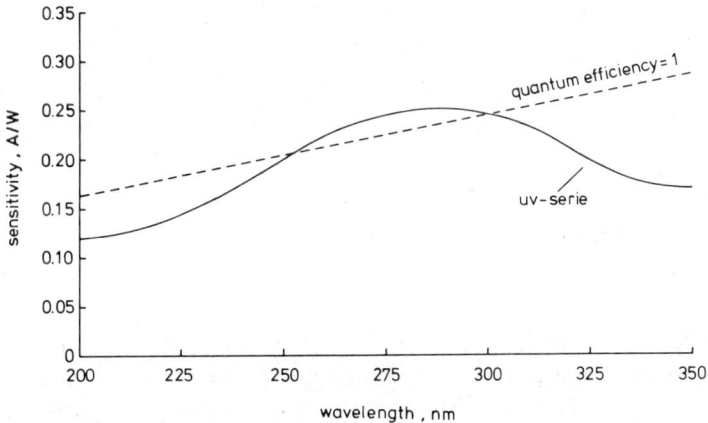

Figure 9.3. Spectral response versus wavelength for a single silicon PD [9].

9.2.2. Optoelectronic Image Detectors (OIDs)

All OIDs simultaneously detect the radiation in various channels, but the readout of the information collected in each channel occurs sequentially. Therefore, OIDs consist of three basic components: a radiation sensitive element which converts the photon image into an electrical one, a storage device for storing this electrical image, and a readout system that permits convenient display of the information. Many types of OIDs, differing in the three basic components, are developed and employed in different fields of spectroscopy. Excellent reviews of this topic have been given by Talmi [12, 13]. In this section, only the two principal OIDs—the self-scanning PD array and the silicon vidicon tube—will be treated, while other OIDs such as image dissector tubes, charge coupled devices (CCDs), will be left out of consideration.

9.2.2.1. *Self-scanning Photodiode Arrays (SPDs)*

The most commonly used SPDs, manufacturered by EG&G/Reticon [14], consist of 1024 single PDs (pixels) in a one-dimensional (linear) array. Such an array, together with the necessary controlling logic and amplifying analogue circuit, is shown schematically in Fig. 9.4. The anode of each PD is connected to the output line (video line of the array) by a FET switch controlled by a single bit, which is shifted through a shift register, the latter being also integrated in the array. The PDs work in the charge storage mode, which is indicated in Fig. 9.4 by capacitors switched parallel to the PDs and which corresponds to the pn-junction capacitances. The reverse bias charge can be discharged by recombination of a hole with an electron in the p-type region.

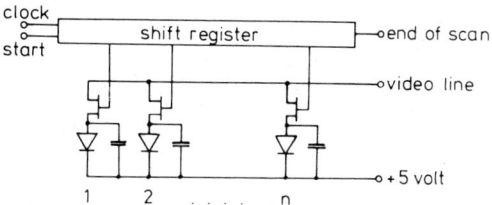

Figure 9.4. Schematical diagram of a self-scanning PD array.

This hole, which moves by diffusion from the n-type wafer, is formed here as an electron-hole pair generated by absorption of a photon, or thermally. During periodic scans, the FET-switches sequentially recharge the capacitors to their initial values. Each recharging pulse is proportional to the sum of the currents contributed by the radiation and the dark current integrated over one-scan period.

The dynamic range of the output pulses related to the noise of the measuring and controlling electronic device amounts to about 500:1 without correction of fixed-pattern noise; when the latter is allowed for, a dynamic range of about 1000:1 can be achieved.

The desired properties of a spectroscopic system discussed in Section 9.1 have been investigated for the SPD, partly in connection with special analytical measurements, by several authors [15–25]. According to data given by the manufacturer, the PD array shows a quantum efficiency of $\geq 30\%$ in the spectral region between 500 and 900 nm. The spectral response is shown in Fig. 9.5.

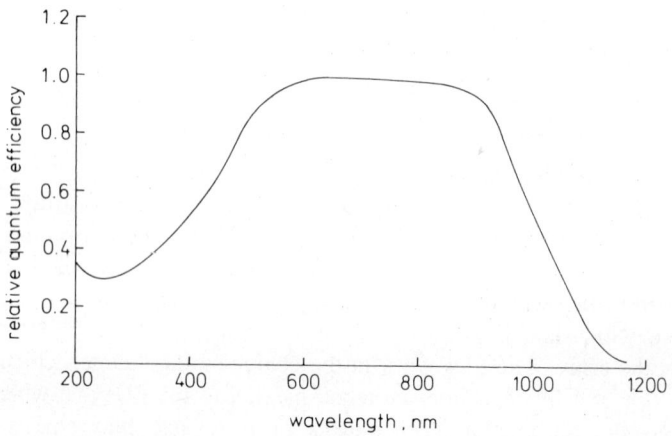

Figure 9.5. Quantum efficiency versus wavelength for a self-scanning PD array according to [14].

9.3. ELECTRONIC MEASUREMENT SYSTEMS

9.2.2.2. Silicon Vidicon Tube (SV)

SVs and two image-intensified devices based on the SV have been developed during the past 20 years and are now commercially available from several manufacturers. The fabrication of complete vidicon scanning spectrometers (Princeton Applied Research Corp., Tektronix) is in particular a reason why SVs have been more often explored in optical spectroscopy than other OIDs. The tube target consists of an n-type silicon wafer covered by p-type regions and forming with them a PD array. Similarly, to the self-scanning PD arrays, the PDs of an SV work in the charge storage mode. The reverse bias charge of each diode can be discharged as described in Section 9.2.2.1. The pn-junction capacitor is recharged by scanning an electron beam over the p-type side of the PDs. The recharging current of the beam is proportional to the sum of the currents generated by the radiation and dark current.

In contrast to the self-scanning PD arrays, the recharging by means of a single scan can be incomplete if the pn-junction capacitor of a PD is strongly discharged by intense radiation, which leads to a memory effect known as lag.

Since the essential part of the SV is the PD array, properties such as quantum efficiency, spectral response (see Fig. 9.6), and dynamic range are similar to those of SPDs. The geometrical dimension of a standard target having 400 by 500 elements is 9.5×12.7 mm^2. The spatial resolution is normally dictated by the diameter of the electron beam (~ 25 μm).

9.3. ELECTRONIC MEASUREMENT SYSTEMS

9.3.1. Introduction

Generally, in optical emission spectroscopy, radiation is detected in one of the ways described in Section 9.2. At the output of the detector, a signal is generated that contains the required information. This signal, hereafter referred to as net signal, is obtained in the form of an electric voltage or current. Normally, the net signal is weak and overlapped by interference signals (e.g., dark current noise) so that its direct display on a readout unit (e.g., oscilloscope, recorder, A/D display) does not provide the desired, correct information. The aim of the electronic measurement system is to (1) amplify the net signal, and (2) simultaneously reduce interference signals. Depending on the kind and intensity of net and interference signals, one will have to decide on one of the three modes of measurement: dc measurement, ac measurement, or pulse measurement (photon counting).

In Section 9.3.2 we shall discuss two important concepts essential to the understanding of the principles of measuring: noise bandwidth and dynamic range, the latter with respect to the net signal that the electronic measurement

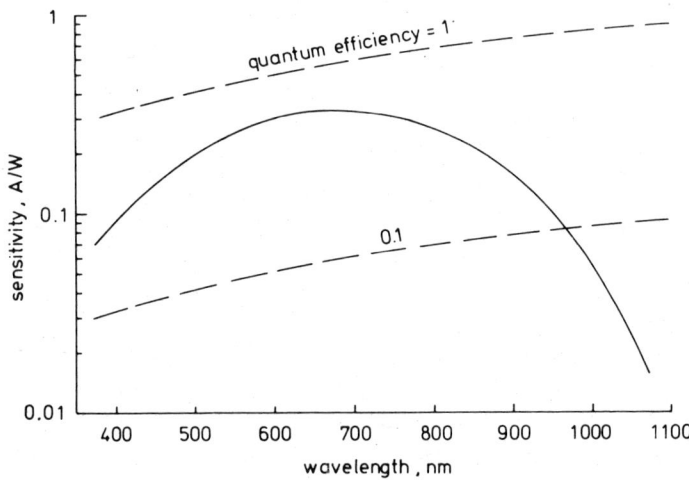

Figure 9.6. Spectral response of a silicon vidicon tube (Hamamatsu N 736/N 747).

system can handle without changing the amplification of the detector. In Section 9.3.3 several common realizations of the three modes of measurement will be dealt with.

9.3.2. Modes of Measurement and Their Characteristics

9.3.2.1. DC Measurement

The dc measurement represents the simplest and, therefore, often applied measuring mode. The net signal can be amplified in a simple manner by using an operational amplifier. Unfortunately, the interference signals occurring at the detector output are amplified in the same manner as the net signal; furthermore, undesired interference signals such as $1/f$ noise and drift occur in a system only employing an operational amplifier. Therefore, one will try (1) to minimize any of these additional interference signals by a suitable selection of the amplifier, and (2) to reduce frequency-dependent interference signals by narrowing the bandwidth to such a degree as is permissible without loss of information in the net signal. This can be achieved, for instance, by the use of an amplifier that operates as an active low-pass filter or an amplifier that acts as or is followed by a dc integrator (see Sections 9.3.3.1 and 9.3.3.2). Thus, the noise bandwidth is determined by the bandwidth of the low-pass filter or the integration time of the integrator.

The dynamic range is defined as the range extending from the noise voltage

Table 9.2. Specifications of Two FET-Input Operational Amplifiers

Models	AD 515 L (Analog Devices)	3527 AM (Burr–Brown)
Open loop gain		
$R_L \geq 10$ kΩ	10^5	4×10^5 (112 dB)
Output Characteristics		
Voltage (V) at $R_L = 2$ kΩ	±12	±12
Short circuit current (mA)	25	20
Load capacitance (nF)	1	1
Frequency response		
Unity gain (MHz)	0.35	1
Full power response (kHz)	16	14
Slew rate (V/μs)	1	0.9
Settling time (μs)	16	45
Input offset voltage (mV)	0.4	0.2
Versus temperature (μV/K), max.	25	10
Versus power supply (μV/V)	100	75
Input bias current (pA)	0.05	2
Input impedance		
Differential (Ω)	10^{13}	10^{12}
Common mode (Ω)	10^{15}	10^{13}
Input noise		
Voltage		
(μV(p-p)), 0.3–10 Hz	3	2.6
(nV Hz$^{-1/2}$), f = 10 Hz	75	75
(nV Hz$^{-1/2}$), f = 100 Hz	55	35
(nV Hz$^{-1/2}$), f = 1 kHz	50	30
Current		
(fA(p-p)), 0.3 Hz–10 Hz	2	15
(fA(rms)), 10 Hz–10 kHz	10	60

to the maximum permissible signal voltage at the output of the operational amplifier. In order to maintain the largest dynamic range, the input noise (current and voltage) must be kept low. A dynamic range of approximately 10^4 at a bandwidth of 1 Hz can be attained, if a detector with a low dark current and low-noise amplifiers are applied. As an example of such an amplifier, the specifications of two of them are given in Table 9.2. On the whole, dc measurement has the disadvantage, in particular at high amplification, that additive interference signals, such as $1/f$ noise and drift, indicate their largest amplitudes just at $f = 0$ Hz; this disadvantage can be avoided to a large extent by the ac mode of measurement.

9.3.2.2. AC Measurement ("Lock-In" Measurement)

An ac measuring system for the frequency f_R presumes a signal that is modulated with this frequency. Additive dc signals that are added to the ac signal after modulation, will be eliminated and other frequency-dependent interference signals can be reduced by the proper choice of f_R. Best conditions prevail if the net signal only is modulated for 100%. In many cases this can be reached by modulation of the radiant beam at the source or with a chopper positioned in the radiation path between the source and the dispersive element or the detector.

A further, very refined method is wavelength modulation, which can be applied if the measurement should yield the difference of two spectral intensities at two different wavelengths. In this approach, a refractor plate in front of the exit slit of the monochromator is moved in such a way that, for example, initially the spectral line plus background is measured at λ_0 and subsequently the background only at $\lambda_0 + \Delta\lambda$ or $\lambda_0 - \Delta\lambda$. With constant spectral background (continuum radiation, stray light) the background is not modulated and therefore eliminated.

Multiplicative interference signals due to changes in detector sensitivity, instabilities in the electric parameters of the radiation source, or fluctuations in the gas or solution flows multiplicatively overlap an ac signal and cannot be reduced by the application of either an ac or a dc measurement system.

The same operational amplifiers used in dc measurement are used in ac measurement but the ac signal is taken capacitively from the amplifier output and fed into a readout unit. In order to reduce the interference signals, the capacitive coupling should operate as a band filter at the resonance frequency f_R. However, it is shown by the characteristic of a simple passive band filter described by the transmitting function that not only the bandwidth is broad but that also the amplitude of the net signal at resonance frequency f_R is reduced. Distinctly narrower noise bandwidths and a correspondingly larger dynamic range can be reached by employing active band filters. Depending on the location of the resonance frequency, the dynamic range lies between 10^4 and 10^5. Since the most common readout units display dc signals only, an ac/dc converter is needed between ac amplifier and readout unit. One of the possible realizations is described in Section 9.3.3.5.

A decisive disadvantage of the ac mode of measurement is the loss of "information intensity" that occurs through modulation of a dc radiation, for example, by the use of a chopper. If the chopper running on the frequency f_R interrupts the radiation for $1/(2f_R)$ s, only one-half of the radiation bearing the information reaches the measuring device. This leads to a decrease in the SNR, which has a negative effect, particularly in view of the measurement of low spectral intensities.

9.3.2.3. Pulse Measurement ("Photon Counting")

Here we will discuss photon counting, a method not only suitable for measuring very low radiation intensities, but also excellently suited for reaching a large dynamic range. It can be applied if the counting time is long enough to keep negligible the error in the counting rate due to the overlapping of the pulses at the detector output. Allowing an error of 0.1% in the mean count rate, that is, if on average, 2 out of 1000 pulses are not time-resolved, the maximum count rate can be calculated in accordance with the Boltzmann distribution as

$$r_{max} \simeq \frac{1}{8\Delta t_r} \text{ Hz} \qquad (9.3)$$

where Δt_r (s) represents the pulse width that can be resolved by the measuring system. For the still resolvable pulse width, one may take the width at 10% of the peak value, which can be calculated as

$$\Delta t_r \approx \Delta t_{10\%} \cong 2.3 R_L C_S \qquad (9.4)$$

where R_L (Ω) is the load resistance and C_S (F) the stray capacitance, in other words, Δt_r is determined by the time constant of the system, which is of the order of 0.1 μs. Therefore, the maximum counting rate is of the order of 10^6 Hz and with a dark pulse rate of 1 Hz, the dynamic range also is of the order of 10^6.

One of the typical photon counting systems, taken from the work of Franklin et al. [26], is represented in Fig. 9.7. The bandwidth for signals must be sufficiently wide so that even high counting rates may be used. The noise band-

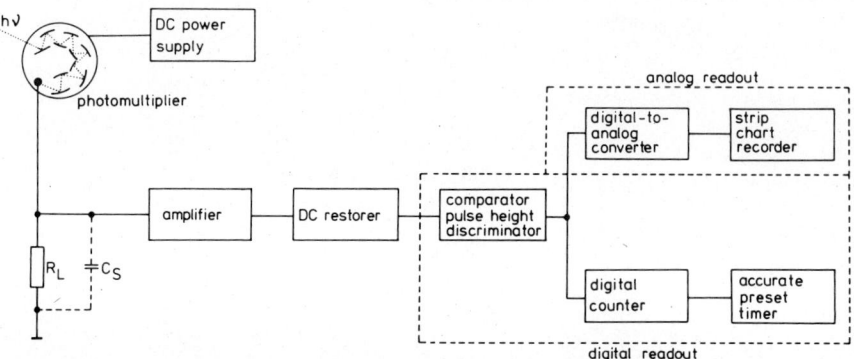

Figure 9.7. Block diagram of the practical photon counting system [23]; R_L, load resistor, C_S, stray capacitance.

width is mainly determined by the pulse height discriminator which suppresses too high pulse voltages (due to cosmic radiation, for instance) and which also suppresses too low pulse voltages (due to the dark current of the PMT, for instance). However, as pulses caused by the relevant photoelectrons vary in pulse height—the mean peak voltage at the amplifier input is $\overline{U} = Ge/C_s$, where G is the PMT gain and e the elementary charge—some of these pulses are discriminated, too.

This, in respect to a *fixed pulse rate*, leads to a constant RSD which cannot be diminished by longer measuring times. Any other additive interferences such as drift or $1/f$ noise can be suppressed to a great extent.

9.3.3. Some Electronic Measurement Devices

9.3.3.1. DC Amplifier with Low-Pass Filter

Table 9.2 lists characteristic data of two integrated operational amplifiers. They are low-noise amplifiers that can be used in all three measurement modes. As illustrated in Fig. 9.8, the amplifier operates as an electrometer-amplifier and simultaneously acts as third order low-pass Bessel-filter (as well as an impedance converter). The dc amplification is

$$A_0 = 1 + \frac{R_F}{R_1} \tag{9.5}$$

The amplification or bandwidth can only be changed by alteration of the feedback resistor R_F or the three capacitors C_1, C_2, and C_3.

Another possibility to alter the bandwidth of a dc amplifier is provided by active filter units (e.g., DATEL INTERSIL, Mansfield, USA; Burr–Brown, Tucson, USA), the bandwidths of which are adjustable by means of external resistors and capacitors.

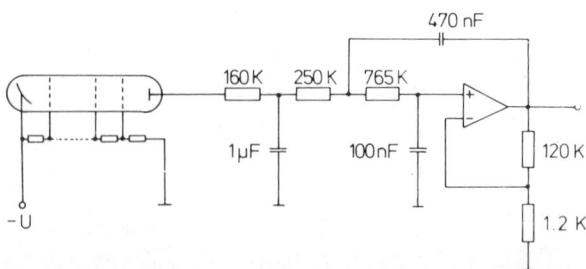

Figure 9.8. Electrical circuit of an electrometer-amplifier acting simultaneously as third-order low-pass Bessel filter. Signal bandwidth $\Delta f_S = 1$ Hz, gain $A_0 = 100$.

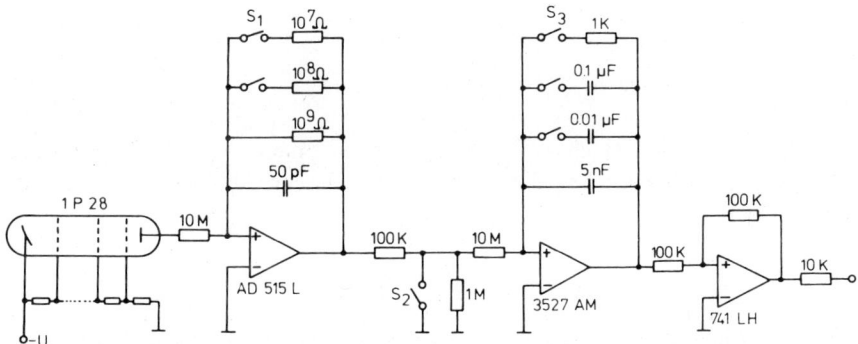

Figure 9.9. Electrical circuit of a dc integrator (dc amplifier and integrator).

9.3.3.2. DC Integrator

The bandwidth of a dc amplifier can be easily changed by introducing a dc integrator. However, the dc integrator necessitates an additional expenditure of digital electronics. Generally positioned at the output of the dc amplifier is a further operational amplifier operating as dc integrator[1].

Figure 9.9 shows the practical realization of this measuring mode. The voltage at the output of the integrator is

$$\overline{U}_i = \frac{1}{\tau_i} \int_0^{t_i} [U_{dc} + U_{ac}(t)]\, dt \qquad (9.6)$$

where $\tau_i = R_i C_i$ is the time constant of the integrator, U_{dc} the dc component and $U_{ac}(t)$ the ac component of the input voltage, and t_i the integration time. If $U_{ac}(t)$ is identical to so-called white noise, Eq. 9.6 reduces to

$$\overline{U}_i \approx U_{dc} \frac{t_i}{\tau_i} \qquad (9.7)$$

where $\overline{U}_i$ is the mean output voltage. For the output noise voltage u_i of the integrator, we then have

$$u_i = \frac{2\sqrt{2}}{\tau_i} u'_n \sqrt{t_{i'}} \qquad (9.8)$$

where u'_n is the effective noise voltage in $\text{VHz}^{-1/2}$ at the integrator input.

---

[1] A direct coupling of the integrator to the output of the PMT is possible, but the disadvantage is that the influence of interferences (e.g., caused by FET-switches) at a very low current of the PMT can bias the results. Therefore, an additional amplifier is introduced which acts as current-to-voltage converter.

The noise bandwidth,

$$\Delta f_n = \frac{1}{2t_i} \qquad (9.9)$$

can easily be varied by changing the integration time t_i. Similarly, the output voltage can easily be changed by using another integration capacitor C_i as can be seen in Fig. 9.9.

9.3.3.3. Boxcar Integrator

A boxcar integrator consists of a dc amplifier and an integrator in which either the amplifier entrance or the integrator entrance is gated electronically for a particular time interval of a normally periodically repeated procedure. The gate is opened exactly as long as the signal of interest is present. The boxcar integrator permits it to eliminate interference signals (due to dark current, stray light, noise, etc.) that are generated beyond the periods in which the signal of interest is predominant. At first glance, this technique does not appear to be of great value in ICP–AES methods; however, it is applied in radiation measurements using, for instance, a self-scanning PD array (see Section 9.2.2.1.).

Figure 9.10 schematically presents the measuring device of a boxcar integrator. Figure 9.11 shows the voltage variations U_A at the output of the dc amplifier, U_I at the output of the integrator, and U_H at the output of the hold unit as well as the necessary controlling pulses. The part of the measuring device presented within the frame of broken lines is commercially available (EG&G/Reticon, Fairchild). As a readout cycle of all n single diodes of an array ($n \leqslant 2048$ at a linear array) occurs during a time of about 4 μs, an oscilloscope is necessary as a fast display unit.

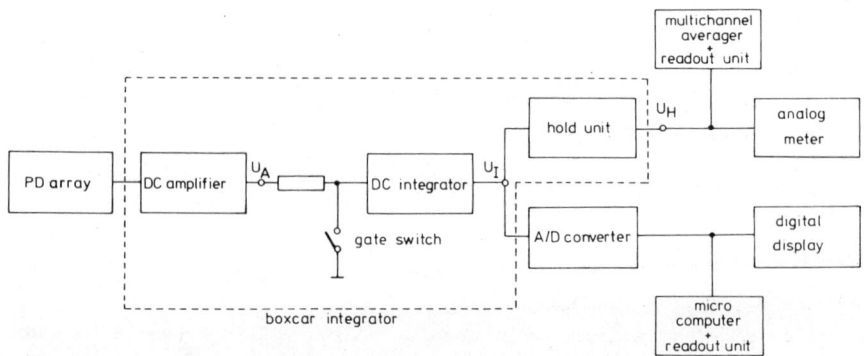

Figure 9.10. Schematical diagram of a boxcar integrator.

9.3. ELECTRONIC MEASUREMENT SYSTEMS

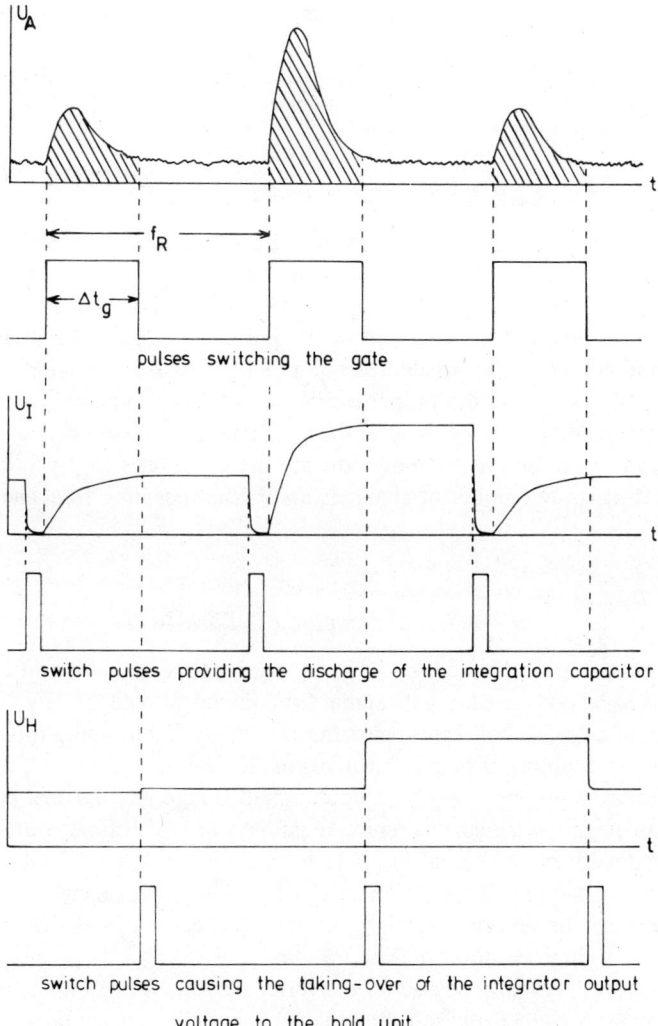

Figure 9.11. Voltage courses of a boxcar integrator: U_A, output voltage of the amplifier, U_I, output voltage of the integrator, U_H, output voltage of the hold unit.

Fast data handling and processing and slow data output can be combined by the use of a microcomputer or a multichannel averager. Therefore, after throughput of K cycles (e.g., the average of K cycles) the data output may be obtained.

The signal bandwidth is $\geq (\Delta t_g)^{-1}$, where Δt_g represents the time interval of the opened gate, while the noise bandwidth amounts to

$$\Delta f_n = \frac{1}{2K\Delta t_g} \qquad (9.10)$$

9.3.3.4. Multichannel Averager

If, for example, the signals of the single PDs of a PD array should not be displayed on a screen of an oscilloscope but should be integrated, a multichannel averager can be used. Here, in each cycle, the signal of each PD is digitized and added to the sum of signals from the previous cycles. The multichannel averager serves as a kind of device comprising n boxcar integrators. It can, therefore, be stated that the properties of the averagers with respect to signal and noise bandwidth are identical to those of a boxcar integrator.

The signal and the noise bandwidths are the same ones as given in Section 9.3.3.3. If n_c is the number of channels and f_R the repeating frequency, $\Delta t_g = (n_c f_R)^{-1}$.

9.3.3.5. AC Amplifier ("Lock-In")

As was mentioned in Section 9.3.2.2, the ac signal at the output of an ac amplifier has to be converted to a dc signal for convenient readout. This is accomplished with an ac/dc converter operating in either a free-running (phase-insensitive) or a synchronous (phase-sensitive) mode.

The free-running type may be a double-path bridge rectifier that passes any positive or negative signal unaltered, regardless at which time it arrives. The noise bandwidth of the signal is given by the time constant of the rectifying circuit—in the simplest case the product of the output impedance and the capacitance—but the dc offset is high. A consequence of this is that high noise signals lead to high dc noise offsets.

The use of a synchronous ac/dc converter having, for example, the bridge rectifier paths switched phase-synchronously, entails a lower dc noise offset. The converter requires a synchronous reference signal. Figure 9.12 shows an arrangement consisting of a single PD, a current-to-voltage converter and a passive band filter, a synchronous rectifier, and a "hold" unit. As the synchronous rectifier reacts with equal sensitivity to all odd harmonics of the basic frequency, the band filter should possibly attenuate the harmonics.

The noise bandwidth of the arrangement presented is given by the band filter,

$$\Delta f_n = \frac{\pi}{2} f_R \qquad (9.11)$$

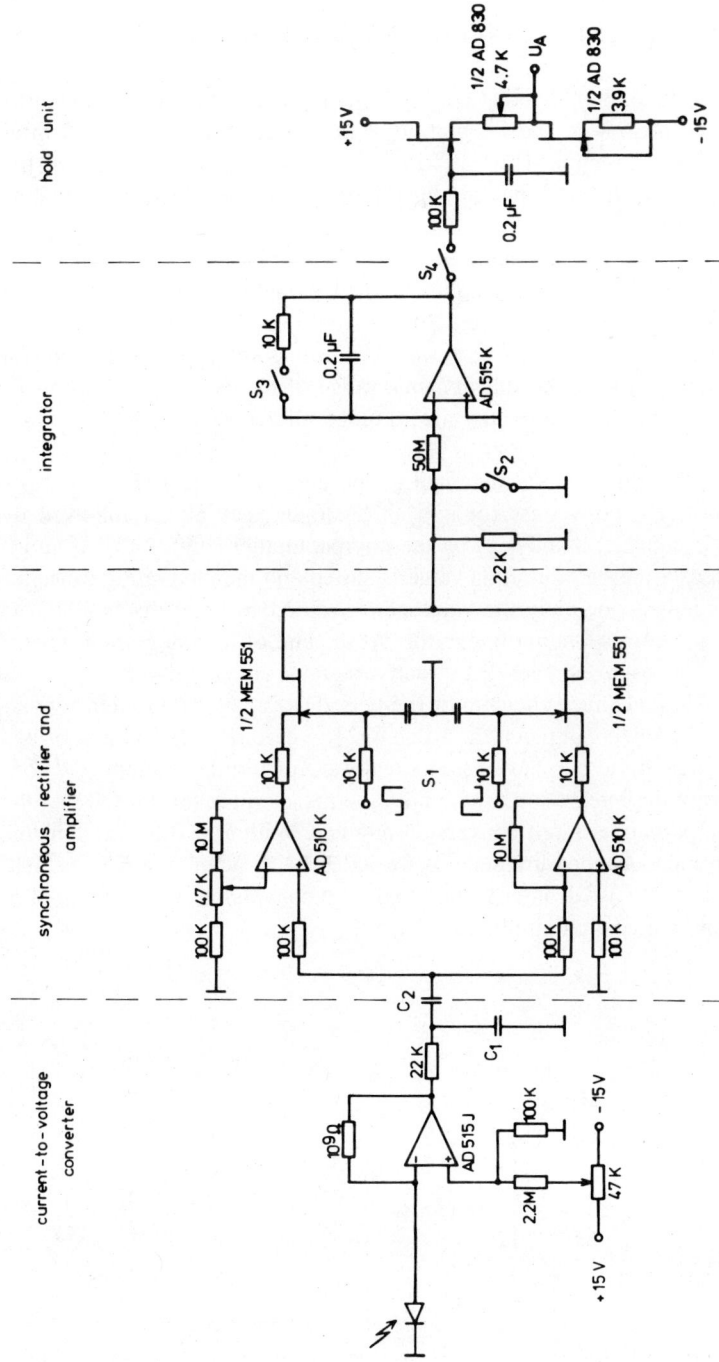

Figure 9.12. Electrical circuit of an ac amplifier consisting of a single PD, current-to-voltage converter, passive band filter, synchronous rectifier, integrator, and hold unit.

9.4. DATA PROCESSING AND CONTROL OF ELECTRONIC CIRCUITS

We mentioned in Section 9.3 that modern integrated amplifiers are well suited for both direct integration and integration of the anode current of a PMT after preamplification to a signal level at which the noise and the influence of the temperature coefficient of the integrating amplifier and other electronic elements can be neglected. The data processing and the control of the electronic circuit is done nowadays by microcomputers. This is cheaper in construction and maintenance, and much more easily adaptable to new parameters than the equivalent analogue data processing and circuit control would be.

To make this approach familiar, we will discuss three possible ways for processing data that save the full dynamic range of intensity measurement offered by PMTs and the analytically useful range offered by the ICP.

1. The preamplified and integrated anode current of the PMT is fed periodically into A/D-converters (or into an analogue multiplexer followed by only one A/D-converter) and read by the microcomputer (Fig. 9.13). If an intensity exceeds a preset value, this value is stored and the integrating capacitor C_I is discharged; if not, only the value zero is stored without discharging the capacitor of the corresponding integrator. At the end of the integration time of the analysis, the voltages reached by each integrator are stored.

For each PMT channel, the sum of the periodically converted intensity values and the end value can reach the value $\delta n 2^\alpha$, where α is the number of bits of the converter, n the number of periodically occurring integrations and conversions during the integration time, and δ the preset value for discharging the capacitors in parts of the full conversion number. With $\delta = 0.5$, the dynamic range of the intensity measurements is extended by a factor of $n/2$. The analogue integration of weak intensities avoids the conversion noise and nonlinearity of small conversion numbers.

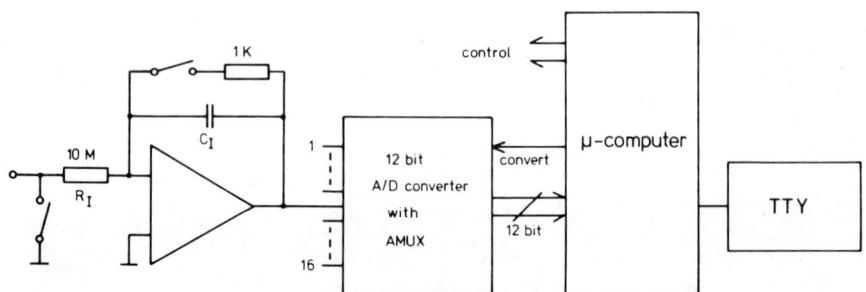

Figure 9.13. Electronic circuit with signal integration and data processing by a microcomputer.

Sampling of the separate intensities measured during each integration interval permits statistical evaluations of the separate signals and the measurement of the time dependence of the signals. Both can be used to control the precision of the analytical results. In spite of the large dynamic range, small programs should control the feedback resistance of the preamplifiers to ensure that their output voltages are within the ratings of the amplifier type. Further control is necessary for the integrators.

2. By adding a comparator to each integrator (Fig. 9.14), a discharge current for this integrator can be switched by a bistable multivibrator. The current is precisely controlled by the microcomputer which counts the number of discharges for each integrator. At the end of the integration time, the voltages (charges) of each integrating capacitor are converted to digital values read by the microcomputer and added to the value derived from the number of discharges in the relevant channel.

Adding a second comparator with multivibrator to the output of the preamplifiers enables the microcomputer to control the feedback resistors and, thus, to ensure that the output voltages of the preamplifiers remain within the rated values of the amplifier type. This measuring principle does not require a "premeasuring time" to set the appropriate feedback resistors and integrating capacitors; it also does not entail "dead time" for the periodical A/D-conversion or for the access to digital results within the integration time. However, it is not possible to obtain the time dependence of the signal and data for statistical analysis.

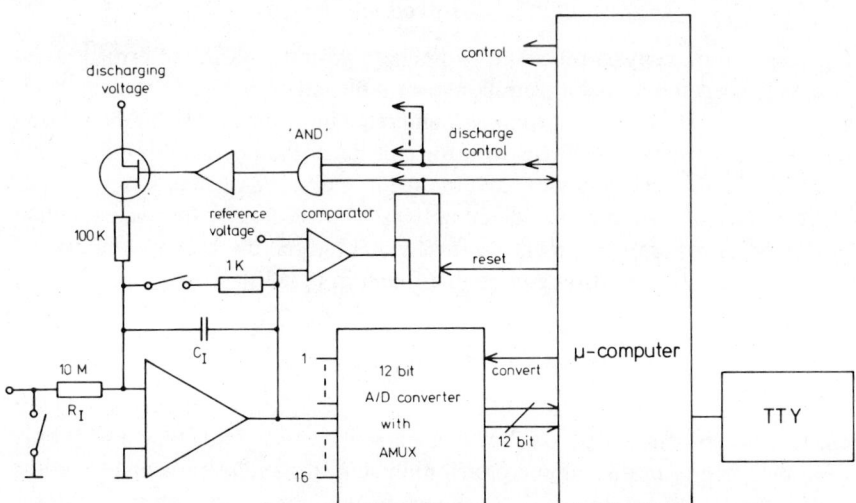

Figure 9.14. Electronic circuit with signal integration, data processing and controlling of the discharge current by a microcomputer.

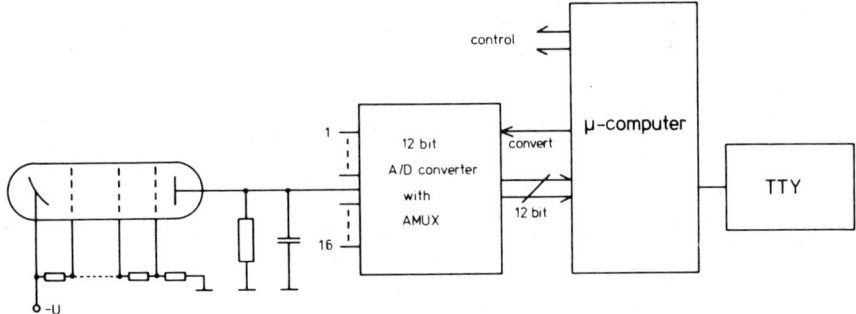

Figure 9.15. Electronic circuit for high repetition A/D-conversion.

3. It is possible to omit preamplifiers and integrators and to smooth the output voltages of the PMTs by small capacitors parallel to the output resistors of the PMTs (Fig. 9.15). These output voltages are digitized with high repetition frequency. But by using this method, no time dependence or statistical data can be collected. Weak intensities suffer from nonlinearities of the A/D-converter and from its conversion noise.

9.5. ERROR AND NOISE CONSIDERATIONS IN DETECTION AND MEASUREMENT DEVICES

9.5.1. Introduction

Each step in an analysis procedure is liable to various sources of error located in sample preparation, nebulization, atomization and excitation, wavelength adjustment and detection, measurement, and readout of the signal. In the following sections, just those sources of error that lie in the operational field of the detector, the measuring device, and the readout unit are treated, while various other sources of error are considered only as far as necessary for understanding the remedies for error treatment. A review and tutorial discussion of noise and SNRs in analytical spectrometry can be found in [27–30].

9.5.2. Systematic Errors

Systematic errors are generally difficult to recognize, and an assessment of their effects can frequently lead to faults in accuracy. Errors of this type do not only have their origin in the sample preparation step, the radiation source, and the optical part of the equipment, but can also be produced at the detector or in the measuring system in the form of dark current, offset voltage, or incorrect calibration of the readout unit.

Mostly, however, analytical procedures only use relative measurements so that systematic errors in detection and measurement of radiation do not lead to systematic errors in the analytical results; therefore, these types of errors will not be further discussed here.

9.5.3. Random Errors

Random errors are due to random fluctuations of the intensities about their long-term temporal average. The frequency spectrum of these fluctuating intensities may comprise a single frequency or several frequencies or be continuous. Interferences showing a discrete frequency spectrum—50 Hz (60 Hz) and weaker harmonics—originate at the power line; in the case of the ICP they may have their origin in the generator frequency or in the transmitter unit in the vicinity. Often, they can be eliminated or considerably reduced, by a suitable choice of the measuring frequency, by introducing notch filters or by carefully screening measuring devices. Interferences in the operational field of the detector and the measuring system are the Nyquist noise of the resistors (continuous frequency spectrum), the flicker noise ($1/f^2$ noise) stemming from the power supply, and the $1/f$ noise and drift of amplifiers. Owing to their frequency dependence, flicker noise and noise of amplifiers may be reduced by shifting the measuring frequency to a higher frequency region; apart from that, the effect of all noise components can be reduced by decreasing the noise bandwidth or increasing the measuring time. However, an extension of the measuring time is frequently hampered by limiting factors, such as the amount of sample available and insufficient long-term stability of the measuring system.

9.5.3.1. Detector Noise

9.5.3.1.1. Noise of Photomultiplier

The two fluxes of electrons originating at the cathode of a PMT are the dark current I_{DK}, caused by thermionic emission and the photoelectron current $I_K = \eta e \Phi$, caused by the photon flux Φ (s^{-1}), where η represents the quantum efficiency in electron per photon. These two currents generate an rms shot noise current i_K

$$i_K = \sqrt{2e \, (I_K + I_{DK}) \, \Delta f_n} \qquad (9.12)$$

Amplification of the secondary electron emission leads to additional shot noise due to the statistical variation in the secondary emission coefficient at the dynodes. This increase is mainly effected by the first two or three stages where small numbers of electrons are handled and can be taken into account by introducing a noise enhancement factor a. Thus, the noise current at the anode is

$$i_D = \sqrt{2ea^2G(I_A + I_{DA})\Delta f_n + \frac{4kT\,\Delta f_n}{R_L}} \qquad (9.13)$$

where G is the total gain, I_A the anodic current due to the photon flux, and I_{DA} the anodic dark current of the PMT. The last term is the Nyquist noise of the load resistor R_L. Low values of a can be obtained by operating the tube at a high first-stage gain. Typical values of a lie between 1.2 and 1.4; the smallest values are achieved when the operating conditions as quoted by the manufacturer are used. This applies in particular to the recommended stabilized voltage between cathode and first dynodes.

A reduction of the noise current can be achieved by lowering the dark current, for instance, by cooling the cathode of the PMT; the thermal dark current at the cathode obeys Richardson's law,

$$I_{DK} = 1.2 \times 10^2 \, QT^2 \exp\left(-\frac{1.16 \times 10^4 P}{T}\right) A \qquad (9.14)$$

where T is absolute temperature in Kelvins, Q the emitting area of the cathode in square centimeters and P the thermal work function in electron-volts of the cathode material.

9.5.3.1.2. Noise of Photodiode

Since in analytical AES only frequencies below 1 kHz must be normally measured, the PD should be operated in the photovoltaic mode. In this case, the shunt resistor appears to be the predominant source of noise. Therefore, the total noise current of a photovoltaic diode is

$$i_D = \sqrt{\left(m\frac{4kT}{R_{Sh}} + 2eI_A\right)\Delta f_n} \qquad (9.15)$$

where I_A is the photocurrent at the output of the PD due to the photon flux and m ($1 < m < 2$) is a correction factor applied in view of the discrepancy between the simple Shockley theory and the experimental behavior of the diodes [31]. If the PD must be operated in the photoconductive mode, the term in Eq. (9.15) due to Johnson noise of the resistor R_{Sh} has to be extended by an expression including the bias voltage U_R, viz.

$$i_D = \sqrt{m\left\{\frac{4kT}{R_{Sh}}\left[1 + \exp\left(\frac{eU_R}{mkT}\right)\right] + 2eI_A\right\}\Delta f_n} \qquad (9.16)$$

9.5.3.2. Noise of Measuring Systems

9.5.3.2.1. Noise of Conventional Measuring Systems

By a conventional measuring system, as often used in optical emission spectroscopy, we denote a system comprising a PMT (or single PD) in connection with a dc amplifier. For such a system, presented in Fig. 9.16, we shall derive a noise expression using the symbols defined in the caption of Fig. 9.16.

By the application of basic laws for electrical circuits, one obtains the following equation for the effective noise voltage at the amplifier output:

$$u_0^2 = \left| \frac{AZ_T Z_F}{Z_F - AZ_T} \right|^2 \left\{ i_D^2 + i_F^2 + i_A^2 + \frac{u_A^2}{|Z_T|^2} \right\} \quad (9.17)$$

where $1/Z_T = 1/Z_A + 1/Z_D + 1/Z_F$. Resistors of high values show an additional noise term to that of the Johnson noise which can best be represented by introducing a frequency-dependent flicker noise terms $c(\omega)$. Thus, the noise current i_F^2 of the feedback resistor is represented by the expression

$$i_F^2 = \frac{4kT}{R_F} [1 + c^2(\omega) \Phi^2] \Delta f_n \quad (9.18)$$

Inserting Eq. 9.18 into Eq. 9.17, and taking into account that the open loop gain of amplifiers often used are of the order of 10^5, we find the total noise voltage at the output to be

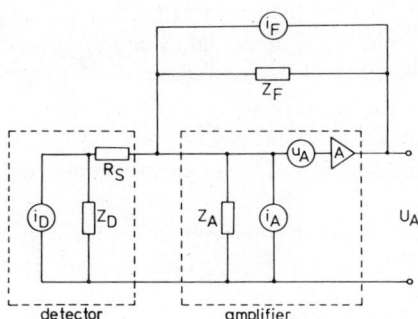

Figure 9.16. Electrical equivalent circuit for a detector/dc amplifier device: u_A, i_A, input noise voltage, input noise current of the amplifier; i_F, noise current of the feedback resistor; i_D, noise current of the detector; $Z_F = R_F - j\omega C_F$, feedback impedance; $Z_A = R_A - j\omega C_A$, input impedance of the amplifier; $Z_D = R_D - j\omega C_D$, effective output impedance of a detector; in the case of a PMT: $R_D = R_L$ (R_L, load resistor), $C_D = C_S$ (C_S, stray capacitance), in the case of a PD: $R_D = R_{Sh}$ (R_{Sh}, shunt resistor), $C_D = C_J$ (C_J, pn-junction capacitance); A, open loop gain of the amplifier.

$$u_0 = \frac{R_F}{\sqrt{1 + \omega^2 R_F^2 C_F^2}} \left\{ i_D^2 + \frac{4kT\Delta f_n}{R_L}(1 + c^2(\omega)\Phi^2) + i_A^2 \right.$$
$$\left. + u_A^2 \left[\left(\frac{1}{R_A} + \frac{1}{R_D} + \frac{1}{R_F}\right)^2 + \omega^2(C_A + C_D + C_F)^2 \right] \right\}^{1/2} \quad (9.19)$$

if we also consider that both impedances Z_F and Z_T are nearly of the same magnitude, and AZ_T is much greater than Z_F.

Equation 9.19 may serve to calculate the lower limit of the dynamic range of a detection and measuring system (DMS) or to compare different systems; it may also be used to determine the SNR which is given by

$$\text{SNR} = \left(\frac{I_k}{2ea^2\Delta f_n}\right)^{1/2} \left(1 + \frac{i_{\text{DMS}}^2}{2ea^2 G^2 I_k \Delta f_n}\right)^{-1/2} \quad (9.20)$$

where the noise current i_{DMS} is

$$i_{\text{DMS}}^2 = 2ea^2 G^2 I_{DK}\Delta f_n + \frac{4kT\Delta f_n}{R_D}(1 + c^2(\omega)\Phi^2) + i_A^2$$
$$+ u_A^2 \left[\left(\frac{1}{R_A} + \frac{1}{R_D} + \frac{1}{R_F}\right)^2 + \omega^2(C_A + C_D + C_F)^2 \right] \quad (9.21)$$

Normally in ICP-AES, the measurements are shot-noise limited—even at low concentration levels—owing to the relatively high background intensity, that is $i_{\text{DMS}}^2 < 2ea^2 G^2 I_k$. Therefore, the noise level as occurring in an analysis with ICP lies higher than the lower limit of the DMS noise level, for instance, the analytically available dynamic range is smaller than the range of the DMS. The opinion that the dynamic range of the DMS can be increased by changing the voltage supplied to the PMT or the amplification of the operational amplifier is not valid, but those changes may serve to fit the output signal to the readout unit (A/D-converter, card-recorder, etc.) in order to make use of its full range. Therefore, it sometimes seems as if the dynamic range could be expanded at its upper end.

This noise expression is currently applicable for both dc and ac measurement. If a filter stage is inserted between the detector and the amplifier, as in ac measurement, the detector noise is multiplied by the transmission function $T(\omega)$ of the filter. An electronic circuit, for instance, an integrator following the amplifier, generally does not contribute any additional noise.

In principle, Eq. 9.20 even holds for pulse counting when the current terms are converted to equivalent pulse rates, that is, $I_K = e\tau_K r_K$ and $i_{\text{DMS}}^2 = 2e^2\tau_{DK}r_{DK}$, where τ_K and τ_{DK} are transmission factor for a photoelectron or a dark electron pulse, respectively, passing undisturbed the measuring device (PMT, amplifier, discriminator), r_K and r_{DK} are the pulse rates for photoelec-

9.5. ERROR AND NOISE CONSIDERATIONS

trons or dark electrons, respectively. With regard to $\Delta f_n = (2 t_o)^{-1}$, where t_o is the measuring time, Eq. 9.20 can be rewritten as

$$\text{SNR} = (\tau_K r_K t_o)^{1/2} \left(1 + \frac{\tau_{DK} r_{DK}}{\tau_K r_K}\right)^{-1/2} \quad (9.22)$$

The opinion that pulse counting should only be applied if low intensities are to be measured is not valid. At high intensities it is normally necessary to attenuate the intensities to avoid pulse overlapping. This can best be done by using optical narrow-band filter by which no additional disturbances will be caused since the photon statistics of the radiation is not influenced. A positive side effect is that such filters additionally reduce the stray light.

Necessarily, the above expressions given for noise and SNR only take account of the noise sources of a DMS; other noise expressions have to be additionally introduced if the entire analytical equipment is considered.

Ingle and Crouch [32] have compared the total noise of dc measurement and pulse counting and concluded that under equivalent conditions, photon counting may yield a 5 to 22% higher SNR under shot noise limited conditions, if pulse overlap is negligible.

9.5.3.2.2. Noise of Optoelectronic Image Detector Systems

Optoelectronic image detector systems produce a noise output voltage analogous to Eq. 9.19. However, this noise voltage is normally covered by fixed-pattern noise, lag, and blooming [33, 34]. Fixed-pattern noise arises from the diode-to-diode variation in dark current and sensitivity. As lag is denoted as that fraction of charge which stays on the sensing element after one readout cycle and results in a memory effect. Blooming occurs through cross talk which means that charge spreading to adjacent sensing elements takes place.

The self-scanning photodiode arrays (SPDs) show a strong, but very regular fixed-pattern noise in both dark current and sensitivity, caused by the feedthrough from the clocking signals, irregularities in the monolithic silicon crystal wafer, and a high dark current, which essentially limits the possibility of measuring weak spectral lines by use of long integration times. The fixed-pattern noise as well as the dark current can be reduced by cooling the array below room temperature. Another possibility is the correction of the previously mentioned faults. This can be done only by a computer data acquisition system. The SPDs show no lag and their blooming, extending at least over two adjacent diodes, does not seriously reduce their capability for AES measurements.

The silicon vidicon tubes (SV) exhibit likewise a fixed-pattern noise and a high dark current level. These properties can be corrected in the same manner as already described. Beyond it, the SVs show lag as well as blooming, which is perhaps the most serious problem for application of SVs in analytical AES. The blooming effect grows with increasing spectral intensity and, under certain circumstances, the entire sensor may be bloomed out by intense radiation, which therefore should be kept away from the sensor. The lag characteristic depends on the intensity of the radiation. The image on the sensing elements generated by intense radiation may be erased to about 90 % in one readout cycle, while the image of weak radiation is only erased to a much smaller extent [35].

REFERENCES

1. P. Fellgett, Ph.D. thesis, Cambridge University, Cambridge (1951).
2. J. D. Winefordner, R. Avni, T. L. Chester, J. J. Fitzgerald, L. P. Hart, D. J. Johnson, and F. W. Plankey, *Spectrochim. Acta* **31B**, 1 (1976).
3. J. P. Keene, *Rev. Sci. Instr.* **34**, 1220 (1963).
4. P. W. J. M. Boumans, R. J. McKenna, and M. Bosveld, *Spectrochim. Acta* **36B**, 1031 (1981).
5. Photomultiplier tubes (brochure ref.: P001/fP70), EMI Electronics Ltd., Hayes, United Kingdom.
6. M. Cole and D. Ryer, *Electro Optical Systems Design*. Milton S. Kiver, June 1972, pp. 16–19.
7. W. E. R. Davies, *Rev. Sci. Instr.* **43**, 556 (1972).
8. R. P. Nanavati, *An Introduction to Semiconductor Electronics*, McGraw-Hill, New York (1963).
9. EG&G Inc., Salem, USA, Data sheet EG 75101.
10. P. W. J. M. Boumans, R. F. Rumphorst, L. Willemsen, and F. J. de Boer, *Spectrochim. Acta* **28B**, 227 (1973).
11. H. Bubert, W.-D. Hagenah, and K. Laqua, *Spectrochim. Acta* **34B**, 19 (1979).
12. Y. Talmi, *Anal. Chem.* **47**, 658A (1975).
13. Y. Talmi, "Optoelectronic Image Detectors in Chemistry, An Overview" in Y. Talmi, ed., *Multichannel Image Detectors*, American Chemical Society, Washington DC (1979).
14. EG&G/Reticon, Sunnyvale, USA, Data sheet 87202.
15. G. Horlick, *Appl. Spectrosc.* **30**, 113 (1976).
16. H. Bubert, W.-D. Hagenah, and D. Stüwer, *Spectrochim. Acta* **34B**, 289 (1979).
17. W. G. Nunn, R. E. Dessy, and W. R. Reynolds, "A Dual-Beam Photodiode Array Spectrometer System for Liquid Chromatographic Separation Methods Development" in Y. Talmi, ed., *Multichannel Image Detectors*. American Chemical Society, Washington DC (1979).
18. M. L. Franklin, C. Baber, and S. R. Koirtyohann, *Spectrochim. Acta* **31B**, 589 (1976).
19. T. E. Edmonds and G. Horlick, *Appl. Spectrosc.* **31**, 536 (1977).
20. Y. Talmi and R. W. Simpson, *Appl. Opt.* **19**, 1401 (1980).

21. S. W. McGeorge and E. D. Salin, *Spectrochim. Acta* **38B**, 633 (1983).
22. R. R. Williams and G. N. Coleman, *Spectrochim. Acta* **38B**, 1171 (1983).
23. S. W. McGeorge and E. D. Salin, *Prog. Analyt. Atom. Spectrosc.* **7**, 387 (1984).
24. J. M. Keane, D. C. Brown, and R. C. Fry, *Anal. Chem.* **57**, 2526 (1985).
25. S. W. McGeorge and E. D. Salin, *Spectrochim. Acta* **41B**, 327 (1986).
26. M. L. Franklin, G. Horlick, and H. V. Malmstadt, *Anal. Chem.* **41**, 2 (1969).
27. C. Th. J. Alkemade, W. Snelleman, G. D. Boutilier, B. D. Pollard, J. D. Winefordner, T. L. Chester, and N. Omenetto, *Spectrochim. Acta* **33B**, 383 (1978).
28. G. D. Boutilier, B. D. Pollard, J. D. Winefordner, T. L. Chester, and N. Omenetto, *Spectrochim. Acta* **33B**, 401 (1978).
29. C. Th. J. Alkemade, W. Snelleman, G. D. Boutilier, and J. D. Winefordner, *Spectrochim. Acta* **35B**, 261 (1980).
30. E. Voigtman and J. D. Winefordner, *Prog. Analyt. Atom. Spectrosc.* **9**, 7 (1986).
31. R. Müller, in W. Heywang and R. Müller, eds. *Grundlagen der Halbleiter-Elektronik,* 2nd ed. *Halbleiter-Elektronik*, Vol. 1, Springer-Verlag, Berlin, Heidelberg, New York (1975).
32. J. D. Ingle and S. R. Crouch, *Anal. Chem.* **44**, 785 (1972).
33. R. W. Simpson, *Rev. Sci. Instrum.* **50**, 730 (1979).
34. E. D. Salin and G. Horlick, *Anal. Chem.* **52**, 1578 (1980).
35. H. L. Felkel, Jr. and H. L. Pardue, "Simultaneous Multielement Determinations by Atomic Absorption and Atomic Emission with a Computerized Echelle Spectrometer/Imaging Detector System" in Y. Talmi, ed., *Multichannel Image Detectors*, American Chemical Society, Washington DC (1979).

INDEX

AAS (atomic absorption spectrometry), 2
  principle, 2f
  using ICP, 2
Absorption bands, O_2, 381
Absorption spectrometry, *see* AAS, atomic absorption spectrometry
AC amplifier, 554f
AC arc, *see* Arc, AC arc
Accuracy, ICP-AES, 85, 176f
AC measurement, 548
Addition technique, analyte, 21
  buffer, 21f
Additive interference, 22
Aerosol breakdown, 298
Aerosol carrier gas, *see* Carrier gas, ICP
Aerosol cooling, organic solvents, 222f
Aerosol desolvation, 322f, 324f
Aerosol droplet size, 302, 312f, 322
  Nukijama-Tanasawa equation, 302
Aerosol gas, *see* Carrier gas, ICP
Aerosol generation:
  efficiency, 202, 302, 311, 322f
  electrothermal, *see* ETV (electrothermal vaporization)
  using arc, 344f
    ETV, 328f
    laser, 346f
    spark, 344f
  see also Nebulization
Aerosol transport, 299f
AES:
  arguments, 86
  atomic emission spectrometry, 2
  literature guide, 34f
  multielement capability, 12
  principle, 2f
AFS (atomic fluorescence spectrometry), 2
  basic setup, 86f
  HCL-ICP-AFS, 86f
  ICP-AFS, 86f
    detection limits, table, 159

ICP-ICP-AFS, 88
LEAF, 88
  principle, 2f
Air-Ar ICP, 153f, 384
Air ICP, 153f, 246, 384
Al continuum, 368f
Alkalis:
  Ar ICP, detection limits, 146f
    interferences, 212f
  as buffer, 21f
  and DC arc, 46f
  and DCP, 46f
Alternating current arc, *see* Arc, AC arc
Ambipolar diffusion, 48
Amplifier:
  AC-, 554f
  DC-, 550
  lock-in-, 554f
  operational, specifications, 547
Analysis line(s):
  program for polychromator, 182
  see also Prominent line(s)
Analysis signal, *see* Line intensity
Analytical calibration function, 18
Analytical evaluation function, 18
Analytical range, 85, 181f
Angstrom, 4
a-parameter, Voigt profile, 397
Appearance of ICP, 74f
Arc:
  AC arc, 4. *See also* DCP
  cascade arc, 51
  copper arc, NBS, 360
  DC arc, 4, 24f
  DC arc methods, 15, 24f
  disk-stabilized arc, 46f
  flat top technique, 26
  gas-stabilized arc, 46f
  globule arc, 24
  Gordon arc, 26
  graphite arc technique, 26

Arc (*Continued*)
  for liquid analysis, 26
  NBS copper arc, 360
  powder insufflation technique, 26f
  powder sifting technique, 26f
  rotating arc, 49
  sampling, and ICP, 343f, 344f
  Scheibe-Rivas technique, 26
  seeded arc, 52
  for solids analysis, 24f
  vacuum cup technique, 26
  wall-stabilized arc, 46f
Ar ICP:
  and ETV, 328f
  H lines, 374
  injection methods, PN, 316f
  molecular bands, 373, 375f
  vs. N_2-Ar ICP, 71, 236f
  O lines, 374
  spectra from organics, 379f
Ar lines:
  Stark broadening, 374
  Stark shifts, 374
Ar-purged spectrometers, *see* VUV
Ar spectrum, 373f
Asundi bands, *see* CO bands, Asundi system
Atlas:
  AC and DC arc, 394. *See also* Wavelength table(s)
  Grimm glow discharge, 394
  ICP, 386, 395f, 406f
    coincidence profiles, 407f
    for geological purposes, 410
    spectra of multiple charged ions, 394
Atomic absorption, *see* AAS (atomic absorption spectrometry)
Atomic emission, *see* AES
Atomic fluorescence, *see* AFS (atomic fluorescence spectrometry)
Atomic spectra:
  ICP, infrared, 10, 146, 361, 411
  MIP, near-infrared, 61
  VUV, *see* VUV
  See also Atlas; Wavelength table(s)
Atomization, 2f
Auxiliary gas, ICP, 73

Babington nebulizer, *see* Nebulizer, Babington

Background:
  blank- *vs.* off-peak-, 110
  complex-, 414f
  gross- *vs.* net-, 110
  line overlap:
    correction, 505f, 509f
    noise, 441f
  from line wings, 368f, 487
  off-peak correction, 413, 506f
  on-peak correction, 413, 509f
  organic solvents, suppression, 222
  simple-, 222
  sloping-, 413, 506f
Background continuum, 186f, 363f, 486.
  See also Continuum
Background correction, 411f, 505f
  errors, 179f, 190f
  novel approach, 417f
  quantitative spectral analysis, 417f
  wavelength modulation, 507f
Background enhancement, 191, 486
Background equivalent concentration, 110, 409
Background noise, *see* Noise; RSDB
Background RSD, *see* RSDB
Background shift, *see* Background enhancement
Background signal, 19, 103, 190
Background spectrum, ICP, 190f, 363f, 486
Background subtraction, *see* Background correction
Bandpass, *see* Bandwidth
Band spectra:
  fluorides, 386
  oxides, 386
Bandwidth, 427f, 474f
  and aberrations, 428f
    detection limit, *see* Detection limit
    physical line width, 435
  grating *vs.* echelle spectrometer, 436f
  required, 431, 435f, 485
  table, 475
  see also Noise
BEC, background equivalent concentration, 110, 409
Beenakker cavity, 59f
Blank fluctuations, 441f
Blaze, *see* Grating
Boltzmann equation, 6, 198
Boundary layer sampling, ICP-MS, 89

INDEX

Boxcar integrator, 552f
Buffer addition technique, 21f

Ca-Al interference, ICP, 213f
Calibration:
　absolute, 20
　bracketing, 170
　quantitative analysis, 18f
　wavelength-, slew-scan spectrometer, 521f
Calibration curve, *see* Calibration function
Calibration errors, 178
Calibration function, 18f
　curvature, 178f, 190
　linear *vs*. logarithmic, 165
　logarithmic, 19, 179
　photographic emulsion, 123f
　types used in AES, 20
　see also Scheibe-Lomakin equation
Capacitively coupled microwave plasma, *see* CMP
Ca-phosphate interference, ICP, 213f
Carbon, *see* C lines; C_2 bands
Carbon arc, *see* Arc, DC arc
Carbon furnace, *see* Furnace, for AES
Carrier distillation technique, 25
Carrier gas, ICP, 73
　control, 300
　wetting, PN, 304
Carrier gas flow, ICP, optimization, 204f
Cascade arc, 51
Cavity, MIP, 59f
C_2 bands:
　Ar ICP, spectral scan, 382, 383
　　Mullikan system, 381f
　　Swan system, 383f
　Suppression, 222, 384
CCR (critical concentration ratio), 397f, 443
Chloroform, *see* Organic solvents
Chromatography:
　and ICP-AES, 311, 320, 342f
　see also GC
Chromatography detector, 343
Cleanup time, *see* Nebulizer
C lines:
　Ar ICP, 375
　organics, 218, 379
CMP:
　capacitively coupled microwave plasma 4, 56f

　and powder injection, 343
CN bands:
　Ar ICP, spectral scan, 382f
　suppression, 222, 384
CO bands, Asundi system, 384
Codding torch, MIP, 64
Coefficient of variation, *see* RSD
Coincidence, 6, 386f
　noise caused by, 441f
　resultant profile, 390f
　see also Line coincidence; Line overlap
Coincidence profiles:
　atlas, ICP, 407f
　examples, 408
Coincidence table(s), ICP, 395f
　example, 400
Combined sources, 34
Complex spectra, *see* Line-rich spectra
Compromise conditions, 71, 128, 189, 194f
　Ar ICP, *vs*. N_2-Ar ICP, 235
　inorganic samples, 204f
　organics, 217
　rules, 216f
　typical, 219
　and detection limits, 194f
　interferences, 194f
Compromise operating conditions, *see* Compromise conditions
Concentration in source, 6
Concentric nebulizer, *see* Nebulizer
Configuration factor, ICP torch, 259f, 286
Configuration of ICP, 74f
Continuous background, *see* Background; Continuum
Continuum:
　Al, Mg, 368f
　ICP, 186f, 363f, 486
　from sample components, 367f
Convolution, 190
　Doppler-Lorentz widths, formula, 426
　see also Deconvolution
Coolant gas, ICP, 73
Cooling:
　aerosol, *see* Aerosol cooling, organic solvents
　detector, *see* Detector
Copper arc, NBS, 360
Copper spark technique, 27
Cr, spectral scan, 388
Critical concentration ratio, 397f, 443

Cross-flow nebulizer, *see* Nebulizer
Cyclone chamber, 344

Dark current:
  PD, *see* PD, leakage current
  PMT, 539
    cooling, 541, 560
Dark current noise, *see* Noise
Data processing, 556f
DC amplifier, 550
DC arc, *see* Arc, DC arc
DC integrator, 551f
DC measurement, 546f
DCP:
  analytical characteristics, 54
  current-carrying *vs.* current-free, 45
  direct current plasma, 4
  excitation mechanisms, 54
  inverted V configuration, 53f
  inverted Y configuration, 54f
  rotating arc, 49
  three-electrode DCP, 54f
  Valente-Schrenk DCP, 54f
  see also Arc, AC arc
DC plasma, *see* DCP
Deconvolution:
  effective line profiles, 431f
  see also Convolution
Densitometer, 14, 515
Derivative spectra, *see* Wavelength modulation, background correlation
Desolvation of aerosol, 322f, 324f
Detection:
  lock-in, *see* AC measurement
  and measurement, 538f
  phase-sensitive, *see* AC measurement
  photographic *vs.* photoelectric, 162
  sequential *vs.* simultaneous, 189
  synchronous, *see* AC measurement
  see also Measurement
Detection limit:
  and additives, 152f
  bandwidth, *see* Spectral resolution
  limit of determination, 163f, 442f
  matrix, 152f
  spectral interference, 152f, 423f, 438f
  spectral resolution, 129, 137, 148f, 160f, 438f
  alkalis, 146f

breakdown, 129, 137, 148f, 361
  Laqua theory, 160f
  concept, 102f
  data, historical, 127
    Ar ICPs, 127f
  vs. detection power, 102, 446f
  effect of rf frequency, 291f
  function of SBR and RSDB, 118f
  high-efficiency ICPs, table, 278f
  HR profit, 444f, 502f, 504f
  hydride technique, 148f
  vs. PN, 338
  ICP-AES, outline, 84
    Ar ICP, PN, high resolution, table, 138f
    PN, oil, 233
    PN, periodic table, 142
    PN, polychromator, table, 138f
    PN, table, 130f
    USN, table, 138f
    with ETV, 330f
    molecular gas ICPs, 153f
      table, 155f
  ICP-AFS, table, 159
  ICP-MS, table, 160
  laser-ICP, 346f
  mass/volume *vs.* mass/mass, 128
  in presence of blank, 126f, 441f
  real samples, 152f
  slit width optimization, 119f
  statistical interpretation, 103f
  PN *vs.* USN, 127, 324f
  SBR-RSD *vs.* SNR approach, 108
  "true," 153f
  USN, 322f
  VUV lines, 142f
    table, 146
  $z\sigma$, value of z, 102, 106f, 125f
Detection power:
  vs. detection limit, 102, 446f
  see also Detection limit
Detector:
  element-selective-, 61, 343
  element-specific-, 61, 343
  GC, 61, 343
  OID, cooling, 563
  PD, cooling, 563
  PMT, cooling, 560, 561
  requirements, 537f
  see also Photodetector, requirements

INDEX

Detector noise, *see* Noise
Diffraction width, 428
Direct analysis of solids, AES, 24f, 334f, 343f
Direct current arc, *see* Arc, DC arc
Direct current plasma, *see* DCP
Direct reading spectrometer, *see* Polychromator
Direct sample insertion devices, 334f
Direct solids analysis, remote sampling, 344f
Direct solids sampling, 24f, 334f, 343f
Discrete nebulization, *see* Injection methods
Discrete sampling, *see* ETV (electrothermal vaporization); Injection methods
Disk-stabilized arc, 46f
Dispersion, 471f
   angular-, 471f
   linear-, 471f
   reciprocal linear-, 471f
   table, 473
Dissociation energy, 7
   oxides, table, 389
Dissociation equilibria, 197f
DMS (detection and measuring system), 537
   noise, 559
Doppler profile, *see* Line profile
Drift:
   of amplifiers, 559
   in DC measurement, 547
Droplet size, aerosol, 302, 312f, 322f
Dynamic range:
   AC measurement, 548
   DC measurement, 546f
   ICP-AES, 20, 85, 178f, 485
   PD array, 544
   photon counting, 549f
   PMT, 539
   powder injection analysis, 344
   and spectral resolution, 187

Easily ionizable element, *see* EIE
Echelle:
   description, 498f
   vs. grating, bandwidth, 436f
Echelle monochromator:
   for analysis of ICP spectra, 405
   with cross dispersion, 498f

line width measurements, 432f
   with moving detector, 526f
   with predisperser, 118, 498f
   bandwidth, 433
Effective line profile, *see* Line profile
Effective line width, 427
   and detection limit, 150f
EIE:
   as buffer, 51
   and DC arc, 46f
   and DCP, 46f
   easily ionizable element, 46
Electrodeless discharge, *see* MIP
Electrothermal aerosol generation, *see* ETV (electrothermal vaporization)
Electrothermal nebulization, *see* ETV (electrothermal vaporization)
Element coverage, ICP-AES, 83, 188f
Element-selective detector, 61, 343
Element-specific detector, 61, 343
Emission spectrometry, *see* AES
Emission spectrum, *see* Spectrum
Emulsion calibration function, 123f
End-on viewing:
   of ICP, 78, 283
   vs. side-on viewing, ICP, 78
Enthalpy, *see* Heat content, atomic *vs.* molecular gas
Error:
   in measurement, random, 559f
   systematic, 558f
Ethanol, *see* Organic solvents
ETV (electrothermal vaporization), 328f
   filament in furnace atomization, 332f
   filaments, 328f
   graphite rod, 331f
   graphite tube, 331f
   graphite yarn, 331f
   with ICP, detection limits, 330f
   memory effects, 331f
   metal boats, 328f
   tantalum filament, 329f
   wire loops, 328f
Evaporation factor, organic solvents, 227f
Evolution techniques, 342f
Excitation, 2
   and ionization, 197f
   schematic representation, 197
   separate from volatilization, 34
Excitation energy, 6

Excitation potential, 6
Excitation sources:
  for AES, 4
  and sample introduction, 24f
Excitation temperature:
  and frequency, 291
  high-efficiency ICPs, 276f, 288
Exploding conductors, for AES, 4, 32
Exploding wire or foil, 4, 32

FAAS (flame atomic absorption spectrometry), 297
FAES (flame atomic emission spectrometry), 297
Fassel torch, 79
Fe spectrum:
  chart, reproduction, 17
  charts and atlas, 394
  ICP, 410
FET:
  field-effect transistor, 537
  see also Operational amplifier, specifications
FIA (flow injection analysis), 319f
Fiber optics, 162
Filter, band blocking-, 495
Find peak routine, see Peak find routine
Flame, 4, 32f
  multielement capability, 11
  temperatures, 11
Flat top technique:
  arc, 26
  spark, 27
Flicker noise, see Noise; RSD; RSDB
Flow injection analysis, see FIA (flow injection analysis)
Flow injector, 321
Fluorescence spectrometry, see AFS (atomic fluorescence spectrometry)
Fluorides, band spectra, 386
Forward power, see Power
Fourier transform spectroscopy, 410, 529
  line width measurements, 432
Fractional distillation, see Selective volatilization
Free spectral range, 473f
Frequency:
  and ICP characteristics, 290f
  and wavelength, 4

Frequency effect, high-efficiency ICPs, 292f
Fritted disk nebulizer, see Nebulizer
FTT (fast Fourier transform), 166
Furnace, for AES, 4, 32
FWHM, (full width at half maximum), 389

Gas consumption:
  Ar ICP, organics, 219
  vs. N_2-Ar ICP, 81
  molecular gas ICPs, 154
  torches for low-, 273f
  see also High-efficiency ICPs
Gas flows, ICP, terminology, 72
Gases and vapors, introduction into ICP, 341f
Gas-stabilized arc, 46f
GC:
  detector, 61, 343
  gas chromatography, 61
  and ICP-AES, 311, 320, 342f
  see also Chromatography
General survey analysis, arc, 15f, 24
Ghosts, see Grating
Globule arc, 24
Glow discharge, 4, 29
GMK nebulizer, see Nebulizer
Grating:
  apodization, 492
  blaze, 475f
  concave grating mounts, 477f
  equations, 469f
  ghosts, 490f
  grass, 490
  holographic-:
    efficiency, 492
    production, 491f
    properties, 491f
  ruled-:
    efficiency, 476f
    imperfections, 490f
    production, 490f
  satellites, 490
  scatter, 490
  stray light:
    perfect grating, 492
    ruled vs. holographic grating, 487f
Grating spectrometer, see Spectrometer
Greenfield torch, 78
Grimm glow discharge, 29
  spectral atlas, 394

INDEX

Gross line signal, 19, 103
Gross signal, 19, 103

Hagen-Poiseuille law, 302
Halogen emission, reduced pressure ICP, 272. *See also* Nonmetals
Hard line, *see* Spectral line(s)
H_β line, 383
HCD (hollow cathode discharge), 29
HCL (hollow cathode lamp), 87
HCL-ICP-AFS, 87f
He, as ICP support gas, 272
Heat content, atomic *vs.* molecular gas, 236f
Hg determination, evolution technique, 342
High-efficiency ICPs, 273f
   detection limits, table, 278f
   frequency effect, 292f
   interferences, 274f
   organic solvents, 274f
   specifications, 276f
High frequency, *see* rf
High-power ICP, *vs.* low-power ICP, 71, 236f
High resolution, *see* HR
High spectral resolution, *see* HR
History ICP, 69
H line(s):
   Ar ICP, 374
   spectral scan, 375
   stark broadening, 374
Hollow cathode discharge, 4, 29f
Holographic grating, *see* Grating
HR:
   dual polychromator setup, 450f
   echelle spectrometer, 497f
   Fabry-Perot interferometer, 497
   grating spectrometer, 497f, 500f
   high resolution, 441f
   Michelson interferometer, 497
   profit, 444f, 502f, 504f
   and spectral interference, 423f, 438f, 485, 502f
   spectrograph, 500
   and wavelength positioning errors, 452f
Hydride generation, 336f
   interferences, 339f
   vs. PN, detection limits, 338
   recoveries, 339f
Hydride generator, 336f
   flow-cell type, 336f
   out-freezing device, 336
Hydrogen, *see* H line(s)

ICP:
   air ICP, 153f, 246, 384
   analytical performance, outline, 82f, 86
   analytical range, 85, 181f
   appearance, 74f
   and arc sampling, 344f
   chromatography, 311, 320, 342f
   ETV, *see* ETV
   GC, 311, 320, 342f
   powder injection, 344
   spark sampling, 345f
   atlas, *see* Atlas
   background, *see* Background
   basic setup, 72
   configuration, 74f
   core, 75
   for crystal growth, 75
   detection limits, *see* Detection limit
   direct solids sampling, 334f, 343f
   dynamic range, 20, 85, 178f, 485
   element coverage, 83, 188f
   Fe spectrum, 410
   gas flows, terminology, 72
   halogen detection, 272. *See also* Nonmetals
   high-efficiency, *see* High-efficiency ICPs
   high-power- *vs.* low-power-, 71, 236f
   history, 69
   hydride techniques, *see* Hydride generation; Hydride generator
   inductively coupled plasma, 2
   infrared atomic spectra, 10, 146, 361, 411
   introduction of gases and vapors, 341f
   ionic line advantage, 127, 360
   laser ablation, 347f
   laser microprobe, 346f
   limit of determination, *see* Limit of determination
   line coincidence tables, 395f
   literature reviews, 92f
   low-flow ICP, *see* High-efficiency ICPs
   low-power ICP, *see* High-efficiency ICPs
   mixed gas-, 246. *See also* Molecular gas ICP
   molecular gas, *see* Molecular gas ICP
   multielement capability, 83f, 188f

ICP (*Continued*)
  noise power spectra, 167f
  O_2-Ar ICP, 384
  O_2 ICP, 384
  optimization, *see* Optimization
  pencil plasma, 88
  plasma rotation, 167f
  power balance, 264
  precision, 162f
  principles, 72
  reduced pressure ICP, 272f
  sample introduction techniques, 297f
  sample size, 82f
  sample types, 82
  selectivity, *see* Selectivity
  spatial profiles, *see* Spatial profiles of emission
  tail, 75
  tear drop ICP, 70
  torch, *see* Torch
  toroidal, 70f, 75f
    vs ellipsoidal, 70
  VUV spectra, *see* VUV
  wavelength tables, *see* Atlas; Wavelength table(s)
ICP-AAS (ICP atomic absorption spectrometry), 2
ICP-AES (ICP atomic emission spectrometry), 2
  detection limits, *see* Detection limit
ICP-AFS (ICP atomic fluorescence spectrometry), 86f
  detection limits, table, 159
  LEAF, 88
ICP characteristics, and frequency, 290f
ICP-ICP-AFS, 88
ICP lines, physical widths, 431f
ICP-MS (ICP mass spectrometry), 89f
  detection limits, table, 160
  performance, 91
ICP operating conditions, *see* ICP parameters
ICP parameters, 75
  Ar ICP, typical, 219
  molecular gas ICPs, table, 155f
  optimization, *see* Optimization
  values, table, 81
  see also Compromise conditions
ICP power, *see* Power
ICP spectra:
  analysis, 405f

  infrared, 10, 146, 361, 411
  organics, 218
IDES (image dissector emission spectrometry), 12
Image dissector tube, 12
Inductively coupled plasma, 4. *See also* ICP
Infrared atomic spectra:
  ICP, 10, 146, 361, 411
  MIP, 61
Injection methods, 302, 315f
  FIA, 319f
  PN, Ar ICP, 316f
  PN, N_2-Ar ICP, 316f
Instrumental width, *see* Bandwidth
Instrumentation for AES, spectroscopic, 12f, 466f
Integrator:
  boxcar-, 552f
  DC-, 551f
Intensity:
  absolute, 6
  relative, 6
  spectral line, 6
Interelement effects, *see* Interference(s)
Interference(s):
  additive *vs.* multiplicative, 22
  CA-Al, ICP, 213f
  CA- phosphate, ICP, 213f
  classification, 22
  and compromise conditions, 194
  high-efficiency ICPs, 274f
  hydride generation, 339f
  in multicomponent system, corrections, 23f, 510f
  multiplicative, 22, 511
    ICP, 212f
  optimization, 204f
  simplex optimization, 230f
  solute vaporization, ICP, 212f
  spark ablation and ICP, 346
  spectral:
    correction, 509f
    vs. nonspectral, 22
    types, 363f
    see also Spectral interference
  suppression, 20f
  USN, 322f
Interference coefficient, multicomponent corrections, 421
Interferent, 20

INDEX

Interferometry, line width measurements, 431f
Intermediate gas, ICP, 73
  ICP, and carbon deposits, 217, 260
Internal standard, 20
  ICP-AES, 170f
Ionic line advantage, ICP, 127, 360
Ionization:
  equilibria, 197f
  stages, 7
Ionization buffer, 22
Ionization energy, 7
  table, 9
Ions, multiply charged, spectral atlas, 394
Isoformation technique, 22
Isotope analysis, hydrogen, ICP, 272
Isotope dilution, ICP-MS, 91
IUPAC recommendations, nomenclature, 35

Kerosene, *see* Organic solvents
Kranz plasma jet, 48f

Laser ablation:
  and ICP, 343f, 346f
    detection limits, 347f
  of solids, for AES, 30f
Laser evaporation, *see* Laser ablation
Laser-excited atomic fluorescence, 88
Laser microprobe, ICP, *see* Laser ablation
LEAF (laser-excited atomic fluorescence), 88
Level population, 6
Limit of detection, *see* Detection limit
Limit of determination, 84, 152f, 163f, 442f
  HR profit, 444f, 502f, 504f
  and spectral interference, 423f, 438f
  detection limit, 442f
Limit of guarantee for purity, 103f
Limit of identification, 103f
Line broadening, 184f, 423f
  self-absorption, 184f
Line coincidence, 6, 386f
  noise caused by, 441f
  resultant profile, 390f
  see also Line overlap
Line intensity, 6
Line overlap, 6, 386f
  background RSD, 441f
  computer simulation, 444f
  noise, 441f

and peak find routine, 454f
see also Line coincidence
Line profile, 184f, 423f
  Doppler, 426
  effective, 427
  instrumental, 427
  physical, 423f
  Lorentz, 426
  and self-absorption, 184f
  Voigt, 426f
Line-rich spectra, 12
  background, 368f
  ICP optimization, 236, 372f
  line selection and line wings, 371f
  number of lines, table, 386
Line selection, 361f
  apparatus constraints, 12f, 362
  criterion, 448
  and line wing background, 371f
  selectivity, 448
  spectral interference, 190f, 386f, 485
  polychromator, 514
  use of spectral scans, 402
Line shape, *see* Line profile
Line signal, *see* Line intensity
Line width, 423f
  measurements, 431f
  see also Effective line width; Physical line width
Line wing(s), 186, 236, 487
  background, 367f
  implications, 371f
  in line-rich spectra, 368f
  vs. line shoulder, 390
  phenomena, 367f
Liquid analysis:
  arc methods, 26
  spark methods, 26f
Local thermal equilibrium, LTE, 7f
Lock-in amplifier, 554f
Lock-in measurement, 548
Lorentz profile, *see* Line profile
Low-flow ICP, *see* High-efficiency ICPs
Low-power ICP:
  vs. high-power ICP, 71, 236f
  see also High-efficiency ICPs
  LTE (local thermal equilibrium), 6
Lyman ghosts, *see* Grating, ghosts

Magnetohydrodynamic pinch, 48
MAK nebulizer, *see* Nebulizer

Mass action law, 198
Mass spectroscopy, *see* MS (mass spectroscopy)
Matrix effect, *see* Interference(s)
Measurement, 545f
  AC-, 548
  DC-, 546f
  errors, 558f
  lock-in-, 548
  photon counting, 118, 549f
  pulse-, 118, 549f
Medium resolution, *see* MR (medium resolution)
Meinhard nebulizer, *see* Nebulizer
Memory effects:
  ETV, 331f
  USN, 327
Methanol, *see* Organic solvents
Mg continuum, 368
MIBK:
  definition, 218
  see also Organic solvents
Microcomputer, in measuring systems, 556f
Microliter sampling, *see* ETV (electrothermal vaporization); Injection methods
Microphotometer, 14, 515
Micro torch, *see* Torch
Microwave induced plasma, *see* MIP
Microwave plasmas, 45, 56f. *See also* CMP; MIP
MINDAP, definition, 62
Mini torch, *see* Torch
MIP, 4, 58f, 343
  microwave induced plasma, 4, 48f, 343
  near-infrared lines, 61
  surfatron, 63
  toroidal, 64
Mixed gas ICP, 246
Modulation:
  of radiant beam, 548
  see also Wavelength modulation
Molecular bands, 5
  analytes, 386f
  Ar ICP, 373, 375f
    organics, 218
  molecular gas ICP, 384f
  see also Band spectra
Molecular gas ICP, 153f, 240f
  detection limits, 153f

  molecular bands, 384f
  O_2 outer gas, 384
  power requirements, 236f
Monochromator, 471
  characterization, 13
  computer-controlled-, 516f
  Czerny-Turner-, crossed, 520
  predispersing-, *see* Predisperser
  programmable-, 516f
  slew-scan-, 13, 516f. *See also* Spectrometer(s)
Most sensitive lines, 360
  types and wavelengths, 10
  use of, in AES, 6
  see also Prominent line(s)
MR (medium resolution), 441f
MS (mass spectroscopy), 89f
  ICP-MS, detection limits, table, 160
  spark source-, 89
Mullikan system, *see* C_2 bands
Multichannel averager, 554
Multicomponent analysis, multiple interferences, 23f
Multicomponent corrections, spectral interference, 23f, 417f, 510f
Multielement analysis, compromise conditions, 194f
Multielement capability:
  AES, 12
  ICP-AES, 83f, 188f
Multiplicative interference, 22, 511

N_2-Ar ICP, 153f, 236f, 240f
  vs. Ar ICP, 71, 236f
  injection methods, PN, 316f
  sensitivity advantage, 239, 246
  see also ETV (electrothermal vaporization); Molecular gas ICP
N_2^+ bands, spectral scan, 380
  Ar ICP, 375f
NBS (National Bureau of Standards), 128
NBS copper arc, 360
N_2-cooled ICP, *see* N_2-Ar ICP
N determination, evolution technique, 342
Nebulization:
  discrete sampling, *see* Injection methods
  droplet size, 302, 312f, 322f
  efficiency, 302, 311
    and carrier gas flow, 202

USN, 322f
FIA, *see* Injection methods
flicker noise, 167f, 312f
PN, aerosol gas wetting, 304
pneumatic, 138, 299f
spray chamber, 170, 299f. *See also* Spray chamber
stability:
  noise power spectra, 167f
  PN, 303f, 307, 312f
  USN, 327
see also Nebulizer
Nebulization effects, 167f, 302
Nebulizer:
  Babington, 283, 299, 310f
  blockage, 303f, 307
  cleanup time:
    PN, 304
    USN, 327
  concentric-, 299, 303f
    and ETV, 332
  cross-flow-, 299, 305f
  drift, *see* Nebulization, stability
  fritted disk-, 299, 311f
  GMK-, 310f
  MAK-, 308f
  Meinhard-, 303f
  pneumatic-, 299f
  PN:
    assessment, 314f
    vs. USN, detection limits, 324f
    tolerance, salt content:
      PN, 303f, 307f
      USN, 327f
  USN, assessment, 327f
  V-groove-, *see* Nebulizer, Babington-
  washout time:
    PN, 304
    USN, 327
Net line signal, 19, 103
Net signal, 19, 103
NH band(s)
  Ar ICP, 375f
  for N determination, 342
Nitrogen, *see* N determination, evolution technique; N_2-Ar ICP; N_2-cooled ICP; N_2 purged spectrometers
NO bands:
  Ar ICP, 375f
  spectral scan, 379

Noise:
  of amplifiers, 559
  blank sample, 441f
  dark current-, PMT, 118, 539
  detector-, 113f, 559f
  $1/f$-, 167, 546, 559
  $1/f^2$-, 559
  fixed pattern-, 563f
  flicker-, 113f, 115f, 559, 561
    line overlap, 441f
    nebulizer effects, 312f
  high-frequency-, *see* Noise, flicker-,
  image detector system, 563f
  Johnson-, 114, 560
  low-frequency-, *see* Noise, $1/f$-,
  in measurement, 558f
  of measuring system, 561f
  Nyquist-, 114, 559f
  OID, 563f
  PD, 560f
  photon-, 113f, 559f
  PMT-, 113f, 559f
  proportional-, *see* Noise, flicker-,
  resistor-, 559f
  shot-, 113f, 559f
  thermal-, 114, 560
  total-, 114, 561
  whistle-, 167
  white-, 167
  see also Noise power spectra; RSDB
Noise analysis, 167f
Noise bandwidth, 113
  AC amplifier, 554
  boxcar integrator, 554
  DC integrator, 552
Noise power spectra, 166f
Nonmetals:
  AES detection, 10, 61, 143f, 146, 272f, 411
  infrared ICP lines, 10, 146, 361, 411
  VUV lines, 10, 143f
Norm temperature, 195f
N_2-purged spectrometers, *see* VUV
Nukijama-Tanasawa equation, 302
Number density, 6

O_2 absorption bands, spectral scan, 381
O_2-Ar ICP, 384
Observation height, 202
  optimization, 204f

OES (optical atomic spectrometry), 2
Off-peak background correction, 413, 506f
OH band interference:
   Al line, spectral scans, 456f
   Bi lines, spectral scans, 460f
   V line, spectral scans, 377f
OH bands:
   Ar ICP, 375f
   spectral scans, 377f, 380, 456f, 460f
O_2 ICP, 384
OID:
   blooming, 563f
   cooling, 563
   lag, 563f
   noise, 563f
   optoelectronic image detector, 543f
   see also PD; SV (silicon vidicon tube)
Oil analysis:
   Ar ICP, 218
   detection limits, 233
   simplex optimization, 232
   spark rotrode technique, 271
O lines, Ar ICP, 374
On-peak background correction, 413, 509f
Operating conditions:
   optimization, see Optimization
   see also ICP parameter
Operational amplifier, specifications, 547
Optical emission spectrometry, see OES
       (optimal atomic spectrometry)
Optimization:
   Ar ICP, 202f
      inorganic samples, 204f
      organics, 217
      organic solvents, 202f, 218f
   compromise conditions, 194f, 235f, 246f
   guidelines, 195
   N_2-Ar ICP, 235f, 240f, 246f
   samples with line-rich spectra, 236, 372f
   simplex-, 217, 229f
      Ar ICP vs. N_2-Ar ICP, 232f
      composite response function, 230
      interferences, 230f
      MAK torch, 289
      N_2-Ar ICP, 230f, 235f, 240f
   single-element-, 194
   see also Compromise conditions
Optoelectronic image detector, OID, 543f
Order sorter, spectrograph, 432

Organics:
   aqueous solution, optimization, Ar ICP, 217
   carbon deposits in torch, 217, 260
Organic solvents:
   aerosol cooling, 222f
   Ar ICP:
      optimization, 202f, 218f
      plasma solvent load, 222f
      sample feed rate, 219f
   carbon deposits in torch, 217, 260
   evaporation factor, 227f
   high-efficiency ICPs, 274f
   oxygen addition, 222, 384
   sample feed rate, 303
   spectra, Ar ICP, 218, 379f
Outer gas, ICP, 73
Oxides:
   band spectra, 386
   dissociation energies, table, 389
Oxygen, see O_2 absorption bands, spectral scan; O_2-Ar ICP: O_2 ICP; O lines, Ar ICP

Parameters, ICP, see ICP parameters
Particle size:
   nebulization, 302, 312f, 322f
   powder injection analysis, 344
   spark ablation, 345f
Particulate analysis, laser-ICP, 347
Partition function, 6
PD:
   alkali detection, 147
   array, 528, 543f
      dynamic range, 544
      quantum efficiency, 544
      self-scanning, 543f
   cooling, 563
   leakage current, 542
   noise, 560f
   photoconductive mode, 542
   photodiode, 12, 541f
   photovoltaic mode, 542
   single-, 541f
   triptych, 147
Peak find routine, 392, 521f
   and line overlap, 454f
      spectral resolution, 505
   moving window, 522

peak area fitting, 522f
Savitsky-Golay fitting, 524
side line indexing, 522
simple, 521f
Peak search routine, *see* Peak find routine
Pencil plasma, ICP, 88
Peristaltic pump, *see* Pump, peristaltic
Persistent lines, 359
Phase-sensitive detection, *see* AC measurement
Photodetector, requirements, 537f
Photodiode, *see* PD
Photographic detection, 14f, 515
    Fe spectrum, plate, 17
Photographic parameter P, 123
Photographic *vs.* photoelectric detection, 162
Photomultiplier, *see* PMT
Photon counting, 118, 549f
Photon noise, *see* Noise
Physical line width:
    and detection limit, 150f
    measurements, 431f
    numerical values, 187, 431f
Pinch:
    magnetohydrodynamic-, 48
    thermal-, 48
Plasma:
    definition, 5
    transferred, 46
    UHF plasma, 57
Plasma gas, ICP, 73
Plasma jet, 46f
    Kranz plasma jet, 48f
    see also DCP
Plasma solvent load, Ar ICP, organic solvents, 222f
Plasma sources:
    definition and types, 11
    multielement capability, 33
Plasma stability curve, torch design, 286f
Plasmatron, 46, 343
Plastic cup technique, spark, 27
PMT, 539f
    dark current, 539
        cooling, 541, 560
    dynamic range, 539
    noise, 113f, 559f
    photomultiplier tube, 10

    properties, table, 540
    quantum efficiency, 539
    solar blind PMT, 496
PN (pneumatic nebulization), 299f
Pneumatic nebulization, *see* Nebulization, PN, aerosol gas wetting
Pneumatic nebulizer, *see* Nebulizer, PN
Polychromator, 470f
    characterization, 13
    conventional mounts, 512f
    Echelle mounts, 513f
    line selection, 514
    moving detector, 515, 525f
    n + 1 channel, 515
    program of analysis lines, 182
    stability, 513f
    wavelength scanning, 514, 515, 525f
    see also Spectrometer
Population, *see* Level population
Porous cup technique, spark, 27
Powder analysis:
    using laser-ICP, 346f
        spark-ICP, 346
    see also Arc; Powder injection
Powder injection, arc, 26f
    dynamic range, 344
    ICP, 343f
    particle size, 344
    segregation, 344
Powder insufflation, *see* Powder analysis; Powder injection
Powder sifting technique, arc, 26f
Power:
    high *vs.* low power, 71, 236f
    optimization, 204f
    and RSDB, 119
    torches for low power consumption, 273f
    see also High-efficiency ICPs
Power balance, ICP, 264
Power of detection, *see* Detection power
Power requirements:
    Ar ICP, organic solvents, 232
    Ar ICP *vs.* N_2-Ar ICP, 81
        organics aqueous solution, 219
    atomic *vs.* molecular gas ICP, 236f
Precision, 162f
    effect of signal integration period, 166f
    ICP-AES, 85f, 162f
    internal standard, 20, 170f

Precision (*Continued*)
 nebulizer stability, 312f
Predisperser, 499f
Predispersing monochromator, 499f
Profile, *see* Line profile; Spatial profiles of emission
Prominent line(s), 359f
 alkalis, 147
 infrared, 10, 146, 361, 411
 listings, 360f, 406
 tables in this work, 130f, 138f, 146, 147, 155f
 VUV region, 146
 wavelength distribution, 144f
 see also Most sensitive lines
PTFE (polytrifluorethylene), 309
Pulse measurement, *see* Photon counting
Pump, peristaltic, 300
Purged spectrometers, *see* VUV

Qualitative analysis, 6, 15f
 photographic recording, 14
Quantitative analysis, 6
 calibration, 18f
 interferences, 18f
Quantitative spectral analysis, background correction, 417f
Quantum efficiency:
 PD, 542
 PD array, 544
 PMT, 539

Radiation detection, 538f
 photographic *vs.* photoelectric, 162
Radio frequency, *see* rf (radio frequency)
Raies ultimes, 359
Readout, *see* Measurement
Recombination continuum, *see* Continuum
Rectification, 554
Reentry spectra, 492f, 520
Reference element technique, 20f, 170f
Reference sample, 18
Reference signal, 20f, 170f
Refractor plate:
 in polychromator, 514
 for wavelength shift, 507f
Remote sampling, direct solids analysis, 344f
Resolution:
 high-, *see* HR

medium-, *see* MR
see also Bandwidth; Spectral resolution
Resolving power:
 experimental-, 474f
 theoretical-, 474
 see also Bandwidth; Spectral resolution
Reviews, ICP literature, 92f
rf current, 70
rf frequency, *see* Frequency
rf plasmas, 45
rf power, *see* Power
rf (radio frequency), 45
Rotating arc, 49
Rotating disk technique, spark, 27
Rotating platform technique, spark, 27
Rotation of ICP, 167f
Rotrode technique, spark, 27f
Rowland circle, 478
Rowland ghosts, *see* Grating, ghosts
RSD:
 of background signal, *see* RSDB
 of blank, 441f
 in case of line overlap, 441f
 of net signal, 163f, 312f, 441f
 relative standard deviation, 84
 see also Noise
RSDB, 110f
 detector noise, 113f, 559f
 flicker noise, 113f, 115f, 441f, 559f
 and ICP power, 119
 photographic detection, 123f
 PMT, 110f
  numerical values, 112, 115f
  theoretical relationships, 113f
 RSD of background, 108f
 shot noise, 113f, 115f, 559f
 trade-off with SBR, 119
Ruled grating, *see* Grating

Saha equation, 198
Salt content tolerance, nebulizers, *see* Nebulizer
Sample feed rate:
 control, 203
 organic solvents, 219f, 303
Sample introduction, 5f, 296f
 and excitation source, 24f
 see also Direct solids sampling

Sample introduction techniques:
  comparison, 349f
  ICP, 74, 296f
Sauter droplet diameter, 302
Savitsky-Golay fitting, 524
SBR:
  and bandwidth, see Detection limit, breakdown
  signal-to-background ratio, 110
  and spectral resolution, see Detection limit, breakdown
  trade-off with RSDB, 119
Scheibe-Lomakin equation, 18, 179f
Scheibe-Rivas technique, arc, 26
S determination, evolution technique, 342
Seeded arc, 52
Segregation, powder injection analysis, 344
Selective volatilization, 25f
  arc-ICP vs. spark-ICP, 345
Selectivity:
  AFS, 86
  ICP-AES, 83f
    and line selection, 448
      multielement capability, 189
      spectral interference, 423f, 438f
Self-absorption, 179
Self-absorption broadening, 181, 184f
Self-reversal, 184f
Semiquantitative analysis, 15f
Sensitivity, 19
Sensitivity advantage, N_2-Ar ICP, 239, 246
Sequential vs. simultaneous detection, 189
Shot noise, see Noise; RSD; RSDB
Signal detection, see Detection
Signal measurement, see Measurement
Signal-to-background ratio, see SBR
Silicon vidicon tube, 545
Simplex optimization, see Optimization
Simulation technique, 21
Simultaneous vs. sequential detection, 189
Skin depth, 264f
  numerical values, 267
Skin effect, rf current, 70, 75, 264f
Slew-scan monochromator, 13, 516f
  wavelength positioning errors, 452f
Slew-scan spectrometer, see Spectrometer

SNR:
  of DMS, expressions, 562
  nebulizer effects, 312f
  signal-to-noise ratio, 12
  theory, 167
Soft line, see Spectral line(s)
Solids, direct analysis, AES, 24f, 334f, 344f
Solute vaporization interference, ICP, 212f
Source flicker noise, see Noise; RSD; RSDB
Sources:
  combined-, 34
  hybrid-, 34
Spark, 4
  flat top technique, 27
  gliding, 10
  graphite spark technique, 27
  for liquid analysis, 26f
  for metal analysis, 26f
  plastic cup technique, 27
  porous cup technique, 27
  rotating disk technique, 27
  rotating platform technique, 27
  rotrode technique, 27f
Spark ablation, and ICP, 345f
Spark elutriation, and ICP, 345f
Spark erosion, and ICP, 345f
Spark sampling, and ICP, 343f
Spark source MS, 89
Spatial profiles of emission:
  DCP, 55
  ICP, 201f, 208f
SPD (self-scanning photodiode array) 543f
Speciation, 343
Spectra:
  complex, see Line-rich spectra
  halogens, reduced pressure ICP, 272. See also Spectra, infrared atomic
  infrared atomic:
    ICP, 10, 146, 361, 411
    MIP, 61
  line-rich, see Line-rich spectra
  line-richness, table 385
  noise power-, 166f
  organics, 218
  reentry-, 492f, 520
  VUV, see VUV
  see also ICP spectra; spectrum
Spectral bandwidth, see Bandwidth

Spectral interference:
  AFS, 66
  and detection limit, 152f, 423f, 438f
    limit of determination, 423f, 438f
    line selection, 190f, 386f, 485
    resolving power, 423f, 438f, 485, 502f
    selectivity, 423f, 438f
    spectral resolution, 423f, 438f, 485, 502f
  classical tables, 393f
  correction, 417f, 509f
  ICP-MS, 91
  minimization, 392f
  multicomponent system, correction, 417f, 510f
  vs. multiplicative, 22
  recent tables and atlases, 394f
  types, 363f
Spectral line profile, see Line profile
Spectral line(s), 4
  "hard" and "soft" lines, 195f
  intensity, 6
  line wing, 367f
  see also Spectrum
Spectral overlap, see Line overlap
Spectral resolution:
  and analytical performance, 423f, 438f, 485, 444f, 450f, 502f
  and detection limit, 152f, 423f, 438f
    dynamic range, 187
    peak find routine, 505
    SBR, see Spectral resolution, detection limit
    spectral interference, 423f, 438f, 485, 502f
Spectral response:
  PD, 542
  PD array, 544
  see also Quantum efficiency
Spectral scan(s):
  for background correction, 412f
  C_2 bands, 382f
  computer manipulations, 402f
  Cr, 388
  H line, 375
  for line selection, 402
  N_2-Ar ICP vs. O_2-Ar ICP, 384f
  N_2^+ bands, 379f
  NO bands, 379f
  O_2 absorption bands, 381
  OH band interference, Al line, 456f
  Bi lines, 460f
  V line, 377f
  seventy elements, 386f, 406f
Spectral slit, resultant-, 428
Spectral window, 190
Spectrochemical buffer, 21f
Spectrograph, 470, 515
  characterization, 13
  electronic, 528
  HR, 500
  line width measurements, 432
  order sorter, 432
  see also Spectrometer(s)
Spectrography, 14f
  atlas, 394
Spectrometer(s):
  aberrations, 479f, 481f, 483f. See also Monochromator; Polychromator
  for AES, 12f, 466f
  astigmatism, 479f, 483f
  basic setup, 467f
  characterization, 13f
  coma, 481f, 483f
  commercial, characteristics, 503
  concave grating mounts, 477f
  curved slits, 480f, 483f
  Czerny-Turner mount, 472f, 481f
  direct reading, 511f. See also Polychromator
  Ebert mount, 472, 481f
  equations, 469f
  Fastie-Ebert mount, 481f
  field curvature, 480f, 483f
  Fourier transform-, 529
  high resolution-, 497f
  for ICP-AES:
    choosing, 483f, 527f
    desiderata, 487
    new directions, 528f
    updatings, 15
  intelligent-, 529f
  interferometer, 529
  Michelson interferometer, 529
  mountings, 477f
  multichannel detection, 528f
  multiplex detection, 529
  Paschen-Runge mounting, 479f
    moving detector, 515, 525f
  plane grating mountings, 480f

Rowland circle, 478
scanning:
  direct drive, 517f
  magnetical drive, 520
  sine bar drive, 517
Seya-Namioka mounting, 480f
slew-scan-, 13, 516f
  encoders, 521
  moving detector, 515, 525f
  peak find routine, 521f
  wavelength calibration, 521f
wavelength range, 10, 12, 362
Spectrum, 5f
  Ar, 373f
  atomic, 7f
  ionic, 7f
  and type of emitting species, 7f
  see also Spectra
Spray chamber, 299f
  cyclone-, 344
  noise, 170
  recirculation-, 315
  thermostatted-, 303, 314
Standard, 18
Standard deviation, see RSD
Standard sample, 18
Stark broadening, 374, 487
Stark effect, 374, 487
Stark shifts, Ar lines, 374
Statistical weight, 6
Stray light, 366f, 487f
  double monochromator, 496
  elimination, 493f
  establishing presence, 187
  post exit slit-, 494f
  reduction by filters, 494f
  ruled vs. holographic grating, 487f
  see also Spectrometer(s), mountings
Surfatron, microwave plasma, 63
SV (silicon vidicon tube), 545. See also OID; PD
Swan system, see C_2 bands
Synchronous detection, see AC Measurement

Tear drop ICP, 70
TE mode, transverse electric mode, 59
Thermal conductivity, atomic vs. molecular gas, 236f
Thermal pinch, 48

Thermochemical reactions, 25
Three-electrode DCP, see DCP
TM_{010} cavity, 59f
TM mode, transverse magnetic mode, 59
Top-down viewing, ICP, see End-on viewing
Torch:
  air-cooled, 284f
  alignment control, spacers, 260, 261
  carbon deposits, organics, 261
  Codding-, MIP, 64
  configuration factor, 259f, 286
  DCP, 64
  demountable, 261
  demountable vs. prealigned, 269f
  extended-, 144f, 270f
  Fassel-, 79
  Greenfield-, 78
  laminar-flow-, 289f
  MAK, 287f
  micro-, 281f
  mini-, 281f
  radiatively cooled-, 286
  for reduced pressure ICP, 272f
  reproducibility, 272
  streamlined-, 80, 260f
  water-cooled-, 283f
Torch design, 258f
  annular spacing, 261f, 281f, 287
  carrier gas speed, 268
  frequency effect, 268, 292f
  minimum diameter, 267f
  plasma stability curve, 286f
  spacers, 261
  swirl velocity, 259f, 281f, 289f
  turbulent flow vs. laminar flow, 289f
  vortex vs. axial flow, 289f
  vortex stabilization, 289f
Torches:
  conventional, 79, 259f
  for low gas consumption, 272f
  for low power consumption, 272f
  MIP, 64
  special-purpose-, 269f
  for VUV spectrometry, 144f, 271f
Toroidal ICP, 70f, 75f
Transition probability, 6

UHF (ultra high frequency), 57
UHF plasma, 57

Ultimate lines, 359
Ultrasonic nebulization, see USN
Ultrasonic nebulizer, see Nebulizer, USN, assessment; USN
Universal analysis, 15f, 24
Uptake rate, see Sample feed rate
USN:
  detection power, 322f
  devices, 322f
  interferences, 322f
  memory effects, 327
  microliter sampling, 328
  nebulization efficiency, 322f
  noise power spectra, 167f
  salt content tolerance, 327
  stability, 327
  ultrasonic nebulization, 321f
UV (ultraviolet), 2

Vacuum cup technique, arc, 26
Valente-Schrenk DCP, 54f
Venturi-effect, 299f
V-groove nebulizer, see Nebulizer
Vidicon, see SV (silicon vidicon tube)
Voigt profile, 397
  a-parameter, 397. See also Line profile deconvolution, 432
Volatilization, 5
  selective, 25
  separate from excitation, 34

Volatilization curves, 26
Vortex stabilization, ICP, 73, 289f
VUV:
  vacuum ultraviolet, 6
  wavelength tables, 145f, 393f. See also Detection limit
VUV prominent ICP lines, 142f
VUV spectrometers, 143f
VUV spectroscopy, 10

Wall-stabilized arc, 46f
Washout time, see Nebulizer
Wavelength, and frequency, 4
Wavelength modulation, background correction, 507f
Wavelength range, spectrometer, 10, 12f, 362
Wavelength scan, see Spectral scan(s)
Wavelength setting errors, and spectral resolution, 452f
  slew-scan monochromator, 452f
Wavelength table(s):
  classical, 393f
  ICP, 394, 405f, 409f
  VUV, 393f
  see also Atlas; Coincidence tables
Wing, see Line wing(s)

Xylene, see Organic solvents

(*continued from p*)

Vol. 67. **An Introduction to Photoelectron Spectroscopy.** By Pradip K. Ghosh
Vol. 68. **Room Temperature Phosphorimetry for Chemical Analysis.** By Tuan Vo-Dinh
Vol. 69. **Potentiometry and Potentiometric Titrations.** By E. P. Serjeant
Vol. 70. **Design and Application of Process Analyzer Systems.** By Paul E. Mix
Vol. 71. **Analysis of Organic and Biological Surfaces.** Edited by Patrick Echlin
Vol. 72. **Small Bore Liquid Chromatography Columns: Their Properties and Uses.** Edited by Raymond P. W. Scott
Vol. 73. **Modern Methods of Particle Size Analysis.** Edited by Howard G. Barth
Vol. 74. **Auger Electron Spectroscopy.** By Michael Thompson, M. D. Baker, Alec Christie, and J. F. Tyson
Vol. 75. **Spot Test Analysis: Clinical, Environmental, Forensic and Geochemical Applications.** By Ervin Jungreis
Vol. 76. **Receptor Modeling in Environmental Chemistry.** By Philip K. Hopke
Vol. 77. **Molecular Luminescence Spectroscopy—Part 1: Methods and Applications.** Edited by Stephen G. Schulman
Vol. 78. **Inorganic Chromatographic Analysis.** Edited by John C. MacDonald
Vol. 79. **Analytical Solution Calorimetry.** Edited by J. K. Grime
Vol. 80. **Selected Methods of Trace Metal Analysis: Biological and Environmental Samples.** By Jon C. VanLoon
Vol. 81. **The Analysis of Extraterrestrial Materials.** By Isidore Adler
Vol. 82. **Chemometrics.** By Muhammad A. Sharaf, Deborah L. Illman, and Bruce R. Kowalski
Vol. 83. **Fourier Transform Infrared Spectrometry.** By Peter R. Griffiths and James A. de Haseth
Vol. 84. **Trace Analysis: Spectroscopic Methods for Molecules.** Edited by Gary Christian and James B. Callis
Vol. 85. **Ultratrace Analysis of Pharmaceuticals and Other Compounds of Interest.** Edited by S. Ahuja
Vol. 86. **Secondary Ion Mass Spectrometry: Basic Concepts, Instrumental Aspects, Applications and Trends.** By A. Benninghoven, F. G. Rüdenauer, and H. W. Werner
Vol. 87. **Analytical Applications of Lasers.** Edited by Edward H. Piepmeier
Vol. 88. **Applied Geochemical Analysis.** by C. O. Ingamells and F. F. Pitard
Vol. 89. **Detectors for Liquid Chromatography.** Edited by Edward S. Yeung
Vol. 90. **Inductively Coupled Plasma Emission Spectroscopy—Part I: Methodology, Instrumentation, and Performance.** Edited by P. W. J. M. Boumans